The Invisible Hand

The Invisible Hand

Neurocognitive Mechanisms of Human Hand Function

Matthew R. Longo

The MIT Press
Cambridge, Massachusetts
London, England

The MIT Press
Massachusetts Institute of Technology
77 Massachusetts Avenue, Cambridge, MA 02139
mitpress.mit.edu

© 2025 Massachusetts Institute of Technology

This work is subject to a Creative Commons CC-BY-NC-ND license.

This license applies only to the work in full and not to any components included with permission. Subject to such license, all rights are reserved. No part of this book may be used to train artificial intelligence systems without permission in writing from the MIT Press.

The MIT Press would like to thank the anonymous peer reviewers who provided comments on drafts of this book. The generous work of academic experts is essential for establishing the authority and quality of our publications. We acknowledge with gratitude the contributions of these otherwise uncredited readers.

This book was set in Stone Serif and Stone Sans by Westchester Publishing Services. Printed and bound in the United States of America.

Library of Congress Cataloging-in-Publication Data

Names: Longo, Matthew R. (Matthew Ryan), author.
Title: The invisible hand : neurocognitive mechanisms of human hand function / Matthew R. Longo.
Description: Cambridge, Massachusetts : The MIT Press, [2025] | Includes bibliographical references and index.
Identifiers: LCCN 2024046259 (print) | LCCN 2024046260 (ebook) |
  ISBN 9780262551878 (paperback) | ISBN 9780262382588 (pdf) |
  ISBN 9780262382595 (epub)
Subjects: MESH: Hand—physiology | Perception | Hand—pathology | Nervous System Physiological Phenomena | Somatosensory Cortex—physiology | Biological Evolution
Classification: LCC QP334 (print) | LCC QP334 (ebook) | NLM WE 830 |
  DDC 612.9/7—dc23/eng/20241218
LC record available at https://lccn.loc.gov/2024046259
LC ebook record available at https://lccn.loc.gov/2024046260

10 9 8 7 6 5 4 3 2 1

EU product safety and compliance information contact is: mitp-eu-gpsr@mit.edu

For my mother

Contents

Preface: The Visible and Invisible Hands   ix
Acknowledgments   xi
List of Abbreviations   xiii

1   **The Visible Hand**   1
2   **The Sensing Hand**   11
3   **The Moving Hand**   45
4   **The Acting Hand**   71
5   **The Exploring Hand**   91
6   **The Embodied Hand**   105
7   **The Hand in Space**   127
8   **The Expert Hand**   145
9   **The Diseased Hand**   167
10   **The Extended Hand**   193
11   **The Evolving Hand**   221
12   **The Developing Hand**   257
13   **The Bilateral Hand**   279
14   **The Dominant and Nondominant Hands**   305
15   **The Communicating Hand**   321

   Epilogue: The Future of the Invisible Hand   343

Notes   345
References   347
Author Index   469
Subject Index   489

Preface: The Visible and Invisible Hands

The human hand does not lack for superlatives. Sir Isaac Newton is quoted as claiming that "in default of any other proof, the thumb would convince me of the existence of a God" (Dickens, 1863, p. 346). The Greek philosopher Anaxagoras claimed that humans were the most intelligent of animals because of their hands. To Immanuel Kant, the hand was the outer part of the brain (Katz, 1925, p. 28). To Aristotle, it was the "tool of tools" (1986, Book III, Chapter 8), "talon, hoof, and horn, at will. So too it is spear, and sword, and whatsoever other weapon or instrument you please" (1882, Book IV, Part 10).

The hand has been assigned a central role in both human evolution and human cultural development, in our *descent* and in our *ascent*. Charles Darwin (1871), in *The Descent of Man*, considered that "Man could not have attained his present dominant position in the world without the use of his hands, which are so admirably adapted to act in obedience to his will" (p. 173). A century later, Jacob Bronowski (1975), whose history of human scientific and technological achievement *The Ascent of Man* deliberately parallels Darwin, again gives the hand central position in the development of culture: "The hand is the cutting edge of the mind. Civilisation is not a collection of finished artefacts, it is the elaboration of processes. In the end, the march of man is the refinement of the hand in action" (pp. 115–116).

The story of the human hand is all of human evolution, all of human culture. It is human art and technological accomplishment. It is human life. It is Giotto drawing a perfect circle freehand, a child tying their shoelaces for the first time. It is Glenn Gould playing the *Goldberg Variations*, Jacqueline du Pré playing Elgar's *Cello Concerto*. It is a newborn's feeble grasp of their mother's breast, the skilled touch of a master potter. It is a bride and groom exchanging rings, the forefingers of the longbowmen at Agincourt. It is the outstretched hand of God, whether through the intermediary of Michelangelo in the Sistine Chapel or of Diego Maradona in the Estadio Azteca.

Yet, there is a paradox about the human hand. For while it is lauded as the pinnacle of evolutionary perfection, the physical hand itself is largely unexceptional. The eminent anatomist Frederic Wood Jones (1920) went so far as to call the human hand "the absolute bed-rock of mammalian primitiveness" (p. 17). The human hand in its overall physical aspect departs little from the basic mammalian template. Other primates, in fact, have far more specialized hands, in the sense of deviation from the basal mammalian form.

This paradox is more apparent than real. As Wood Jones (1920) himself noted, "It is not the hand that is perfect, but the whole nervous mechanism by which movements of the hand are evoked, co-ordinated and controlled" (p. 236). Similarly, according to the anatomist John Napier, who literally wrote the book on *Hands* (1993), "the hand itself is derived from yeoman stock but the factor that places it among the nobles is, as it were, its connections—its connections with the higher centers of the brain" (p. 10).

To understand the significance of the hand to human achievement, we need to look not at the hand as a physical object—what I call the *visible hand*. Instead, we need to understand the *invisible hand*. By this term, I refer collectively to the set of neural mechanisms underlying the interpretation of sensory signals from the hand, the sending of motor commands, and everything in between. The invisible hand inserts itself into the visible hand like a glove, quickening and animating the hand, and investing it with the full suite of perceptual, motor, and cognitive abilities of the human brain.

The hand has been the focus of an enormous amount of research from a dizzying range of disciplines, from anatomy to evolutionary biology, from psychology to neurology, archaeology, and neuroscience. The invisible hand provides an organizing concept that helps to integrate and contextualize findings from these disparate fields. In this book, I will introduce the concept of the invisible hand and discuss its key features. I will argue that the neurocognitive mechanisms comprising the invisible hand are central to understanding a wide range of phenomena, including basic sensory and motor function, space perception, the self, tool use, handedness, and gesture. More generally, an overarching argument will be that the extraordinary abilities of the hand arise from the complementary nature and tight integration of the visible and invisible hands.

Acknowledgments

I started work on this book at the start of my first sabbatical in the 2021–2022 academic year. Perhaps unsurprisingly, my ambition to complete the book during this period was unsuccessful. I am finishing the draft of the book nearly two years later while a visiting researcher at the Center for Information and Neural Networks (CiNet) at Osaka University in Japan. I wish to thank my colleagues at Birkbeck, University of London, for covering my teaching and administrative responsibilities, which made my sabbatical possible. I would also like to thank Nobuhiro Hagura, Tamami Nakano, and my other colleagues at CiNet for their hospitality to me and my family.

Researchers in many fields have made important contributions to understanding hand function. The writing of this book has therefore taken me far outside of my own fields of experimental psychology and cognitive neuroscience into disciplines as diverse as neurology, evolutionary biology, physiology, and archaeology. I am acutely aware of my lack of expertise in many of these fields. I hope my colleagues will feel that I have done justice to their disciplines and beg their forbearance when I have failed to do so. However, my self-consciousness at not being an expert in all the fields I discuss is tempered by an awareness that nobody else is an expert in every one of these fields either. Despite this, I believe that there is value in this type of integrative approach, and I hope that colleagues working in several disciplines find the book has value to their work and is a useful entry point to literatures they may have been only dimly aware of.

While the range of topics I cover is diverse, it is not comprehensive. I have discussed issues that I consider to be especially important to understand hand function. But space limitations have meant that I have needed to be selective in my coverage. These choices have been shaped by my interests and inclinations as well as, no doubt, by many accidents of what I have come across in my research. I am thus well aware of significant topics that I have neglected. For example, readers familiar with my own experimental work may be surprised that there is not more discussion of distortions of mental body representations. Other indisputably important topics, such as pain, which I had originally intended to cover, have, in the end, been dropped. My discussion of the nervous system also focuses heavily on the cerebral cortex, giving less coverage than is probably warranted to subcortical structures, such as the thalamus, cerebellum, and spinal cord. The lack of discussion of a specific topic should not be taken to indicate that I consider it unimportant. This book is most certainly not intended as the last word on the subject of the hand.

Just as it is not the last word, this book is not the first word either. There is a distinguished lineage of books on the hand, from such illustrious scientists as Sir Charles Bell (1833), Frederic Wood Jones (1920), David Katz (1925), John Napier (1993), and Vernon Mountcastle (2005). Although the focus of this book is different from each of these other works, I have been deeply influenced by each of them. I hope that this book contributes, in its modest way, to this tradition, and that it may inspire other researchers to study the human hand.

Several colleagues have provided helpful feedback on drafts of chapters. I particularly wish to thank Denise Cadete, Christos Ganos, Lúcia Garrido, Ryo Ishibashi, Ori Ossmy, Luigi Tamè, and two anonymous reviewers. I also wish to thank Ellen Blythe for creating figure 8.3, Raffaele Tucciarelli for help making figure 10.3, Guido Orgs for assistance with German language translation, and Adrian Alsmith for wise advice and encouragement at several stages of this project. Of course, such errors that remain are entirely my responsibility.

Finally, the writing of a book like this can take over life at home. I therefore especially wish to thank Lúcia and Sofia for putting up with me for the past two years.

Matthew Longo

August 2023

Osaka, Japan

List of Abbreviations

2PDT	Two-point discrimination threshold
ADM	Abductor digiti minimi
AIP	Anterior intraparietal area
aSMG	Anterior supramarginal gyrus
BOLD	Blood-oxygen-level-dependent signal
CCE	Crossmodal congruency effect
CM	Corticomotoneuronal
CT	Computed tomography
dPMC	Dorsal premotor cortex
DTI	Diffusion tensor imaging
EEG	Electroencephalography
EMG	Electromyography
FEF	Frontal eye fields
FPL	Flexor pollicis longus
fMRI	Functional magnetic resonance imaging
GABA	Gamma-aminobutyric acid
GP_e	Globus pallidus external segment
GP_i	Globus pallidus internal segment
IPS	Intraparietal sulcus
LIP	Lateral intraparietal area
LOC	Lateral occipital complex
LRN	Lateral reticular nucleus
M1	Primary motor cortex
MDS	Multidimensional scaling
MEG	Magnetoencephalography
MEP	Motor evoked potential

MIP	Medial intraparietal area
MRI	Magnetic resonance imaging
MVPA	Multivoxel pattern analysis
OFC	Optimal feedback control
PET	Positron emission tomography
PMC	Premotor cortex
PPC	Posterior parietal cortex
PPFM	Pli de passage fronto-pariétal moyen
pRF	Population receptive field
PRR	Parietal reach region
RA1	Rapidly adapting type 1
RA2	Rapidly adapting type 2
RF	Receptive field
rTMS	Repetitive transcranial magnetic stimulation
S1	Primary somatosensory cortex
S2	Secondary somatosensory cortex
SA1	Slowly adapting type 1
SA2	Slowly adapting type 2
SEP	Somatosensory evoked potential
SNr	Substantia nigra pars reticulata
SMA	Supplementary motor area
STN	Subthalamic nucleus
TMS	Transcranial magnetic stimulation
tDCS	Transcranial direct current stimulation
V1	Primary visual cortex
VBM	Voxel-based morphometry
VIP	Ventral intraparietal area
vPMC	Ventral premotor cortex

1 The Visible Hand

On his death in 1829, Francis Henry Egerton, 8th (and last) Earl of Bridgewater, left £8,000 to the Royal Society of London to be paid to people nominated by the society's president who would "write, print, and publish one thousand copies of a work On the Power, Wisdom, and Goodness of God, as manifested in the Creation" (Topham, 1998, p. 234). This directive was dutifully implemented, and between 1833 and 1840, eight books were published, known collectively as the *Bridgewater Treatises* (Topham, 1998). These books cover a diverse range of topics, from astronomy to chemistry, geology to botany. They share a perspective, however, shaped by the "natural theology" of the English clergyman William Paley—a view that would now be called "intelligent design." In 1802, Paley advanced the famous "watchmaker" analogy, which was memorably challenged by Richard Dawkins (1986) nearly two centuries later.

The fourth of the *Bridgewater Treatises* was written by the Scottish surgeon and anatomist Sir Charles Bell. Bell was widely respected for his scientific work and remains known today for his description of Bell's palsy and his discovery of the segregation of sensory and motor nerves in the spinal cord (the so-called Bell–Magendie law). The overall argument of Bell's book, admirably summarized by its title *The Hand: Its Mechanism and Vital Endowments as Evincing Design*, is unlikely to be taken seriously by scientific readers today. Nevertheless, it is a work of careful scholarship, which describes the anatomical structure of the human hand and places it into comparative context with a rich and thoughtful discussion of the limbs of many other animals. Bell's interpretation of the hand as evidence for design was based on his analysis of the efficiency of its operation:

> Seeing the perfection of the hand, we can hardly be surprised that some philosophers should have entertained the opinion with Anaxagoras, that the superiority of man is owing to his hand. We have seen that the system of bones, muscles, and nerves of this extremity is suited to every form and condition of vertebrated animals; and we must confess that it is in the human hand that we have the consummation of all perfection as an instrument. (1833, p. 209)

Even if we do not wish to follow Bell in interpreting the consummate perfection of the human hand as evidence for divine creation, we can still recognize the extraordinary elegance of its construction. In this initial chapter, I will briefly introduce some key features of the visible hand, which will set the context for the discussion of the invisible hand that will occupy the remainder of the book. When I refer to the "visible hand," I mean not only the hand as we can see it from the outside but also what is visible to anatomists, including the bones, muscles, and tendons.

A Tour of the Human Hand

Pentadactyly and the Mammalian Archetype

The standard mammalian forelimb has thirty bones: one bone forming the upper arm, two bones forming the forearm, eight carpal bones forming the wrist, five metacarpal bones forming the main body of the hand, and fourteen phalanges forming the digits. In all mammals, the phalanges occur in the formula 2, 3, 3, 3, 3, with the digit on the radial side of the hand (i.e., the thumb) having only two bones and the others having three. Remarkably, a clearly homologous bone structure is seen even in mammals that lack fingers entirely and instead have appendages ending in hooves (e.g., horses), flippers (e.g., dolphins), and wings (e.g., bats). The discovery of this precise homology was an immensely important scientific discovery in the nineteenth century (Desmond, 1982), as will be discussed further in chapter 11. The Victorian paleontologist Richard Owen (1849) argued that each individual skeleton reflected a common *archetype*, which was common not only to mammals but also to all vertebrates. Charles Darwin (1859) in *On the Origin of Species* highlighted this archetype and, of course, went on to provide an elegant explanation for it: "What can be more curious than that the hand of a man, formed for grasping, that of a mole for digging, the leg of the horse, the paddle of the porpoise, and the wing of the bat, should all be constructed on the same pattern, and should include the same bones, in the same relative positions?" (p. 434).

In the case of limbs, the archetype consists of a sequence of, first, a single long bone (the humerus in the forelimb, the femur in the hindlimb), then a pair of bones (the ulna and radius in the forelimb, the tibia and fibula in the hindlimb), then a cluster of irregularly shaped bones (the carpals in the forelimb, the tarsals in the hindlimb), and finally five digits. As Shubin (2009) pithily put it, Owen's limb archetype consists of "one bone–two bones–lotsa blobs–digits" (p. 33).

The most obvious feature of human hands is the presence of five long, thin peninsulas—that is, fingers—jutting off it. This five-fingered condition, known as "pentadactyly," is older even than mammals, having arisen early in the evolution of the tetrapods—the group of fishes that moved onto land, giving rise to all living amphibians, reptiles, birds, and mammals. The earliest known fully terrestrial tetrapod, the 340-million-year-old *Casineria*, had clearly pentadactyl limbs (Paton et al., 1999). Intriguingly, several earlier animals that appear transitional between fish and tetrapods had more than five digits (Coates & Clack, 1990), such as *Tulerpeton* with six, *Ichthyostega* with seven, and *Acanthostega* with eight. Nevertheless, the pentadactyl pattern emerged early in tetrapod evolution more than 300 million years ago and has formed the default pattern ever since.

Individual humans, of course, may have more than five fingers—a condition that has been described since antiquity. Pliny the Elder (1855) in his *Natural History* writes of two six-fingered daughters of C. Horatius, known accordingly by the name "Sedigitae." More recently, a family has been studied who have a supernumerary finger on each hand, which they can use in highly dexterous ways (Mehring et al., 2019). The presence of extra fingers is not particularly rare, occurring in around one in five hundred human births (McCarroll, 2000), and is even more common in other mammals such as cats (Danforth, 1947). Nevertheless, despite being clearly hereditary within individual families, polydactyly does not appear to have become the standard pattern for any species of mammal. Apparent exceptions to this rule have proved illusory. For example, the so-called thumb of pandas, which appears to be a sixth digit, is, in fact, an extension of one of the wrist bones rather than a true digit (Gould, 1980). While this reflects a remarkable adaptation to the panda's mode of life in bamboo forests, it nevertheless occurs within the standard plan of a mammalian limb. Similarly, while the Siamang gibbon appears to have only four digits on each foot, in fact, it retains all the bones of other primates, with a web of skin linking the second and third digits (Napier, 1993).

Movements of the human hand are facilitated by several types of joint (Kapandji, 2019). Two types of joint are found in the arms, which greatly expand the range of motion of the hand. First, the long bone of the upper arm (the humerus) attaches to the torso via a complex arrangement of five joints. The most critical of these is the glenohumeral joint, which is a typical ball-and-socket joint, allowing the hand to be transported to a wide region of external space, especially when combined with flexion and extension of the elbow joint. Second, the existence of two distinct bones in the forearm (the radius and ulna) allows the entire forearm to be rotated in the movements of pronation and supination. These rotations are supported by pivot joints at both ends of the forearm, the superior and inferior radioulnar joints, which function mechanically like a system of ball bearings. Rotations of the forearm, known as pronation (with the thumb pointing toward the midline of the body) and supination (with the thumb pointing away from the midline of the body), allow the hand to be not only brought to the correct location in space but also oriented appropriately for a specific action.

Rotation of the forearm is especially powerful in combination with movements of the wrist. The wrist consists of two distinct joints: the radio-carpal joint, linking the carpal wrist bones to the forearm, and the mid-carpal joint, between the two rows of carpal bones. Collectively, the movements of these joints combined with forearm rotation allow the hand to be positioned at nearly any angle to grasp or manipulate objects. The wrist is thus an example of what is known as a "universal joint" (Kapandji, 2019), a device originally used to suspend compasses and other instruments to protect them from rolling and pitching on moving ships. In modern engineering, universal joints are often used to transmit rotational forces between rods with different orientations. In the case of the wrist, it allows the large forces generated by the arm and torso muscles to be transferred to the hand regardless of orientation.

There are also several types of joint within the hand itself. The phalanges within each finger form hinge joints with each other, allowing each finger to flex and extend. Each finger attaches to its metacarpal with a biaxial joint, which allows the finger not only to flex and extend but also to move from side to side. Finally, the joint linking the metacarpal of the thumb to the wrist, the trapeziometacarpal joint, has a unique "saddle" shape, which allows the thumb to rotate across the rest of the hand, facilitating its precise opposition with the other fingers (Napier, 1993).

The Hand Musculature
Movements of the hand are controlled by an extraordinary system of muscles and tendons that are crammed into a remarkably small space. Most humans have twenty-one muscles within the hand itself, which are collectively known as the "intrinsic hand muscles" (Lemelin & Diogo, 2016). In addition, another system of muscles, the "extrinsic hand muscles," located in the forearm and upper arm, is also involved in controlling the hand. To a first approximation, the intrinsic hand muscles provide the precise dexterity of hand movements, while the extrinsic muscles provide the force, as shown by studies that measured muscular activities during performance of different types of grasp (Long et al., 1970). As will be discussed further in chapter 8, the development of expertise in a specific task, such as playing the piano, is frequently associated with a shift from the use of intrinsic muscles to extrinsic muscles (Furuya & Kinoshita, 2008; Kay et al., 2003).

There are no muscles controlling the thumb that are actually specific to humans, as every muscle can be found in other primates (Richmond et al., 2016). For example, two major extrinsic muscles controlling the thumb in humans, the flexor pollicis longus (FPL) and extensor pollicis brevis, are normally absent in other great apes (Diogo et al., 2012). However, these muscles are present in gibbons (i.e., lesser apes). Similarly, in humans and gibbons—but not in great apes—the FPL is fully separated from the flexor digitorum profundus that attaches to the other fingers (Diogo et al., 2012), allowing more individuated movements of the thumb. Conversely, one intrinsic thumb muscle, known as the first volar interosseous muscle of Henle (Susman, 1994) or the accessory adductor pollicis (Diogo et al., 2012), is present in humans and great apes—but not in gibbons. Thus, the musculature of the human hand is unique in its combination of features, but the individual features themselves are not unique. Rather, the human thumb is unique in having a larger total number of muscles than that of any other individual primate (Diogo et al., 2012), and the proportion of the total hand muscular devoted to the thumb is higher than in other species (Tuttle, 1969b).

Although the FPL muscle appears to exist in gibbons and so is not unique to humans, it is thought to have played a key role in the emergence of characteristically human hand skills, such as tool use (Hamrick et al., 1998; Susman, 1994). Hamrick et al. (1998) used electromyography (EMG) to record electrical activity from the FPL while participants performed several tasks, including the manufacture and use of stone tools. Strong activity was found in the FPL across all of these tasks, indicating that this muscle is widely involved in tool use.

The Skin and Sensory Receptors
The skin is the boundary between our body and the external world. As such, a major role of the skin is to keep water and potentially harmful things outside and blood and other bodily tissues inside. The human hand is covered by two distinct types of skin. The hand dorsum is covered by hairy skin, which is similar to the skin on the arm. The palm, in contrast, is covered by glabrous skin, which lacks hair entirely and is found on only a few other places on the body, such as the soles of the feet. The skin is much more, however, than just a waterproof sack. The skin is highly flexible and stretches to allow the hand to move freely. At the same time, the skin is also tightly bound to the underlying tissues in places, such as the center of the palm, which allows objects to be held tightly without the skin slipping and sliding around. Another feature of the skin that aids grasping is the presence of papillary ridges on the glabrous skin—permanent thickenings of the skin that increase the friction between the skin and held objects, much like the treads on car tires. It is the papillary ridges that comprise the familiar whorls and loops of our fingerprints.

The skin is also a remarkable sensory device, with several types of sensory receptors providing a variety of information about the hand and objects it is touching (Mountcastle, 2005). The glabrous skin of a human hand contains around seventeen thousand mechanoreceptive end organs (Johansson & Vallbo, 1983)—specialized units in the skin that are sensitive to physical stimulation and provide the raw inputs for the sense of touch. These end organs come in four main types, each associated with a distinct type of nerve fiber carrying information to primary sensory neurons in the dorsal root ganglia, which in turn send information to the spinal cord (Handler & Ginty, 2021; Johnson, 2001).

These four systems differ in their properties and sensitivity to mechanical stimuli, and they appear to be specialized for different aspects of tactile perception. They differ in two main dimensions. First, some respond continuously in a sustained fashion to the presentation of a stimulus—a property known as "slow adaptation." Others, in contrast, respond rapidly at the onset of a stimulus and then stop responding—a property known as "rapid adaptation." Together, these properties provide information about both the stable features of objects we touch and changes that may require rapid attention. Second, the systems differ in their degree of spatial sensitivity, with some showing highly localized responses and others generalizing across much of the hand. Each of these systems is briefly described below:

- The slowly adapting type 1 (SA1) afferent system is associated with Merkel cell end organs in the epidermis. These organs densely innervate the glabrous skin, with approximately a hundred Merkel cells per square centimeter of skin in the fingertips of monkeys and humans. This system shows very high spatial sensitivity and responds well and continuously to stable object features such as points, edges, and curvature. This system is likely involved in tactile perception of form and texture.

- The rapidly adapting type 1 (RA1) system is associated with Meissner corpuscle end organs located superficially in the epidermis at the apex of the dermal papillae—protrusions of the dermis into the epidermis that form the papillary ridges. This system also has a high innervation density, with as many as 150 Meissner corpuscles per square centimeter of skin in the fingertips. This system responds rapidly to dynamic changes in skin deformation but shows no response to static skin deformation. Thus, it is well suited for tactile perception of object motion. As discussed above, the papillary ridges increase the friction of the fingertips against objects. Via the RA1 system, these ridges also provide rapid information about changes in friction resulting from an object slipping against the skin. Accordingly, this system is likely important for rapid feedback control of grip.

- The slowly adapting type 2 (SA2) system is associated with Ruffini corpuscle end organs located deep in the dermis. This system shows poor spatial sensitivity and little sensitivity to indentation of the skin. In contrast, however, it shows strong sensitivity to the skin itself stretching. This likely has two important perceptual functions. First, it can provide information about the direction and force of object motion against the skin. Second, it can provide information about the posture of the hand, since hand movements produce systematic changes in skin stretch across the hand and forearm.

- Finally, the rapidly adapting type 2 (RA2) system is associated with Pacinian corpuscles, which, like Ruffini corpuscles, lie deep within the dermis. This system has virtually no spatial resolution but is exquisitely sensitive to high-frequency vibrations. Because these responses do not depend on the spatial location of the stimulus, they carry information about vibrations applied to objects held in the hand, such as tools, as will be discussed further in chapter 10.

The Dexterous Movements of the Hand

Given the large number of bones making up the hand, and the equally large number of muscles controlling it, there are a vast number of possible hand movements. The space of potential hand movements includes at least twenty independent degrees of freedom. As will be discussed in chapter 3, simplifying this complexity is a central problem in motor control. In an influential paper, John Napier (1956) argued that all prehensile actions of the hand could be classified into just two types: power grips and precision grips. Importantly, this is not to suggest that there are only two things that we can do with our hands. Rather, each of these grip types encompasses an extraordinary range of actions, allowing us to perform a dazzling array of skilled behaviors.

Nearly all prehensile grips involve flexion of the fingers against other parts of the hand. In power grips, the flexed fingers form a clamp pinching the object against *the palm of the hand*, as shown in the left column of figure 1.1. This allows the object to be held securely and for large forces to be applied using the extrinsic hand muscles of the arm. Power grips are what we use for hammering, chopping, pulling open a heavy door, or throwing a ball. In contrast, precision grips involve flexion of the fingers against *the thumb*. In its most canonical form, precision grips involve the opposition of the index finger against the thumb as shown in the top right panel of figure 1.1. But precision grips can also involve larger numbers of fingers as well, as shown in the center right and lower right panels of figure 1.1. This allows objects to be held and manipulated precisely and dexterously using the intrinsic hand muscles. Precision grips are what we use for holding a pencil to write, threading a needle, or picking up a paper clip.

It is important to emphasize that the distinction between power and precision grips is not about the nature of the object but rather about what one intends to do with the object. The same object can be held with either type of grip, as shown in the middle row of figure 1.1. Indeed, different grips may be used sequentially at different stages of a complex action. For example, when unscrewing a jar, a power grip may be used initially, since a large amount of force may be required to dislodge the lid (figure 1.1, bottom left). Once the lid is loose, we may switch to a precision grip to quickly unscrew the lid (figure 1.1, bottom right). Similarly, when using a screwdriver, we may start with

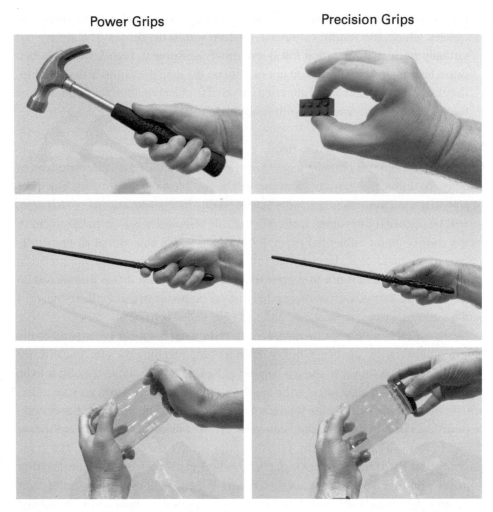

Figure 1.1
Napier's distinction between power and precision grips. *Top row*: Canonical power grips (top left) and precision grips (top right). *Center row*: The same object can be held using either a power grip (center left) or a precision grip (center right), depending on what one is doing. *Bottom row*: Power grips (bottom left) and precision grips (bottom right) may be used sequentially—for example, when opening a jar.

a precision grip to rapidly thread a screw into a hole. At the end, however, we may switch to a power grip to firmly drive the screw into place.

Napier's distinction between power and precision grips has been enormously influential in many fields and will come up repeatedly throughout this book. Numerous research groups have suggested different classifications of hand grips, although these have generally not undermined Napier's basic distinction. One important extension comes from the work of the anthropologist Mary Marzke and her colleagues who classified the grips used during experimental manufacture and use of stone tools (for a review, see Marzke, 2013). Marzke argued that there are three main grips used in these tasks, which combine elements of precision and power grips. These grips will be discussed in greater detail in chapter 10 when we discuss Paleolithic toolmaking. For now, I will give just one example. Suppose you are peeling a potato. The potato is held securely against the palm of the hand to withstand the force of the peeler pressing against it—a classic power grip. At the same time, we may use the fingertips to continuously rotate the potato to expose different areas of its surface to be peeled. This rotation involves opposition of the thumb against the fingers—a feature of a precision grip. Marzke refers to this as a "cradle grip."

This chapter has briefly surveyed the remarkable properties of the physical hand—the visible hand. Although we may, like Sir Charles Bell, admire the consummate perfection of the hand itself, we must also keep in mind Wood Jones's (1920) dictum that the true perfection of the hand is found in "the whole nervous mechanism" (p. 236) underlying hand function. We turn now to this nervous mechanism—to the invisible hand.

2   The Sensing Hand

On April 25, 1903, a curious operation was performed at the London home of Dr. Henry Percy Dean, assistant surgeon at the London Hospital. Assisted by Dr. James Sherren, Dr. Dean made a 16.5 cm incision near the left elbow of their colleague, the eminent neurologist Dr. Henry Head. They proceeded to cut the radial and external cutaneous nerves in Head's arm and to tie the severed ends with silk thread. This procedure led to an immediate loss of sensation in a wide region around the thumb and index finger of Head's left hand (Head et al., 1905).[1] Over the next four years, Head underwent extensive testing at regular intervals by his colleague William Rivers to map the gradual reemergence of sensitivity (Rivers & Head, 1908).

Touch and the various bodily senses are ubiquitous. In vision, we can close our eyes, wear a blindfold, or go into a dark room. In audition, we can cover our ears, wear earplugs, or seek a quiet room. But touch is less easily blocked. We can stand motionless, but even then, we feel the pressure on the soles of our feet, the gentle pressure of our clothes, the posture of our limbs, and the temperature of the air surrounding us. We can sit awkwardly on our hand or foot, making the limb "fall asleep." While this may stop touch, it is characterized as much by the emergence of novel sensations (tingling, throbbing) as the absence of usual ones. That tactile sensations could be impaired by damage to the nerves, to the spinal cord, or to the brain was well known (indeed, it was Head's clinical work with patients suffering from nerve injury that inspired him to volunteer for this study). Nevertheless, the immediacy and specificity of Head's tactile loss is striking.

While the details of Rivers and Head's interpretations of the study have been superseded by time, the experiment stands as a landmark in the study of human hand sensation, inspiring subsequent research and public fascination. For example, the study forms a central metaphor in the acclaimed *Regeneration* trilogy of historical novels by Pat Barker (1991), which focus on the relationship between Rivers and the war hero, anti-war protestor, poet, and fox-hunting man Siegfried Sassoon. In this chapter, I will discuss the

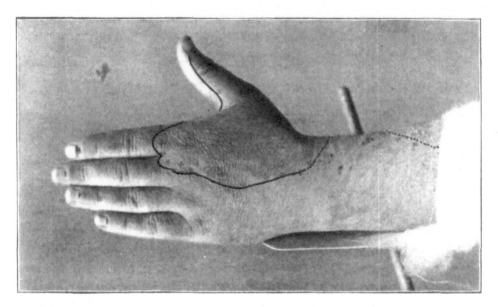

Figure 2.1
Area of sensory loss in Henry Head's hand shortly after his radial nerve was cut at the elbow. From Rivers and Head (1908).

sensing hand, starting with an overview of the cortical somatosensory system, before discussing higher-level cognitive processes supported by this system—what my colleagues and I (Longo et al., 2010) have called "somato*perception*."

The Cortical Somatosensory System

A central feature of the invisible hand is the use of neural maps to represent the hand. Maps are everywhere in the brain, from cognitive maps of the external world (Epstein et al., 2017) to maps of high-level semantics (Huth et al., 2016), and within sensory systems, from retinotopic maps of visual space (Silver & Kastner, 2009) to tonotopic maps of sounds (Dick et al., 2017). The most conspicuous maps in the brain, however, are the maps of the body itself, to which I now turn.

Somatotopic Maps of the Body
Before discussing the somatosensory cortex, we first need to introduce the motor cortex, which historically was studied first. Studies in the latter part of the nineteenth century showed that the cortex just in front of the central sulcus (i.e., the precentral

gyrus), separating the frontal from the parietal lobe, was critical for the control of movements on the contralateral side of the body. In one of the most influential studies in the history of neuroscience, Gustav Fritsch and Eduard Hitzig (1870, 2009) applied electrical currents to the exposed cortex of dogs, showing that stimulation of the posterior part of the frontal lobe evoked twitches of specific parts of the contralateral side of the animal's body—that is, stimulation of the left frontal lobe evoked twitches on the right side of the dog's body and vice versa. This region of the brain, the precentral gyrus, is now known as the primary motor cortex or M1.

Similar results were found in a variety of animals by David Ferrier (1873, 1876), including frogs, fish, rabbits, cats, jackals, and monkeys. Ferrier applied longer trains of stimulation than had Fritsch and Hitzig, and he described more complex, ecologically meaningful behaviors—a point I will return to in the following chapter. Ferrier further found that ablation of regions of the motor cortex eliminated movements of the same body parts (Ferrier & Yeo, 1884). Later research by Leyton and Sherrington (1917) described detailed maps of motor responses evoked by stimulation of the motor cortex in great apes, including gorillas, orangutans, and chimpanzees. Like Ferrier, they found good correspondence between the body part in which movement was evoked on stimulation of a particular location and the site of weakness and disuse following ablation of that brain area. Eventually, analogous results were also reported in which electrical stimulation of the exposed human brain caused similar movements (Bartholow, 1874; Foerster, 1926).

Just behind M1, on the opposite side of the central sulcus, is a region known as the primary somatosensory cortex or S1. The stimulation studies in nonhuman animals described above all investigated M1, not S1. This is a result of the fact that stimulation of M1 produces overt behaviors that can be observed, but stimulation of S1 (usually) does not. These animals may very well have also experienced tactile sensations following stimulation of S1 but were unable to report these to the researchers. As Cushing (1909) notes, "it is a matter which requires on the part of the subject not only an analysis of his sensory impressions, but their transference by descriptive speech to the observer, who in turn must interpret this description" (p. 45). For this reason, comparable studies of humans were critical to investigate somatosensory maps in the brain. Cushing was the first to stimulate the postcentral gyrus in awake humans undergoing surgery for treatment of epilepsy. His two patients both reported clear sensory experiences, including of "warmth," "touch," and "stroking." One patient described the sensation as feeling "as though someone had touched or stroked the finger" (p. 50).

More detailed maps of touch in S1 were first reported in the seminal studies of Wilder Penfield and his colleagues (Penfield & Boldrey, 1937; Penfield & Rasmussen, 1950). Penfield was a neurosurgeon, operating on patients to treat epilepsy and brain tumors.

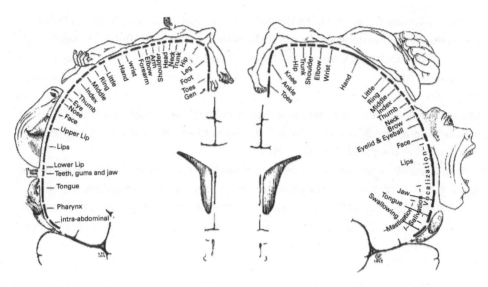

Figure 2.2
The "homunculi" in S1 (left panel) and M1 (right panel) mapped using intracranial electrical stimulation in human patients undergoing surgery for epilepsy. From Penfield and Rasmussen (1950).

As many of these surgeries were conducted without general anesthesia, the patient was awake and able to report their experiences. The first report of this work (Penfield & Boldrey, 1937) aggregated data from 126 patients tested over a number of years. As in the studies described above, Penfield showed that electrical stimulation of M1 produced twitches of muscles on the contralateral side of the body, as shown in the right panel of figure 2.2. Analogously, stimulation of S1 led the patients to report feeling touches or tingling at specific locations on the contralateral side of the body, as shown in the left panel of figure 2.2. Moreover, these maps in M1 and S1 were in approximate alignment with each other.

These maps of the body in M1 and S1 have become known as the motor and somatosensory "homunculi" because it was as if there was a little person spread across the surface of the cortex.[2] There are two important features of these homunculi. The first feature is somatotopy, which refers to the fact that the spatial layout of the body itself is preserved (at least to some degree) in the spatial layout of the representation in the brain. This means that adjacent parts of the body are represented by adjacent parts of the cortex. In Penfield's studies, as in earlier research on M1 in monkeys (Horsley & Schäfer, 1888) and apes (Leyton & Sherrington, 1917), the representations of the feet were most medial, at the top of the brain, followed by the torso, upper arms, forearms, hands, and then the face and head. Indeed, this general pattern is seen in virtually all placental mammals. At least in S1, there is also clear somatotopy at a finer spatial scale, as will be discussed below. In M1, the fine somatotopic organization is less clear, as will be discussed in more detail in the next chapter.

A second important feature of both the motor and somatosensory maps is the phenomenon of cortical magnification. This refers to the fact that the amount of cortical territory devoted to each part of the body is not related to its physical size but instead to its dexterity (in the motor cortex) or sensitivity (in the somatosensory cortex). This feature was depicted in the famous visualization of the homunculus, first provided by Penfield and Boldrey (1937) and later revised by Penfield and Rasmussen (1950), as shown in figure 2.2. The size of each body part is drawn in proportion to the amount of cortical territory representing that body part, resulting in a figure of a person with oversized fingers and lips, and tiny legs and torso. The Penfield homunculus has become an iconic image, shown in virtually all textbook discussions of sensorimotor function (Catani, 2017). As a depiction, the homunculus has also been widely criticized for oversimplifying a more complex representation, with some level of overlap between body parts and differences between individuals (Desmurget & Sirigu, 2015; Farrell et al., 2007). The homunculus is doubtless a simplification, as Penfield himself was well aware. To some extent, of course, this type of simplification is exactly the purpose of such scientific visualizations.

Recently, Roux and colleagues have revisited the homunculus, testing a large number of patients using modern stimulation and mapping methods to construct detailed maps of somatotopy in both the somatosensory (Roux et al., 2018) and motor (Roux et al., 2020) cortices. Overall, these results provide striking confirmation of the basic organization described by Penfield. Nevertheless, some relatively modest differences with Penfield's reports were identified, notably somewhat more overlap between body parts in the motor homunculus—a point I will return to in the next chapter. Another recent study involved implanting two microelectrode arrays into S1 of a tetraplegic patient with a spinal cord injury (Flesher et al., 2016). Whereas Penfield's stimulation using relatively large electrodes produced generally rather diffuse and unnatural sensations, such as tingling and pins and needles, the patient described by Flesher and colleagues reported largely natural feeling sensations, which could be precisely localized to specific segments of single fingers.

As noted above, studies using direct cortical stimulation are unable to map the somatosensory cortex in nonhuman animals because of obvious limitations in the ability of animals to report their subjective sensations of touch, although later in the book, we will encounter some creative ways that researchers have devised to circumvent this problem, such as the demonstrations of the unusual tactile experiences allochiria (Mott, 1892) and synchiria (Eidelberg & Schwartz, 1971) in monkeys, discussed in chapter 13. Studies using microelectrode recordings to map the receptive fields (RFs) of individual neurons in S1, however, revolutionized understanding of the organization of the somatosensory cortex in monkeys and other animals. By using electrophysiological recordings rather than stimulations, the researchers could apply touch to different locations on the animal's body to map the region of skin that each neuron responds to—what is known as the RF of the neuron.

For example, figure 2.3 shows the somatotopic organization of two subregions of S1—Brodmann's areas 3b and 1—in the owl monkey (Merzenich et al., 1978). The detailed somatotopic organization is beautiful. Of particular note is the clear and detailed somatotopic maps of the hand, with an orderly progression of the five fingers (at least for the glabrous skin of the palmar side of the hand). It is also worth noting the good correspondence between these maps in New World monkeys and those of Penfield in humans. In both cases, there is a similar progression, with the feet at the top of the map, followed by the legs, torso, arms, hands, and face. There is thus a high level of similarity in these maps both across species (humans vs. monkeys) and across measurement techniques (verbal reports following electrical stimulation vs. RF mapping with single-unit recordings).

Multiple Somatotopic Body Maps
Another conspicuous feature of the maps shown in figure 2.3 is the fact that there are two distinct maps of the body: one in area 3b and the other in area 1. Indeed, in a landmark study, Jon Kaas et al. (1979) showed that there are four distinct maps of the body in S1 of monkeys: one in each of the four cytoarchitectonic areas (i.e., Brodmann's areas 3a, 3b, 1, and 2). While these maps are closely aligned, with representations of each body part adjacent in different maps, they are also clearly distinct, showing mirror reversals of the body at the boundary of each area. This reversal is clear in the hand maps shown in figure 2.3, with the representations of the palm in areas 3b and 1 adjacent to each other, but with two distinct sets of finger representations, one in each area.

This reversal is shown more precisely for areas 1 and 2 in figure 2.4, which shows the exact location on the fingers of RFs for neurons recorded at different places within areas 1 and 2. In area 1, the fingers are oriented pointing toward the back of the brain; in area 2, this is reversed. Subsequent research in humans using functional magnetic resonance imaging (fMRI) has identified similar reversals (Blankenburg et al., 2003; Sánchez-Panchuelo et al., 2014). Interestingly, such reversal of topographic maps between adjacent brain areas is also characteristic of retinotopic maps of visual space in the visual cortex (Sereno et al., 1995). These reversals may provide important insight into the evolutionary origins of these distinct cortical regions, as will be discussed in chapter 11.

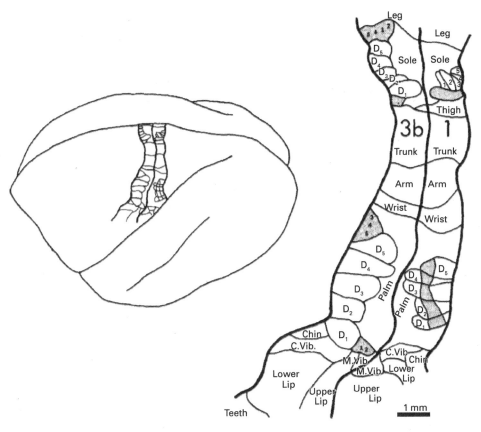

Figure 2.3
Somatotopic maps in the primary somatosensory cortex (areas 3b and 1) of the New World owl monkey. For the hand and feet, the glabrous surfaces of the palm and sole are shown in white, while the hairy skin of the dorsal surfaces is shown with gray crosshatching. While only areas 3b and 1 are shown here, there are two additional somatotopic maps in areas 3a and 2. From Merzenich et al. (1978).

An elegant study by Sur et al. (1980) quantified the distortions of these maps. They calculated a measure of magnification of different body parts, defined as the ratio of the area of the cortical representation of a skin region and the area of that skin region itself. Because cortical area and skin area are both defined in centimeters squared, this appears in both the numerator and denominator of this measure and cancels out, leaving a dimensionless measure of magnification. In areas 3b and 1 of owl monkeys, magnification for the hand was on the order of 10^{-2}, meaning that the cortical representation of the hand was about one hundredth the area of the hand itself. In contrast,

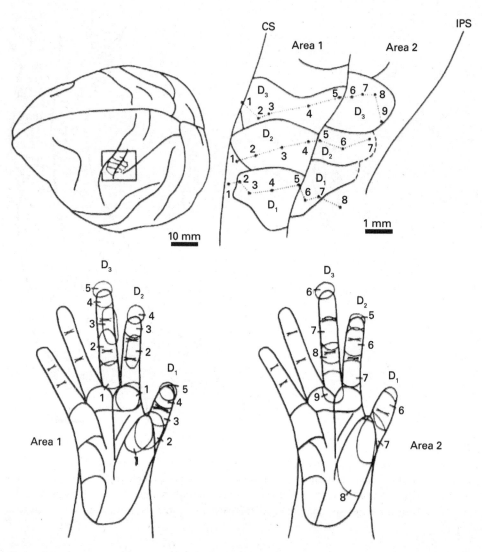

Figure 2.4
Reversal of finger orientation in different maps within S1 of a macaque monkey (Kaas et al., 1979). RFs on the skin are numbered, matching the recorded locations with the monkey's brain. In area 1, the fingers are oriented pointing away from the central sulcus. For example, for the middle finger (D3), recording site 1 is closest to the central sulcus and at the base of the finger, while recording site 5 is farthest from the central sulcus. In area 2, however, this pattern is reversed, with the fingers pointing toward the central sulcus. CS, central sulcus; IPS, intraparietal sulcus. From Kaas et al. (1979).

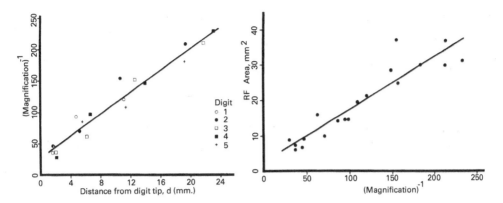

Figure 2.5
Relation between magnification and RF size in the study by Sur et al. (1980). *Left panel*: Magnification in area 3b as a function of distance from the fingertip. Magnification was highest at the fingertip and declined progressively toward the base of each finger. *Right panel*: RF area as a function of magnification for the same nineteen skin regions. Areas with high levels of magnification had correspondingly small RFs. Similar results were also found in area 1. Adapted from Sur et al. (1980).

magnification on the torso was about 10^{-4}, meaning the cortical representation was one ten-thousandth the size of the actual skin. Compared to the torso, the hand is thus magnified a hundred times!

Even this, however, understates the differences across the skin, since the number for the hand includes regions that themselves vary in magnification. Sur et al. (1980) showed a highly systematic relation between the magnification of a skin region and the RF size of neurons. They divided the representation of the fingers into nineteen regions (three on the thumb, four on each of the other fingers). As shown in the left panel of figure 2.5, magnification in area 3b was highest at the fingertip and decreased progressively toward the base of the finger. The right panel of figure 2.5 shows the relation between magnification and RF area for these same nineteen regions. There is an almost perfect linear relation between these variables. Areas, such as the fingertips, with high levels of magnification (i.e., with many neurons representing them) have correspondingly small RFs. As Sur et al. (1980) note, this allows for consistent tiling of the skin surface by a population of neurons in a somatotopic map, such that areas of high magnification are represented by a high density of neurons with small RFs, while areas of low magnification are represented by a smaller number of neurons with large RFs. The amount of overlap between the RFs of adjacent neurons as a proportion of their diameter is thus approximately constant across the body. This relation between magnification and RF size again closely parallels the organization of retinotopic maps in the visual cortex (Hubel & Wiesel, 1974).

One interesting feature of somatotopic maps is that the RFs of neurons are generally oval shaped rather than circular. Specifically, they appear elongated along the proximodistal axis of limbs. Such elongation has been found in both cats and monkeys in S1 and other areas (Mountcastle, 1957). Brown et al. (1975) found that RFs of neurons representing the legs of cats in the dorsal horn of the spinal cord were on average two to five times as large in the proximodistal limb axis as in the mediolateral axis at different locations on the leg. This pattern suggests that the spatial resolution of touch should be higher across than along the limbs, as has indeed been found for both tactile acuity (Cody et al., 2008; Schlereth et al., 2001) and precision of tactile localization (Margolis & Longo, 2015).

Studies have also used fMRI to identify and characterize somatotopic maps in humans. An elegant series of studies by Sereno and Huang (Sereno & Huang, 2006; Huang et al., 2012; Huang & Sereno, 2007) used an MRI-compatible compressed-air device to apply tactile stimulation to different parts of the body using "phase-encoded" mapping methods originally developed to map retinotopic representations in the visual cortex. The key feature of this method is to apply a "traveling wave" of stimulation across the body surface. Voxels that include somatotopic maps should respond at the frequency at which stimuli sweep across the skin. The specific body part represented by each voxel can then be estimated by the phase of this response. Sereno and Huang used this method to identify numerous somatotopic body maps throughout the parietal and frontal lobes. They found remarkable alignment between some of these tactile maps with visual maps responding to stimuli presented near the body, as will be discussed in more detail in chapter 7. Similar methods have been used to explore detailed somatotopic maps of the fingers in S1 (Kolasinski et al., 2016; Mancini et al., 2012; Sánchez-Panchuelo et al., 2010).

Another approach was used by Duncan and Boynton (2007) to measure the magnification of the fingers (D2–D5) in S1. Each pair of fingers was contrasted directly by stimulating the fingers in alternating blocks of twenty seconds. By averaging the responses from the three blocks including each finger, Duncan and Boynton estimated the magnification of each finger by calculating the number of voxels representing each one. Magnification decreased progressively from the index finger to the little finger—a pattern that has also been found in microelectrode studies of area 3b in owl and squirrel monkeys (Merzenich et al., 1987). Intriguingly, this change in magnification is paralleled by similar changes across the fingers in tactile acuity, which is highest at the thumb and index finger and declines across the hand to the little finger, both in Duncan and Boynton's own data and in other studies (e.g., Vega-Bermudez & Johnson, 2001).

Numerous studies have provided evidence for somatotopic maps in a wide variety of brain regions, including the premotor cortex (PMC; Raos et al., 2003), the supplementary motor area (SMA; Fried et al., 1991), the cerebellum (Manni & Petrosini, 2004), the basal ganglia (Zeharia et al., 2015), the posterior parietal cortex (PPC; Seelke, Padberg, et al., 2012), the secondary somatosensory cortex (S2; Disbrow et al., 2000), and the thalamus (Kaas et al., 1984). This suggests that somatotopic maps reflect a widespread pattern of organization and not something idiosyncratic to S1 and M1. There is evidence, however, that the nature of these maps and the magnification of different body parts varies. Saadon-Grosman et al. (2020) applied tactile stimuli across the body and used fMRI to map cortical magnification factors for different body parts across the brain. In S1, they found patterns of magnification similar to those reported by Penfield, with large magnification of the hand and lips and low magnification of the legs and torso. In other areas, however, this pattern was somewhat different. In the PMC, the lips were more magnified than in S1 and the hands less so. In more medial regions such as the cingulate cortex, in contrast, there was higher magnification of the torso and legs. The spatial specificity of this analysis is fairly crude in terms of both brain areas and body parts. Nevertheless, this study provides important insight into the ways that magnification changes in body maps across the brain.

Categorical Representations of Body Parts
While the cortex is a continuous sheet, there is evidence that discontinuities within somatotopic maps correspond to categorical boundaries between discrete body parts. This phenomenon has been studied in most detail in the so-called barrel cortex of rodents, in which the representation of each whisker in the somatosensory cortex is surrounded by a region of comparatively low myelinization known as a "septum" (Woolsey & Van der Loos, 1970). These septa are not merely empty buffers but appear to be involved in larger-scale spatial integration of information across whiskers and overall guidance of behavior (Alloway, 2008; Kim & Ebner, 1999). There is also evidence for similar septa separating the representations of each finger in S1 of monkeys as well as another septum separating the representations of the hand and face (Jain et al., 1998).

There is also evidence for such septa in the sensorimotor cortex in humans. More than a century ago while studying human cadavers, the German neuroanatomist Paul Flechsig (1920) identified a septum of low myelinization cutting across both M1 and S1 and separating the representations of the hand and face. Flechsig's ideas have recently been revisited by Kuehn et al. (2017), who used MRI-based myelin mapping to investigate the presence of septa in living humans. Kuehn and colleagues identified an S-shaped septum separating both M1 and S1 into regions representing the hand and face. The structural borders, identified on the basis of myelinization, also aligned closely to the borders between discrete regions identified on the basis of resting-state functional connectivity. Intriguingly, in prosimians such as lemurs and lorises, which lack a central sulcus entirely, another sulcus (the coronal sulcus) separates the hand and face representations in the sensorimotor cortex (Radinsky, 1975). Thus, it may be that the septa described by Flechsig and by Kuehn and colleagues functionally replaced the categorical boundary previously served by the coronal sulcus as the central sulcus evolved in early monkeys, presumably linked to the evolution of distinct primary somatosensory and motor cortices. This point will be discussed further in chapter 11.

Cortical Columns and Modularity in the Somatosensory Cortex

Another important feature of the organization of the somatosensory cortex is the presence of cortical columns—modular units of cortex in which similar sensitivity is found in contiguous groups of neurons spreading vertically across all six layers of the cortical sheet (Buxhoeveden & Casanova, 2002; Mountcastle, 1997). Columns vary in width between 0.3 and 1.0 mm (Mountcastle, 2005) and generally have a hexagonal shape, being surrounded by six other columns with different features. Intriguingly, the size of columns is broadly consistent across animals whose brains vary in size over several orders of magnitude (Bugbee & Goldman-Rakic, 1983; Manger et al., 1995). Evolutionary changes in brain size are thus associated with changes in the number of columns in brain areas, not the size of individual columns. As Mountcastle (2005) notes, the modular structure of columns allows for an "intermittently recursive mapping," (p. 274) which allows for multiple stimulus variables to be mapped onto the two-dimensional cortical surface.

On Mountcastle's (1997) account, there are two defining characteristics of columns in S1: place specificity and modality specificity. Place specificity is seen in the high overlap in the RF location of neurons within a column. For example, the left panel of figure 2.6 shows the location of fifteen recordings of neurons in a vertical section through the cortical sheet in area 3b of cats (Favorov et al., 1987). The right panel shows the location of the RFs of these neurons, which line up almost perfectly.

There is also evidence that place specificity coexists with larger-scale spatial integration. While the RFs of most S1 neurons are highly focal, there is also evidence that responses can be modulated by simultaneous tactile stimuli at more distant locations (Laskin & Spencer, 1979a; Reed et al., 2008). Reed et al. (2010), for example, quantified the relation between neural firing and the distance of stimuli for the location of peak activation in area 3b of owl monkeys. While most neurons show a rapid drop off in activation when stimuli are presented even 0.5 mm from the peak activation site, a minority of neurons show broader spatial tuning, responding to stimuli as much as 3 mm away.

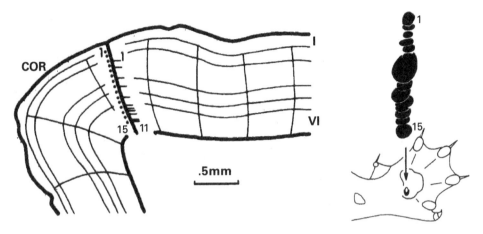

Figure 2.6
Place specificity within cortical columns in area 3b of S1 in cats. Adapted from Favorov et al. (1987).

The highly local response pattern of most neurons, combined with broader spatial integration among a minority of neurons, shows that, generally, modular organization can coexist with more widespread spatial integration.

The modality specificity of S1 columns has been more controversial. Mountcastle (1957) found that neurons in S1 of cats were highly selective in their responses to specific types of stimulation. He described neurons specific to pressure on the skin, to movement of hairs, to stimulation of deep tissues, and to movement of joints. Pressure-sensitive neurons also differed in the time course of their response to a tactile stimulus. Some neurons responded continuously throughout the period of stimulation, while other neurons responded only at the onset and offset of the stimulation. Mountcastle classified these neurons as slowly adapting (SA) and rapidly adapting (RA), respectively, and drew a direct link between the adaptation properties of these neurons and the characteristics of SA and RA mechanoreceptors, discussed in chapter 1. The idea was that SA neurons received inputs from SA1 (Merkel) afferents and RA neurons received inputs from RA1 (Meissner) afferents. As a single cortical column would include only neurons of one of these types, Mountcastle argued that columns should be thought of as computational modules for processing a specific type of afferent input at a specific spatial location on the body.

Many subsequent single-unit studies reported similar specificity of responses in S1 columns, both in cats (Dykes & Gabor, 1981; Sretavan & Dykes, 1983) and in monkeys (Paul et al., 1972; Sur et al., 1981, 1984). Sur et al. (1981, 1984), for example, showed that there are alternating bands of cortex in area 3b with SA and RA neurons. These alternating bands cut across the overall somatotopic plan of the hand, resulting in columns with each type of response on each digit. More recent research has provided similar evidence using intrinsic optical imaging, which measures changes in the light spectrum of blood, depending on brain activity (Chen et al., 2001; Friedman et al., 2004). Romo et al. (1998, 2000) showed that monkeys can behaviorally discriminate between trains of electrical pulses in the flutter range applied to area 3b, but only when RA columns are stimulated.

Other studies, however, have found less strict segregation of sub-modalities in S1 neurons. Some neurons show characteristics of both SA and RA responses (Hyvärinen & Poranen, 1978; Pei et al., 2009) or information about both touch and limb movement (Cohen et al., 1994; Darian-Smith et al., 1982). One study of neurons in S1 found both SA responses that showed a high level of orientation selectivity and RA responses with no orientation selectivity (Bensmaia et al., 2008). Another study suggested that distributed representations of stimulus amplitude and frequency were multiplexed onto the same neurons (Harvey et al., 2013). Saal and Bensmaia (2014) argue that the different responses of S1 columns should be thought of in terms of differential functional roles, or selectivity for ecologically relevant stimulus features, rather than specific classes of afferent inputs.

Hierarchical Processing of Somatosensory Information

As mentioned above, while broadly aligned somatotopic maps are apparent in all four cytoarchitectonic areas of S1 (i.e., areas 3a, 3b, 1, and 2), these maps are not identical. These regions were first distinguished as cytoarchitectonically distinct by Brodmann (1909) in his seminal studies using Nissl stains. This parcellation has been confirmed by more recent studies using observer-independent analyses of the density of darkly staining cell bodies (Geyer et al., 1997, 1999; Grefkes et al., 2001). MRI studies in humans have revealed other structural features differentiating these areas, including thickness of the cortical sheet and levels of myelinization (Sánchez-Panchuelo et al., 2014; Wagstyl et al., 2015).

There is evidence that these different maps in distinct subregions of S1 are related to processing of different types of sensory inputs (Mountcastle, 2005). Area 3a, for example, lying immediately posterior to the primary motor cortex, is dominated by inputs from non-cutaneous receptors related to proprioception, notably from muscle afferents (Tanji & Wise, 1981). Area 3b, in contrast, predominantly receives inputs from cutaneous receptors, although with a more modest input from deep receptors (Hyvärinen & Poranen, 1978; Tanji & Wise, 1981). Area 1 also receives predominantly cutaneous inputs, although with a greater input from deep receptors and a more complicated RF structure (Hyvärinen & Poranen, 1978). Finally, the major input to area 2 consists of deep receptors related to muscles and joints, although with a substantial input of cutaneous responses as well (Hyvärinen & Poranen, 1978; Merzenich et al., 1978). Thus, overall, areas 3b and 1 appear largely specialized for the processing of cutaneous inputs, and areas 3a and 2 seem to be more specialized for the processing of deep inputs relating to proprioception. At the same time, the proportion of cutaneous inputs appears to decrease gradually from area 3b to 1 to 2, with a corresponding increase in deep inputs (Iwamura et al., 1993).

The different regions of S1 differ not only in the inputs they receive but also in the amount of complexity and level of spatial integration they show. For example, a seminal series of studies by Yoshiaki Iwamura and his colleagues has shown progressive enlargement of RF size from rostral to caudal regions of S1 (Iwamura, 1998). In area 3b, neurons have RFs limited to one finger and indeed limited to a single functionally distinct part of that finger (e.g., the fingertip; DiCarlo et al., 1998; Iwamura et al., 1983b). In area 1, many neurons show increases in size, showing spatial integration either across regions of a single finger or within a region across fingers (Iwamura et al., 1983a, 1985). In area 2, many neurons show even larger spatial generality, frequently having RFs that cover multiple entire fingers (Iwamura et al., 1980, 1985). Interestingly, such cross-finger generalization is more common among the ulnar fingers (D3–D5) and less common for the thumb and index finger (Iwamura et al., 1983a; Shoham & Grinvald, 2001). This may relate to the critical role of these latter fingers in precision grips. There is also evidence that neurons in area 2 also generalize across the left and right sides of the body, showing bilateral RFs (Iwamura et al., 1994)—a point that will be discussed in more detail in chapter 13. A recent study has found that neurons in area 2 appear to code for kinematics of the arm as a whole rather than just for individual joints (Chowdhury et al., 2020).

A similar progressive increase in spatial generality and complexity across areas 3b, 1, and 2 has also been found in humans using population RF (pRF) modeling of fMRI data (Schellekens et al., 2021). This method estimates the average RF properties across the set of neurons within a single fMRI voxel. Consistent with the microelectrode recordings in monkeys described in the previous paragraph, there was a progressive increase in the size of pRFs from area 3 to area 1 to area 2. Interestingly, in each of these areas, there was also a progressive increase in the size of pRFs from the thumb to the little finger, mirroring results on magnification and tactile acuity described above.

Posterior to S1, even higher levels of integration have been reported. Duffy and Burchfiel (1971) recorded from neurons in posterior parietal area 5 of macaque monkeys, finding complex responses to movements of multiple joints. Other cells responded to both tactile and joint movement stimuli, while others had bilateral responses or even responses to all four limbs. Complex patterns of both excitatory and inhibitory convergence were reported, which were not random but appeared to reflect patterns

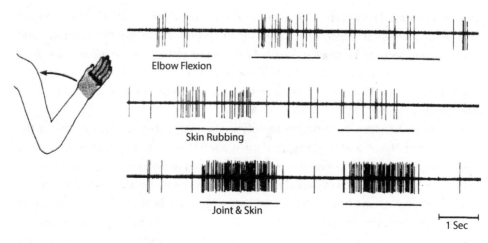

Figure 2.7
Integration of tactile and kinesthetic information in posterior parietal area 5 of a macaque monkey. This neuron shows modest responses to flexion of the elbow (top row) and to tactile stimulation of the hand dorsum (middle row). When both stimuli are presented simultaneously (bottom row), there is a vigorous response, much stronger than the sum of responses to each stimulus individually. From Sakata et al. (1973).

characteristic of ecologically relevant approach and withdrawal behaviors. Similar results were reported by Sakata et al. (1973), who also reported neurons coding complex combinations of tactile and kinesthetic information. For example, the neuron shown in figure 2.7 integrates information about flexion of the elbow joint and tactile stimulation of the hand dorsum. While the neuron responds at low levels to joint movement or to touch alone, robustly over-additive responses are seen when both stimuli are presented at the same time. We can only speculate on the precise function of such responses in the animal's behavior. The combination of tactile cues with information about limb position and movement, however, may be involved in localization of touched objects in external space (i.e., tactile spatial remapping, discussed below) and in haptic object recognition, as will be discussed in chapter 5.

In an influential study, Randolph and Semmes (1974) surgically removed different subregions of S1 in monkeys and investigated the consequences for a range of tactile tests. Following removal of area 3 (both 3a and 3b), monkeys were globally impaired on all tactile tests. In contrast, more selective deficits were seen following removal of areas 1 and 2. Monkeys with damage to area 1 showed the largest deficits in tasks requiring perception of texture, such as discrimination of hard versus soft or of rough versus smooth objects. Monkeys with damage to area 2 showed the largest deficits in tasks requiring perception of geometric form, such as discrimination of convex versus concave surfaces or of squares versus diamonds. These results are consistent with a serial organization in which area 3 provides separate inputs to areas 1 and 2. This serial model has been supported by subsequent studies showing that removal of area 3 abolishes responses in area 1 (Garraghty et al., 1990). In contrast, removal of area 1 had no apparent effect on the responses of area 3b neurons.

There is also evidence for serial organization of S1 and S2, a region on the surface of the Sylvian fissure. Surgical removal of the S1 representation of a specific body part leads to the complete elimination of responses in S2 on that body part, without apparent effects on other body parts (Pons et al., 1987). In contrast, removal of S2 has no apparent effect on S1 responses, providing strong evidence for serial organization of these areas. Subsequent work by the same group showed a striking degree of specificity of these effects (Pons et al., 1992). Removal of areas 3b and 1, which predominantly receive cutaneous inputs, abolished responses in regions of S2 responsive to cutaneous stimulation but not in regions responsive to deep stimulation. Conversely, removal of areas 3a and 2, which receive more inputs from deep receptors, abolished responses in regions of S2 responsive to deep inputs but not to cutaneous inputs.

Beyond S1, there is evidence for two broad processing streams. One "dorsal" stream progresses from S1 to areas 5 and 7 of the PPC. The other "ventral" stream progresses from S1 to S2 and then on to the posterior insula. How these streams should be characterized, however, remains uncertain. Mishkin and colleagues (Mishkin, 1979; Pons et al., 1987) argue that the S1–S2–insula pathway should be thought of as a corticolimbic pathway serving perceptual learning and memory. This perspective is related to Mishkin's work on the bifurcation of the visual system into two broad processing streams (Ungerleider & Mishkin, 1982): a ventral stream terminating in the temporal lobes and supporting semantic object recognition and conscious visual experience, and a dorsal stream terminating in the PPC and supporting spatial vision. The corticolimbic, ventral tactile pathway would thus be analogous to the ventral visual stream and would link the somatosensory cortex with regions of the medial temporal cortex involved in learning and memory formation. Dijkerman and de Haan (2007) suggested that these pathways reflect touch for conscious perception versus touch for action, in direct parallel to Milner and Goodale's (2006) influential reinterpretation of the visual ventral and dorsal pathways as reflecting vision for perception versus action.

Another perspective on the function of the two somatosensory pathways comes from the distinction made by Per Roland and his colleagues between perception of the microgeometry versus the macrogeometry of felt objects. Using positron emission tomography (PET), Roland et al. (1998) and Bodegård et al. (2001) found activations in

S2 for judgments of object texture or roughness, which they call "microgeometry," and activations in the PPC for judgments of object size and shape, which they call "macrogeometry." These results were consistent with a large study of ninety-four patients who had had brain regions surgically removed to treat a variety of diseases (Roland, 1987). Other patient studies have reported similar results (Bohlhalter et al., 2002; Hömke et al., 2009). On this view, the pathway terminating in the insula is related to the perception of microgeometrical object properties, while the pathway terminating in the PPC is related to perception of macrogeometrical properties.

Body Representations and Somatoperceptual Information Processing

In 1911, eight years after volunteering to have his own nerve cut, Henry Head, together with his colleague Gordon Holmes, published an extraordinary account of deficits in higher-level somatosensory function in stroke patients (Head & Holmes, 1911). Loss of basic somatosensory function on one side of the body, hemianesthesia, is a common consequence of strokes, as will be discussed further in later chapters. The patients described by Head and Holmes, however, did not have hemianesthesia, as they could generally detect when they had been touched. Their problems appeared to be with other aspects of touch, which appeared to be coded at stages of processing subsequent to their basic detection. For example, the patients might be able to detect *that* they had been touched, but they were uncertain about *where* on the body the touch had occurred.

Head and Holmes argued that sensory signals coming from the body needed to be integrated with central representations of the body in the brain—what they called "schemata." This concept of schemata has been highly influential in experimental psychology, notably in the work of Sir Frederic Bartlett (1932) on memory. Within somatosensation, it has also been central to more recent theories that have emphasized the role of mental body representations in the brain in the interpretation of primary sensory signals (Longo et al., 2010; Medina & Coslett, 2010; Schwoebel & Coslett, 2005; Tamè et al., 2019). In the second part of this chapter, I will discuss several domains of higher-level processing of somatosensory information.

Tactile Localization on the Skin

The ability to localize where on the skin a touch occurs is among the most fundamental of all tactile abilities. One source of evidence comes from studies using a technique known as microneurography, in which an electrode is placed directly into a peripheral nerve of an awake human, allowing both recording from and stimulation of individual nerve fibers. Remarkably, electrically stimulation of even single-nerve fibers in the human median nerve can produce sensations that are clearly localized on the skin and which are accurate in the sense that they align with the RF of that fiber (Ochoa & Torebjörk, 1983; Schady et al., 1983).

Head and Holmes (1911) described patients who had a seemingly intact ability to detect that they had been touched but were unable to tell where on the skin the touch had been located—a condition sometimes called "atopognosia." One patient described their experience: "I feel you touch me, but I can't tell where it is; the touch oozes all through my hand" (p. 139). In some cases, a touch to the hand was judged as being on the face or torso. Head (1918) described other patients who could tell which finger had been stimulated but had no idea where on that finger the touch had been located, as if the touch had "spread out" to cover the entire finger. Head and Holmes argued that localizing touch on the skin requires that immediate sensory signals be integrated with a schema—what has since come to be called the "superficial schema."

Disruption of tactile localization on the skin following brain damage has since been described by numerous other researchers (Birznieks et al., 2016; Critchley, 1953; Denny-Brown et al., 1952; Halligan et al., 1995; Rapp et al., 2002; White et al., 2010). In some cases, the patterns of mislocalization are highly systematic and preserve the overall somatotopic pattern of body parts, even as the mapping between actual and perceived location is distorted. A common finding is that localization errors on the hand tend to involve shifts of location proximal to the true location (Critchley, 1953; Halligan et al., 1995; Rapp et al., 2002). In some cases, such proximal shifts for stimuli at relatively distal locations on a body part coexist with distal shift of more proximal locations (Denny-Brown et al., 1952; Rapp et al., 2002), as if perceived locations were being funneled toward the center of the limb or body part. For example, figure 2.8 shows the pattern of mislocalizations of patient R.S.B., who suffered extensive damage to his left parietal lobe (including both S1 and S2 as well as more posterior regions) following a stroke (Rapp et al., 2002). Touches applied to the fingertips were perceived as being applied to the middle of the fingers, while touches on the wrist were perceived farther up the hand. The whole structure of the hand appears compressed. Critically, however, despite these distortions, the *relative* location of touches with respect to each other is preserved. The mislocalizations preserve the somatotopic structure of the hand but systematically misplace each point onto the skin surface. This suggests that R.S.B.'s deficit involves not somatotopic representation as such but rather the way in which locations within such maps are related to the body surface.

Poeck and Orgass (1971) wrote critically about the concept of the superficial schema, saying, "It is difficult to see the difference between the superficial schema as an organized model and the 'homunculus' of the cortical sensorimotor representations" (p. 258).

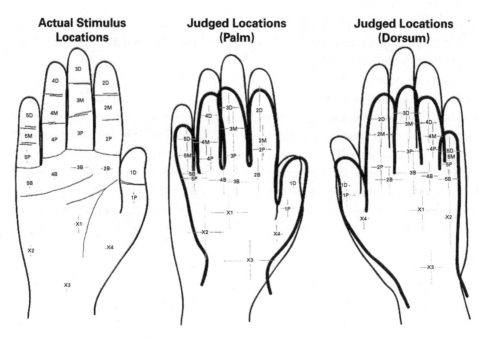

Figure 2.8
Tactile mislocalizations in patient R.S.B. Adapted from Rapp et al. (2002).

However, there are both theoretical and empirical reasons for differentiating mechanisms underlying tactile detection and localization. First, the localization of a stimulus within a somatotopic map is not in itself sufficient to localize the stimulus on the body surface, as there is no hardwired link between these locations (Longo et al., 2010; Medina & Coslett, 2010). If there *were* such a hardwired linkage, plastic changes to somatotopic maps, such as those seen following peripheral trauma (Merzenich et al., 1984; Pons et al., 1991) or learning (Elbert et al., 1995; Pascual-Leone & Torres, 1993), would invariably produce mislocalization of tactile stimuli—that is, if a set of neurons had RFs on the middle finger before training but on the index finger after training, touch applied to the index finger that activates these neurons should produce an experience of being touched on the middle finger. While examples of such referred sensations have been reported following massive cortical reorganization in limb amputation (Halligan et al., 1993; Ramachandran et al., 1992) or plasticity induced by simultaneous tactile co-activation of multiple fingers (Schweizer & Braun, 2001), most instances of plasticity do not produce mislocalizations. For example, while mislocalizations between fingers have been reported for three-finger braille readers, this does not occur for one-finger braille readers (Sterr et al.,

1998), although large plastic changes in the somatosensory cortex have been found in these individuals (Pascual-Leone & Torres, 1993).

Many of the patients described above who had impaired ability to localize stimuli also had a range of other somatosensory impairments, including some (if not total) impairment in stimulus detection (Anema et al., 2009; Birznieks et al., 2012, 2016; Halligan et al., 1995; White et al., 2010). For example, patient A.H. was unable to localize stimuli that he was able to detect, but he was only able to detect 33 percent of delivered stimuli in the first place (Halligan et al., 1995). A recent case report of patient D.S., in contrast, has provided strong evidence that tactile localization can be selectively impaired despite normal detection ability (Liu et al., 2020). D.S. had a stroke that affected the subcortical white matter in the right hemisphere. D.S.'s tactile detection thresholds were measured at the middle fingertip and were virtually identical between the two hands and within the normal range. In contrast to his intact detection abilities, D.S. showed striking mislocalizations of tactile stimuli on his contralesional left hand. Specifically, he perceived touches as shifted leftward of their true location, relative to the proximodistal hand axis. Notably, this leftward bias means that the reference frame of mislocalizations is not somatotopic. Rather, touches were mislocalized toward the ulnar (i.e., little finger) side of the hand when the hand was held palm down but toward the radial (i.e., thumb) side of the hand when it was held palm up. When D.S.'s hand was rotated ninety degrees relative to the rest of the body, these mislocalizations remained constant relative to the hand, showing that they were defined in a hand-centered reference frame. This pattern of impaired localization together with intact detection indicates that the process of localization is distinct from the initial stages of somatosensory processing on which detection depends.

Further evidence that tactile localization relies on mechanisms distinct from detection comes from brain stimulation and neuroimaging studies in healthy humans. One study used a method known as transcranial magnetic stimulation (TMS) in which magnetic pulses are delivered into the brain by a coil placed on the scalp, allowing reversible disruption of specific neural processes. Pulses applied over the sensorimotor cortex produced more severe and longer-lasting effects on localization than on detection (Seyal et al., 1997). Another study found that tactile localization—but not detection—was selectively disrupted by pulses of three magnetic pulses delivered to S1 starting 150 ms (but not 300 ms) after stimulus onset (Porro et al., 2007). Finally, an fMRI study comparing judgments of stimulus orientation versus location found that localization judgments produced selective activations of regions in the right temporoparietal junction (Van Boven et al., 2005). One recent study has investigated the neural processes underlying tactile localization using an electroencephalography (EEG) repetition-suppression paradigm (Miller et al., 2019). Two successive touches were applied to the forearm, either at the same location or at different locations. When the second location was the same, clear suppression was found in S1 and M1 from around 50 ms and in the PPC from around 80 ms.

Other studies have described patients who are unable to detect tactile stimuli but are nevertheless strikingly accurate when forced to guess where the stimulus was applied (Paillard et al., 1983; Rossetti et al., 1995). Paillard and colleagues called this condition "numbsense"—in analogy to the better-known condition of blindsight in which seemingly blind patients can nevertheless localize and act on visual stimuli they deny being able to see (Weiskrantz, 2009). For example, patient J.A. experienced profound somatosensory loss, which affected the entire right side of his body following a stroke that damaged the thalamus (Rossetti et al., 1995). Sensitivity to all tactile and proprioceptive stimuli on the right side was absent, including to pressure, temperature, vibration, pain, and passive movement. In striking contrast, when J.A. was touched at one of six locations on his right hand and encouraged to point to the stimulated location with his left hand, he pointed to the correct location on 45 percent of trials—well above chance (i.e., $1/6 = 16.7$ percent). Similarly, when asked to point to one of eight stimulated locations on his forearm, he pointed correctly 43 percent of the time, again well above chance (i.e., 12.5 percent). Remarkably, however, when asked to indicate which location was touched verbally or by marking the corresponding location on an arm silhouette, J.A.'s performance was at chance levels. A similar pattern has been reported in other patients (Aglioti et al., 1996). Other studies have provided evidence for numbsense related to properties other than location, including stimulus motion (Aglioti et al., 1998; Brochier et al., 1994), object size (Maravita, 1997), object identity (Berti et al., 1999), and shape and texture (Hanada et al., 2021).

These findings have proved controversial. Harris et al. (2004, 2006) have argued that the apparent dissociation between detection and localization in putative cases of numbsense could reflect differences in the nature of the two tasks. Specifically, the yes/no nature of detection is susceptible to bias in the sense that patients have to decide how liberal or conservative a criterion to use. In contrast, for localization, no such criterion is involved. Thus, it is possible that these patients have a conservative response criterion for detection rather than an actual complete absence of awareness of stimulus presence. Harris and colleagues tested healthy participants and showed that when tactile detection and localization performance were both quantified using bias-free signal-detection measures, that sensitivity for detection was always higher than for localization. Their results did replicate numbsense in healthy participants in the sense that they found above-change localization performance on trials in which participants denied having felt any stimulus. However, they argued that this was an artifact of participants using a conservative response criterion for detection. Harris and colleagues concluded that once response biases are taken into account, the evidence suggests that tactile detection is a necessary prerequisite for tactile localization, with detection and localization occurring as successive, serial processing stages.

This view has been challenged by a recent study by Ro and Koenig (2021), who reported numbsense in healthy participants induced by TMS to the primary somatosensory cortex. They applied near-threshold electro-cutaneous stimuli to the left index or ring finger, although on half the trials, no stimulus was presented at all. On each trial, participants made forced-choice judgments of whether they felt a stimulus and of which finger had been stimulated. Single-pulse TMS applied to the hand area of right S1 led to clear reductions of detection sensitivity (d'). Critically, however, TMS did not appear to affect response bias. Ro and Koenig showed that on the TMS trials in which participants reported not feeling a touch that was actually present, their subsequent guesses about which finger had been stimulated were significantly above chance levels. They interpret these results as evidence for numbsense in healthy participants, suggesting that tactile localization can be driven by neural pathways bypassing S1. While these results are striking, it is not clear that they fully address the concerns raised by Harris and colleagues. Even if TMS didn't *change* the decision criteria that participants used for detection, participants may have adopted a relatively conservative criterion in all conditions, such that above-chance localization could be supported by modest levels of conscious detection that nevertheless remained below the respond threshold for the detection task.

Proprioceptive Localization of the Hand in Space
Ian Waterman, also known as patient I.W., was nineteen years old and working as a butcher on the island of Jersey in the English Channel when he became sick. Although it seemed at first that he might have the flu, he eventually found that he was unable to feel anything from his body. He had suffered a viral infection that destroyed most of the afferent fibers in his spinal cord (Cole, 1995, 2016). Although he can still feel some level of temperature and pain from his body, he receives no tactile or proprioceptive signals. Remarkably, however, his efferent motor fibers remain physiologically intact, allowing him to move his body. Despite this, in the initial period after his illness, he was almost entirely unable to produce skilled actions. Remarkably, over a period of years, I.W. was able to regain an impressive range of motor abilities by using focused visual monitoring to replace the effortless and automatic monitoring his brain had done before his illness. His actions thus require constant attention and lack the automaticity of normal everyday

actions. His neurologist Jonathan Cole (1995) has thus referred to I.W.'s life as a "daily marathon."

I.W.'s case highlights the importance of the perceptual ability known as "proprioception," which refers to our ability to tell the posture and location of our limbs in space, even when we can't see them. As noted above, Head and Holmes (1911) reported numerous patients whose primary deficit following a stroke was an inability to perceive the location of body parts in space in the absence of vision. Indeed, these authors wrote that the "inability to recognize the position of the affected part in space is the most frequent sensory defect produced by lesions of the cerebral cortex" (p. 157). In addition to the superficial schema underlying tactile localization, Head and Holmes proposed another schema, now generally known as the "postural schema" or "body schema," which keeps track of body posture by incorporating information from a variety of relevant sensory signals: "By means of perpetual alterations in position we are always building up a postural model of ourselves which constantly changes. Every new posture or movement is recorded on this plastic schema, and the activity of the cortex brings every fresh group of sensations evoked by altered posture into relation with it" (p. 187).

Although his illness had not affected his brain, patient I.W., like the stroke patients described by Head and Holmes, could not tell where the different parts of his body were in space when he couldn't see them. Gallagher and Cole (1995) interpreted this as a near-complete loss of the body schema. Other studies of patients without proprioception have similarly found that this produces dramatic impairments in the ability to produce skilled actions. Rothwell et al. (1982) studied patient G.O., who, like I.W., suffered an acute illness that appeared to selectively damage the sensory nerves within the spinal cord. G.O.'s ability to perform simple motor behaviors was intact. He could tap his thumb successively against the other fingertips and trace shapes in the air with his fingers. Nevertheless, he was severely impaired with regard to more complex everyday activities, such as using a knife and fork, writing, or buttoning his shirt. In these cases, the movements cannot be made in a purely feedforward manner but rather must be continuously monitored and updated—a point that will be discussed further in the next two chapters.

It is difficult to precisely localize the brain areas involved in proprioceptive localization of the hand, given that deficits in this ability result from a diverse range of brain lesions. Single-unit recording studies in monkeys have shown many neurons in posterior parietal area 5 that encode arm postures (Duffy & Burchfiel, 1971; Graziano et al., 2000; Sakata et al., 1973). Studies in humans have implicated the superior parietal lobe in maintaining postural representations of current body state, using fMRI both in healthy participants (Filimon et al., 2009; Pellijeff et al., 2006) and in brain-damaged patients (Wolpert et al., 1998). There is also evidence for laterality effects, with such deficits more severe, on average, following lesions of the right parietal lobe than the left (Sterzi et al., 1993; Vallar et al., 1993).

This older literature has generally implicated the PPC in proprioceptive representations. One recent study, however, found rich representations of hand postures in both S1 and M1 of macaque monkeys (Goodman et al., 2019). The authors used chronically implanted electrodes to record simultaneously from large populations of neurons in areas 3a and 2 in S1 as well as in M1. Neural responses were recorded while monkeys made naturalistic grasping actions of twenty-five different objects of a wide range of shapes and sizes. Simultaneously, a system of fourteen cameras was used to track the movement and orientation of the joints in the monkeys' hands. In all three regions, neurons showed sensitivity to complex postures, formed by combinations of the positions of several joints. For example, on average, eight joints were needed to account for 90 percent of the explainable variance in a single neuron's response. Moreover, neurons appeared to encode the posture of the entire hand in the sense that the joints coded by a single neuron did not tend to be spatially contiguous (e.g., multiple joints from the same finger) but were widely spread across the hand. There was thus no evidence for a hierarchical arrangement across brain regions, with the complexity of proprioceptive representation changing progressively across successive processing steps. This suggests that the nature of proprioceptive representations may differ qualitatively from that for touch, as described earlier in this chapter.

As described in the previous chapter, several types of sensory signal provide information about body posture, including receptors in joints, skin, and muscles (Proske & Gandevia, 2012). Each of these signals can provide information about joint *angles*—that is, the relative flexion or extension of each joint. However, none of them provide information analogous to a GPS signal, which would allow determination of the absolute position of a body part in space. To determine the absolute location of a body part in space, joint angles must be combined trigonometrically with information about the length of segments between joints. This length information is not specified by immediate afferent signals, suggesting that it comes from a stored representation of body size and shape (Longo et al., 2010).

Longo and Haggard (2010) developed a procedure to measure this stored representation of body size and shape. As shown in figure 2.9, participants placed their hand underneath an occluding board and used a long baton to judge the location of different landmarks on their hand (i.e., the knuckle and tip of each finger). By comparing the relative judged location of each landmark to each other, implicit perceptual maps of hand size and shape could be constructed and compared to actual hand form. The key idea underlying this paradigm is that whereas the perceived location of any individual

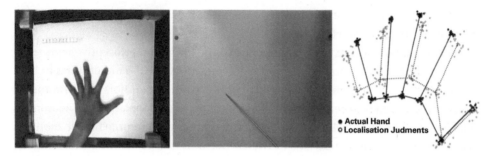

Figure 2.9
The proprioceptive hand mapping procedure of Longo and Haggard (2010).

landmark will be affected by numerous factors, most of these will affect multiple landmarks in the same way. For example, if a participant misperceives their elbow as farther forward than it actually is, this will produce a shift in the perceived location of both the knuckle and tip of each finger. By focusing on the relative location of judgments relative to each other (ignoring their location relative to the true location of each landmark), the representation of the internal metric configuration of the hand can be estimated.

Typical results from this paradigm are shown in the right panel of figure 2.9. As can be seen from the figure, the hand maps are massively distorted in a highly consistent way across people.

A limitation of this method is that it relies on the presence of discrete landmarks with known verbal labels. Other studies have applied a similar logic based on localization judgments cued by touch or by a visual cue shown on an image of a hand (Longo et al., 2015). As with localization of hand landmarks, these maps show strong overestimation of hand width compared to length.

That the distortions seen in these maps come from central representation of the hand is shown by the fact that very similar distortions are found when participants are asked merely to imagine that their hand is lying flat underneath the board (Cocchini et al., 2018; Ganea & Longo, 2017). Similarly, patient I.W., who lacks proprioceptive afferent signals entirely, showed similar distortions (Miall et al., 2021), as did a woman who was born without a left arm (Longo et al., 2012). Similar distortions are also found when participants respond while blindfolded (Longo, 2014) or by giving verbal instructions to an experimenter to move the baton (Longo, 2018), suggesting that they do not arise from biases in vision or active pointing. Nevertheless, there is also evidence for plasticity of these maps, which differ in expert magicians (Cocchini et al., 2018), elite baseball players (Coelho et al., 2019), and people with experience of using sign language (Mora et al., 2021).

A series of studies by Peviani et al. (2020) and Peviani and Bottini (2018) has provided evidence that these distortions also affect active movement of the hand. In these studies, participants were asked to move different parts of their unseen hand to the location of a visual cue, allowing construction of implicit hand maps underlying active movement. Similar distortions were found as in more perceptual tasks, although their magnitude was somewhat reduced, providing evidence for partial correction of distortions for active movements. The precise role of this distorted map in skilled action remains unclear (for recent discussion of this issue, see Bassolino & Becchio, 2023, and Longo, 2023).

Tactile Localization in External Space
When we hold a beach ball with the fingertips of both hands, we receive tactile inputs on ten noncontiguous regions of skin. Our experience, however, is of perceiving a single—and contiguous—ball. This point was made forcefully by James Gibson (1962), who labels this phenomenon the "unity of phenomenal experience," and it raises fundamental questions about the relation between our subjective experience of touch and the brain mechanisms underlying somatosensation. Specifically, it highlights the fact that, in most everyday situations, we interpret tactile sensations as being about the objects we are touching, not about our skin or body itself.

An intriguing exception to Gibson's point about phenomenal unity comes from an illusion generally referred to as "Aristotle's illusion." The main feature of this effect is the illusory perception of two distinct objects when a single object is placed between crossed fingers. In his *Metaphysica*, Aristotle (1908a) writes that "touch says there are two objects when we cross our fingers, while sight says there is one" (p. 1011). Similarly, in Part 2 of *On Dreams*, Aristotle (1908b) writes that "when the fingers are crossed, the one object [placed between them] is felt [by the touch] as two; but yet we deny that it is two; for sight is more authoritative than touch. Yet, if touch stood alone, we should actually have pronounced the one object to be two." (p. 460).

If you have not experienced the illusion already, it is worth trying it yourself. Take your little finger and cross it underneath your ring finger. Close your eyes, then roll a small ball (such as a marble) between your crossed fingertips. In most people, this produces a vivid experience of two objects, exactly as Aristotle described more than two thousand years ago. The effect works equally well on the tip of your nose. Fabrizio Benedetti (1986) labeled this tactile duplication "diplesthesia" in analogy to diplopia, or duplication of visual objects.

Various interpretations of the Aristotle illusion have been presented over the years. Schilder (1935), for example, suggested that people have difficulty perceiving touch when the body is in an unusual posture. However, beyond vagueness about what counts

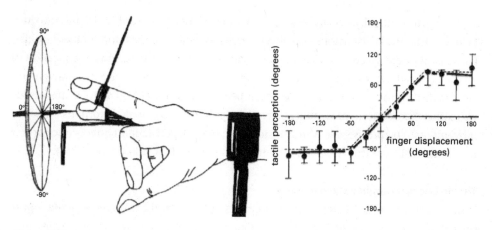

Figure 2.10
Benedetti's (1985) experiment showing that the location of touch is constrained by the limits of voluntary movement of the finger. Adapted from Benedetti (1985).

as "unusual," it is easy to identify postures that are unusual in the sense of being uncommon in daily activities but which nevertheless do not result in illusory experiences. Tastevin (1937), in contrast, emphasized the fact that crossing the fingers is not merely unusual but is artificial in the sense that it cannot be achieved through voluntary muscular action. Tastevin suggested that the crossed fingers are perceived to be in the most extreme position that could be voluntarily produced by ordinary movement of the fingers and that forcing the fingers beyond this limit produced no further change in their perceived location.

Evidence broadly consistent with Tastevin's interpretation was provided by an elegant experiment by Benedetti (1985). As shown in the left panel of figure 2.10, Benedetti used a string to pull the participant's right middle finger to different positions relative to the ring finger. After tactile stimuli were applied to the tip of both fingers, participants judged the angular location of the touch on the middle finger relative to the ring finger. Within the central range of postures (i.e., those that can be produced by ordinary movement of the fingers), perceived location of touch scaled almost perfectly with actual location, as shown in the right panel of figure 2.10. In striking contrast, for more extreme postures, which cannot be achieved voluntarily, no further displacement of touch was observed. Rather, touch on the middle fingertip was perceived as being located at the extreme limit of voluntary movement of the finger, even when the finger was moved beyond that limit by the experimenter.

In terms of location, we do not usually think about where touch is on the skin, but rather we implicitly combine this information with proprioceptive information about body posture to perceive the location of objects in external three-dimensional space. Suppose you wake up in the middle of the night while staying in an unfamiliar hotel room. Trying to find your way to the bathroom, you grope in the dark to find the light switch, eventually feeling your fingertip brush across it. You would not think, "Aha, there's the switch, right there on my left hand." This interpretation of the location of the switch in terms of which bit of skin it had stimulated is almost comically inappropriate in this situation. We conceive of the light switch as a distinct object, with a location within the three-dimensional space of the room.

A striking demonstration of this point comes from a study of tactile neglect by Moscovitch and Behrmann (1994). Patients with neglect fail to notice stimuli on the left side of space, whether in vision or touch, especially when there are competing stimuli on the right side of space. With the patient's eyes closed, Moscovitch and Behrmann applied simultaneous touches to the right wrist on both the thumb and little finger sides of the hand. When the patient's hand rested palm down on the table, they frequently failed to notice the stimulus near their thumb, classic neglect of the left side of space. The critical question was what would happen when the hand was flipped upside down with the palm facing up. In this posture, it is the little finger side of the hand that is toward the left. Indeed, in this posture, patients failed to notice touches near the little finger. These results show that the impairment of spatial attention in these patients is not for specific location on the skin but rather is related to a representation of stimulus location in which tactile location on the skin is integrated with proprioceptive information about the posture of the hand in space. This process of integrating tactile and proprioceptive information to perceive the location of touched objects in external space is known as "tactile spatial remapping."

Yamamoto and Kitazawa (2001a) developed an elegant paradigm for investigating this process. On each trial, tactile stimuli were presented to the participant's right and left hands, and the participant judged which hand had been touched first by wiggling the index finger on that hand. By systematically varying the delay between the touches on the two hands, Yamamoto and Kitazawa measured the precision with which participants could discriminate the temporal order of the stimuli. The critical manipulation was when participants were asked simply to cross their arms over each other, so that the left hand was in the right side of space, and the right hand in the left side of space. It is important to note that this change in posture is logically irrelevant to the task that the participants are performing, which concerns only which hand is touched and not where the touch is in external space. Nevertheless, performance was dramatically impaired with crossed hands. Indeed, even with 300 ms difference between the two stimuli, participants' performance was only modestly above chance.

The study by Yamamoto and Kitazawa (2001a) also provides clues about the time course of tactile spatial remapping. In the uncrossed posture, performance approaches ceiling when the delay between touches is about 100 ms. The time course of this process was measured in an elegant experiment by Azañón and Soto-Faraco (2008), who asked participants to make rapid judgments about the spatial position (top/bottom) of visual stimuli that appeared on a board covering the participants' hands. They were instructed to ignore task-irrelevant tactile stimuli applied to either hand. When the hands were uncrossed, a standard visuo-tactile cueing effect was present (Spence et al., 2000), such that participants responded faster when touch was applied to the hand immediately underneath the light that came on. In striking contrast, when the hands were crossed such that right hand was underneath the left set of lights and vice versa, the nature of the cueing effect reversed, depending on the temporal relationship of the touch and light. When touch was delivered several hundred milliseconds before the light, a cueing effect was found similar to the uncrossed posture. However, when touch was delivered briefly before (e.g., <60 ms) the light, the cueing effect was reversed—that is, responses were enhanced to visual stimuli not on the side of space where the touched hand *actually* was but rather to the side of space where the hand *usually* is. These results suggest that touch is rapidly remapped based on the initial feedforward sweep of neural activity into the somatosensory cortex, which relies on a stored, standard representation of body posture. The later cueing effect, in contrast, reflects recurrent processing from the PPC incorporating real-time information about current body posture.

Research using TMS has implicated the PPC in the process of tactile spatial remapping (Azañón et al., 2010). Participants' left arms were held at different heights with their fingers pointing upward. On each trial, they received a touch on their left forearm and on their face, and they had to judge whether the touch on the arm was higher or lower than the touch on the face. TMS to the right PPC (but not at a control location on the top of the head) reduced the precision of these judgments (i.e., the slope of the psychometric function). Critically, no such effect of TMS was found when participants judged the location of touch on the skin or the proprioceptively experienced location of the arm. This indicates that remapping is a distinct computational process from tactile localization on the skin or proprioceptive localization of the body in space.

Interestingly, while the crossed-hands deficit occurs in the absence of immediate visual inputs (e.g., with the eyes closed), there is evidence that a developmental history of visual experience is crucial for the effect. Congenitally blind individuals, who have never had any visual experience, show no crossed-hands deficit at all (Röder et al., 2004). In contrast, individuals who became blind later in life show a crossed-hands deficit indistinguishable from that of sighted individuals, even after decades of blindness. This indicates that visual experience early in life is critical for shaping the reference frames involved in tactile spatial remapping. H.S. was born blind due to dense bilateral cataracts, which were removed when he was two years old (Ley et al., 2013). Tested at the age of thirty-three, H.S. showed no hint of a crossed-hands deficit, despite more than three decades of visual experience. Similarly, L.M. had her cataracts removed at the age of seven, and she also showed no deficit when tested a decade later (Azañón et al., 2018). These results indicate the existence of a sensitive period during the first few years of life for acquiring normal tactile remapping. In contrast, however, a group of individuals who had had cataracts removed in the first six months of life showed a normal crossed-hands deficit (Azañón et al., 2018), indicating that this sensitive period does not include the earliest phase of infancy.

Tactile Distance Perception
Illusions of tactile distance perception have been described since the very first psychophysical investigations of the sense of touch in the nineteenth century. Ernst Weber (1834) reported a fascinating illusion in which the distance between the two points of a compass moved across the skin feels larger when applied on a relatively sensitive region of skin (e.g., the palm of the hand) than when applied on a less sensitive region (e.g., the forehead). This illusion is now generally known as "Weber's illusion" and has been replicated by several subsequent studies that have shown a systematic relation between the spatial sensitivity of skin surfaces and the perceived distance between pairs of touches (e.g., Cholewiak, 1999; Miller et al., 2016). Because, as discussed above, tactile spatial acuity is itself linked to cortical magnification factors (Duncan & Boynton, 2007), Weber's illusion can be thought of as a perceptual echo of the distortions of the Penfield homunculus.

Other studies have shown that there are large anisotropies of tactile distance perception on the hand, with the distance between touches oriented across hand width felt as substantially larger than equivalent distances oriented along hand length (e.g., Green, 1982; Fiori & Longo, 2018; Longo & Haggard, 2011). In one study, for example, distances across the hand dorsum were perceived as about 40 percent bigger than those along the hand (Longo & Haggard, 2011). Qualitatively similar biases are also apparent on the palm, although they are smaller in magnitude (Longo, 2020).

Illusions that alter the perceived size of body parts produce corresponding changes in perceived tactile distance (de Vignemont et al., 2005; Tajadura-Jiménez et al., 2012; Taylor-Clarke et al., 2004). Taylor-Clarke and colleagues used a video camera together with magnifying and reducing mirrors to give participants the visual experience of seeing their hand reduced in size by half, while their forearm was doubled. Participants judged whether the perceived distance between pairs of touches on the index finger was bigger or smaller than the distance on the forearm. At baseline, the classic Weber's illusion was found, with identical distances judged as larger on the finger than on the forearm. After one hour of distorted vision, however, the magnitude of this effect was significantly reduced, suggesting that visual experience of altered arm size had produced corresponding alterations of perceived tactile distance. Another study by Tajadura-Jiménez and colleagues used an audio-tactile mismatch to alter the perceived length of the arm. Participants tapped a surface with their right hand extended off to their side. A set of speakers was used to create apparent auditory feedback indicating that the taps had occurred farther away, as if the participant's arm was longer than its actual length. Participants judged the relative distance between pairs of touch on the affected right arm compared to the contralateral left arm. When the auditory feedback indicated that the arm was double its true length, there was a clear increase in the perceived distances of touches on the right arm compared to the left arm. Other studies have found modulation of tactile distance perception following tool use (Canzoneri et al., 2013; Miller et al., 2014), as will be discussed in more detail in chapter 10.

There is also evidence for categorical perception of tactile distance across body-part boundaries, such as joints (de Vignemont et al., 2008; Le Cornu Knight et al., 2014, 2017, 2020). De Vignemont and colleagues asked participants to make absolute judgments of the distance between touches oriented in the proximodistal axis of the arm, either using verbal estimates or by comparison with visual stimuli. Pairs of touches that crossed the wrist boundary were judged as larger than equivalent pairs entirely on either the forearm or hand. It is possible that this effect could reflect the classic form of Weber's illusion rather than categorical perception, given that tactile acuity itself appears to be increased in the vicinity of the wrist (Cody et al., 2008). To address this, Le Cornu Knight et al. (2014) compared the magnitude of tactile distance anisotropy on the forearm, hand, and wrist. The bias for mediolateral stimuli to be judged as larger than proximodistal stimuli was reduced at the wrist, indicated that it is specifically stimuli that *cross* the wrist that are perceived as larger, not merely those that are near the wrist.

While such results suggest that tactile distance perception can be modulated by relatively high-level body representations, other evidence indicates that it also emerges from basic features of the low-level organization of the somatosensory cortex. For example, extended sensory exposure (adaptation) to large or small tactile distances produces aftereffects on subsequently perceived tactile distances (Calzolari et al., 2017). Participants received adaptation to tactile distances on each hand, one of which was large (4 cm) and the other small (2 cm). On test trials, they received a tactile distance on each hand and judged which one felt larger. Performance on test trials differed systematically as a function of adaptation. Test stimuli were judged as smaller on the hand that had been adapted to the larger distance than on the hand adapted to the smaller distance—a classic adaptation aftereffect. These tactile distance aftereffects show several features that suggest that they arise from relatively low-level aspects of somatosensory processing. For example, clear aftereffects were found when both adapting and test stimuli were oriented across the width of the hand and when they were oriented along the length of the hand. In striking contrast, however, when adaptors were presented in one orientation and test stimuli in another, no aftereffects were found, showing a high degree of orientation selectivity of adaptation. Similarly, the effects of adaptation were specific to the adapted region of the hand and showed no transfer between the palm and dorsum or between the left and right hands. Together, these results indicate that tactile distance is a basic somatosensory feature, likely to be computed at relatively early stages of somatosensory processing.

As I mentioned above, there is a systematic relation between tactile distance perception and the spatial acuity of different skin surfaces. Distances feel larger on skin regions with high sensitivity and correspondingly small RFs. This pattern also holds for the anisotropy of tactile distance perception, given that tactile acuity is also higher across hand width than along hand length (Cody et al., 2008). Recall that there is also a corresponding anisotropy in the shape of tactile RFs in the somatosensory cortex, which tend to be oval shaped rather than circular, being elongated along the long axis of the limbs (Brooks et al., 1961; Mountcastle, 1957).

To account for these relations, Longo and Haggard (2011) proposed what they called the "pixel model" to tactile spatial perception. The central idea of this model is that the RFs within a somatotopic map are treated as the pixels of a two-dimensional array, with perceived distance corresponding to the number of unstimulated RFs between two distinct activation peaks (figure 2.11). This model can account for the classic form of Weber's illusion, since more sensitive skin regions will have a large number of small RFs. Since all RFs are treated equivalently for computations of distance, regardless of their physical size on the skin, this will have the effect of perceptually magnifying perceived distances on sensitive skin regions compared to less sensitive regions, as seen by comparing the left and center panels in figure 2.11. Similarly, the model can account for anisotropies in tactile distance perception in an analogous way. On skin regions with oval-shaped RFs, pixel maps will be stretched in the axis in which RFs are smaller, as shown in the right panel of figure 2.11. In both cases, perceptual distances are expanded where RFs are small and compressed where they are big.

One prediction of the pixel model is that illusions of tactile distance should be geometrically coherent, in the sense that the layout of tactile space should be related to the physical layout of the skin via geometrically simple deformations, such as stretch or compression. Fiori and Longo (2018) tested this prediction by asking participants to judge the size of tactile distances presented at a range of orientations on the hand. If the anisotropy described above reflects a coherent stretch of tactile space, perceived

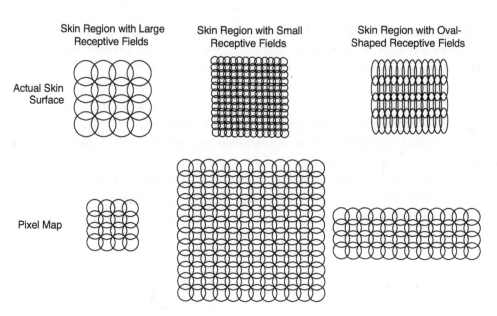

Figure 2.11
The pixel model of tactile spatial perception.

distance should vary systematically as a sinusoidal function of the angle between the actual stimulus orientation and the axis along which tactile space is stretched. This characteristic pattern was clearly apparent on both the dorsum and palm of the hand.

Other recent research has used a statistical method called "multidimensional scaling" (MDS) to construct perceptual maps of the perceived spatial layout of the hand. MDS is a variant of principal component analysis, which reconstructs the latent spatial structure in a matrix of distances or dissimilarities. Longo and Golubova (2017) used MDS to reconstruct the perceptual space of the hand. They made a four-by-four grid of marks on the back of each participant's hand and obtained judgments of perceived distances following stimulation of every possible pair of points. By applying MDS to the resulting perceptual distance matrix, they constructed perceptual maps of tactile space, showing that these were stretched across hand width, consistent with the other results described in this section. Tamè et al. (2021) applied a similar approach to reconstruct maps of the skin from the pattern of representational distances between fMRI activations following stimulation of different locations on the hand dorsum. Organized skin maps could be reconstructed from both M1 and S1, which showed distortions very similar to those of perceptual maps. These results suggest that the distortions of tactile space emerge from the basic organization of the sensorimotor cortex.

3 The Moving Hand

In Book I of his *Philosophical Investigations*, the Austrian philosopher Ludwig Wittgenstein (1953) posed (but did not answer) a question that has haunted the philosophy and neuroscience of action ever since. "What is left over," Wittgenstein asks, "if I subtract the fact that my arm goes up from the fact that I raise my arm?" As Haggard (2009) notes, a key component of an answer is the conscious intention of willing the arm to move. Indeed, Wittgenstein's question is typically discussed in relation to volition and the conscious experience of agency.

This answer, however, is incomplete. If we subtract the movement of the physical arm, what remains are all of the processes in the nervous system that contribute to producing arm movements. While some of these processes are accessible to conscious awareness (such as our conscious intention to raise our arm), others are inaccessible to conscious introspection (such as the efferent motor command sent along the corticospinal tract). There are many nonconscious processes involved in controlling arm and hand movements, and these would remain just as surely as conscious intentions. Indeed, Wittgenstein's question provides a natural way of thinking about the central concepts of this book. If we subtract the movements of the visible hand, we are left with the operations (both conscious and nonconscious) of the invisible hand.

Hierarchical Coding

In this chapter, I will focus on the sensorimotor control of concrete and immediate actions we perform with our hands. It is important to remember, however, that motor programs are themselves nested seamlessly within larger goals and plans that are more abstract and cognitive in nature. Since the earliest work in cognitive psychology in the 1950s and 1960s, researchers have emphasized the importance of hierarchical, recursive nesting of successively more concrete goals as part of a process of "means-end" reasoning (Miller et al., 1960). A recognizably motor goal, such as REACHING TO PICK UP A MUG OF

COFFEE, may be only a small part of a larger goal, such as EATING BREAKFAST or even GETTING READY TO GO TO WORK. The specification of a motoric action reflects the lowest level within this hierarchy—the output of an essentially cognitive process of planning and reasoning. Our representation of actions and their associated goals thus exists at many levels of abstraction and timescales. Some of our actions are based on immediate goals, such as scratching an itch, while others exist on the timescale of minutes, days, years, or even an entire lifetime.

This point was put well by Karl Lashley (1951) in his classic statement of what he called the "problem of serial order in behavior":

> There is a series of hierarchies of organization; the order of vocal movements in pronouncing the word, the order of words in the sentence, the order of sentences in the paragraph, the rational order of paragraphs in a discourse. Not only speech, but all skilled acts seem to involve the same problems of serial ordering, even down to the temporal coordination of muscular contractions in such a movement as reaching and grasping. (pp. 121–122)

From the perspective of a high-level cognitive plan, an action such as picking up a coffee mug may seem concrete and well specified. From the perspective of the motor system, however, it remains high level and abstract. Indeed, we have no introspective access at all to the details of which muscles are contracted or extended in performing this action. As Lashley noted, a seemingly simple act such as picking up a coffee mug must itself be decomposed into nested components—for example, a transport phase in which the arm is moved to bring the hand near the mug, a grasp phrase in which a power grip is used to lift the mug, and continuous monitoring of contact forces of the fingers and palm on the mug to prevent it from slipping. As we will see, each of these steps is associated with distinct patterns of muscular contraction, sensory feedback, and underlying neural machinery.

Motor Equivalence
Early writers such as Lashley (1930) and Hebb (1949) emphasized that, in contrast to basic spinal reflexes, skilled and purposive actions tend not to be performed rigidly in the same way upon repeated performance. Rather, behaviors that are recognizably the "same" action are instantiated using a range of patterns of muscular contractions and kinematic patterns. Successive reaches for a coffee mug may differ in their exact trajectory and kinematics, given changes in overall body posture, slightly different locations of the mug, or the presence of a cup of juice next to the mug. The significance of such *motor equivalence* is that the internal representation of an action reflects the final state or goal to be achieved rather than the details of kinematic motion or muscular contraction required to implement the goal.

In a seminal series of studies, the Soviet neurophysiologist Nikolai Bernstein (1967) developed optical methods for tracking natural movements in three-dimensional space. This allowed him to track the trajectories of repetitive movements, such as hammering a nail. Bernstein observed that while the end position of the action was consistently highly precise, there was substantial trial-to-trial variability in the path traced at intermediate points on the hand trajectory. Such trial-to-trial variability has also been described for precise opposition movements between the thumb and index finger (Cole & Abbs, 1986). Across repeated movements, virtually every kinematic property of the joints in the finger and thumb varied substantially. The one invariant feature was the precise final contact between the pulps of the fingertips.

One classic and dramatic example of the same motor program being implemented in different manners is in writing. For example, Merton (1972) noted that if we write the same text using the dominant hand on both a sheet of paper and on a blackboard at much larger scale, our personal writing style appears in a nearly identical way, despite totally different patterns of hand and arm movement to produce them. Kadmon Harpaz et al. (2013) asked participants to write three letters with a stylus at two different spatial scales while in an MRI scanner. The identity of the letter could be classified independent of scale from the distributed pattern of brain activation in both the primary motor cortex and in the anterior intraparietal sulcus (IPS).

Even more dramatic is the generalization of writing ability, and personal style, to effectors other than the hand. Almost all of us have learned to write using just one preferred hand. Nevertheless, we can grasp a pencil with our non-preferred hand, our mouth, or even our foot and use it to write legibly (if not elegantly). Textbook discussions of motor control almost invariably include a figure showing text written by a single person using different effectors—for example, the dominant hand, nondominant hand, foot, and mouth—an effect first demonstrated by Lashley (1942), as shown in figure 3.1. While the writing may be smoothest and neatest with the dominant hand, overall style remains remarkably constant across effectors. While most of us only have experience of writing using our dominant hand, the motor programs we use to control writing movements appear to be easily transferrable to other body parts.

Rijntjes et al. (1999) recorded fMRI while participants signed their name by moving either their hand or foot through the air. This signing condition was contrasted with a condition in which participants made up-and-down zigzag movements with their hand or foot. They conducted what are called "conjunction" analyses to identify voxels in the brain that became active during writing, regardless of which body part was used. Both writing conditions activated common regions of the PPC and occipitotemporal junction, which were not activated in either zigzag condition. Moreover, both signing

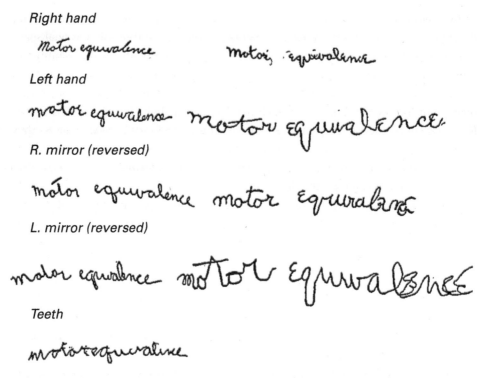

Figure 3.1
Motor equivalence in handwriting. Each column shows text from one participant using either hand to write normally or mirror reversed (these have been flipped for comparison with non-mirrored writing). One participant also used their teeth. In each case, the individual style of handwriting is preserved. From Lashley (1942).

conditions activated a set of brain regions that were activated during zigzag movements of the hand—but not the foot—consistent with the interpretation that writing movements performed with the foot co-opt motor programs developed and generally used for the hand.

Numerous subsequent studies have found that movement and preparation-related activity in premotor and posterior parietal cortices generalizes, at least partly, across the specific effectors used, such as between the two hands (Gallivan et al., 2013), the hand and mouth (Castiello et al., 1999), or the hand and foot (Cunningham et al., 2013; Heed et al., 2016). These studies have revealed a complex pattern of effector-specific and effector-independent representations.

Motor Synergies

One basic problem in the hierarchical control of action concerns how the brain goes from a high-level plan formulated at a cognitive or semantic level and converts this into a specific pattern of commands to muscles. Recall from chapter 1 that the hand has twenty-seven bones and is controlled by more than thirty muscles, producing a vastly complex space of possible hand movements. This raises what Bernstein (1967) referred to as the "degrees of freedom problem": There are a vast number of ways in which any given action might be performed, so how does the brain choose just one? A potential solution to this problem is for the brain to construct complex behaviors through the linear combination of a modest set of motor synergies, each of which involves the combined activity of several muscles. This simplifies the process of selecting actions, since there is a lower-dimensional space of possible actions to select from.

One approach to measuring motor synergies is to investigate the statistical structure of naturalistic actions. Santello et al. (1998) asked participants to wear a special glove known as a "CyberGlove," which continuously recorded the angles of fifteen joints within the hand while participants made the postures they would use when grasping fifty-seven familiar objects, such as an egg or a hammer. They applied a statistical method known as "principal component analysis" to this data and found that it could be well characterized by a small number of "effective" degrees of freedom. The first two components accounted for more than 80 percent of variance in the data, and the value of including additional dimensions plateaued after around five or six dimensions. Similar results occur for hand movements performed in more specific expert contexts, such as piano playing (Furuya et al., 2011) and sign language (Jerde et al., 2003). Other studies apply similar logic to electromyographic signals recorded from peripheral muscles, again showing that only a modest number of dimensions are needed to describe patterns of muscular activation during grasping (Weiss & Flanders, 2004) and reaching (D'Avella et al., 2006).

Thus, the effective dimensionality of human hand behavior is much smaller than the theoretical dimensionality based on a purely anatomical consideration of joints and muscles. Despite the vastly complex space of possible hand actions, the real behavioral states of the human hand appear to occupy a smaller-dimensional subspace. The dimensions of this subspace correspond to the motor synergies, and the position of any behavior within this hypothetical space represents the degree of activation of each synergy. This has the effect of greatly simplifying Bernstein's degrees-of-freedom problem.

There is increasing evidence that such motor synergies are represented in M1. One set of studies has shown that stimulation of M1 does not produce idiosyncratic activations of individual muscles but rather complex correlated patterns across multiple muscles. Such findings have been found using both intracranial microelectrode stimulation of individual neurons in monkeys (Overduin et al., 2012) and TMS in humans (Gentner & Classen, 2006). Indeed, as will be discussed further below, longer-duration stimulation of M1 can elicit complex, ethologically meaningful actions (Graziano, 2016). Similar conclusions come from studies recording M1 activity, whether using microelectrode recording in monkeys (Gallego et al., 2018; Saleh et al., 2012) or fMRI in humans (Leo et al., 2016). For example, Saleh and colleagues developed a statistical model to characterize the firing rates of M1 neurons in monkeys during reaching and grasping actions based on forty features of hand and arm joint angles and velocities, each modeled with eight potential time lags. Despite the data falling within a 320-dimensional space, principal component analysis on the data across M1 showed that the variance explained started to plateau after five dimensions, and by thirty dimensions, it accounted for more than 80 percent of variance.

These results suggest that the general patterns of reaching and grasping actions are implemented through the combination of a quite modest number of motor synergies, corresponding to the first few principal components of behavioral (Santello et al., 1998), electromyographic (Weiss & Flanders, 2004), or neural (Saleh et al., 2012) data. At the same time, they suggest that there may also be more subtle fine-tuning of actions corresponding to higher principal components. In a recent study, Yan et al. (2020) used video-based tracking to measure continuous hand posture with twenty-nine degrees of freedom while participants grasped familiar objects or produced American Sign Language signs. As in previous studies, a large proportion of the variance in both tasks could be characterized by a modest number of principal components (three to five to account for 80 percent, eight to eleven to account for 95 percent). Indeed, these low-dimension spaces were nearly identical for grasping and sign language. Critically, however, Yan and colleagues showed that the higher principal components did not merely reflect noise but were structured in systematic ways between conditions. For example, the values of higher-dimensional components were more similar for two grasps toward the same object (e.g., a coffee mug) than to different objects (e.g., a mug and a stapler). These higher dimensions thus appear to reflect subtle refinements of movements by which the overall structure of hand movements implemented by motor synergies is fine-tuned to reflect the specific features of the action being performed.

Direct Corticomotoneuronal Connections

The main pathway along which the cortex sends motor commands to the periphery is known as the corticospinal (or pyramidal) tract. Projections from M1 related to the hand terminate on spinal interneurons (propriospinal neurons) located in vertebrae C3 and C4 in the neck. This system of spinal interneurons then sends motor commands to motoneurons at lower levels of the spinal cord that directly control muscles. This propriospinal system likely reflects the construction of the overall form of hand movements by motor synergies, which are converted to detailed signals to each individual muscle by the machinery of the spinal cord. In humans and some other primates, however, there is a second type of projection that allows the motor cortex to send commands directly to spinal motoneurons, bypassing the spinal interneuron system (Bernhard et al., 1953; Porter & Lemon, 1993). These direct corticomotoneuronal (CM) connections appear closely linked to skilled control of the hand and may correspond to the fine-tuning of action that is not captured by motor synergies. As Lemon (1993) argues, "the skilful forms of movement that characterize the primate hand require a radically different form of control than those originally evolved for the control of posture and locomotion" (p. 264).

Several lines of evidence implicate CM projections specifically in fine-motor control of hand movements (Lemon, 1993). First, CM connections are most dense for motoneurons controlling the distal muscles of the hand (Kuypers, 1960). Indeed, in macaques, these neurons have been specifically linked to precision grasps (Muir & Lemon, 1983) and active tool use (Quallo et al., 2012). Second, lesions of the corticospinal tract in adult macaque monkeys disproportionately affect fine movements of the hand and fingers compared to other motor behaviors (Lawrence & Kuypers, 1968). Third, lesions of the corticospinal tract in infant macaques prevent the development of fine-motor control of the hand in the first place (Lawrence & Hopkins, 1976), while the developmental trajectory of CM projections parallels development of fine-motor control of the hand during the first eight months of the monkey's life (Galea & Darian-Smith, 1995). Finally, comparative studies across species show that the presence and strength of CM projections are related to manual dexterity (Bortoff & Strick, 1993; Heffner & Masterton, 1975, 1983), as will be discussed in chapter 11.

What is the advantage of direct CM connections for manual dexterity? When the motor cortex controls movements indirectly by way or propriospinal neurons, actions are constructed from the collection of motor synergies stored in the spinal cord. The motor repertoire of the spinal cord, however, is limited, highly stereotyped, and inflexible. By allowing the cortex to control muscular contractions directly, an unprecedented level of flexibility became available in shaping behaviors. As Sherrington (1906) put it, the shift to cortical control transformed the hand "from a simple locomotor prop to a delicate explorer of space" (pp. 352–353). Kuypers (1978), similarly, argues that the CM system is critical for producing the highly selective, "fractionated" movements of the fingers, critical to fine-motor control.

As an analogy, consider getting a new smartphone preloaded with a number of apps. These apps may do useful things, but you are limited to the functionality that has been preprogrammed into them. Now suppose you are given an interface to program your own apps from scratch or to fine-tune the functionality of existing apps. This opens new possibilities, allowing your phone's functionality to be precisely tailored to your specific needs. This is what the CM pathway allows, letting the full computational power of the cortex be brought to bear in sculpting highly precise and flexible motor programs, going beyond mere linear combinations of motor synergies. At the same time, for many routine everyday functions, it may be easier and less error prone simply to use the built-in apps that come with your phone rather than reinventing the proverbial wheel.

The Primary Motor Cortex
One interesting feature of the hand representation in the primary motor cortex is that it is located on a specific anatomical landmark in the middle of the central sulcus, described as a "knob" (Yousry et al., 1997). The exact shape of this landmark varies across individuals, frequently shaped like a Greek letter omega or epsilon. The association of this landmark with the hand motor representation has been demonstrated both using neuroimaging to record brain responses when participants make finger and wrist movements (Boling et al., 1999; Yousry et al., 1997) and brain stimulation studies with TMS (Boroojerdi et al., 1999) and direct intracortical electrical stimulation (Boling et al., 2008).

Intriguingly, the cause of this bulge in the precentral gyrus is the presence of an additional gyrus buried in the depth of the central sulcus known as the "pli de passage fronto-pariétal moyen" (PPFM), which was first described by Paul Broca in the nineteenth century (Boling et al., 1999). As shown in figure 3.2, this gyrus is formed by a folding of the cortical sheet within the central sulcus, linking the precentral and postcentral gyri. This folding appears to be specific to the great apes (Hopkins et al., 2014) and may have the function of increasing the cortical magnification of the hand in M1 and S1 without disrupting the overall somatotopic organization of these regions. Indeed, the portions of the PPFM bordering on the pre- and post-central gyri have been linked to motor (Boling et al., 1999) and somatosensory (Boling & Olivier, 2004) functions, respectively. More broadly, a recent study provided strong evidence that the boundaries between different body parts in M1 are closely linked at the individual subject level to specific morphological features of the central sulcus, such as the PPFM (Germann et al., 2020).

Another important feature of M1 concerns its somatotopic organization. As discussed in the previous chapter, S1 features an exquisitely detailed somatotopic organization. In M1, in contrast, this somatotopy is coarser. It has been clear since the first stimulation studies of M1 in the nineteenth century that there is a broad

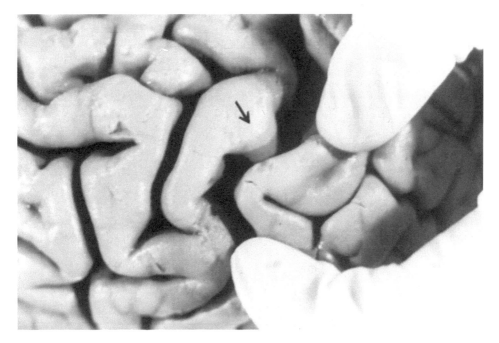

Figure 3.2
The PPFM in a human brain. The fingers hold the postcentral gyrus, spreading open the central sulcus. The PPFM is a gyrus formed by folding of the cortex inside the central sulcus. This folding has the effect of increasing the cortical magnification of the hand, without disrupting the overall somatotopic organization of the sensorimotor cortices. Morphologically, it creates a "knob," frequently shaped like a Greek letter omega or epsilon on the precentral gyrus. From Boling et al. (1999).

somatotopic organization, with the foot and leg represented at the top of M1, the arm and hand in the middle, and the face at the bottom (Ferrier, 1876; Horsley & Schäfer, 1888), as shown in the maps in the top row of figure 3.3. This overall pattern is aligned with that of S1, reflecting the fact that these two brain areas work closely together. At a finer spatial scale, however, the detailed somatotopy seen in S1 breaks down in M1. For example, studies using microelectrode recordings in monkeys (Schieber & Hibbard, 1993) and neuroimaging in humans (Sanes et al., 1995) have shown that representations of the five fingers in M1 are overlapping and spatially intermingled, as shown in the bottom panel of figure 3.3.

There is also evidence that there may be two distinct representations in the motor cortex, which may have strikingly different evolutionary histories, as will be discussed further in chapter 11. Studies in both monkeys (Strick & Preston, 1982) and humans (Geyer et al., 1996) have identified distinct representations within M1, one more rostral

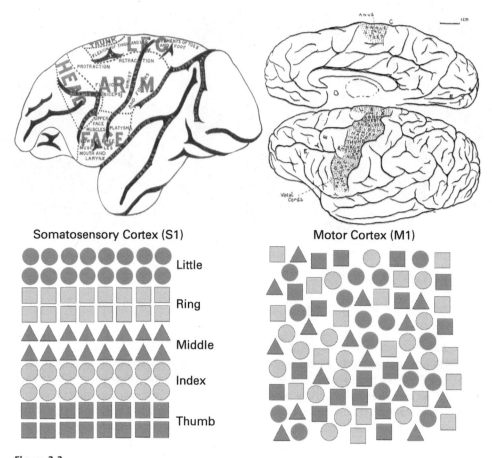

Figure 3.3
Top row: Maps of the motor cortex based on electrical stimulation in monkeys (top left; from Horsley & Schafer, 1888) and a gorilla (top right; from Leyton & Sherrington, 1917). *Bottom row*: Comparison of the fine-grained somatotopic organization of the fingers in S1 (left panel) and M1 (right panel). Following Strick et al. (2021).

near the crest of the precentral gyrus and another more caudal in the depth of the central sulcus.

Rathelot and Strick (2006) investigated these representations by injecting the rabies virus into individual hand muscles of macaque monkeys. This virus is transported between neurons exclusively in the retrograde direction (i.e., in the direction opposite to the flow of the signals themselves) and at a relatively constant speed. This allowed Rathelot and Strick to identify which regions of the motor cortex send projections to each muscle and, critically, also how many intermediate synapses the signals pass through. Direct CM projections (i.e., ones that make only one synapse between the cortex and the muscle) appeared only in the more caudal region of M1, buried in the central sulcus. More rostral parts of M1, in contrast, showed only multisynaptic connections to muscles.

In a subsequent study, Rathelot and Strick (2009) argued that these regions represented two distinct primary motor cortices: an "old" M1, which projects to spinal interneurons, and a "new" M1 with direct CM projections. On this view, old M1 uses the repertoire of spinal motor programs to construct actions, while new M1 constructs actions more directly, allowing greater flexibility and control over precise manipulative actions.

Another approach to investigating CM connections is to stimulate neurons in M1 and measure the latency with which spinal motoneurons become active. Witham et al. (2016) stimulated both old and new M1 and replicated the finding that only new M1 had direct CM connections. More importantly, they found that even in new M1, the majority of neurons project to interneurons rather than directly to motoneurons.

Thus, rather than construct actions entirely from scratch, new M1 uses CM connections to apply fine adjustments to existing spinal motor programs. This interpretation is consistent with the finding that many CM neurons show directional tuning preferences different from those of the specific muscles that they project to (Griffin et al., 2015). This finding indicates that these CM neurons are involved not in the initiation of the movement itself but rather in fine-tuning and "sculpting" the action in a precise way. As Rathelot and Strick (2009) put it, "the direct access to motoneurons afforded by CM cells enables New M1 to bypass spinal cord mechanisms and sculpt novel patterns of motor output that are essential for highly skilled movements" (p. 918).

New and old M1 also appear to differ in their patterns of connectivity with the premotor and somatosensory cortices. Dea et al. (2016) found in capuchin monkeys that old M1 is more densely connected to the premotor cortices, while new M1 is more densely connected with S1 and the PPC. Moreover, within each region of M1, they identified lateral and medial subregions that themselves had different patterns of connectivity. For example, in old M1, the more medial region had stronger connections to the dorsal PMC (dPMC), whereas the more lateral region had stronger connections to the ventral PMC (vPMC). Similarly, within new M1, the more medial region had stronger connections to PPC area 5, whereas the more lateral region had stronger connections to area 2 in S1. These results suggest that there may be a more modular organization of hand actions, even within new and old M1s, with subregions potentially involved in specific types of actions. For example, Dea and colleagues suggest that, within new M1, the more lateral region is involved in grasping, whereas the more medial region is involved in reaching and defensive movements.

Muscle and Movement Representations

Since the earliest neurophysiological work on the motor system in the nineteenth century, there has been debate about what features of movement the primary motor cortex codes. Fritsch and Hitzig (1870, 2009), who first showed that movements could be elicited by direct electrical stimulation of the motor cortex, suggested that the motor cortex represented contractions of individual muscles. In contrast, David Ferrier (1873) applied longer duration stimulation to the motor cortex of monkeys and reported behaviors that he interpreted as more complex and purposeful than simple muscle twitches. In the past 150 years, this debate has still not been fully resolved.

One influential view on M1 function came from Edward Evarts (1968), who recorded from M1 neurons while monkeys made reaching movements using a manipulandum to which different weights could be applied. Evarts found that the responses of neurons graded with the amount of force required to make movements and argued that M1 codes the amount of muscular effort required for actions rather than more abstract features such as direction in space.

A different perspective came from studies by Apostolos Georgopoulos et al. (1982), who showed that neurons in the primary motor cortex code the direction of an arm movement. They trained monkeys to perform a task in which they held a manipulandum at the location of a center light surrounded by a circular array of eight other lights. On each trial, one of the peripheral lights would come on, and the monkey's task was to move the manipulandum to that location. Of more than six hundred neurons tested, nearly 75 percent showed preferential responses to movements of the arm in specific directions. Moreover, the firing rates of neurons decreased smoothly for directions of movement progressively different from the preferred direction, showing a sinusoidal tuning function.

While any individual neuron will thus only specify broadly the direction of arm movement, Georgopoulos et al. (1986) argued that a population of such neurons could specify the direction of motion with a high degree of precision. They modeled the activity of each neuron as a vector, with vector direction determined by the neuron's preferred direction of motion and vector length determined by the change in the neuron's firing rate from baseline. The vector sum of these individual vectors—the population vector—showed a high degree of coherence with the actual direction of motion of the monkey's arm. By drawing random samples from the population of neurons, Georgopoulos et al. (1988) showed that the activity of a population of as few as 100–150 neurons could be sufficient to specify movement direction with high precision. More recent studies have shown an orderly mapping of preferred orientation across the cortical sheet in the primary motor cortex, with neurons having similar preferred directions arranged in mini columns interleaved with neurons having orthogonal preferences (Georgopoulos et al., 2007), mirroring the modular structure found in other cortical areas, such as the somatosensory cortex (Mountcastle, 1957), as discussed in the previous chapter.

Other studies in monkeys have combined the approaches of Evarts and Georgopoulos and asked how the responses of direction-selective neurons are affected by opposing loads. Kalaska et al. (1989) found separate populations of neurons in M1, some being strongly affected by the amount of force required to move in a specific direction and others appearing to code movement irrespective of force. In contrast, however, neurons in the PPC are almost completely unaffected by such loads (Kalaska et al., 1990), suggesting that the parietal motor regions code actions at a more abstract level than the frontal motor regions.

An elegant approach to exploring the features coded in motor cortices was pioneered by Kakei et al. (1999). They trained monkeys to move a manipulandum in different directions using three different hand postures (figure 3.4). In the pronated posture, the monkey grasped the manipulandum with an overhand grip. In the supinated posture, the monkey grasped the manipulandum with an underhand grip. Finally, the mid posture was intermediate between pronation and supination. The differences in forearm rotation across these conditions allowed Kakei and colleagues to dissociate the direction of motion of the movement in three-dimensional space from the direction of joint movement with respect to the arm itself. For example, when the hand is pronated, upward motion is produced by wrist extension; conversely, when the hand is supinated, upward motion is produced by wrist flexion. By mapping the direction selectivity of M1 neurons across postures, Kakei and colleagues determined whether each neuron coded the direction of motion in external space (extrinsic neurons) or the flexion of the wrist joint or muscular activity (intrinsic neurons). Substantial populations of both types of neurons were found, with 50 percent of neurons tested showing extrinsic-like responses and 32 percent showing intrinsic-like responses. In a follow-up study, Kakei et al. (2001) recorded from neurons in the vPMC using the same paradigm. In striking contrast to the mixed pattern of response in M1, 94 percent of premotor neurons coded for direction of movement in external space. These results showed not only that combinations of both extrinsic and intrinsic reference frames are used to code actions, but also that there is a progression from predominantly extrinsic coding in the PMC and predominantly intrinsic coding in M1.

A recent study in humans combined fMRI and magnetoencephalography (MEG) to investigate movement and muscle representations in M1 during a wide range of naturalistic hand actions (Kolasinski et al., 2020). Interestingly, these appeared to map onto

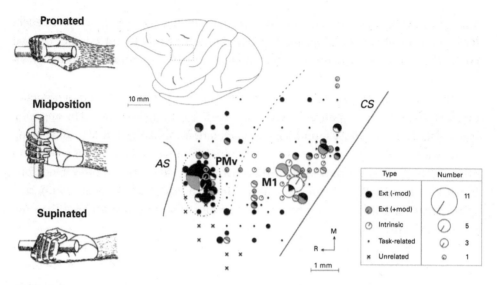

Figure 3.4
Coding of movement in extrinsic and intrinsic reference frames by neurons in the vPMC and M1. Adapted from Kakei et al. (2001).

new and old M1, respectively. Caudal regions, buried in the central sulcus, corresponding to new M1 carried information about direction of motion regardless of which muscles were involved, and they were active as much as 200 ms before movement onset. In contrast, more rostral regions, on the surface of the central sulcus, corresponding to old M1 became active much later and corresponded more directly to the specific muscles involved.

A different perspective on M1 came from a remarkable series of studies by Michael Graziano and his colleagues (for review, see Graziano, 2009, 2016). As already discussed, much of the original understanding of the motor cortex came from studies measuring the muscle twitches evoked by brief (~50 ms) electrical stimulation of M1 in animals (Fritsch & Hitzig, 1870) and humans (Penfield & Boldrey, 1937). While there were some suggestions in the literature that longer durations of stimulation might evoke more complex actions (Ferrier, 1873; Penfield & Welch, 1951), these had not been systematically studied. Graziano applied electrical stimulation to M1 in monkeys for substantially longer (~500 ms) and found that he could reliably evoke complex actions, which had obvious ethological significance and were representative of the animal's actual behavioral repertoire. For example, stimulation might evoke movements of the contralateral hand to the mouth or movements of the arm to defend the face.

The actions evoked by stimulating M1 appeared to be organized into zones, each containing an ethologically relevant type of action. This action map covers not only M1 but also parts of the PMC. This is obviously a very different way of conceiving of the organization of M1 from the traditional homunculus. Nevertheless, the two types of maps are not necessarily inconsistent. For example, the top of M1 has traditionally been associated with movements of the feet (Penfield & Rasmussen, 1950). Even where Graziano finds hand (rather than foot) movements evoked by stimulation of this area, they tend to be hand movements directed *toward* the feet. By the same token, the bottom of M1 is traditionally associated with the face, and Graziano finds that stimulation of this region evokes movements of the hand to the mouth as well as movements such as chewing. Thus, the traditional homunculus and Graziano's action maps may be complementary rather than alternative forms of organization. The issue may be that motor behavior varies along many dimensions, whereas the cortical sheet is two-dimensional. The multiplexing of broad somatotopy and ethological action maps may be an optimal form of organization in structuring a complex and diverse behavioral repertoire onto a limited two-dimensional cortex (Graziano & Aflalo, 2007).

The studies discussed so far in this section have focused on the responses of single neurons. The increasing ability to record simultaneously from many neurons has opened new perspectives on this issue. Churchland et al. (2012) used ninety-six-electrode arrays to record from large numbers of M1 neurons while monkeys prepared and then executed reaches. They used a form of principal component analysis to explore the temporal evolution of population-level responses among these neurons. Remarkably, there was a strong oscillatory response at the population level, although there was nothing rhythmic about the reaching movements themselves. The population dynamics of M1 during reaching thus result in an essentially circular path through the state space of neural activations across time, ending back at the initial starting location—a phenomenon known as "rotational dynamics." Churchland et al. suggest that preparatory activity may specify parameters of movement by specifying the initial state of the system, with movement execution controlled by the dynamic progression of M1 activity through state space, ending back at approximately the same state as it started.

On this account, and in striking contrast to the idea of the population vector, the response patterns of individual neurons may not be representative of the larger-scale computational processes happening among the full population of M1 neurons. Russo et al. (2018) found that individual M1 neurons can show "muscle-like" responses, even as the larger neural population in which they are embedded shows rotational dynamics with no direct relation to either individual muscles or abstract direction in space. At the same time, rotational dynamics may be specific for reaching rather than a general characteristic of M1 function. A recent study (Suresh et al., 2020) found clear evidence for rotational dynamics during reaching but not during grasping. This suggests that these two aspects of hand function may operate very differently—an issue that will occupy the second half of this chapter.

The Grasping and Reaching Networks

Schemas for Reaching and Grasping

In one of the first systematic studies of the nature of hand movements, Woodworth (1899) showed that goal-directed movements to a target include two distinct phases. Woodworth asked participants to move the tip of a pencil toward targets on a sheet of paper attached to a rotating drum, which allowed him to analyze the kinematics of movement during the course of the action. The initial phase of movement was rapid and relatively uncontrolled but was followed by a slower and more deliberate final stage as the hand reached the target.

In a seminal series of studies, Marc Jeannerod (1981, 1984, 1988) extended this idea, arguing that reaching and grasping reflect the operation of distinct yet coordinated visuomotor channels. One channel focuses on the extrinsic properties of objects, such as their location in space, and operates to transport the arm and hand to the appropriate location in three-dimensional space. The other channel focuses on the intrinsic properties of objects, such as their size and shape, and operates to shape the fingers and hand to precisely grasp and lift the object. There is evidence that these channels rely on distinct parieto-frontal networks, as will be discussed in more detail below.

Jeannerod (1981) asked participants to reach and grasp a range of manipulable objects, including spheres, boxes, and cylinders. Using frame-by-frame analyses of movies of these reaches, he showed that they consisted of two broad phases: an initial high-velocity transport stage, in which the hand is quickly transported to the vicinity of the object, and a later low-velocity grasp stage, in which the fingers and hand contact and lift the object. These two phases were modulated by different features of the context and the object to be grasped. Extrinsic features of the object, such as its location in space, affected the transport component but not the grasping component. When objects were farther away from the participant, the peak velocity of arm movement increased, but no changes were found in the pattern of grip aperture. Conversely, intrinsic features of the object, such as its size, affected only the grasp component. For example, the size of maximum grip aperture varied systematically with object size, which had no effect on the transport component.

The transport phase lasts for around 70 percent of the total reach duration, from the onset of movement to the point of maximum deceleration. During the transport phrase, the position of the fingers is already being adjusted, the grip aperture between them systematically overestimating the necessary size to grasp the object. The point at which grip aperture reached its maximum value temporally coincided with the end of the transport phrase. The pre-shaping of grip aperture during the transport phase indicates that the transport and grasp components are not purely serial but are activated simultaneously. Moreover, the temporal coordination of peak aperture with hand deceleration suggests, in Jeannerod's (1984) words, "a common program which achieves the timing of coordination" (p. 253).

The PPC as a Command Apparatus

The sensory functions of the parietal lobe are well known. The anterior parietal lobe houses the primary somatosensory cortex, as discussed in the previous chapter. Meanwhile, the PPC is the terminus of both the hierarchical somatosensory processing stream (Iwamura, 1998) and the dorsal visual stream (Milner & Goodale, 2006; Ungerleider & Mishkin, 1982). But the parietal lobe also has motor functions. In a seminal study, Vernon Mountcastle et al. (1975) showed that individual neurons in posterior parietal areas 5 and 7 also respond when monkeys perform actions with the arm and hand.

One group of neurons, the arm-projection neurons, fire when the monkey moves its arm in space. In contrast, hand-manipulation neurons are silent during the transport phase of reaching but fire intensely as the monkey configures the hand for grasping the object and while it manually explores it. As Mountcastle et al. (1975) put it, "Cells of the first set discharge at high rates when the waking animal projects his arm into the immediate extrapersonal space that surrounds him; those of the second when he manipulates within it" (p. 881).

Critically, these same neurons were not activated when the monkey performed other movements that involved the same muscles, suggesting that they are not related to the low-level implementation of the action. Instead, they appear to correspond to the animal's intentions and behavioral goals. Mountcastle et al. (1975) continue:

> This activity occurs only if the projected movement of the manipulation is aimed at securing for the animal an object he desires, such as food when he is hungry ... The large majority of these cells ... is silent during other active movements, such as those of an aggressive or aversive nature in which the arm and hand may be manipulated in the same zones and with the same muscles. (p. 881)

Mountcastle et al. (1975) concluded that the PPC contains "a command apparatus for operation of the limbs, hands and eyes within immediate extrapersonal space"

(p. 871)—a proposal that contrasted with more traditional concepts of the PPC as a visual/somatosensory association area. Other evidence also suggests a direct motor role for the PPC. Fleming and Crosby (1955) showed that electrical stimulation of areas 5 and 7 in macaques continued to elicit movements of the hands and other body parts, even following surgical removal of the primary motor cortex. This result indicated that the pathway terminating in the primary motor cortex was not the only way for motor commands to be sent from the cortex. A recent study by Rathelot et al. (2017) provided evidence that a lateral region in area 5 is able to control hand movements directly. First, they demonstrated that electrical stimulation of this region reliably evoked movements in the contralateral hand. Then, they used a combination of anterograde and retrograde tracers to show that this region connects directly with regions of the medial dorsal horn in the spinal cord that communicate directly with both somatosensory inputs from the hand and motoneurons controlling hand movements.

Recent research has provided further evidence for the motor functions of the parietal cortex. Baldwin et al. (2018) applied long-train intracortical stimulation to several areas of the parietal lobe in macaque monkeys. Movements were elicited by stimulation of both S1 (areas 3b, 1, and 2) as well as the PPC (areas 5 and 7), including hand movements corresponding to both precision and power grasps. There were large clusters of response in both regions that involved only the thumb and index finger, presumably reflecting the behavioral importance of precision grasps to these monkeys. There was, at least broadly, a correspondence between the location of tactile RFs and motor responses, indicating that motor and somatosensory coding in these areas is tightly integrated. Intriguingly, while movements could be elicited from a wide area of cortex, the stimulation threshold required to produce such movements varied systematically. Thresholds were lowest in M1, intermediate in S1, and highest in the PPC. This pattern suggests that the PPC represents actions at a more abstract level than other, more classically sensorimotor, regions like S1 and M1.

The suggestion that the PPC sits atop a hierarchy of motor abstraction is consistent with recent fMRI research by Turella et al. (2020), who used a multivariate analysis technique called "multivoxel pattern analysis" (MVPA) to investigate three levels of hierarchical encoding for hand actions. While in the MRI scanner, participants made reaches to grasp objects that varied along three dimensions: (1) whether they used a precision or a power grip, (2) whether the wrist was oriented in a pronated or a supinated posture, and (3) whether the reach was made with the right hand or the left hand. By analyzing how neural responses generalized across these three dimensions, Turella and colleagues aimed to differentiate brain areas coding actions at different levels of abstraction. In M1, there was a combination of responses that were specific to each individual action and others that generalized across wrist orientations, consistent with neurophysiological studies in monkeys (Kakei et al., 1999). In contrast, responses in the PPC showed a range of responses across all three levels of abstraction. Turella and colleagues suggest that the PPC may thus function as "a central hub for moving across representations along the hierarchy, from concrete action specification to goals within and across different effectors" (p. 2933).

Consistent evidence comes from studies of motor imagery in patients with brain damage. Motor imagery refers to our ability to imagine performing actions without actually moving the body. In healthy adults, there is a close correspondence between the duration of real movements and of imagined movements, suggesting that the same neural processes are used in both cases (Decety et al., 1989). Patient C.P., who had damage to the right M1, showed large increases in the time she took to produce sequences of hand movements with her contralesional left hand, but also in the time it took for her merely to imagine performing the same sequence (Sirigu et al., 1995). In striking contrast, she was unimpaired at both real and imagined movements of her ipsilesional right hand. A similar pattern occurs in patients with basal ganglia dysfunction in Parkinson's disease (Dominey et al., 1995). In both cases, patients show correlated increases in duration of real and imagined movements. This suggests that the damaged M1 or basal ganglia machinery is being used to implement motor imagery. But the process of imagery itself remains intact and indeed continues to accurately reflect the patient's motor abilities. In contrast, Sirigu et al. (1996) found that PPC lesions disrupted the ability to imagine hand movements at all, resulting in no apparent relation between the duration of actual and imagined hand movements. Thus, while several regions of the motor system may be involved in motor imagery, it is the PPC that coordinates and directs this process. It seems to be the PPC where the idea of the action is implemented—a point that we will return to in the discussion of ideational apraxia in chapter 9.

A similar conclusion comes from studies of direct brain stimulation in humans. Recall that Penfield had mapped M1 by measuring twitches induced by applying direct electrical stimulation to the brain of patients undergoing brain surgery. Similar twitches can be induced noninvasively in healthy humans by using electrical currents (Merton & Morton, 1980) or TMS (Bestmann & Krakauer, 2015). In each case, participants deny that they performed the action themselves. The action is evoked without a corresponding intention to perform it. The same appears to be true for stimulation of the PMC (Desmurget et al., 2009). In contrast, however, Desmurget and colleagues found that stimulation of the inferior parietal cortex produced the desire to move, or even the experience of having moved, without eliciting any actual movement or electromyographic activity in muscles.

The Grasping Network

The finding of arm-projection and hand-manipulation neurons by Mountcastle et al. (1975) has obvious relevance to the behavioral findings of Jeannerod (1988) of distinct schemas for reaching and grasping. An influential view on the distinction between grasping and reaching is that these abilities are mediated by largely distinct parietal-premotor circuits (Jeannerod et al., 1995; Matelli & Luppino, 2001). As these networks appear to be separate subdivisions of the dorsal visual pathway, they are sometimes referred to as the "dorsolateral pathway" and the "dorsomedial pathway," but I will refer to them as the "grasping network" and the "reaching network." Grasping, on this account, is mediated by a network including the anterior IPS (anterior intraparietal area [AIP]) and the vPMC—regions that show strong anatomical interconnections (Borra et al., 2008). Interestingly, and unusually among parietal regions, AIP is also connected with inferotemporal regions of the ventral visual pathway (Borra et al., 2008), which makes sense, given the importance to grasping of intrinsic features of objects, such as size and shape, as well as learned semantic knowledge about object function—a point that will be discussed further in the next chapter. Notably, AIP also has connections to both the basal ganglia and cerebellum (Clower et al., 2005) and is thus widely linked with broader parts of the motor system as well as the ventral visual pathway.

Neurophysiological studies in monkeys show that both AIP (Murata et al., 2000; Sakata et al., 1995) and vPMC (Murata et al., 1997; Rizzolatti et al., 1988) contain neurons sensitive to shaping the hand and grasping manipulable objects, as shown in figure 3.5. An important characteristic of this network is that many of these neurons responsive to specific actions also show purely visual responses when the monkey sees an object that affords that very same action (Murata et al., 1997; Sakata et al., 1995)—what are sometimes called "canonical visuomotor neurons." Neuroimaging studies in humans have identified similar motor and visual responses in broadly similar regions (Binkofski et al., 1998; Fang & He, 2005; Konen & Kastner, 2008).

Evidence showing that this network is causally involved in controlling grasping comes from studies that have disrupted activity in AIP or vPMC. In monkeys, the gamma-aminobutyric acid (GABA) agonist drug muscimol can be used to temporarily inactivate a specific brain area. Injection of muscimol into both AIP (Gallese et al., 1994) and vPMC (Fogassi et al., 2001) produces rapid problems with forming appropriate grasping movements in the contralateral hand. Critically, however, more proximal movements of the arm involved in transporting the hand toward objects are spared. Similarly in humans, TMS has been used to disrupt processing in this network, resulting in impairment in several features of grasping, including shaping of hand posture to match object size, scaling grip force to object weight, and online adjustment of grasp in response to object

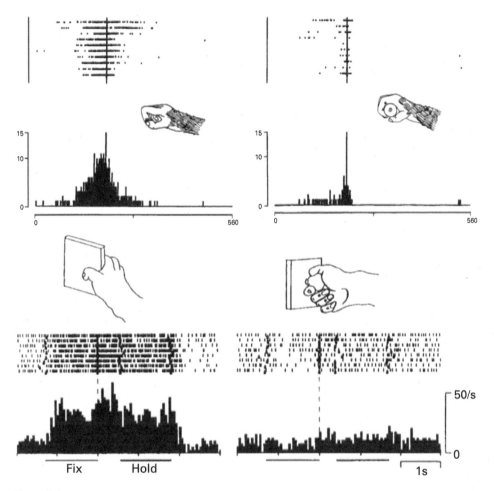

Figure 3.5

Motor responses of neurons in the macaque vPMC (area F5; top row) and AIP (bottom row). *Top row*: A neuron in vPMC coding for precision grip (Rizzolatti et al., 1988). The neuron shows a strong response in the lead-up to a precision grip of a raisin (top left) but shows little response to a power grip. *Bottom row*: A neuron in AIP coding vertically oriented grasps of thin objects (Murata et al., 2000). The neuron shows a strong response to grasping a vertical plate (bottom left) but little response to grasps of a small sphere (bottom right) or several other types of objects (not shown here). Top row adapted from Rizzolatti et al. (1988). Bottom row adapted from Murata et al. (2000).

movement (Davare et al., 2006; Tunik et al., 2005). Finally, studies of human patients with damage to AIP have revealed highly specific deficits in grasping (Binkofski et al., 1998; Jeannerod et al., 1994), as will be discussed in the next chapter in the context of optic ataxia.

Several studies have investigated the flow of information within the grasping network. Schaffelhofer et al. (2015) recorded from a large number of neurons in AIP, vPMC, and M1 while monkeys grasped a wide range of objects. During the planning phase of movements, before the start of muscular activity, the grip type could be decoded from neural responses in both AIP and vPMC, but not from M1. In contrast, during grasp execution, grip type could be decoded from M1 and vPMC, but not AIP. This suggests that AIP is involved early in grasp planning and M1 is involved later in grasp execution, with vPMC being involved continuously and linking planning and execution. Evidence consistent with this largely serial model of information flow from AIP to vPMC to M1 comes from a study using intracranial electrical stimulation to evoke grasping movements in monkeys (Stepniewska et al., 2014). At baseline, grasps could be elicited from stimulation of all three regions. Inactivation of M1 using muscimol disrupted the ability of stimulation in vPMC and AIP to evoke movements. In contrast, inactivation of vPMC blocked effects of stimulation to AIP, but not to M1, while inactivation of AIP did not affect stimulation to vPMC or M1.

Studies using TMS have provided further insight into the flow of information in the grasping network in humans. Davare et al. (2008) used a paired-pulse design in which a motor evoked potential (MEP) from stimulation of the M1 hand area was preceded by a conditioning TMS pulse applied to vPMC. This allowed them to measure the influence of vPMC on M1, depending on the type of action being performed. At baseline, vPMC stimulation had an inhibitory influence on M1, reducing the amplitude of MEPs. When TMS was applied during execution of a power grip, this inhibition did not occur, and during execution of a precision grip, it converted to excitation. This shows that the influence of vPMC on M1 differs, depending on the type of grip being used. A subsequent study from the same group used theta-burst TMS to disrupt processing in AIP (Davare et al., 2010), finding that this blocked the grasp-specific modulation from vPMC to M1.

Together, these studies of both monkeys and humans suggest that AIP lies at the top of the motor hierarchy within the grasping network. This interpretation is reinforced by studies that have investigated levels of motor equivalence in different areas. Many neurons in the vPMC show generality across response effectors—for example, firing during precision grasps using either the fingers or lips (Rizzolatti et al., 1988), or of the contralateral or ipsilateral hand (Tanji et al., 1988), as will be discussed further in chapter 13. Motor equivalence in the grasping network was studied explicitly by Michaels and

Scherberger (2018), who recorded from a large number of neurons in AIP and vPMC of macaque monkeys that performed a delayed grasping task. During the preparation phase in which the monkey knew which grasp it would have to perform but before it received the "go" signal, there was only weak differentiation in AIP between planned movements of the contralateral and ipsilateral hands. In contrast, laterality dominated responses in vPMC. There thus appears to be a hierarchy of abstraction, with AIP coding grasping actions in the most abstract manner, M1 coding the details of action implementation, and vPMC performance the coordinate transformations between these representations.

At the same time, this hierarchy of abstraction does not imply an exclusively one-way flow of information. A recent study used EEG to quantify connectivity between AIP and vPMC in children and adults performing a difficult precision grip task in which forces needed to be precisely controlled (Beck et al., 2021). Performance on this task was related more closely to "backward" coupling from vPMC to AIP than coupling in the reverse direction. Thus, as with many other domains, control of the hand in the grasping network relies on complex dynamics of feedforward and feedback signals.

The Reaching Network
Parallel to the grasping network, another set of regions appears to correspond more closely to the transport phase of reaching. This reaching network includes a complex of areas known as the "parietal reach region" (PRR) as well as dPMC (Andersen et al., 2014; Matelli & Luppino, 2001).

One difficulty in studying reaching responses in monkeys is that monkeys tend to look where they reach, meaning that observed neural activity could reflect either motor control of reaching or of eye movements. Even if the monkey is trained to maintain fixation during reaching, they could still be planning a saccade that they never execute. Snyder et al. (1997) used a clever method to discriminate between saccade-related and reach-related activity by using a dissociation task in which the monkey made reaches and saccades to different locations. The monkey was trained to reach after a delay toward the location in which a red light appeared and to saccade toward the location in which a green light appeared. On some trials, the monkey made only a reach or only a saccade, while on other (dissociation) trials, the monkey made a saccade and a reach in opposite directions. This allowed Snyder and colleagues to determine whether a given neuron coded for movement of the eyes or the arm. They found two areas in the PPC that showed strong activity during the delay period, suggesting they were related to the animal's intention to move. One area, the lateral intraparietal area (LIP), showed strong selectivity for eye movements, while another, the PRR, showed strong selectivity for hand movements. A subsequent study showed the same distinction between responses in LIP and PRR in a task where monkeys freely chose on each trial whether to make a saccade or a reach (Cui & Andersen, 2007), showing that the neurons code the monkey's motor intentions to move and not responses to stimulus cues.

The PRR appears to be a complex composed of several distinct brain areas (Andersen et al., 2014). In monkeys, the PRR includes the medial intraparietal area (MIP)—a region on the IPS located medial and posterior to AIP, as well as area V6a, even more posterior at the boundary between the parietal and occipital cortices. Both of these regions are directly connected to dPMC (Bakola et al., 2017).

Processing within this reaching pathway appears to be linked to transformation of information from eye-centered to hand-centered reference frames. Pesaran et al. (2006) measured responses from neurons in both MIP and dPMC while tracking the relative positions of the monkey's eyes, hand, and the object being reached for. Using a reference frame dissociation task, similar to that of Snyder and colleagues, Pesaran and colleagues found that while area MIP codes the location of the target object relative to the eyes, dPMC appears to code the three-way relation between the locations of the eyes, hand, and object. This suggests that the reaching network is involved in the transformation of visual information from the intrinsically retinotopic reference frame in which it begins into a hand-centered frame of reference required for implementing actions. This progression can be seen even among the areas within the PRR. V6a shows stronger connections to the visual cortices, while MIP shows stronger connections to motor and somatosensory regions (Bakola et al., 2017).

Several neuroimaging studies in humans have tried to identify a human homologue of the PRR. Connolly et al. (2003) compared fMRI activity while human participants planned either points or saccades to different spatial locations. A region in the medial part of the superior parietal lobe was consistently activated during planning of points compared to saccades, making it a potential human homologue of the PRR. Filimon et al. (2009) compared reaches made with visual feedback, reaches made without vision, and saccades to identify reach-selective regions as well as areas coding visual feedback from the hand. They identified a region at the boundary between the parietal and occipital lobes that responded more strongly during reaches performed under visual guidance than without vision. They also identified another reach-sensitive region in the medial parietal lobe that responded regardless of visual feedback. The human PRR, like the monkey's, thus appears to include multiple regions with distinct functions. The two regions described by Filimon and colleagues likely correspond to macaque areas V6a and MIP, respectively.

Other studies have found different forms of spatial sensitivity across regions of the reaching network. For example, reaching selective regions in the parieto-occipital boundary such as V6a show wide sensitivity across the visual field, suggesting a specialization for peripheral vision (Pitzalis et al., 2006; Prado et al., 2005). In contrast, medial intraparietal regions and dPMC respond more strongly when reaches are performed in foveal vision (Prado et al., 2005). This pattern suggests that the more posterior regions function to transport the arm into broadly the right location, while more anterior regions serve more precise guidance toward the end of a reach.

Interactions between the Grasping and Reaching Networks
Thus far, I have discussed the grasping and reaching networks as if they were completely modular and distinct. In reality, however, these networks are closely linked, as might be suspected from the fact that reaching and grasping themselves are closely linked actions. Microelectrode recording studies have found evidence for reach (as well as grasp) selectivity in both AIP and vPMC of the grasping network (Lehmann & Scherberger, 2013) and grasp (as well as reach) selectivity in area V6a of the reaching network (Fattori et al., 2010). Similarly, Takahashi et al. (2017) found reach and grasp responses in both dPMC and vPMC. In capuchin monkeys, both the dPMC and vPMC are densely interconnected with the hand representation in M1 (Dum & Strick, 2005). Dum and Strick suggest that dPMC, vPMC, and M1 form a densely interconnected network for the control of hand function.

Similar results have been found in fMRI studies in humans. Fabbri et al. (2014) used MVPA to determine whether different regions contained information specifying the direction of reaching or the type of grasp used. Some regions contained information about both grasping and reaching, such as AIP and vPMC. Other regions, however, contained exclusively information about reaching, such as portions of dPMC and the posterior part of the superior parietal lobule. Thus, the two networks are overlapping but at least partly distinct. Rather than fully distinct modules, there appears to be a gradient across the parietal cortex, with reaching represented more posterior and medially and grasping more anterior and laterally (Konen et al., 2013).

4 The Acting Hand

In his chapter on "The Will" in *The Principles of Psychology*, William James (1890) describes a problem he faced at the dinner table: "I sit at table after dinner and find myself from time to time taking nuts or raisins out of the dish and eating them. My dinner properly is over, and in the heat of the conversation I am hardly aware of what I do; but the perception of the fruit, and the fleeting notion that I may eat it, seem fatally to bring the act about" (p. 522). We can all recognize the phenomenon James describes. In this chapter, I will discuss the real-time control of action. As we will see, the mechanisms underlying this ability provide rich insight into James's problem.

Internal and External Circuits

James Gibson (1979) famously described the environment we perceive in terms of the *affordances* or opportunities for action provided to us by objects in the world. The concept of affordance provides a rich way of thinking about the actions we perform with our hands and their link to the objects we interact with. For example, objects of moderate weight and with elongated handles afford wielding. Rigid objects with sharp edges afford cutting or scraping, and elongated elastic objects afford knotting or weaving. Affordances, by their fundamental nature, reflect an intimate link between the physical properties of the world and the capabilities for action of an organism.

Gibson's concept of affordance has been enormously influential within perceptual psychology and has interesting links with a broader set of concepts in biology on the links between the body and the physical world. The British biologist J. B. S. Haldane (1927), for example, in a wonderful essay titled "On Being the Right Size," emphasized that different physical forces operate differently at different spatial scales. For humans, gravity is a dominant force, while the influence of surface tension on our lives is negligible. For a fly, however, the situation is reversed. A fly walks happily up a vertical wall but must, at all costs, avoid getting wet, for the film of water that surface tension leaves on its body weighs several times its body weight. The dangers—and opportunities—of the environment are fundamentally different for animals at different spatial scales to the point that there is little use in even calling it the same environment at all.

The intimate linkage between perception and action has led to various proposals within experimental psychology that perception of object affordances will automatically elicit those actions. Some observers of early infant reaching behaviors suggested that these were automatically elicited by objects rather than being a result of deliberate attempts to grasp the object (Baldwin, 1891; McGraw, 1941). Others linked such automaticity to what they considered less developed mental abilities, such as Kurt Koffka's (1935) famous claim: "To primitive man each thing says what it is and what he ought to do with it: a fruit says, 'Eat me'; water says, 'Drink me'; thunder says, 'Fear me,' and woman says, 'Love me'" (p. 7).

Magnetic Apraxia and Utilization Behaviors

In the 1950s, the neurologist Derek Denny-Brown (1958) argued for a dissociation between the effects of lesions to the parietal lobes or to the frontal lobes. Patients with medial frontal lesions showed compulsive tendencies to reach for and grasp seen objects—a condition that Denny-Brown termed "magnetic apraxia." In contrast, patients with parietal lesions showed an opposite tendency to retract their limbs and withdraw from stimuli—what he termed "repellent apraxia." "Apraxia" refers to deficits in the control of skilled action, as will be discussed in chapter 9, while "magnetic" and "repellent" emphasize the idea that the behaviors are driven directly by stimuli in the environment rather than by the patient's desires and intentions. On Denny-Brown's view, these two forms of motor response form opposing behavioral poles of approach and avoidance:

> The magnetic, exploratory aspect of behavior in relation to the environment is managed by the cortex of the parietal lobe, and released by frontal and temporal lesions. Conversely, the repellent, negative bias to behavior is determined by a strip of cerebral cortex in the premotor, cingulate and hippocampal region and released by parietal lesions. We believe that these two types of response represent two areas of normal organization, positive and negative, of the tropisms to the environment managed by the cerebral cortex. (p. 22)

The "magnetic" behaviors described by Denny-Brown consist largely of groping toward and grasping of objects—a pattern reported repeatedly in brain-damaged patients (Adie & Critchley, 1927; Seyffarth & Denny-Brown, 1948) as well as in young infants (Twitchell, 1965), as will be discussed in chapter 12. In other cases, however, more complex goal-directed actions can be seen that occur independent of—and often contrary to—the patient's conscious goals and intentions. In the "anarchic hand" sign, one hand performs complex actions that the patient denies performing themselves and is unable to control or inhibit (Della Sala et al., 1991; Goldberg et al., 1981). For example, when Della Sala and his colleagues took their patient, G.P., to dinner with her family, "out of the blue and much to her dismay, her left hand took some leftover fish-bones and put them into her mouth" (Della Sala, 2005, p. 606). The anarchic hand has been repeatedly linked to damage to regions of the medial PMC, such as the SMA (Della Sala et al., 1994; Goldberg, 1985; Goldberg et al., 1981). Della Sala et al. (1994) reviewed thirty-nine cases of anarchic hand reported in the literature and identified the SMA as being damaged in most of them. Similarly, removal of the SMA in monkeys has been linked to forced grasping and groping behaviors (Fulton et al., 1932).

In some cases, the two hands may work in opposition to each other—a condition known as "diagonistic dyspraxia" (Akelaitis, 1945; Tanaka et al., 1996). For example, when the patient described by Tanaka and colleagues was asked to place a toothbrush in front of a mirror, he took the brush in his right hand to do so, but his left hand "snatched it from the right hand and put it back where it was" (p. 861). Bogen (1993) similarly reports a patient whose anarchic hand was undoing the buttons on his shirt even as his healthy hand was buttoning them.

Finally, other patients show apparently compulsive tendencies to approach and use objects around them—a condition known as "utilization behavior" (Lhermitte, 1983). This differs from conditions such as magnetic apraxia and the anarchic hand in that it is the entire body producing the actions, not only one limb. For example, the patient shown in figure 4.1 uses each object shown to him in its conventional manner, irrespective of whether this is situationally appropriate. Despite already wearing glasses, he immediately puts on a second pair when it is placed in front of him. Such reports raise natural questions about the demand characteristics of the situation: How should the patient interpret his doctor placing familiar objects right in front of him? Critically, similar behaviors have also been reported in situations in which the object is not placed directly in front of the patient and where attention has not been directed toward it (Shallice et al., 1989). Unlike patients with anarchic hands who deny responsibility for the actions performed, when asked about their behaviors, patients with utilization behavior confabulate reasons for their actions, although these are often strained and unconvincing.

The conditions described in this section differ in the level within the motor hierarchy at which actions are elicited. The grasping and groping of magnetic apraxia involve low-level sensorimotor mechanisms, which appear to be innately present at birth (McGraw, 1940), as will be discussed further in chapter 12. The anarchic hand involves a higher level of motor organization in which coordinated, goal-directed actions of the entire limb are implemented, likely involving the reaching and grasping networks described in

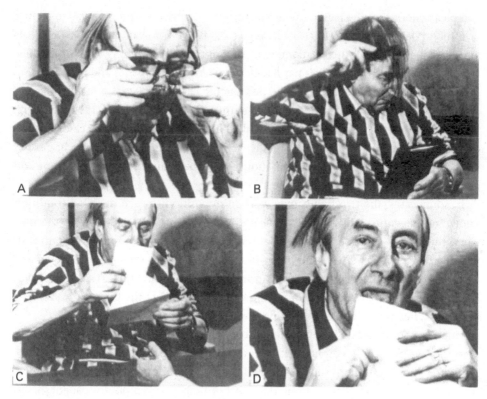

Figure 4.1
Utilization behavior in a patient with a right frontal tumor. Presented with familiar objects, he immediately uses them in their conventional manner, including (A) glasses, (B) a comb, (C) a sheet of paper, and (D) an envelope. From Lhermitte (1983).

the previous chapter. Finally, utilization behaviors appear to involve a still higher level of motor organization, possibly above the sensorimotor level entirely and involving more cognitive processes of planning and decision-making. In each case, however, the behaviors of neurological patients, while exaggerated, bear more than a passing resemblance to William James's observation of his own reaching for the after-dinner fruit.

Priming of Actions from Object Affordances

If conditions such as magnetic apraxia and utilization reflect exaggerated or disinhibited forms of response tendencies that are continuously operating in all of us, then similar effects should be measurable in the lab. Tucker and Ellis (1998) used a simple stimulus-response compatibility paradigm to measure automatic motor activations from task-irrelevant affordances of everyday objects. Participants judged whether familiar objects, such as a frying pan or teapot, were right side up or upside down by pressing buttons with either their left hand or their right hand. Across trials, the objects also varied in terms of whether their handles were pointed toward the right side of the image, affording being grasped with the right hand, or pointing toward the left side of the image, affording being grasped with the left hand. Although this latter variable was irrelevant to the task, it nevertheless produced clear effects on reaction time. Responses were faster when they involved the hand on the side toward which the handle was oriented, suggesting that the affordances of the object had automatically primed motor responses in the hand. A control experiment showed that this effect did not occur when participants responded using two fingers of a single hand, showing that it is not due to more abstract priming of spatial locations (i.e., a Simon effect).

Other research from the same lab has shown that precision and power grips can be similarly primed by seeing objects of the appropriate size (Tucker & Ellis, 2001). Participants made judgments about whether seen objects were manufactured (e.g., a key) or natural (e.g., a banana) by pressing one of two buttons held in their dominant hand. One of the buttons was held between the thumb and index finger and required a precision grip to press. The other button was held between the palm and the other fingers and required a power grip to press. Orthogonal to the manufactured/natural dimension, the objects also varied in size such that some, such as the key, were appropriate to be picked up with a precision grip, whereas others, such as the banana, were appropriate to be picked up with a power grip. Despite the size of the objects being completely irrelevant to the task that the participants were instructed to perform, it nevertheless affected responses in a systematic way. Precision grip responses were faster to small objects than to large ones, and power grip responses were faster to large objects than to small ones.

Gentilucci (2002) used motion tracking to measure kinematic effects of task-irrelevant distractor objects. He embedded identical wooden dowels into the top of an apple and a strawberry. Participants reached to make precision grips of the dowel. Although the size of the object the dowel was embedded in was not relevant to the task, it nevertheless altered several kinematic features of reaching and grasping. For example, the maximum aperture of the grasp during the approach to the target was larger for the dowel in the apple than for the one in the strawberry.

McBride et al. (2012) used a paradigm similar to that of Tucker and Ellis (1998), but they had participants respond by squeezing one of two bars held in each hand, depending on whether the seen object belonged in a kitchen (e.g., a frying pan) or a toolbox (e.g., a screwdriver). The handle of each object was oriented to be easily graspable by either the left or the right hand. In addition to showing a congruency effect, as in previous studies, the pressure applied to each bar was measured continuously. This allowed McBride and colleagues to show that even on trials on which the correct response was made, there was frequently a small initial increase in pressure applied by the incorrect hand, which was quickly corrected.

There is also evidence that this sort of automatic priming of motor actions from object affordances underlies the anarchic hand sign (McBride et al., 2013; Riddoch et al., 1998). McBride et al. (2013) tested S.A., a seventy-two-year-old woman with an anarchic right hand, using the paradigm described above. S.A. showed a substantially larger affordance congruency effect for her anarchic right hand than for her non-anarchic left hand. Indeed, the magnitude of the congruency effect for S.A.'s anarchic hand (76 ms) was several times as large as that found for healthy participants in the same task (16 ms on average). S.A.'s anarchic hand thus appeared to reflect an exaggeration of a response tendency that is present in everyone.

Evidence consistent with these effects comes from neurophysiological studies in monkeys and humans. Recall from chapter 3 that regions of the grasping network such as the AIP and vPMC contain canonical visuomotor neurons that fire both when the monkey makes a grasping movement and when it merely sees objects that afford that same action (Sakata et al., 1995). Studies using fMRI have found evidence for broadly similar activations in humans as well (Binkofski et al., 1998; Fang & He, 2005).

Cisek has argued for what he calls the "affordance competition hypothesis" (Cisek, 2007; Cisek & Kalaska, 2010). On this view, sensory information about environmental affordances is continuously processed to prepare and specify the kinematic details of appropriate actions. Simultaneously, competitive winner-take-all interactions among representations of potential actions implement the process of action selection. Cisek links these competitive interactions to the parallel parieto-frontal reaching and grasping networks discussed in the previous chapter. This process of decision-making can also be modulated by signals from cortical regions (e.g., the prefrontal cortex) and subcortical structures (e.g., the basal ganglia) involved in higher-level planning and decision-making processes. Importantly, the claim is not that every possible action we might perform is prepared and specified. Rather, attentional selection operates to ensure that only the most promising and relevant potential actions in the current situation will be processed.

Distinct Premotor Representations of Externally Elicited and Internally Generated Actions

As discussed above, damage to the medial PMC has been repeatedly linked to disorders of volitional movement such as the anarchic hand (Della Sala et al., 1994; Goldberg, 1985). Such findings reinforce the distinction coming from monkey neurophysiology that the lateral and medial premotor cortices comprise two distinct systems involved in externally elicited and internally generated movements, respectively (Passingham, 1993). Passingham and colleagues have reported clear double dissociations in the motor deficits produced by damage to these regions. In tasks in which the monkey has to make different movements depending on which stimulus is presented, monkeys are impaired following lesions of the lateral but not medial PMC (Halsband & Passingham, 1985). Conversely, in a task in which the monkey is rewarded with food every time it raises its arm without any external cue, monkeys are impaired following medial but not lateral PMC lesions (Thaler et al., 1995).

Additional evidence supporting this distinction was provided by Mushiake et al. (1991), who trained Japanese macaques to perform a sequential reaching task requiring them to respond by pressing sequences of four touch pads arranged in a diamond shape in front of them. The monkeys responded either on the basis of immediate visual cues or on the basis of a remembered sequence. In the visually guided task, each reach was cued by an LED on the touch pad illuminating. In the internally determined task, in contrast, the monkey was given an auditory cue to initiate a memorized sequence of reaches. Mushiake and colleagues recorded the activity of neurons in M1, the SMA (a region of the medial PMC), and the lateral PMC during preparation for and execution of the reaches, and they classified each neuron according to whether it responded more strongly to visually guided or internally determined reaches. While M1 neurons responded indifferently in both conditions, there was a dissociation between lateral PMC and SMA neurons. While a range of response types was found in both regions, SMA neurons showed a clear tendency to respond more during the internally determined task, and lateral premotor neurons showed a tendency to respond more during the visually guided task.

Neuroimaging studies in humans have also supported this distinction. For example, Playford et al. (1992) asked participants to freely choose which direction to move a joystick on each trial, while trying to be as random as possible. PET scanning showed that this free choice condition produced robust activations in the SMA and medial premotor regions. As will be discussed further in chapter 9, these activations appear to be disrupted in patients with Parkinson's disease, mirroring the difficulty these patients have in initiating self-generated movements. Intriguingly, direct stimulation of the SMA in human patients undergoing brain surgery has been reported to sometimes produce the "urge" to perform an action rather than eliciting an overt movement (Fried et al., 1991).

Online Control Programs

We are all familiar with the experience of getting into our car and finding ourselves at work, with little recollection of the act of driving, or getting off the bus and finding ourselves at home, with no detailed memory of the route we walked along. Such examples are like William James's experience with the after-dinner fruit, albeit with one important difference. James was struck by the fact that he was performing actions that he had never consciously decided to perform. In contrast, we are most likely aware of our decision to leave the house for work in the morning. Our lack of awareness instead concerns the continuous monitoring of ongoing behaviors. In this section, we will discuss the mechanisms underlying this online control of hand action.

Optic Ataxia and the PRR

In a classic report, the Hungarian neurologist Rezső Bálint (1909, 1995) described a patient with bilateral damage to the parietal lobes following a stroke. This patient was able to describe and recognize objects placed in front of him, showing no evidence of visual agnosia. In contrast, however, when he tried to reach out for objects with his right hand, he was only able to grope in their general direction. Remarkably, however, when asked to reach with his left hand, he was able to do so accurately, indicating that he did in fact know where the object was located. Moreover, he was also able to point to parts of his body using his right hand with his eyes closed, showing that he also did not have a purely motor impairment in controlling his right hand. Instead, his problem appeared to involve the real-time use of visual information to control online actions of his right arm. Bálint named this condition "optic ataxia," although he claimed it occurred as just one component of a larger condition, now generally known as "Bálint's syndrome."

Many subsequent patients with similar patterns of impairment have been described over the past century. Optic ataxia can result from unilateral damage to the parietal lobe in just one hemisphere, in which case the impairment is usually restricted to reaching toward objects in the contralateral visual hemifield (Cole et al., 1962; Ratcliff & Davies-Jones, 1972; Riddoch, 1935). Indeed, the deficit can even be restricted to a single quadrant of the visual field (Ross Russell & Bharucha, 1984). Frequently, reaching deficits occur only toward objects seen in peripheral vision, whereas reaches to foveated targets are accurate (Buxbaum & Coslett, 1997; Jackson et al., 2009; Rondot et al., 1977). This is consistent with neuroimaging results in healthy participants showing different parietal networks underlying reaching in central and peripheral vision (Prado et al., 2005). Patients with optic ataxia also show impairments in other aspects of visually guided action, including grasping objects effectively (Damasio & Benton, 1979), scaling grip aperture to object size (Jakobson et al., 1991; Jeannerod, 1986), and orienting the hand to match the orientation of the object (Perenin & Vighetto, 1988). Some patients also show analogous deficits for hand and foot movements (Evans et al., 2012; Rondot et al., 1977), consistent with the principle of motor equivalence. Importantly,

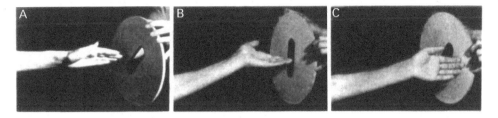

Figure 4.2
Errors of hand orientation (B) and placement (C) in optic ataxia using the "slot" task of Perenin and Vighetto (1988).

such deficits in reaching and grasping can occur independently of corresponding problems in making eye movements to the same targets (Trillenberg et al., 2007).

For example, Perenin and Vighetto (1988) used a simple task in which the patient had to move their hand through a narrow slot in a circular board held in front of them. Figure 4.2 shows performance on this task by a patient with a tumor in the right parietal lobe. When reaching with her left hand into the right hemifield (figure 4.2A), her reaches were accurate; the hand has been transported directly in front of the slot and is oriented appropriately to go through it. In contrast, when reaching into the left visual field, she made errors of both hand orientation (figure 4.2B) and location (figure 4.2C).

Despite the word "optic" in the condition's name, these patients also have difficulties with reaching in the dark under proprioceptive guidance. For example, Blangero et al. (2007) asked two patients to point in the dark with their ataxic hand toward their intact hand. Despite there being no visual component to this task, both patients were impaired. This proprioceptive deficit showed a similar pattern to deficits in visually directed reaching in that it showed both a "field effect," with less accurate reaching when the intact hand was in the contralesional than in the ipsilesional hemispace as well as a "hand effect" in that deficits were seen for reaches with the ataxic hand into both hemispaces. Thus, optic ataxia can be thought of as a general deficit in reaching under sensory guidance.

Other research has shown that the deficit in visually directed action in optic ataxia can be dissociated from the patient's conscious, semantic understanding of the objects. Jeannerod et al. (1994) described patient A.T., who suffered bilateral parietal lobe damage. While A.T. could reach to the correct location, her ability to configure her hands for grasping was severely impaired. Her attempts at precision grasping failed to end with the object in contact with the fingertips, and her grip aperture correlated poorly with actual object size. In striking contrast, however, when A.T. was asked to use her thumb and index finger to match the size of seen objects, her estimates correlated well with actual object size. This shows that her deficit was not in the visual perception of object size itself, nor in basic motor control of the hand. Rather, her deficit appeared specific to the online control of object-directed grasping actions.

The role of disrupted online control of action in optic ataxia can also been seen in some remarkable and counterintuitive improvements in reaching in these patients when sensory information is withdrawn. For example, Milner et al. (1999) asked patient A.T. to point to the location of lights that lit up at various points in reaching space. Typically for optic ataxia, she made large errors in the location of her reaches. In another condition, however, A.T. waited for five seconds after the light turned off until cued to respond. Remarkably, these delayed points were highly accurate. A.T.'s ability to maintain the location of the light in working memory was intact, as was her ability to use conscious, effortful control of her arm movements. She was impaired only when using sensory information in real time. Similarly, Jackson et al. (2005) used a set of spectacles with a rapidly closing shutter to remove vision from their patient at exactly the moment he was cued to start reaching. In this condition, the errors he made with the presence of vision were substantially reduced. Another patient showed poor scaling of grip aperture to object size when grasping actual objects but much better performance when asked to pantomime such grasps toward the remembered locations of objects (Milner et al., 2001). In each of these cases, the errors that patients make when relying on immediate sensory signals disappear when they rely on memory instead. Indeed, Rossetti et al. (2005) found that some patients with optic ataxia have learned to rely on remembered rather than online information. They asked patients and controls to point to the remembered locations of objects after a five-second delay. On some trials, however, a conflicting stimulus came on as they were cued to reach. While the points of control participants were strongly affected by these latter cues, the responses of the patients were much less affected, suggesting a reduced role of immediate sensory information in guiding their actions.

Studies in monkeys have provided consistent support for the role of the PPC in the control of hand movements. Ferrier (1890) found that following bilateral PPC lesions monkeys "are unable like normal monkeys to take things offered them delicately with the fingers, but make grabs at them with the whole hand" (p. 1414). Subsequent studies confirmed this observation, showing that reaches made by the contralateral arm were inaccurate and slow following PPC lesions (Ettlinger & Kalsbeck, 1962; Faugier-Grimaud et al., 1985; LaMotte & Acuña, 1978) For example, Faugier-Grimaud and colleagues surgically produced unilateral lesions in area 7 of the PPC in four monkeys, showing clear impairments in both the accuracy and speed of reaching with the contralateral hand.

Traditionally, optic ataxia has been linked to lesions of the superior parietal lobe or areas around the IPS (Perenin & Vighetto, 1988). A lesion-overlap study comparing stroke patients with and without optic ataxia linked the condition to lesions of the medial PPC, near the border with the occipital lobe (Karnath & Perenin, 2005). This is highly consistent with the location of the PRR, described in the previous chapter as a key part of the reaching network. The link between the PRR and optic ataxia was confirmed by Hwang et al. (2012), who used the drug muscimol to temporarily inactivate the PRR in monkeys, showing that this rapidly produced deficits highly similar to human optic ataxia, with deficits in visually guided reaches but not in saccades.

Blindsight and the Dorsal Visual Pathway

Optic ataxia shows a selective *impairment* in online control of hand actions. These mechanisms can also be selectively *spared* in people with dramatic impairments in other aspects of vision. For example, some patients with blindness in portions of their visual field nevertheless show evidence of processing information from within their scotoma (i.e., the blind region of their visual field)—a condition known as "blindsight" (Weiskrantz, 2009).

Notably, evidence for blindsight in cortically blind patients is generally substantially higher for manual reaching performance than for eye movements (Perenin & Jeannerod, 1978; Weiskrantz et al., 1974)—a phenomenon known as "action blindsight." For example, figure 4.3 shows data on the ability of the well-studied patient D.B. to make hand and eye movements to the location of objects in his blind visual hemifield (Weiskrantz et al., 1974). The location of his reaching responses correlates almost perfectly with the true location of the objects (figure 4.3, right panel), although he claims not to be able to see them. In contrast, his eye movements toward these objects have little relation to their actual location (figure 4.3, left panel). Critically, this is not due to poor control over eye movements per se, as his ability to direct eye movement to seen targets in his intact visual field is preserved. Similarly, patient T.N., with complete blindness following bilateral damage to the primary visual cortex (V1), was 75 percent accurate at pointing to targets placed on either his left or right but at chance when asked to make a discrete judgment of location by simply lifting his left or right hand (Buetti et al., 2013).

Other studies show that in addition to basic information about object location, some patients can also use more detailed information about the size and shape of objects in their scotoma to guide action. For example, patient P.J.G., described by Perenin and Rossetti (1996), reported having no visual experiences from the right side of his visual field. Nevertheless, he was able to place his hand through an oriented slot on the right side but was unable to indicate the orientation verbally. Similarly, he showed appropriate scaling of grip aperture when reaching to pick up objects of different size but was unable to indicate the size of objects either verbally or by matching the size using his thumb and index finger. Patient S.J. showed a similar pattern (Whitwell et al., 2011). Intriguingly, when

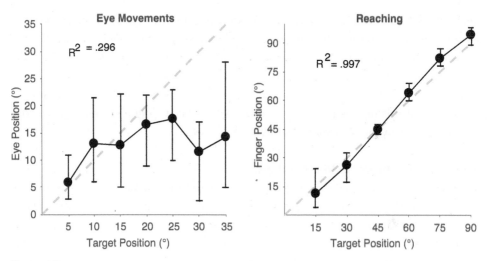

Figure 4.3
Object localization performance of blindsight patient D.B. tested by Weiskrantz et al. (1974).

a two-second delay was introduced between S.J. seeing the object and being allowed to start her grasp, she was unable to scale grip aperture. Whitwell et al. (2020) refer to this unconscious ability to scale grip to object size as "grip constancy" in contrast to the well-known principle of "size constancy" in conscious vision.

Other studies have shown similar sparing of online action abilities in patients with damage to higher levels of the visual system, who, while not blind, show visual object agnosia. For example, the famous patient D.F. suffered from profound visual agnosia as a result of brain damage from carbon monoxide poisoning (Goodale & Milner, 2004). Although not blind, she was unable to name or recognize everyday objects, such as apples, books, or sailboats. Despite these profound difficulties, D.F. was strikingly unimpaired in her everyday behaviors. She was able to walk around confidently, even on rough mountain terrain. When the experimenters held up a pencil, D.F. was completely unable to say what the object was (aside from that it was yellow) or how it was oriented. When asked to reach out and grasp the object, however, D.F. did so effortlessly, orienting her hand appropriately from the start of the reach. In order to test this ability formally, Goodale et al. (1991) used a slot task similar to that used with optic ataxic patients (figure 4.2) through which D.F. was asked to post a card she held in her hand as if she were placing a letter into a mailbox. Her ability to do this was well above chance levels and similar to control participants. In contrast, when asked simply to orient the card to match the orientation of the slot, D.F. performed at chance levels. Thus, despite being unable to tell consciously how the slot was oriented, D.F. was able to use visual information about the slot's orientation to guide actions as well as anyone else.

Milner and Goodale (2006) influentially interpreted visual object agnosia, as in D.F., as constituting a double dissociation with optic ataxia. On this interpretation, these conditions result, respectively, from damage to the ventral visual pathway for conscious visual perception and the dorsal visual pathway for guidance and control of action. Other researchers have questioned this simple dissociation and suggested that processing within the dorsal stream should be further subdivided. For example, some authors have suggested that the dorsal stream should be thought of as two distinct streams (Binkofski & Buxbaum, 2013; Pisella et al., 2006): a dorsal-dorsal stream (broadly consistent with the reaching network described in the previous chapter), damage to which results in optic ataxia, and a ventral-dorsal stream (broadly consistent with the grasping network), damage to which results in conditions such as apraxia, which will be discussed in chapter 9. Other research has suggested that there may be comparable dissociations between processing streams for other sensory modalities, such as audition (Bizley & Cohen, 2013).

Online Action Correction and the "Automatic Pilot"
In the previous sections, we have discussed cases in which the ability to use real-time sensory information to control movements is selectively impaired or spared. These processes are, of course, constantly shaping our everyday behaviors outside of our conscious awareness. In a seminal study, Goodale et al. (1986) showed rapid and unconscious corrections to reach trajectory. Participants started each trial with their hand on the location cued by a light and made reaches toward a new location when the light moved at the start of each trial. Participants saw their arm through a semi-silvered mirror, which allowed Goodale and colleagues to remove visual information about the hand as soon as the reaching movement started. In half the trials, the light remained at the location where it had first moved to (single-step trials). On other trials, however, the target made a second jump (double-step trials), which happened as the participant's saccade to the first location reached peak velocity.

Remarkably, the accuracy of reaching was equivalent in single-step and double-step trials, showing that the reach had not been programmed ballistically and that visual information was continuously being used to adjust the trajectory of the reach. Moreover, the duration of hand movements increased linearly with the distance traveled and was not systematically longer for double-step trials. Indeed, there was no apparent discontinuity in the velocity profile of reaches in double-step trials to indicate the initiation of any special process of correction. Rather, Goodale and colleagues argue that the updating of reach trajectory on double step trials reflects "nothing more than the normal updating of the motor programming that occurs at the end of the first saccade on an ordinary trial" (p. 749).

Intriguingly, despite the smooth and rapid corrections to reach trajectory on double-step trials, participants remained totally unaware of the second jump of the stimulus and, indeed, of the difference between single-step and double-step trials. In control experiments in which participants were asked to discriminate between single-step and double-step trials, they were unable to do so. This provides striking evidence that visual information is continuously being monitored to fine-tune our actions and that this is happening outside of our conscious awareness. It also, incidentally, provides further evidence for hierarchical organization of motor control: participants were perfectly aware of reaching to the target (i.e., a high-level goal) but remained entirely unaware of corrections to the precise trajectory of reach, which was dealt with effectively and automatically by lower-level processes operating outside of conscious awareness.

Castiello et al. (1991) investigated this dissociation between participants' awareness of the action and the automatic adjustment of reaching responses using larger and more qualitative changes in the reach target that could be noticed. Participants were asked to reach for one of three objects that was illuminated with a light. On 20 percent of trials, the light shifted to a different object immediately as the participant's hand started to move toward the first object. Participants were instructed both to adjust their reach to grasp the new object and to indicate verbally when they became aware of the new target. Adjustment of the actual reach trajectory was observed around 100 ms after the change in light location. In contrast, vocal responses indicated awareness of the change more than 300 ms later. In general, participants reported seeing the light shift between objects at the very end of their reach, although it occurred, in fact, at the exact moment their reach began.

The corrections of reach trajectory following target displacement also show appropriate coordination between reaching and grasping motor programs. Visual displacement of the target results in the duration of reach being increased by approximately 100 ms (Paulignan et al., 1991). Critically, this delay is reflected in a correspondingly altered time course of grip aperture, which decreases following the target displacement and then peaks closer to the time of contact.

Intriguingly, such corrections occur even when participants are instructed to treat the movement of the target as a stop signal (Pisella et al., 2000). This suggests that they happen automatically and outside of volitional control. Pisella and colleagues call this the "automatic pilot" for the hand. According to Pisella and colleagues, online control is the most central function of the entire PPC, and disruption of this automatic pilot is the core feature of optic ataxia. Cressman et al. (2006) replicated this effect and showed that no such automatic corrections occur for changes cued by a change in object color—a feature unlikely to be directly processed by the dorsal stream. Subsequent studies have shown that these automatic responses can be modestly reduced by cognitive control, but even then, a strong automatic and uncontrollable component remains (Cameron et al., 2009; Day & Lyon, 2000; Striemer et al., 2010).

Such rapid online correction of hand actions is not specific to visual inputs. Pruszynski et al. (2016) showed that tactile information can also be used in this way. They asked participants to make reaches with the right hand to grasp a ball at the end of a rod. On half the trials, the rod rotated after the start of the reaching movement. The thumb of the participant's left hand rested on the other end of the rod, providing tactile cues about the rod's rotation. Participants were able to use this tactile information to rapidly update their reach trajectory. Comparison of updating based on tactile and on visual cues showed that, in both cases, muscular changes associated with the correction started about 100 ms after the start of the rotation. Similar corrections can be based on proprioceptive (Crevecoeur et al., 2016), auditory (E. O. Boyer et al., 2013), or vestibular (Bresciani et al., 2002) signals.

The Long-Latency Stretch Reflex
The studies described so far have involved alterations to the perceived location of the target object. Another set of studies have shown rapid motor adjustments in response to perturbations of the position of the reaching limb itself. Extremely rapid corrections within 20–50 ms, known as the "stretch reflex," were described in cats by Liddell and Sherrington (1924). They showed that this response is controlled by monosynaptic connections between muscles and the spinal cord. While this was classically taken as an example of a purely spinal reflex without involvement of the central nervous system, an elegant study in humans by Hammond (1955) showed that in addition to this rapid correction, there was a second, longer-latency muscular response, occurring between 50 and 100 ms after perturbation. This effect is striking because its latency appears too slow to reflect classical monosynaptic spinal reflexes and yet too rapid to reflect a voluntary response. More dramatically still, Hammond (1956) showed that the magnitude of the long-latency response—but not the short-latency response—was modulated by verbal instructions to the participant to "resist" or to "let go." Thus, while the long-latency response was not directly under voluntary control, it was nevertheless influenced by the goals and mental "set" of the participant, blurring the traditional distinction between reflexes and voluntary behavior.

Many studies over the last seventy years have replicated these results and shown that the intentions, goals, and situational context modify long-latency stretch reflexes

(Pruszynski et al., 2008; Rothwell et al., 1980). In one classic example, Marsden et al. (1981) applied perturbations to the left arm of kneeling participants while they supported themselves by holding onto a table with the right hand. There were clear long-latency contractions of the right triceps muscle in response to these perturbations of the left hand, which helped to stabilize the participant. These responses in the right arm vanished when the participant let go of the table, demonstrating that they were being controlled flexibly to take into account the configuration of the body and its relation to other objects. Most remarkably, the modulation of the corrective response was not merely related to the fact that the participant was grasping something with their right hand. The same adjustment that will help stabilize the body when holding onto a stable object such as a table would be counterproductive when holding a different type of object such as a teacup. Indeed, the corrective response to perturbation of the left arm when holding a teacup was opposite to that when holding the table, resulting in relaxation of the triceps and contraction of the biceps muscles in the right hand, preventing the teacup from tipping over.

Similarly, in the so-called sherry-glass experiment (Traub et al., 1980), disturbances were applied to the participant's wrist as they held their thumb and index finger along the rim of a wine glass, which elicited rapid responses in the forearm muscles stabilizing the glass. This response could not be suppressed by conscious will, suggesting that it was operating at an automatic level. However, no such responses were seen when the fingers were holding onto a block of wood rather than a glass, and remarkably they were also absent when the participant was told not to worry about knocking over the glass. This is a remarkable dissociation: the participant cannot suppress the response directly but can do so indirectly by approaching the task of holding the glass using a different mental "set." Despite being implemented at an automatic, unconscious level, the sherry-glass response nevertheless reflects and implements the participant's conscious intentions. These results show that the long-latency corrective responses are modulated in a sophisticated way—a point we will return to below when we discuss the concept of the "control policy" in optimal feedback control (OFC).

Similarly, Nashed et al. (2012) asked participants to reach for either circular or bar-shaped targets. Mechanical perturbations of the arm away from the target resulted in different types of correction for the two shapes. When reaching for circular targets, responses rapidly mirrored the shift in target, resulting in similar grasps of the objects. In contrast, when reaching for the bar, smaller corrections were made, resulting in grasps of the bar at locations closer to the new location of the hand.

Beyond their role in shaping real-time motor responses, there is also emerging evidence that these rapid motor adjustments also provide signals used for subsequent learning and adaptation to altered states of the body (Albert & Shadmehr, 2016; Maeda et al., 2020). It is also important to keep in mind that in the same way that the hierarchical organization of motor actions themselves fits within a more cognitive process of planning and means-end reasoning, these automatic processes of action monitoring and correction are only one aspect of a broader set of processes for recognizing and correcting errors at multiple levels of abstraction within the motor hierarchy. This includes behavioral evidence for slowing of subsequent responses in the aftermath of errors (Rabbitt & Rodgers, 1977) and neural mechanisms in the anterior cingulate cortex involved in error detection and cognitive control (Carter et al., 1998).

Computational Models of Feedback Control

Traditionally, reflexes have been treated as stereotyped and rigid, in stark contrast to the flexibility and context specificity of skilled voluntary action. While motor behaviors certainly vary along this dimension, the sophistication and speed of automatic corrective responses such as the long-latency stretch reflex has blurred this distinction somewhat (Scott, 2016). Over the past two decades, OFC has become increasingly influential as a model of motor control (Diedrichsen et al., 2009; Scott, 2004; Todorov, 2004; Todorov & Jordan, 2002). OFC proposes a complex and continuous linkage between sensory feedback and motor planning, as shown in figure 4.4. As such, it builds on the idea of internal models, which describe computational frameworks for combining motor and sensory information for effective planning and control of action (Wolpert et al., 1995). A key aspect of this framework is the idea of inverse and forward models. An inverse model transforms a desired goal state into the required pattern of motor commands to implement the action. In contrast, a forward model transforms a set of potential motor commands into a prediction of the resulting state and its sensory consequences. Within OFC, a state estimator performs a Bayesian integration of predictions from a forward model and immediate sensory feedback to come up with an optimal estimate of the current state of the body and world.

The critical novel feature of OFC is the idea of a control policy used to generate motor commands, which minimizes some criterion of performance. This criterion is defined by a cost function, which incorporates the high-level goal of the action, the behavioral context, and the current state of the body and assigns to each possible movement a cost. The generation of motor commands, whether the initiation of an action or adjustments of an ongoing action, can then be implemented by finding the commands that minimize this cost function.

Consider the sherry-glass experiment described above (Traub et al., 1980). When we hold a wine glass, we need to adjust the movements of our hand and arm to protect the

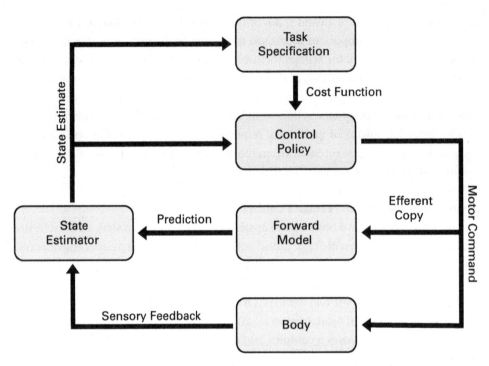

Figure 4.4
A schematic depiction of OFC. Following Scott (2004) and Diedrichsen et al. (2009).

fragile glass itself and prevent it breaking or wine from sloshing out. These considerations can be expressed in terms of the cost function, in which costs can be assigned to changes in the orientation of the glass or contact of the glass with other objects that would not make sense if we were holding, for example, an apple. This cost function then determines the control policy that will remain active while we are holding the glass. This control policy, in turn, will shape the way in which automatic corrective movements such as the long-latency stretch reflex become expressed in behavior. This provides an elegant system in which high-level behavioral intentions and semantic knowledge can shape the expression of automatic control processes that themselves are outside of conscious awareness and happen more quickly than would be possible for volitional movements.

Various models have been proposed for what performance criterion such cost functions would optimize, such as minimizing the jerk (i.e., the derivative of acceleration; Flash & Hogan, 1985), minimizing changes in torque (Uno et al., 1989), or metabolic costs of movement (Berret et al., 2011). Whatever the control policy aims to minimize (a factor that may, itself, vary depending on context), the important point is that the balance between a control policy and cost function provides a principled basis for selecting among the multitude of potential kinematic patterns that could implement a desired goal state. Given that all neural systems are inherently noisy, such models can account for why repeated goal-directed actions show trial-to-trial variability and yet succeed at implementing the desired goal, thus solving Bernstein's degrees-of-freedom problem discussed in the previous chapter. Moreover, in solving this problem, it produces movements that are smooth, fluid, and appropriately sensitive to the current context.

One key aspect of OFC is that movement errors will only be corrected if they are actually relevant to the action goal—the so-called minimum intervention principle. Errors in orthogonal directions will not be penalized by the cost function and so will not be corrected.

Brain Networks Underlying Online Action Correction
Several lines of evidence support a direct causal role of the PPC in the online control of actions in humans. Desmurget et al. (1999) asked participants to point with their right hand to visual targets that appeared in their peripheral vision. On some trials, the location of the target changed while the participant was making a saccade. As in previous research, although participants were unaware that targets had changed location, they were nevertheless able to smoothly update their reach trajectory. Single-pulse TMS applied to the left PPC, however, disrupted this updating. Importantly, when reaches were instead made with the left hand (ipsilateral to the TMS), no effect of TMS was found, indicating that TMS affected updating of reaches with the contralateral hand and not visual perception of stimulus location per se. Subsequent studies have replicated this general pattern (Reichenbach et al., 2011) and reported similar TMS-induced disruption of adjustments related to other features of an object, such as its orientation (Tunik et al., 2005) and size (Glover et al., 2005).

Similar conclusions come from studies of patients following damage to the PPC. For example, Gréa et al. (2002) tested patient I.G. with bilateral PPC lesions. I.G. was able to reach effectively to grasp objects, showing that she did not have an impairment in primary motor planning or execution. On trials, however, in which the object to be grasped made an unexpected jump, I.G. was unable to update her ongoing reach. Similar results were reported in two other patients by Rossetti et al. (2005), who suggest that the effects of PPC damage result in a shift from "fast visuomotor" to "slow cognitive" control. These patients can still perform actions, but only by approaching the task in a different way and using effortful and deliberate processes to produce them. In this sense, these patients may be similar to patient I.W., described in chapter 2, who lost tactile and proprioceptive signals from below the neck.

In the case of the long-latency stretch reflex, the perturbation is applied to the body itself and not to the target object. In this case, S1 and M1 are thought to be involved.

In an early series of studies, Evarts recorded from neurons in M1 while macaque monkeys performed arm movements in which the trigger to respond was the perturbation of arm position itself (Evarts, 1973; Evarts & Tanji, 1976). Like in the stretch reflex recorded from muscles themselves, M1 neurons showed both short-latency (20–25 ms) and long-latency (40–50 ms) responses to perturbation. Moreover, long-latency—but not short-latency—M1 responses were modulated depending on whether the monkey's task was push or to pull a lever, similar to the modulation of the long-latency stretch reflex described above (Hammond, 1956). More recent studies have confirmed that M1 responses to perturbation of the arm are modulated by the monkey's intention (Pruszynski et al., 2014) and involve integration of postural responses across multiple joints (Pruszynski et al., 2011).

Cheney and Fetz (1984) linked the long-latency stretch reflex not merely to M1 but more specifically to the direct CM neurons discussed in the previous chapter. The CM neurons linked with "new M1" (Rathelot & Strick, 2009) provide a natural mechanism for mediating rapid and flexible responses. First, because these CM neurons project monosynaptically onto motoneurons, they should produce rapid responses. Second, because they bypass spinal mechanisms, they allow responses to be flexible and context dependent. The same features that make CM neurons an important substrate for fine manual dexterity thus also make these neurons ideal for fast and flexible feedback responses.

Other research has also implicated a wider range of brain regions in fast-feedback responses. For example, some studies have shown responses in the dentate nucleus of the cerebellum consistent with this role (Strick, 1983). Omrani et al. (2016) showed using electrophysiological recordings in macaque monkeys that somatosensory feedback information spreads rapidly to several parieto-frontal brain areas, especially when it is task relevant to the monkey. Intriguingly, the earliest responses were found in superior parietal area 5, consistent with the role of this area in controlling adjustments based on visual information about target movements discussed above.

In a recent study, Takei et al. (2021) transiently deactivated area 5 and dPMC in macaque monkeys by cooling each area. On the basis of computational modeling, they attempted to link the patterns of deficits observed when each area was cooled to specific processes in OFC. Cooling of dPMC produced impairments consistent with disruption of the control policy. In contrast, cooling of area 5 produced impairments consistent with disruption of state estimation. This latter finding is consistent with other results showing that neurons in the PRR indicate the direction of reaching movements with 0 ms lag (Mulliken et al., 2008), suggesting that they reflect the output process of a forward model, as well as with studies of human patients linking damage to this region and impaired state estimation of the arm (Wolpert et al., 1998).

5   The Exploring Hand

As a field surgeon in the German army in World War I, Hermann Krukenberg developed an extraordinary procedure for helping soldiers who had lost their hands. In the so-called Krukenberg procedure, the two bones of the forearm (the radius and ulna) are converted into a pair of claw-like pincers (Kleeman & Shafritz, 2013). The forearm muscles are repurposed to allow independent control of each pincer, allowing them to be used like two fingers. While this procedure is not widely used today, at least partly on cosmetic grounds, the Krukenberg "hand" can be used dexterously for many daily activities.

The procedure was widely used in the Soviet army in World War II and studied extensively by Leont'ev and Zaporozhets (1960). In the weeks and months following the surgery, these patients were unable to recognize objects felt using their Krukenberg "hand," despite having intact basic tactile and motor function. Patient Ukr, for example, underwent the Krukenberg procedure on his right arm after losing both of his hands in 1942. Handed a wooden cylinder, he comments: "This is a very difficult shape to make out but it is tall. It's a smooth thing—I can feel it well, but what it is I have no idea" (p. 41). Ukr can clearly feel the object and can tell its texture and basic features, but he cannot recognize what the object is. He has tactile agnosia. Just having tactile sensations was insufficient to recognize the object; something else was needed. Describing these patients, Leont'ev and Zaporozhets observe: "The general impression is created that the hand is blind; that although in contact with the object it does not reflect its properties in its movements; they lose their wisdom and, as it were, the inspiration which specifically characterizes the movements of the hand" (p. 11).

Some patients never managed to use their Krukenberg hands effectively, despite having intact tactile sensations and ability to move the hand. With training and experience, however, other patients regained the "wisdom" of their hands. Patient Mus, who was jokingly called "Professor Krukenberg" by the other patients, was outstandingly successful at learning to control his Krukenberg hand for all manner of daily activities and was able to recognize objects by touch as well as two-handed participants. In general, the ability of patients to recognize objects by touch with their Krukenberg hand was not related to their basic tactile sensitivity but rather to their ability to act effectively with the hand in daily life.

In previous chapters, somatosensory and motor functions have been discussed separately. The work of Leont'ev and Zaporozhets, however, highlights the close relationship between complex tactile judgments such as recognizing objects and motor control. In this chapter, we will discuss the active use of the hand for tactile perception—what is known as "haptics," which Gibson (1966) defined as "the sensibility of the individual to the world adjacent to his body by the use of his body" (p. 97).

Haptic Perception

Haptic Object Recognition

In his extraordinary book *The World of Touch*, David Katz (1925) almost single-handedly initiated the study of haptics. Katz emphasized that much of the information that we acquire through touch is only available when we actively move our hands across objects. As Lashley (1951) observed, "the shape of an object impressed on the skin can scarcely be detected from simultaneous pressure, but the same shape can readily be distinguished by touch . . . when explored by tactile scanning" (p. 128). Gibson (1962) experimentalized Lashley's observation, asking participants to judge which of six cookie cutters they felt on each trial. When the stimuli were passively applied to the palm of the hand, judgments were correct on 49 percent of trials. In striking contrast, when participants could actively feel the shapes, they were correct 95 percent of the time.

An elegant series of studies by Roberta Klatzky and Susan Lederman provided dramatic evidence for the power and accuracy of haptic object recognition. In an initial study (Klatzky et al., 1985), blindfolded participants felt a hundred familiar everyday objects from eight categories, such as foods (e.g., carrot, potato), tools (e.g., wrench, screwdriver), and office supplies (e.g., pencil, stapler). Participants were asked to verbally identify each object as quickly as possible using their hands. Performance was extremely high, with correct classifications made on more than 95 percent of trials. Moreover, most errors involved using superordinate category labels (e.g., "vegetable" instead of "carrot") or were category related (e.g., "sock" instead of "sweater"), showing some level of accurate recognition. Equally impressive, classifications were made quickly, typically in just one or two seconds.

Such object recognition can be achieved even from quite minimal information. Klatzky and Lederman (1995) asked participants to make brief contact with objects using only their fingertips—what they call a "haptic glance." While performance was not as high as with unconstrained object exploration, participants were able to classify objects at substantially better than chance levels even with exposure times of as little as 200 ms. This was particularly true when they were given information about which superordinate category the object belonged to (e.g., "Is this container top a cork?").

These findings reinforce the argument of Katz (1925) and Gibson (1962) that the hand should be thought of as a highly effective sensory organ. Without question, the specialized sensory machinery of the hand contributes to this ability. But at the same time, the physical characteristics of the hand may not be the most critical factor. After all, the findings of Leont'ev and Zaporozhets (1960) described above show that even the Krukenberg hand can be used to effectively recognize objects once it has become well integrated into the sensorimotor system. In several studies, Katz (1925) also showed that healthy participants are surprisingly good at making haptic discriminations using other body parts, such as the lips, teeth, and toes. A more recent study by Lawson (2014) showed that while object recognition was not as accurate for the feet as for the hands, people could still recognize objects with their feet more than half the time. Moreover, performance on the two body parts was correlated across participants. Such results generalize the concept of motor equivalence to the haptic domain.

Haptic Exploratory Procedures

It is not, of course, merely the fact that the hand is moving that makes haptic object recognition effective. The hand can be moved in specific ways that optimize the information obtained. As Katz (1925) notes, however, when we haptically explore an object, it is very difficult to introspect about the specific movements we use. The features of the object being felt are immediately and vividly apparent; the hand itself fades into the background.

Lederman and Klatzky (1987) proposed that the hand movements involved in recognizing objects and their physical properties could be described in terms of a relatively small set of exploratory procedures, as shown in figure 5.1. To investigate how exploratory procedures were used to judge different object properties, they used a haptic match-to-sample task in which participants were first asked to feel a "standard" object, followed by three "test" objects, and to judge which test object was most similar to the standard along one specific dimension. The objects varied across nine dimensions (e.g., texture, hardness, volume). By coding the video-recorded hand movements that participants used to explore the objects, Lederman and Klatzky showed that people systematically use specific exploratory procedures in order to obtain information about different object properties. For example, when asked to compare stimuli in terms of their texture, participants spent on average 76 percent of their time using the lateral motion exploratory procedure. In contrast, when asked to compare stimuli in terms of hardness, they spent 64 percent of their time using the pressure exploratory procedure. These results show that people have clear and consistent preferences about which hand movements to use to discriminate different object properties. A subsequent study in which participants were instructed on each trial which hand movement to use showed that these intuitive choices that people make are near optimal in terms of the accuracy and speed of discriminations.

Figure 5.1 depicts six of the most studied exploratory procedures:

- Lateral motion: The fingertips move quickly from side to side across the surface of an object. This provides rich information about object *texture*.
- Unsupported holding: An object is held in the palm of the hand, frequently with hefting movements of the wrist or forearm. This provides information about object *weight*.
- Pressure: The fingertips poke or press against the surface of an object that is supported in place. This provides rich information about the *hardness* of the object.
- Enclosure: The hand is molded around as much of the object as possible. Successive periods of enclosure may occur separated by shifts of the object in the hand. This provides information about the *global shape* and *volume* of the object.
- Static contact: The palm is held flat against the surface of an object. This provides information about object *temperature*.
- Contour following: The hand or fingertips move smoothly and continuously along the contour of an object. This provides information about the *exact shape*, *volume*, and *thickness* of an object.

Tactile Agnosia

In vision, there has been extensive research on higher-level disorders of visual recognition, visual agnosias, in which patients are not blind but are unable to recognize what objects are, as in the case of patient D.F. described in the previous chapter (Goodale & Milner, 2004). There is also evidence for selective impairments of haptic object recognition, known as "tactile agnosia" or "astereognosis." For example, patient Ukr, described at the start of this chapter, was able to feel touch with his Krukenberg hand and could recognize basic tactile features such as size, smoothness, and curvature, but he could not use this information to recognize what the object was (Leont'ev & Zaporozhets, 1960).

Reports of such impairments go back to the mid-nineteenth century and have been controversial ever since (Mauguière et al., 1983; Reed et al., 1996; Reed & Caselli, 1994). Wernicke (1895) described two aphasic patients who also showed impairments in recognizing objects haptically. Wernicke distinguished between primary and secondary forms of haptic object-recognition deficit. A primary deficit in haptic object recognition, or

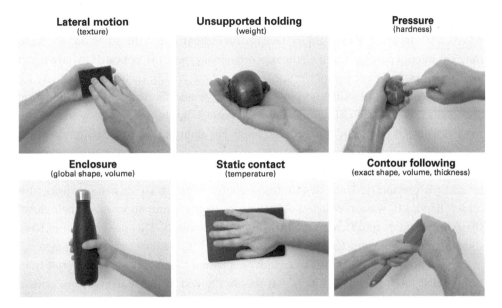

Figure 5.1
Exploratory procedures used for haptic exploration of objects. Following Lederman and Klatzky (1987).

astereognosis, would result from impairments in primary tactile processing of features such as size, shape, and texture. A secondary deficit, in contrast, a true tactile agnosia, would involve impaired recognition of an object's identity or semantic meaning but preserve elementary sensory processing of its basic features. This directly parallels the distinction in vision between apperceptive and associative agnosia.

Large studies of patients with parietal lobe damage resulting from stroke (Roland, 1976) and surgical excision (Corkin et al., 1970) linked impaired haptic object recognition to the primary somatosensory cortex (i.e., the postcentral gyrus). Consistent with this localization, patients with tactile object recognition problems also showed substantial deficits in more basic aspects of somatosensation, suggesting that their problems with object recognition were secondary to basic sensory problems. The link between tactile object deficits and impairments of elementary sensory processes led many researchers to question whether true tactile agnosia, as opposed to astereognosis, existed at all (Critchley, 1953; Teuber, 1965).

Other studies have, however, reported selective deficits in haptic object recognition in patients with seemingly preserved basic somatosensory processing (Caselli, 1991; Gerstmann, 1918; Platz, 1996; Reed et al., 1996; Semmes, 1965; Veronelli et al., 2014). For example, Gerstmann (1918, 2001) described the case of patient J.H., a soldier in the

Austro-Hungarian army in World War I who received a gunshot wound to the head. Based on the location of J.H.'s injury, Gerstmann estimated that the bullet had damaged the right inferior parietal lobe (i.e., the supermarginal gyrus). J.H. appeared to have intact motor control and basic sensory processing from both hands, including sensitivity for pain, temperature, vibration, and pressure, as well as normal two-point discrimination thresholds (2PDT) and localization ability. In contrast, he showed a striking inability to recognize objects haptically using his left hand. His ability to recognize objects with his right hand was unaffected, as was his ability to recognize objects visually. Intriguingly, he also performed well when given a set of objects to feel with his left hand and judge which he had felt previously. Thus, despite being unable to recognize or name objects with his left hand, J.H. was nevertheless able to retain haptic information about their global shape and features to differentiate them from other objects. This preservation of basic somatosensory processing led Gerstmann to describe J.H. as a pure case of tactile agnosia.

Other authors have suggested that cases of apparent tactile agnosia could result from broader deficits in spatial cognition, not specifically tied to the somatosensory system. Semmes (1965), for example, tested sixty-five patients who had suffered penetrating bullet wounds to the brain in World War II or the Korean War on a range of tests of tactile and spatial function. While most patients who showed deficits on haptic object recognition had severe difficulties on more basic tactile tests, Semmes did identify several patients whose object recognition difficulties were out of proportion with basic sensory abilities. However, Semmes argued that these patients' problems on object recognition were a result of difficulties with more general spatial cognition, which she measured using a task in which patients had to use a map to navigate a path around the testing room.

Evidence for a pure tactile agnosia independent of broader spatial problems was provided in a detailed case study by Reed and colleagues (Reed et al., 1996; Reed & Caselli, 1994). E.C. had a stroke that damaged her left inferior parietal lobe. She complained of numbness and an inability to recognize objects using her right hand. Formal testing of somatosensory abilities showed normal performance in her right hand and no asymmetry with her unaffected left hand. Critically, E.C. also performed normally on a range of tests of spatial cognition, including the same map-following task used by Semmes (1965). Moreover, when E.C. was asked to recognize letters traced onto her skin, she showed a selective impairment on her right hand, which did not appear on her right forearm or cheek. Similarly, her ability to recognize objects using her right foot was similar to her left foot and to the performance of control participants. This high degree of selectivity to her right hand strongly suggests that domain general spatial deficits cannot account for her difficulties. Interestingly, while E.C. was unimpaired at simple metric judgments of object length and simple judgments of object shape, she did show impairment for more complex shape judgments with her right hand. For example, when asked to judge whether two felt shapes had the same aspect ratio of width to length, E.C. showed modest impairment with her right hand. Reed and colleagues thus argued that E.C.'s tactile agnosia reflected a deficit in high-level tactile shape perception rather than a deficit in semantic knowledge about objects.

Recall from chapter 2 that there are two broad processing streams from S1 (Dijkerman & de Haan, 2007; Sathian, 2016). The ventral somatosensory pathway runs from S1 to S2 to the insula and then on to the medial temporal lobe to integrate with the ventral visual pathway. This pathway appears to be closely linked to object recognition. The dorsal somatosensory pathway runs from S1 to the PPC and is likely involved in integrating somatosensory information with the motor functions of the reaching and grasping networks. In the context of haptic object recognition, the dorsal pathway would manage the interplay between motor programming of haptic exploratory procedures and interpretation of the resulting tactile and proprioceptive inputs. Effective haptic object recognition would thus require the effective operation of both the ventral and dorsal somatosensory pathways. Indeed, lesion studies in monkeys have shown that tactile discrimination deficits can result from damage to either area S2 in the ventral somatosensory pathway (Garcha & Ettlinger, 1978) or regions of the PPC linked to the dorsal somatosensory pathway (Moffett et al., 1967). Similarly, studies of human patients have linked tactile agnosia to both S2 (Caselli, 1993) and the PPC (Hömke et al., 2009).

Tactile Apraxia

One interesting aspect of E.C.'s case is that her use of exploratory procedures with her agnosic right hand was similar to her intact left hand and to healthy participants, as described above (Reed et al., 1996). Her difficulty was with the tactile interpretation of the information received from these exploratory procedures. The converse problem has also been reported following damage to the PPC (Pause et al., 1989; Valenza et al., 2001). The patient of Valenza et al. (2001) showed poor object recognition ability from active haptic movements of her left hand following a brain aneurysm. However, her recognition improved substantially when her hand movements were guided by the experimenter actively moving her hand or by giving her verbal instruction on how to move. Thus, her problem appeared to be in the use of hand movements to control haptic exploration rather than in the interpretation of tactile information from the hand.

Remarkably, however, she did not have more general problems in motor control. The left panel of figure 5.2 shows examples of her writing with both hands, showing good ability to make dexterous movements with her impaired left hand (even though she was right-handed). In contrast, the right panel of figure 5.2 shows the trajectory of her index

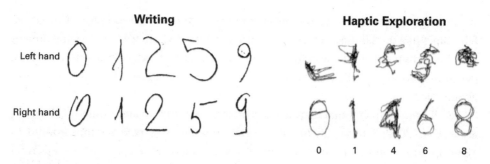

Figure 5.2
Tactile apraxia. This patient is able to control both hands effectively for writing but is impaired at guiding exploratory haptic movements of her left hand. Adapted from Valenza et al. (2001).

finger as she tried to haptically recognize two-dimensional shapes of letters. With her right hand, these movements are clearly in line with the "contour following" exploratory procedure described above. With her left hand, in contrast, her exploratory movements were disorganized and had little apparent relation to the shape being explored.

This patient thus suffers from tactile apraxia, or a deficit in controlling exploratory haptic movements, in the absence of basic deficits in either motor control or somatosensation (Binkofski et al., 2001). Tactile apraxia would thus reflect disruption in the grasping network, preventing the effective generation of haptic exploratory procedures.

Haptic and Visual Object Perception

In most everyday situations, we use vision and haptics simultaneously. Unless a devious psychologist like me is around, the signals we receive from each modality provide consistent information about the world. The correspondence between the seen and felt shape of objects intuitively seems to be extraordinarily natural. Even bumble bees are able to transfer knowledge of object shape between vision and touch (Solvi et al., 2020). At the same time, however, the basic nature of the information reaching the brain from the somatosensory and visual systems is strikingly different. This raises important questions about how the brain combines information across sensory modalities.

In *An Essay Concerning Human Understanding*, John Locke (1690) shares a question he received from his friend and correspondent, William Molyneux of Dublin:

> Suppose a man born blind, and now adult, and taught by his touch to distinguish between a cube and a sphere of the same metal, and nighly of the same bigness, so as to tell, when he felt one and the other, which is the cube, which the sphere. Suppose then the cube and sphere

placed on a table, and the blind man to be made to see; query, Whether by his sight, before he touched them, he could now distinguish and tell which is the globe, which the cube? (p. 94)

Molyneux's question has become a classic thought experiment in discussions of multisensory perception. Both Molyneux and Locke thought that the newly sighted person would be unable to match their visual and tactile experiences and would need to learn to link them. A recent study has provided empirical support for this view. Held et al. (2011) tested five children between eight and seventeen years of age who were born blind due to dense bilateral cataracts. These children were tested within two days of the surgical removal of the cataracts, which allowed them to see for the first time. The children were presented with shapes made from large blocks and tested in a task in which they were first presented with a stimulus and then asked to judge which of two test stimuli matched the original stimulus. Critically, all the children were able to do the task nearly perfectly when the sample and test stimuli were both presented haptically (98 percent correct) or both visually (92 percent correct). This shows that they had clear and reliable experiences from each modality. In striking contrast, however, when the first stimulus was presented haptically and the second visually, performance was near chance (58 percent correct).

Optimal Integration of Vision and Touch

In an influential study, Rock and Victor (1964) created an artificial mismatch between vision and touch. Participants viewed an object through a lens that compressed the apparent size of the object along its horizontal but not its vertical dimension. They simultaneously reached behind the object and grasped it through a black cloth, preventing vision of their hand. Rock and Victor found that judgments of object size were almost completely determined by vision, which appeared to dominate over touch. Although this paper is often cited as definitive evidence that vision dominates over touch, subsequent research has questioned the extent to which this is invariably the case. Heller (1983), for example, showed that when vision is degraded by having participants view stimuli through stained glass, haptic judgments can dominate over vision.

This pattern suggests that rather than one sense or the other being dominant, information from different modalities may be given different weights, depending on the quality of information that each provides in a given context. Ernst and Banks (2002) developed a mathematical framework for quantifying these weights, showing that the statistically optimal approach is to assign weights to each modality in direct proportion to its precision (i.e., the inverse of its variance). Ernst and Banks tested this model using an augmented-reality paradigm, which allowed them to produce mismatches in the size of objects in vision and haptics. This also allowed them to parametrically manipulate the amount of noise in the visual signal. Consistent with the optimal integration model, as the quality of the visual signal declined, correspondingly greater weight was given to haptic cues.

Helbig et al. (2012) used fMRI to investigate the neural mechanisms underlying this optimal integration. Participants made visuo-tactile judgments of object shape while the blurriness of visual signals was manipulated across trials. Increased levels of visual blur led to increases in activation in the primary somatosensory cortex as well as the PPC. Conversely, blur produced reductions in activation of the lateral occipital complex (LOC)—a region of the ventral visual pathway involved in visual object perception. These results suggest that optimal integration involves regulating the gain in visual and somatosensory regions to produce the optimal weighting of each modality.

Common Mechanisms for Tactile and Visual Object Recognition
The results just discussed may give the impression that touch and vision are in an essentially competitive relation to each other. This, however, is an artifact of those paradigms involving devious experimenters deliberately making vision and touch conflict. In real life, we perceive one and the same external world with both modalities. Accordingly, vision and touch work closely together, and indeed there is evidence that common neural processes underlie object perception in both modalities. For example, numerous studies have shown that braille reading in blind people activates the early visual cortex (Büchel et al., 1998; Sadato et al., 1996, 1998). Similarly, disruption of primary visual cortex in blind people using TMS impairs the ability to read braille (Cohen et al., 1997). It is possible, however, that such activations could reflect compensatory plasticity as a result of blindness and thus be unrepresentative of tactile perception in sighted people. Importantly, however, there is also evidence for the involvement of visual cortices in tactile discriminations in sighted people. For example, Zangaladze et al. (1999) found that TMS applied over the primary visual cortex disrupted performance on tactile grating orientation.

Snow et al. (2014) used an fMRI repetition-suppression paradigm to investigate the involvement of V1 in haptic object perception. The logic of the study builds on the well-established principle that neural responses to the second presentation of a stimulus in rapid sequence are reduced relative to the first presentation. In each block of trials, participants haptically explored sixteen objects in sequence. Across blocks, Snow and colleagues varied the number of different objects presented across the sixteen trials. In the "same" condition, for example, the exact same object (e.g., a banana) was presented on each trial. In the "different" condition, in contrast, sixteen different objects were presented. The magnitude of the blood-oxygen-level-dependent signal (BOLD) signal in the

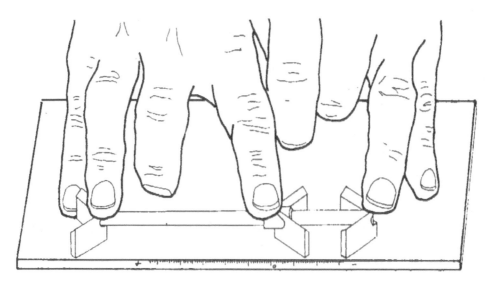

Figure 5.3
The apparatus used by Frisby and Davies (1971) to measure the haptic Müller-Lyer illusion.

primary visual cortex was reduced in the "same" condition compared to the "different" condition, showing clear involvement of V1 in haptic object perception.

Another line of evidence for similar processes underlying haptic and visual object processing comes from similar geometric illusions in both modalities (Révész, 1934). Many of the classic geometric illusions found in vision show up in a very similar way when blindfolded participants explore objects haptically, including the Müller-Lyer (Frisby & Davies, 1971; Over, 1966), horizontal-vertical (Gentaz & Hatwell, 2004; Over, 1966), and Ponzo (Suzuki & Arashida, 1992) illusions. Figure 5.3 shows the apparatus used by Frisby and Davies (1971) to measure the haptic Müller-Lyer illusion. They found a significant correlation across participants in the magnitude of the haptic and visual versions of the illusion, consistent with common underlying processes being involved. Indeed, the haptic Müller-Lyer illusion is found even if the fins are presented visually while the judged line itself is presented haptically (Mancini et al., 2010).

Other evidence for links between tactile and visual form perception comes from studies of repetition priming, in which previous exposure to a stimulus facilitates subsequent identification or reproduction of that stimulus (Tulving & Schacter, 1990). Studies using auditory and visual stimuli showed that while priming was robust within each modality separately, priming was strongly reduced or eliminated when the priming and test stimuli were in different modalities (Roediger & Blaxton, 1987), suggesting that these effects rely on modality-specific mechanisms. In striking contrast, however, studies using haptic and visual stimuli have found robust priming across modalities, with little difference between priming within versus between modalities (Easton et al., 1997; Reales & Ballesteros, 1999). For example, Easton et al. (1997) presented participants with thirty novel shapes comprised of sets of three lines, which could be connected in different ways. Participants were asked to draw the shapes following a brief presentation in either vision or touch. Some of the shapes had previously been presented during a familiarization phase in which participants explored them in vision or touch and then gave descriptions of them. Drawing performance was improved for shapes that had been described during familiarization, even when presented in a different modality. A second experiment using familiar three-dimensional objects similarly showed an increase in the speed of naming for objects that had previously been primed in a different modality.

Striking evidence for such common visuo-haptic mechanisms has come from neuroimaging studies. In a seminal study, Amir Amedi et al. (2001) investigated whether haptic object perception activates regions of the ventral visual pathway known to be involved in visual object recognition, such as the LOC (Grill-Spector et al., 2001). They presented participants with eighteen familiar objects and eighteen textures in the MRI scanner. Each of these stimuli was presented through both vision and touch, and participants were asked to covertly name each object or texture. While the LOC as a whole was clearly more responsive to visual objects, a subregion within the LOC showed robust activations for objects in both modalities. Within each modality, activations in this region were stronger for objects than for textures. A subsequent study from the same group showed that this area was not activated by hearing object-related sounds (Amedi et al., 2002), indicating that it is not a modality-independent representation of object concepts but rather appears specialized for visual and haptic object recognition. Amedi et al. (2002) named this area the "lateral occipital tactile–visual region." This region is likely to be the terminus of the ventral somatosensory pathway discussed in chapter 2, running from S1 to S2 to the insula and then on the LOC.

Many subsequent studies have found comparable activations in the LOC during haptic exploration of objects. These studies have also shown that during haptic exploration, there are changes in the functional connectivity between LOC and regions involved in hand control, which we have already discussed, including S1 (Tal et al., 2016) and area AIP of the grasping network (Monaco et al., 2017; Verhagen et al., 2008). This likely reflects the fact, already mentioned in chapter 3, that AIP is unusual among parietal areas in having direct connections to the ventral visual pathway (Borra et al., 2008). Such connections likely have multiple functions, including utilizing the machinery of the ventral visual pathway for more effective haptic object recognition and also integrating the rich semantic knowledge of objects into skilled hand movements.

A characteristic feature of the occipitotemporal higher-level visual cortex is the existence of a set of brain areas showing category-selective responses for visual perception of ecologically meaningful classes of stimuli (Kanwisher, 2010), such as the fusiform face area, the extrastriate body area, and the parahippocampal place area. There is evidence that each of these areas also show similar category-selective responses to haptic stimuli. Haptic exploration of faces activates regions of the fusiform gyrus (James et al., 2006; Kilgour et al., 2005; Kitada et al., 2009), consistent with the location of the fusiform face area. Similarly, haptic exploration of body parts activates the extrastriate body area (Costantini et al., 2011; Kitada et al., 2009), while haptic exploration of scenes constructed using Lego blocks activates the parahippocampal place area (Wolbers et al., 2011). Pietrini et al. (2004) compared visual and tactile exploration of faces and man-made objects (plastic bottles, shoes). They showed that the distributed patterns of activation produced by each category across the occipitotemporal cortex were correlated between vision and haptics at the individual subject level.

Multimodal Agnosia

Earlier in this chapter, we discussed tactile agnosia. The results discussed in this section raise the question of whether agnosia in the tactile modality would be linked to visual agnosia. Indeed, many patients with tactile agnosia also show visual object agnosia, consistent with common underlying mechanisms (Feinberg et al., 1986; Ohtake et al., 2001; Sirigu et al., 1991). For example, patient F.B. described by Sirigu et al. (1991) was impaired at recognizing objects explored either visually or haptically following bilateral damage to the temporal lobes as a result of herpes encephalitis. Interestingly, he was also impaired at recognizing objects from hearing characteristic sounds. In contrast, F.B. was able to determine whether a haptically felt object was the same shape as a seen object.

Other evidence, however, indicates that haptic and visual object-recognition deficits can be dissociated (Allen & Humphreys, 2009; Riddoch & Humphreys, 1987; Snow et al., 2015). Patient H.J.A. suffered from severe visual object agnosia following bilateral damage to the ventral occipitotemporal cortex (Allen & Humphreys, 2009; Riddoch & Humphreys, 1987). H.J.A.'s ability to recognize objects haptically, however, was intact. Neuroimaging with fMRI showed a region in the dorsal part of the lateral occipital cortex that was activated by haptic object exploration. Given that H.J.A. was unable to recognize objects visually, Allen and Humphreys argued that haptic object perception recruited regions of the ventral visual pathway directly and not secondarily as a result of visual imagery. Conversely, patient E.C., described above, showed deficits of haptic object recognition only for her right hand, with no hint of corresponding visual deficits (Reed et al., 1996).

There is also evidence that the LOC is not necessary for haptic object recognition. Snow et al. (2015) described patient M.C., who had severe visual object agnosia following a stroke that damaged the occipitotemporal cortex (including the LOC) bilaterally. In contrast to her visual impairments, she was seemingly unimpaired at recognizing objects haptically and also showed a normal haptic Müller-Lyer illusion. fMRI results showed a complete absence of LOC activation in response to object perception in either vision or touch. However, brain activations in response to haptic exploration of objects in M.C. were highly similar to those seen in control participants.

6 The Embodied Hand

Fifty meters above Trafalgar Square, in the very heart of London, a statue of a one-armed man looks out along Whitehall toward Big Ben. Horatio, Viscount Nelson, Vice Admiral of the White Squadron, Knight of the Bath, Duke of Bronte, and hero—from the British perspective—of the battles of the Nile, Copenhagen, and Trafalgar, lost his right arm in 1797 leading an amphibious assault on Santa Cruz de Tenerife in the Canary Islands (Southey, 1813). Nelson was wounded by a musket ball, which passed through his right arm just above the elbow, and the arm was immediately amputated by the surgeon of H.M.S. *Theseus* (Gooddy, 1970).

For the rest of his life, Nelson had vivid phantom experiences of his right arm. Nelson is frequently cited as claiming that his phantom constituted "a direct proof of the existence of the soul."[1] This argument has a certain logic. If the experience of the hand continues after the annihilation of the physical hand, why should we think the experience of body as a whole is any different? The concept of the invisible hand provides a natural answer to this question. Although Nelson's *visible* hand had been destroyed, nothing had happened to his *invisible* hand. His brain and central nervous system remained intact and thus so too did all of the mechanisms controlling his hand. From this perspective, the phantom hand simply is the invisible hand.

Phantom limbs are a very pure example of what I will call the "embodied hand." Embodiment includes our conscious experience of the size, shape, and material properties of our body, the conscious body image, as well as the attribution of the limb to the self, the sense of body ownership. In most cases, our body image lines up well with the actual physical structure of our body, such that it is difficult to separate the two things at all. The distinction is driven home clearly by phantom limbs, however, since the physical limb no longer exists. In this chapter, I will discuss our conscious experiences of our hand.

The Conscious Body Image and the Experience of Embodiment

The Experience of Phantom Limbs

The first recorded description of phantom limbs comes from the sixteenth-century French surgeon Ambroise Paré (1634), here in the original seventeenth-century English translation:

> Verily it is a thing wondrous strange and prodigious, and which will scarse be credited, unlesse by such as have seene with their eyes, and heard with their eares the patients who have many monthes after the cutting away of the Legge, grievously complained that they yet felt exceeding great paine of the Leg so cut of. Wherefore have a speciall care least this hinder your intended amputation. (p. 338)

It is striking and curious that while people have lost limbs for as long as there have been people, the first written description of phantom experiences was not published until the sixteenth century. There may be several reasons for this. One obvious possibility is that, before this, people with injuries requiring limb amputation tended not to survive. Indeed, Paré himself made important advances in surgical practice, which increased the likelihood of survival (Finger & Hustwit, 2003). However, there is also clear evidence that at least some people have survived amputation for thousands of years. The Neanderthal known as Shanidar 1, who lived in what is now Iraqi Kurdistan around fifty thousand years ago, survived the loss of most of his right forearm (Trinkaus & Zimmerman, 1982). Similar cases of limb amputation have been described from ancient Egypt (Brothwell & Møller-Christensen, 1963; Messina et al., 2022; Zaki et al., 2010), China (Zhang et al., 2022), Greece (Ganhos & Ariyan, 1985; Hippocrates, 1849), and Rome (Weaver et al., 2000). Recent studies have revealed evidence of successful surgical amputation of limbs in Neolithic communities in France around seven thousand years ago (Buquet-Marcon et al., 2007) and among hunter-gatherers in Borneo as much as thirty-one thousand years ago (Maloney et al., 2022). In Book VII of his *Natural History*, Pliny the Elder (1855) describes the Roman general Marcus Sergius who not only survived the loss of his hand during the Second Punic War but also had quite an adventurous life thereafter:

> In his second campaign he lost his right hand; and in two campaigns he was wounded three and twenty times; so much so, that he could scarcely use either his hands or his feet; still, attended by a single slave, he afterwards served in many campaigns, though but an invalided soldier. He was twice taken prisoner by Hannibal, (for it was with no ordinary enemy that he would engage,) and twice did he escape from his captivity, after having been kept, without a single day's intermission, in chains and fetters for twenty months. On four occasions he fought with his left hand alone, two horses being slain under him. He had a right hand made of iron, and attached to the stump, after which he fought a battle, and raised the siege of Cremona, defended Placentia, and took twelve of the enemy's camps in Gaul. (p. 172)

Given the prevalence of phantom sensations in modern humans, it is virtually certain that Shanidar 1 and Marcus Sergius experienced phantom sensations much like Admiral Nelson's. Another likely possibility is that patients felt that other people would not believe their claims to experience phantoms or might interpret them as possession by evil spirits. Indeed, amputees may have believed this themselves. As detailed by Bourke (2016), during and after World War I, complaints by injured soldiers of pain in phantom limbs were commonly interpreted as reflecting mental weakness or a personality defect—an attitude that lasted well into the 1950s. As Riddoch (1941) put it, "Such a state of affairs was beyond reason and it would not be surprising if the unfortunate patient was regarded as an obstinate, lying fellow or even possessed of the devil. In fact, it was a matter best left alone" (p. 197).

The most influential account of phantom limbs came from Silas Weir Mitchell, a doctor treating soldiers injured in the American Civil War. Mitchell worked with patients at Turner's Lane Military Hospital in Philadelphia, which came to be known as "Stump Hospital" on account of the number of amputations performed (Kline, 2016). Mitchell (1872) found that the majority of his patients experienced vivid phantom sensations: "Nearly every man who loses a limb carries about with him a constant or inconstant phantom of the missing member, a sensory ghost of that much of himself, and sometimes a most inconvenient presence, faintly felt at times, but ready to be called up to his perception by a blow, a touch, or a change of wind" (p. 348).

In a systematic study of ninety amputees, Mitchell (1872) reported phantom sensations in eighty-six (96 percent). This is broadly in line with modern studies, which have found an estimated prevalence of phantom sensations in amputees of 87 percent (Stankevicius et al., 2021). In many cases, the presence of the limb is so vivid that patients mistakenly try to act with it. Mitchell (1871) describes one American Civil War veteran: "A very gallant fellow, who had lost an arm at Shiloh, was always acutely conscious of the limb as still present. On one occasion, when riding, he used the lost hand to grasp the reigns, while with the other he struck his horse. He paid for his blunder with a fall" (p. 566). A similar example from more than a century later comes from patient John McGrath: "When I play tennis, my phantom will do what it's supposed to do. It'll want to throw the ball up when I serve or it will try to give me balance in a hard shot. It's always trying to grab the phone. It even waves for the check in restaurants" (Ramachandran & Blakeslee, 1998, p. 42).

Phantom Limb Movements

A fascinating feature of phantom limbs is that many amputees can voluntarily move them. In a survey of 283 amputees, 132 (47 percent) reported that they were able to move their phantom limb (Giummarra et al., 2010). In other cases, movements occur automatically, without any deliberate intention to move the phantom. These may involve a range of naturalistic behaviors, reaching for objects, movements of individual fingers, breaking a fall, or, as mentioned by John McGrath above, waving for the bill at a restaurant (Scaliti et al., 2020).

In comparison to movements of the intact hand, phantom hand movements are slow and perceived as effortful, but they are otherwise natural, with normal kinematics (de Graaf et al., 2016). Fatigue is commonly felt in the phantom, some amputees describing the arm as feeling "bound" or "restrained in a tube" (de Graaf et al., 2016). Moreover, when amputees try to point with the phantom toward targets of different size, the same slowing of movements with smaller targets (i.e., Fitts's law) is seen as for movements of the intact hand (Saetta, Cognolato, et al., 2020). Studies using fMRI have further shown that movements of individual fingers in the phantom hand activate somatotopically organized finger maps in S1 (Kikkert et al., 2016; Wesselink et al., 2019).

Phantom hand movements also elicit muscular activity in the stump, which can be measured using EMG (Reilly et al., 2006). These EMG patterns are highly stable within an individual, although they are unrelated to the muscle groups normally involved in producing these movements. Complementing this finding, TMS to the primary motor cortex evokes subjective sensations of movements of the phantom hand, including movements the patient claims to be unable to perform voluntarily (Mercier et al., 2006). Using modern multivariate classification methods, Jarrassé et al. (2017) showed that up to fourteen different movements of the phantom hand can be decoded from stump EMG and used to control a multifinger robotic prosthesis. Intriguingly, when a blood pressure cuff was used to induce an ischemic nerve block on the stump, the ability of the patient to voluntarily move the phantom stopped, and no EMG signals were found (Reilly et al., 2006). One amputee described feeling that "I can still feel my phantom hand but it feels as though it is totally frozen in a block of ice" (p. 2220). Such results suggest that even though the physical hand may no longer exist, the experience of movement is nevertheless linked to integration of motor commands telling the hand to move with sensory feedback from the arm.

The importance of efferent motor commands in phantom movements is demonstrated by the distinction between executed and imagined movements of the phantom hand. For an intact hand, this distinction is obvious: when I try to move my hand, my hand moves; when I imagine my hand moving, my hand does not move. In an amputee, the distinction is less visually obvious, as no physical movement occurs in either case. Nevertheless, EMG activity in the stump is found only when amputees actually try to execute phantom hand movements, not when they imagine the phantom moving (Raffin, Giraux, et al., 2012). Neuroimaging studies have also shown that M1 and S1 are activated during execution of phantom hand movements but not during imagery (Raffin, Mattout, et al., 2012). Similarly, organized finger somatotopy is seen in S1 during executed—but not imagined—phantom finger movements (Kikkert et al., 2016).

Another approach to investigating motor representations of the phantom is through bimanual coordination. For example, if you try to move your right hand back and forth in a straight line, the movements you make trace much straighter paths if you move your left hand in straight lines oriented parallel to those of the right hand, rather than in a perpendicular direction. Franz and Ramachandran (1998) showed that similar coupling occurs between movements of the intact and phantom hands. Amputees used their intact hand to trace straight, parallel lines onto a digitizing tablet. At the same time, they were asked to make simultaneous movements of their phantom hand in either the same orientation as the intact hand or in the perpendicular orientation. As in two-handed individuals, the movements of the intact hand were influenced by those of the phantom.

Phantoms in Congenital Limb Absence

One natural interpretation of phantom limbs is as a memory of the previously existing limb. That is, they might reflect the preserved organization of a system that developed in the presence of the intact limb, potentially for years or decades before amputation. From this perspective, phantoms would not be expected in cases of congenital limb absence, which lack this history. Indeed, in contrast to the near ubiquitous experience of phantoms in amputees, most people with congenital limb absence do not report such experiences (Simmel, 1961). Nevertheless, there are numerous cases of such phantoms described in the literature. For example, one eleven-year-old girl who had been born without hands and forearms started to experience vivid phantom sensations of both hands (but not forearms) at the age of six (Poeck, 1964). She was able to move the fingers of the phantom hands and even used her phantom fingers for counting when learning arithmetic at school.

Combining eighteen of their own cases with a review of the literature as far back as the 1830s, Vetter and Weinstein (1967) described a total of twenty-seven cases of congenital phantom limbs. A more recent, large-scale study found that as many as 20 percent of individuals with congenital limb absence may experience phantom limbs (Melzack et al., 1997).

One concern raised about such reports is that many of these cases are of children, who may be less reliable at reporting their symptoms or in differentiating between reality and imagination (Skoyles, 1990). In other cases, however, phantom experiences emerge only in later life following minor injuries or surgery. For example, E.F. was born without a left forearm and hand. At the age of sixteen, she fell while horseback riding and landed on her stump, resulting in a small hematoma. Shortly afterward, she started to have vivid experiences of a phantom forearm and hand and was able to move the phantom fingers (Saadah & Melzack, 1994).

The most detailed study of phantoms in congenital limb absence is the study of A.Z. by Brugger et al. (2000). Tested at age forty-four, A.Z. was born without forearms or legs. Her upper arms consist of stumps, which she is able to use to skillfully grasp objects and perform actions such as typing and using a fork. She has had clear mental images of both arms and legs for as long as she can remember, which she experiences as integral parts of her self-image. Her phantom limbs are present continuously, except when the space they are located in is invaded by a person or object, or when she looks in a mirror. fMRI showed that phantom finger movements produced bilateral activations in the dPMC, the supplementary motor area, and the IPS. TMS applied to the PMC elicited movements of the fingers of the contralateral phantom hand as well as EMG activity in the contralateral stump muscles. In addition, pure sensations of phantom hand movement without corresponding EMG in the stump were elicited following stimulation at one premotor location and two locations in the PPC.

The Rubber Hand Illusion
Another influential method for studying the experience of embodiment is the so-called rubber hand illusion (Botvinick & Cohen, 1998). In this illusion, a prosthetic rubber hand is placed in front of the participant so that it appears in an anatomically plausible posture, while their actual hand is hidden from view. Two paintbrushes are then used to apply temporally synchronous and spatially coherent touch to the rubber hand and to the unseen actual hand. Botvinick and Cohen compared this condition to one in which the touches applied to the rubber hand and the participant's actual hand were temporally asynchronous. The majority of participants report experiencing several striking things. They agree that the touch they felt seemed to occur at the same spatial location as the touch they saw being applied to the rubber hand. Similarly, they experience a direct, causal relation between these two events. Most strikingly, most people report feeling like the rubber hand *actually is* their own hand. These experiences occur specifically when the touch on the two hands is temporally synchronous. Thus, the temporal and spatial correspondence between visual and tactile events is sufficient to induce a vivid experience of ownership over the rubber hand, despite this conflicting with the participant's high-level cognitive understanding of the situation.

Dozens, if not hundreds, of studies over the past two decades have replicated this effect and extended it in a variety of ways. Many studies have used subjective report questionnaires to investigate the conscious experience of embodying the rubber hand (Botvinick & Cohen, 1998; Longo et al., 2008). Another common approach is to measure proprioceptive displacement (or "drift") of the perceived location of the actual hand in the direction of the rubber hand (Tsakiris & Haggard, 2005). While these measures are often interpreted as reflecting a single underlying experience of embodiment, there is also evidence that they can be dissociated (Rohde et al., 2011). A recent meta-analysis (Tosi et al., 2023) has provided strong evidence that there is a correlation across participants between subjective experiences of embodiment and proprioceptive drift, but this correlation is modest, suggesting that these measures may largely reflect distinct aspects of embodiment. Other measures have suggested that the embodiment of the rubber hand is integrated into other aspects of neural functioning. For example, participants show larger skin-conductance responses to the rubber hand being threatened by a knife following synchronous touch to the two hands compared to asynchronous touch (Guterstam et al., 2011).

A fundamental problem with studying ordinary embodied experiences is the fact that the body, as William James (1890) noted, is "always there" (p. 242). The rubber hand illusion provides a valuable experimental tool to investigate embodied experience, since it allows for embodiment over a single object to be varied across conditions. Longo et al. (2008) used principal component analysis to investigate the psychometric structure of the experience of embodiment in the rubber hand illusion. After both synchronous and asynchronous conditions, participants rated their agreement with twenty-seven questionnaire items, covering a range of possible experiences they might have had. In both conditions, four distinct components emerged, which were labeled "embodiment of the rubber hand," "loss of own hand," "movement," and "affect." A secondary analysis of the "embodiment" component showed that it could be further divided into three subcomponents: "ownership", "agency", and "location". The synchronous and asynchronous conditions differed in the extent to which these dimensions of experience were present or absent. Nevertheless, the similar pattern of components indicates a common underlying structure to both experiences. However, one additional component did appear only in the asynchronous condition, which was labeled "deafference." This related to experiences of numbness and pins-and-needles sensations, presumably due to the conflicting information from vision and touch in this condition.

Research has identified several temporal and spatial constraints underlying the rubber hand illusion that provide important clues about the nature of the mechanisms underlying the emergence of limb ownership. First, several studies have manipulated the delay between the touches applied to the rubber hand and actual hand. Delays of around 200–300 ms have been found to be sufficient to disrupt the illusion (Chancel & Ehrsson, 2020; Shimada et al., 2014). Moreover, the magnitude of delay that disrupts the illusion is related to the window of temporal binding—the range of delays between a visual and a tactile event such that they are perceived as happening simultaneously (Costantini et al., 2016). Similarly, the illusion also requires spatial correspondence between vision and touch and is disrupted if the stroking in the two modalities is applied in different directions (Gentile et al., 2013) or to different fingers (Kammers et al., 2009). The illusion is also disrupted if the rubber hand is oriented differently from the actual hand (Tsakiris & Haggard, 2005) or is farther than about 30 cm away (Lloyd, 2007).

Another constraint on the illusion is related to the physical form of the object being embodied. The illusion does not require the rubber hand to be exactly like the participant's actual hand. Ownership can be successfully induced over hands that differ from the participant's hand in skin color (Maister et al., 2013) or size (Bruno & Bertamini, 2010), indicating that the two hands do not need to be identical. However, some degree of similarity in overall structure does appear necessary. The illusion does not work for a block of wood (Tsakiris & Haggard, 2005), for a hand of opposite laterality from the stimulated hand (Guterstam et al., 2011; Tsakiris & Haggard, 2005), or for a rubber foot when the hand is stimulated (Guterstam et al., 2011). In one study, Tsakiris et al. (2010) created a series of stimuli that varied from a featureless box to progressively more hand-like shapes. A clear rubber hand illusion was found only for the actual rubber hand. Together, these results suggest that an external object needs to have the generic shape of a human hand to be embodied but need not be like the participant's specific hand.

In the canonical version of the rubber hand illusion, ownership is driven by a correspondence between visual and tactile cues. However, neither vision nor touch is required to induce the illusion. Similar illusions can be induced without any tactile stimuli at all based on matches between motor commands to move fingers and visual feedback about finger movement (Tsakiris et al., 2006). Conversely, illusions can also be produced in the absence of vision by moving one of the participant's hands across the rubber hand while the corresponding location on the participant's other hand is touched at the same time (Ehrsson et al., 2005). Thus, no single sensory signal is critical for body ownership. What does appear necessary is temporal and spatial congruence between multiple signals, whether vision and touch, touch and proprioception, or motor commands, vision, and proprioception.

Pathological Embodiment and the Experience of Disembodiment

In many cases of brain damage, the experience of embodiment becomes disrupted. Indeed, such disruptions can sometimes be induced experimentally in healthy individuals by distorting sensorimotor signals. These cases of disrupted embodiment, or even disembodiment, are a rich source of information about the cognitive and neural processes underlying our ordinary experience of having a body.

Anosognosia for Hemiplegia

One intriguing condition is anosognosia, in which patients with paralysis affecting one side of their body remain unaware of their deficit, or even actively deny it. This condition was first described by Joseph Babinski (1914, 2014) in the early twentieth century. For example, Babinski (2014) described one patient who was almost entirely unable to move her left arm:

> In contrast with the apparent preservation of intellect, she seemed to be unaware of the existence of the almost complete hemiplegia with which she was affected . . . If she was instructed to move the right arm, she immediately carried out the given command. When she was asked to move the left arm, she remained immobile, keeping silent and behaving as if the question had been addressed to someone else. (p. 6)

Many similar patients have been described subsequently. Indeed, prospective studies of stroke patients with motor impairments following right-hemisphere lesions have found anosognosia in between 15 percent (Baier & Karnath, 2008) and 32 percent (Vocat et al., 2010) of patients. Although less studied than denial of paralysis, patients may also deny somatosensory deficits (Marcel, Tegnér et al., 2004; Spinazzola et al., 2008).

Some evidence suggests that anosognosia may result from a disconnection of sensory predictions from motor intentions and sensory feedback. Some authors (Berti et al., 2007; Frith et al., 2000) have interpreted anosognosia in terms of forward models of motor control (Wolpert et al., 1995), discussed in chapter 4. In these models, sensory predictions are generated on the basis of efferent copies of motor commands and then compared to sensory feedback. If this comparator mechanism was damaged, the predictions generated by motor intentions could dominate over sensory feedback, resulting in anosognosic delusions. Berti and colleagues recorded clear EMG responses in the upper arm muscles of an anosognosic patient with left hemiparesis when trying to move the left hand. This result shows that motor intentions to move the left hand were being produced as usual, even if brain damage prevented the hand from actually moving. Similarly, there is evidence for bimanual coupling between successful actions of the intact hand and attempted actions of the paralyzed hand (Garbarini et al., 2012; Piedimonte et al., 2015). Piedimonte and colleagues showed interference effects in anosognosic patients between intended reaches to two different sized objects by the two hands, even though the paralyzed hand didn't actually move. This again shows that motor commands are being generated and influencing sensorimotor processes. In contrast, no such interference effects were found in hemiparetic patients without anosognosia, suggesting that, in these patients, motor commands may not be sent at all.

A study by Fotopoulou et al. (2008) provided support for this model. They used a rubber hand to provide false feedback about the movement of patients' paralyzed left hand.

Patients judged visually whether the hand had moved in conditions where they had or had not tried to move it. Patients with anosognosia made errors specifically in the case where they had tried to move the hand but where no movement occurred, claiming that they had actually seen the hand move. Such false alarms were not seen in a control group of patients with left hemiplegia but without anosognosia. These results provide striking confirmation of the hypothesis that anosognosia is linked to a failure to recognize mismatch between motor intentions and sensory feedback.

Disturbed Sensation of Limb Ownership

In some patients, a variety of other delusions and misperceptions of the limb occur, which collectively have been called "disturbed sensation of limb ownership" (Baier & Karnath, 2008; Jenkinson et al., 2023). In some cases, patients may personify their paralyzed limb, giving it a separate identity (Critchley, 1955). The stroke patient Mrs. H. referred to her paralyzed left hand as her "poor little withered hand," picked it up, and kissed it (Gilliatt & Pratt, 1952, p. 267). One patient said of her left hand, "That's an old man. He stays in bed all the time" (Nielsen, 1938, p. 555), while another named her hand "zodoquio" (Starkstein et al., 1990). A forty-five-year-old male stroke patient described by Zingerle (1913; Benke et al., 2004) had erotic experiences from his left arm, which he believed belonged to a mysterious woman lying next to him in bed.

In other cases, patients dislike the limb, claiming that the hand has changed, has become "horrid, ugly, unnatural, heavy, like a block of wood or a piece of dead meat" (Critchley, 1955, p. 286). In extreme cases, the hand may be viewed as malicious, even evil, "as endowed with a perverse will of its own" (Critchley, 1955, p. 286)—a condition known as misoplegia. For example, patient E.P. was afraid that her left arm was trying to drown her while she was swimming (Hari et al., 1998). Another ten-year-old stroke patient expressed hatred toward his left limbs as well as a desire to remove them and replace them with his mother's limbs (Moss & Turnbull, 1996). One stroke patient hated his left hand and called it a "jerk" (Starkstein et al., 1990), while another claimed the left side of his body was controlled by the Devil (Nightingale, 1982).

In asomatognosia, part of the body is perceived as missing or absent, seeming "to disappear, or to fall out of corporeal awareness" (Critchley, 1953, pp. 237–238). For example, a fifteen-year-old epileptic patient described by Cushing (1909) described a "funny feeling," a "gone sensation" in his right hand during his seizures. More recently, Arzy et al. (2006) described a patient who experienced a sudden dizziness and felt that her left arm had disappeared. She claimed to be able to see her arm only above the elbow, and to be able to see the table below where she knew her hand must be. She later made a drawing of her experience, shown in figure 6.1. Over several minutes, she experienced her hand

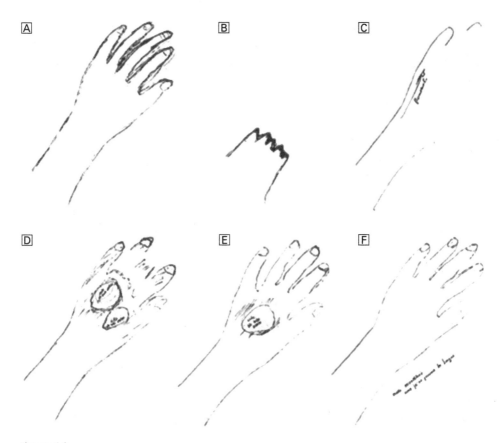

Figure 6.1
Images drawn by the asomatognosic patient of Arzy et al. (2006) showing successive stages of her experience of her left hand: the normal left hand (A); the hand perceived as having disappeared (B); the hand gradually returns, starting from the little finger (C); two holes appear in the hand (D), which then merge (E) and eventually disappear (F).

and arm slowly return, initially with two holes in it. MRI scans showed two small lesions in the right PMC. Another patient who suffered seizures in the right PPC experienced sudden loss of connection or awareness of his left leg upon seizure onset, causing multiple falls and injuries (So & Schauble, 2004).

Perhaps the most dramatic manifestation of disturbed sensation of limb ownership is the phenomenon of somatoparaphrenia, in which patients deny that one of their limbs is their own (Vallar & Ronchi, 2009). This condition was named by Gerstmann (1942), who reported two cases, although he also describes a number of earlier cases from the German and French language literatures. For example, the stroke patient Harriet C. denied ownership of her right hand: "(Is that your hand?)" "It does not look like it, it is too big and swollen, it does not feel like it either and it just won't." (Where is your arm?) "I don't know, you must have it here, I wish you would help me and give me my arm" (Schilder, 1935, p. 311).

Somatoparaphrenia may involve a feeling of non-belonging or disownership of the hand or an explicit attribution of the hand to another person, such as the examining doctor or a family member. One patient, on being shown that the hand she claimed belonged to someone else was connected to her body, replied, "But my eyes and my feelings don't agree, and I must believe my feelings. I know they look like mine, but I can feel they are not, and I can't believe my eyes" (Nielsen, 1938, p. 555). In other cases, disownership appears only in the absence of vision, when the affected hand is held by the other hand—what Brion and Jedynak (1972) called the *main étrangère* (foreign hand): "The patient who holds his hands one within the other behind his back does not recognize the left hand as his own . . . the sign does not consist in the lack of tactile recognition of the hand as such, but in *the lack of recognition of the hand as one's own*" (p. 262).

In its most extreme form, somatoparaphrenia is infrequent but not especially rare. Baier and Karnath (2008) found an incidence of around 8 percent of patients with right-hemisphere stroke. Such numbers may even underestimate prevalence. A recent study found that even when somatoparaphrenia was not apparent in a clinical interview, a substantial proportion (25 percent) showed some level of disownership when assessed using a visual analogue scale to assess how much left-sided body parts felt like they were part of the patient's body (Ronchi et al., 2020). Indeed, some studies have suggested that nearly half of patients with right-hemisphere brain damage may experience some form of disturbed sensation of limb ownership (Jenkinson et al., 2020). While somatoparaphrenia has been reported for the right hand following left-hemisphere damage (D'Imperio et al., 2017; Schilder, 1935), it appears much less frequently than for the left hand. Of the fifty-six cases in the literature identified by Vallar and Ronchi (2009), the left hand was affected in fifty-one cases (91 percent). Romano and Maravita (2019) found an even more extreme pattern, estimating that 97.5 percent of cases involved the left hand. In rare cases, the entire body may be disowned (Smit et al., 2019).

Most cases of somatoparaphrenia involve sensorimotor deficits (Romano & Maravita, 2019; Vallar & Ronchi, 2009). An intriguing example of spared tactile perception comes from patient F.B., who claimed that her left hand belonged to her niece (Bottini et al., 2002). When asked to detect when she felt touch on her left hand, she failed to detect any touches across four testing sessions. Remarkably, however, when asked to indicate when she felt touch in her niece's hand, she correctly detected between 70 and 100 percent of touches applied to her left hand. In virtually every case, however, limb disownership does appear to be linked to impaired proprioception of the affected limb (Romano & Maravita, 2019; Vallar & Ronchi, 2009). This suggests that the perceived spatial position of the limb is a key feature related to body ownership.

Finally, there is the curious condition of body integrity dysphoria, in which apparently healthy people insist that some part of their body does not belong and that they would feel more complete without it (Brugger et al., 2016). In many cases, these people wish for the limb to be surgically amputated, raising important issues of biomedical ethics. This condition is a curious inversion of phantom limbs—a type of "negative" phantom of a body part that does, in fact, exist. It is importantly different from the other forms of disturbed ownership discussed so far in that it is not a result of brain damage but occurs in otherwise healthy people. A recent study nevertheless showed that body integrity dysphoria is linked to lower gray matter density in the PPC and vPMC as well as altered functional connectivity of these regions with wider brain networks (Saetta, Hänggi, et al., 2020). As will be discussed below, damage to similar regions has been linked to disrupted limb ownership following brain damage, suggesting important links between these conditions.

Experimental Inductions of Disembodiment
Some early studies reported confusion between fingers and hands in the face of unusual sensory situations. Jackson and Zangwill (1952) asked participants to perform sequences of finger movements while seeing their hand reflected in a mirror, such that the seen image of the hand was mirror reversed. At debriefing, fifteen of thirty-six participants reported experiences that Jackson and Zangwill interpreted as similar to depersonalization. These included comments that the hand was "isolated," "peculiar," "ill at ease," and "separate from me" and that it "did not belong to anything" or was different from its usual size and shape (p. 8). They called this phenomenon "experimental finger dyspraxia."

Half a century later, McCabe et al. (2005) used the mirror box method to induce sensorimotor conflict more directly. As shown in figure 6.2, participants sat with a mirror aligned with their body midline with their hands positioned on either side of the mirror. The mirror image of the left hand thus appeared to be a direct view of the right hand behind the mirror. When the hands were arranged symmetrically on either side of the mirror (as in the left panel of figure 6.2), vision, proprioception, and motor commands provided consistent information. In contrast, when the hands were arranged asymmetrically, visual feedback conflicted with signals from proprioception and motor commands. After each condition, participants were asked open-ended questions about any changes in the experience of their limbs. The majority of participants reported at least one anomalous sensation, usually (although not exclusively) in the incongruent condition. These

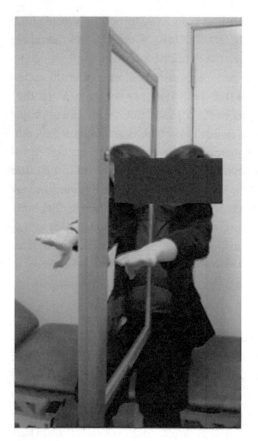 

Figure 6.2
The mirror box paradigm used by McCabe et al. (2005) to induce sensorimotor conflict.

sensations included reports of "pins and needles," "ache," "shooting pain" as well as feeling that the limb had changed weight, altered in temperature, or had disappeared.

Features of disembodiment have been reported in a range of types of multisensory and sensorimotor conflicts. These include conflict between proprioceptive inputs (Longo, Kammers, et al., 2009; Moseley et al., 2006), between vision and touch (Gentile et al., 2013), between vision and proprioception (Otsuru et al., 2014), and between vision and motor commands (Osumi et al., 2018). For example, Osumi and colleagues used a combination of mirrors and computer-controlled video delay to show participants a first-person perspective of their own moving hand with varying delays up to 600 ms. In each condition, participants made a series of flexion and extension movements of the hand

and then rated the experience that "I do not feel as if my hand is my own hand." Such disownership over the hand increased monotonically with increasing delay.

In an elegant series of studies, Newport and colleagues used an augmented-reality system (MIRAGE) that allows them to show near real-time video images of participants' hands that have been digitally altered in various ways. This allowed them to induce disembodied experiences in a variety of ways, including through perceptual illusions of the hand disappearing (Newport & Gilpin, 2011), another object occupying the space where the hand should be (Preston & Newport, 2011), or a finger being detached from the hand (Newport & Preston, 2010). For example, Newport and Gilpin (2011) used the MIRAGE system to create a situation where the participant's right hand seemed to disappear. By moving the hands outward slowly enough that participants did not notice, the right hand was felt to be absent when the left hand reached across and tried to touch it. Following these combined visual and haptic signals of right-hand absence, participants reported feeling that their right hand no longer belonged to their body and also showed reduced skin-conductance responses when a knife threatened either the last seen location of the right hand or its real location.

A recent study by Roel Lesur et al. (2020) used principal component analysis to decompose the experience of disembodiment using a logic analogous to that used previously with the rubber hand illusion to explore embodiment (Longo et al., 2008), described above. Sensorimotor incongruence was induced by showing participants a first-person video image of their body in a head-mounted display, which could be shown in real time or with a systematic delay. Three components emerged from this analysis, which they labeled "disownership," "deafference," and "embodiment." Like in the study by Longo and colleagues with the rubber hand, the "deafference" component only emerged in the asynchronous condition. Consistent with other results, both visuo-tactile and visuomotor mismatches resulted in increased levels of disownership and deafference and lower levels of embodiment, although these effects were stronger for visuomotor mismatch. These results indicate that the experience of disownership is not merely the absence of the ordinary experience of ownership but is a complex experience in its own right.

Neural Mechanisms Underlying Embodiment and Disembodiment
We have discussed a variety of methods for investigating embodiment and disembodiment in both healthy individuals and neurological patients. Much of this research has also provided information about the underlying brain mechanisms underlying these experiences. Neuroimaging studies with healthy human participants have used the rubber hand illusion to identify brain areas that show increased activation when participants

embody the rubber hand. These studies have consistently found activations in the PPC around the IPS and in vPMC (Ehrsson et al., 2004, 2005; Gentile et al., 2013). Other studies have also identified activations in the insula (Tsakiris et al., 2007), consistent with this region's role in processing interoceptive signals from the body. All three of these regions emerged in a recent meta-analysis of the neuroimaging literature on body ownership (Grivaz et al., 2017). Consistent evidence has been found in studies using invasive electrophysiological recordings in both monkeys (Fang et al., 2019) and humans (Guterstam et al., 2019).

Similarly, studies of neurological patients with disembodiment following brain damage have identified the insula (Baier & Karnath, 2008), vPMC (Zeller et al., 2011), and PPC (Martinaud et al., 2017) as relevant lesion locations. The PMC has also been implicated in related conditions such as anosognosia (Berti et al., 2005) and asomatognosia (Arzy et al., 2006). Moro et al. (2023) compared lesion locations in twenty-three right-hemisphere stroke patients with disturbed hand ownership and twenty-six matched patients without ownership abnormalities. Disturbed ownership was associated with lesions of vPMC and the supramarginal gyrus of the inferior parietal lobe. Moreover, it was also associated with damage to white-matter pathways connecting the PMC with both the PPC and the insula.

The experience of ownership is thus closely linked to the parieto-frontal reaching and grasping networks described in chapter 3 and their connections with other networks involved in interoception (e.g., the insula). This suggests that there are no specific brain areas dedicated to experiences of embodiment, such as ownership, but that these arise from basic processes of sensorimotor hand control. Conscious experiences of embodying our body are thus not something extra sitting on top of basic sensorimotor functions but rather are intimately linked to them.

Supernumerary Body Parts

As discussed in chapter 1, the five-fingered pentadactyl limb pattern is ancient and goes back to the earliest tetrapods more than three hundred million years ago (Paton et al., 1999). Nevertheless, it is not rare for individual humans to be born with more than five fingers on each hand (McCarroll, 2000). Such individuals can often show highly dexterous control over these supernumerary body parts (Mehring et al., 2019). This shows that the brain is not innately pre-wired to expect exactly five fingers, and in the developmental presence of extra fingers, sensorimotor brain networks can gracefully accommodate extra digits. There are also reports of adult humans who lack any such developmental history coming to experience supernumerary body parts, either following brain damage or from multisensory illusions of prosthetic augmentation of the body.

Figure 6.3
Self-portraits of three patients who experienced supernumerary phantom limbs following stroke. *Left*: This patient periodically experienced a duplicated right arm extending from the elbow and a duplicated right leg extending from the knee (Staub et al., 2006). These phantom sensations were triggered by attempts to move the paretic right arm or leg. She claimed to be able to perform actions such as putting on her glasses with her phantom hand. *Center*: This patient reported a second left arm extending from her left elbow (Khateb et al., 2009), which she claimed to be able to move freely. In contrast, she labeled her actual plegic left arm as *inutile* (useless). Left hemineglect is apparent in her drawing of her torso and neck. *Right*: Patient E.P. reported periodic experiences of a duplicated left arm and leg (Hari et al., 1998). Objects held in one hand (such as the basket drawn) were triplicated and experienced as held in all three hands.

Supernumerary Phantom Limbs

There are numerous reports of patients with delusional experiences of having extra limbs. Ehrenwald (1930) describes a fifty-nine-year-old patient who had a right-hemisphere stroke and complained of "a nest of hands in his bed" (p. 91), which he requested be amputated. Similarly, Weinstein et al. (1954) reported the case of a woman who suffered right-hemisphere hemorrhage resulting in complete paralysis of her left hand. She claimed to have an extra left hand, which generally lay across her abdomen. Sometimes, she claimed that she herself had two left hands, and on other occasions, she claimed that the extra hand belonged to a friend of hers. Figure 6.3 shows drawings made by three similar patients.

Supernumerary phantom limbs have been described following strokes or tumors affecting several regions of the brain, including the PPC (Vuilleumier et al., 1997) and the medial PMC (Hari et al., 1998). The prevalence of this phenomenon is uncertain. One study of fifty stroke patients showed that while most (27/50) reported misperceptions of limb position, only one reported experiences of an extra limb (Antoniello et al., 2010), suggesting that it may be quite rare. In one case, a patient experienced a supernumerary phantom arm during intracranial electrical stimulation of M1 (Canavero et al., 1999).

One invariant feature of supernumerary phantoms is that they involve perceived duplication of body parts that are injured, paretic, or with sensory loss, and never of healthy limbs. Supernumerary experiences are commonly experienced in conjunction with anosognosia, somatoparaphrenia, or misoplegia for the affected limb. Like somatoparaphrenia, they show a particularly strong association with disrupted proprioception in the affected limb. For example, Riddoch (1941) described an epileptic patient with cortical atrophy in the right parietal lobe who experienced a second set of toes in her left foot during her seizures. While the phantom sensation persisted, she showed a complete loss of postural sensitivity in the toes of her left foot. Similarly, the patient described by Vuilleumier et al. (1997) reported the distressing duplication of both legs, which was associated with weakness in both limbs and proprioceptive loss.

A more complex pattern that also implicates proprioceptive information about limb position comes from patient E.P. (Hari et al., 1998). E.P. reported periodic experiences of a third arm and (less frequently) a third leg following an aneurysm that affected the right medial PMC (see figure 6.3, right panel). She suffered from the anarchic hand syndrome for her left hand with diagonistic dyspraxia. For example, while reading a newspaper, her right hand would turn the page, but her left hand would turn the page right back. She also showed misoplegia, as mentioned above, believing that her left hand wished to harm her. Intriguingly, when her phantom hand appeared, it occupied the location that her real left hand had been in thirty to sixty seconds earlier. In contrast, movements of the phantom hand were evoked by active movements of her *right* hand, although passive movements of her right hand by the experimenter were ineffective. This pattern suggests that mismatches between afferent proprioceptive signals and efferent copies of motor commands play a fundamental role in producing E.P.'s phantom third arm. An fMRI study found evidence for sustained activations within the SMA when E.P.'s third arm was present compared to when it was absent (McGonigle et al., 2002).

The patient described by Khateb et al. (2009) experienced a supernumerary left arm following a right-hemisphere stroke (figure 6.3, center panel). The extra arm seemed to emerge from her left elbow. Interestingly, the supernumerary arm was not always present but emerged specifically when she decided to "trigger" it. When the limb was present, she claimed to be able to see it, describing it as "pale," "milk-white," and "transparent."

She claimed to be able to use the phantom hand to scratch an itch on her head and reported an actual sensation of relief when she did so. Brain imaging using fMRI showed activations of contralateral right premotor and motor cortices when she "scratched" her cheek with her phantom hand, similar to those seen when she tried to do so with her plegic left hand. Additional activations occurred only when she moved the phantom hand in the occipital cortex, which presumably relate to her visual experiences of seeing the limb.

While there are obvious similarities between the phenomenology of supernumerary phantom limbs and the phantoms experienced by amputees, there are fundamental differences between the two phenomena (Brugger, 2003). First, phantom limbs following amputation reflect the ordinary body plan, both of people generally and of the amputee themself before their accident. In contrast, supernumerary phantoms reflect a dramatic deviation from such body plans and thus reflect a *change* in the body image, in contrast to the continuity seen in phantoms of amputated limbs. Second, while pain is commonly linked with phantom limbs following amputation, pain is rarely associated with supernumerary phantoms. Third, while amputees claim to experience their phantom when not looking toward their limb, they do not claim to have visual experiences of actually seeing it. Many patients with supernumerary phantoms, however, report actually seeing it (Critchley, 1953; Halligan, Marshall, & Wade, 1993; Khateb et al., 2009). Finally, while amputees report experiencing phantom limbs, they clearly recognize these as illusory percepts. In contrast, patients with supernumerary phantoms often have delusional beliefs that the extra limbs are real.

Another phenomenon that may be related to supernumerary phantom limbs involves the delusional claim that someone else's limb belongs to the patient's body—what Critchley (1953) referred to as "identification anosognosia." For example, Gerstmann (1942) described patient J.M., who experienced the sudden onset of left hemiplegia, presumably from a stroke. She periodically experienced obsessive feelings that the hand of a nearby person was her own plegic left hand. More recently, Garbarini, Pia, and their colleagues have described numerous patients with similar experiences. These patients report the delusional belief that the arm of an experimenter placed on the table next to their own plegic limb is part of their own body. Several studies have shown that visual stimuli applied to the embodied experimenter's hand produce responses similar to a real hand, including arousal responses (Garbarini et al., 2014), extinction of touch on the contralateral hand (Fossataro et al., 2020), and even referred pain (Pia et al., 2013). Similarly, the embodied hand appears closely integrated with motor representations (Garbarini et al., 2013) and peripersonal space (Garbarini et al., 2015).

A recent study of ninety-six patients (Pia et al., 2020) found that this pathological embodiment requires a set of basic constraints similar to those found in illusions such as the rubber hand illusion. These include the arm being presented from a first-person perspective, in an anatomically correct position, and parallel to the patient's actual arm. Notably, all such patients showed proprioceptive deficits, consistent with other forms of supernumerary phantom sensations. A lesion overlap analysis found the pathological embodiment was linked to damage to white-matter tracts, including the superior longitudinal fasciculus, which carries information between the parietal and frontal cortices.

Experimental Induction of Supernumerary Body Parts
A number of studies have shown that the experience of having supernumerary body parts can also be induced in healthy participants. Some studies have used the rubber hand illusion and shown that participants can experience ownership over two identical rubber hands presented side by side (Fan et al., 2021) or two arms seen in augmented reality (Newport et al., 2010). Similarly, participants can experience ownership over both a rubber hand and their own real hand (Guterstam et al., 2011) or phantom hand (Giummarra et al., 2011). In each case, the participant has the experience of owning three hands.

The perceived structure of the hand can also be altered—for example, by the experience of a supernumerary sixth finger. A study by Hoyet et al. (2016) induced the experience of owning a six-fingered hand in virtual reality. By tracking movements of the participant's actual hand, the fingers of the virtual hand moved accordingly, movements of the sixth finger based on a weighted average of the movements of the ring and little fingers. Participants reported ownership and agency over this extra finger. Other studies have used a mirror box to induce illusions of a sixth finger by creating the illusion that felt touches were caused by a stimulus moving in empty space next to the little finger (Cadete & Longo, 2020; Newport et al., 2016). This sixth finger is perceived as distinct from the little finger and can be perceived as having different properties, such as length (Cadete & Longo, 2022) and curvature (Cadete et al., 2022).

Recent research in engineering has explored a variety of approaches to augmenting manual motor abilities (Eden et al., 2022). It appears remarkably easy for humans to learn to operate supernumerary fingers or limbs and to use them in dexterous ways. For example, one elegant approach involves a robotic third thumb that attaches to the ulnar side of the hand and is controlled by pressure sensors placed on the big toes of each foot, allowing movements with two degrees of freedom (Kieliba et al., 2021). Kieliba and colleagues let participants gain experience using the third thumb for a variety of manual tasks over five days. By the end of this training, participants were able to use the third thumb dexterously. This was accompanied by increased subjective experiences of ownership and agency over the third thumb as well as alterations in somatotopic finger maps in M1 and S1 measured using fMRI.

Such research indicates that the mechanisms involved in sensorimotor function of the hand described throughout this book can be repurposed to control extra body parts. This is an intriguing extension of the principle of motor equivalence discussed in chapter 3. This has important applications in many areas, from the design of prostheses to the rehabilitation of hemiplegia following stroke to laparoscopic surgery.

7 The Hand in Space

The Swiss biologist Heini Hediger became known as the "father of zoo biology" for his work as long-time director of the Basel and Zürich zoos. He pioneered approaches to zoo design that sought to understand and reflect the animals' psychology. Among many important insights, Hediger (1955) found that many animals showed different patterns of behavior when other animals, such as humans, were close to them versus farther away. When an animal, such as a deer, sees a potential predator, it will not immediately flee so long as the other animal is far away. When, however, the predator moves within a specific distance, the flight distance, the animal will immediately run away. Hediger showed that this flight distance is highly consistent and can often be measured to within centimeters, although it may change depending on context, such as the type of terrain or the animal's own fatigue.

The concept of flight distance has been enormously influential, not only in studies of animal behavior but also in anthropological (Hall, 1966) and psychological (Sommer, 1969) studies of how humans manage the spacing between themselves and others. While this work on the representation of the personal space around us in social interactions is beyond the scope of this book, these studies are important in emphasizing that we do not perceive the space around us in a homogenous way. For different purposes, different regions of space are prioritized. This chapter will discuss the representation of the space around the hands, which involves a set of highly specialized mechanisms.

Peripersonal Space

Over the past forty years, a large literature in neuroscience and psychology has revealed a remarkable set of mechanisms for perceiving the space immediately surrounding the body—what has come to be known as "peripersonal space."

Neurophysiological Studies of Peripersonal Space in Animals

Several studies during the 1970s described neurons in area 7 of the PPC with combined visual and tactile responses. For example, Leinonen et al. (1979) recorded from neurons in area 7b (area PF) of the PPC of monkeys. Some of these neurons had tactile RFs, while others had visual RFs. Most intriguingly, some of the neurons responded *both* to visual and to tactile stimuli. These neurons had tactile RFs on the arms and hands (or, less commonly, on the chest or face). The visual RFs of these neurons were defined not with respect to location within the visual field. Rather, they were immediately adjacent to the tactile RF in three-dimensional space. Visual stimuli presented within 5–10 cm of the tactile RF would activate the neuron, as would stimuli within a larger range (~30 cm) that were moving toward the tactile RF. When the monkey's arm was moved, the location of the visual RF moved along with it. The visual responses were also not dependent on the monkey gazing in the direction of the relevant body part, nor even being able to see the body part. For example, because the monkey's head position was fixed, the monkey was unable to see locations on their chest. Nevertheless, neurons with tactile RFs on the chest also responded to visual stimuli moving toward the (unseen) tactile RF. Subsequent research has found similar properties of neurons in a ventral region of the IPS (area VIP; Duhamel et al., 1998).

Rizzolatti et al. (1981) reported similar properties in neurons in the vPMC. These neurons also had both tactile and visual RFs, which were in alignment with each other, as shown in figure 7.1. Rizzolatti and colleagues introduced the term "peripersonal space" to refer to the immediate space around the body that was coded by the visual RFs of these neurons. They distinguished between two basic classes of peripersonal neurons. Pericutaneous neurons respond to visual stimuli extremely close to (within ~10 cm) the tactile RF. For example, the left panel of figure 7.1 shows the RFs of a neuron with tactile RFs covering both hands and with pericutaneous visual RFs that surround the hand like gloves. The center panel shows another neuron that had tactile RFs on both the lips and on the index finger and thumb of the left side of the body. The neuron had a pericutaneous RF adjacent to the lips. Interestingly, this neuron also fired during passive movements of the monkey's arm in the direction of the lip. These features suggest that the neuron is involved in the use of precision grips to bring food to the mouth.

In contrast to pericutaneous neurons, distant peripersonal neurons respond to visual stimuli in a wider range of locations, although always within reaching distance of the monkey's arm. The right panel of figure 7.1 shows an example of a neuron with a tactile RF covering the left hand and a visual RF covering a wide region of the left side of space. Many distant peripersonal neurons had visual RFs that abutted directly onto the

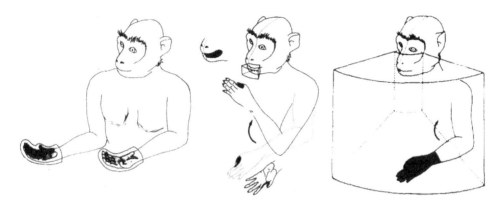

Figure 7.1
RFs of peripersonal neurons. The black shading indicates tactile RFs, and the drawn contours indicate visual RFs. Adapted from Rizzolatti et al. (1981).

monkey's skin. Others, in contrast, had visual RFs that were somewhat detached from the skin, as if specifically excluding the pericutaneous region.

Similar responses were described by Iriki et al. (1996) in a region of the postcentral gyrus, just opposite the IPS from area PF studied by Leinonen et al. (1979). They described two distinct types of neurons coding the space around the monkey's body. One set of neurons, which Iriki and colleagues called "distal type neurons," have tactile RFs on the monkey's hand or forearm and visual RFs immediately surrounding the tactile RF, similar to the pericutaneous neurons described by Rizzolatti and colleagues. These visual RFs follow the arm as it is moved and are unrelated to the direction of the monkey's gaze. Another set of neurons, which they termed "proximal type neurons," have tactile RFs on the shoulder and neck and visual RFs that include the entire space reachable by the arm, similar to distant peripersonal neurons.

Traditionally, the RFs of visually sensitive neurons have been defined as a region of the visual field for which that neuron is sensitive to stimuli. Peripersonal neurons flout this definition, since their visual RFs are not defined with respect to the visual field but rather exist in three-dimensional volumetric space in reference frames defined with respect to the animal's body. Graziano et al. (1994, 1997) systematically investigated the relation of these RFs in the vPMC of monkeys to both arm position and gaze direction. They used an elegant paradigm in which objects approached the monkey along four trajectories while independently manipulating the direction of gaze and the posture of the monkey's arm. This allowed them to show that the visual responses of these neurons were locked to the current location of the monkey's arm, regardless of the direction of gaze.

These recording results have been complemented by lesion studies. Rizzolatti et al. (1983) surgically lesioned either the PMC or the frontal eye fields (FEF). Monkeys with PMC lesions showed visual neglect specific for pericutaneous and distant peripersonal space around the arm and face. In striking contrast, no such neglect was apparent for the space more distant from the face. Conversely, lesions of the FEF produced neglect for far space but not for peripersonal space. This double dissociation shows that the representation of peripersonal space is linked to control of the arm and hand in the PMC, while the representation of far space is linked to eye movements in the FEF.

Neuropsychological Studies of Peripersonal Space in Humans

Insight into the representation of external space in humans has come from studies of patients with hemispatial neglect following right-hemisphere brain damage. Patients with neglect are not blind, but they fail to pay attention to stimuli appearing in the left side of space. Many studies have shown that patients with neglect often have attentional deficits specific to certain regions of space. A first distinction is between neglect for the body surface itself (personal neglect) or for external space (extrapersonal neglect). Bisiach et al. (1986) tested ninety-seven patients with right brain damage on tasks designed to assess personal and extrapersonal neglect. Personal neglect tended to only occur in patients with clear deficits in both motor and somatosensory function as well as visual-field deficits. In contrast, extrapersonal neglect appeared strongly linked to somatosensory deficits and was statistically almost unrelated to the presence of motor deficits.

Even among patients with extrapersonal neglect, there are many reports of selective attentional deficits restricted to either peripersonal space (Berti & Frassinetti, 2000; Brain, 1941; Halligan & Marshall, 1991) or far space (Cowey et al., 1994, 1999; Vuilleumier et al., 1998). For example, patient J.K.P. suffered from left hemiplegia with impaired touch and proprioception of the left hand following a right-hemisphere tumor (Brain, 1941). He additionally showed defective visual localization of objects on the left side of space, which was "limited to objects within arm's length" (p. 255). Other studies have used classic neuropsychological measures of neglect, such as line bisection (figure 7.2). Halligan and Marshall (1991) described patient T.M., who suffered from neglect for the left side of space following a right-hemisphere stroke. When given a line to bisect with a pen, T.M. placed the mark substantially right of center, as if ignoring the left half of the line. In striking contrast, however, when T.M. was given a laser pointer and asked to bisect a line on the wall 2.4 m away, his neglect disappeared. Conversely, Cowey et al. (1994) reported five patients who showed the opposite dissociation, with neglect that was worse in far space than in peripersonal space. Patient H.P., for example, showed

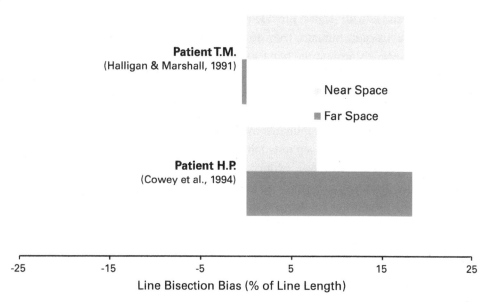

Figure 7.2
Double dissociation of hemispatial neglect in peripersonal (patient T.M.) versus far space (patient H.P).

neglect following a stroke that damaged his right hemisphere. Across two testing sessions eight months apart, he showed strong neglect when he used a laser point to bisect lines 2.4 m away. In contrast, however, he showed substantially less bias when bisecting lines 45 cm away, whether with a pen or a laser pointer.

Other studies using different measures of neglect have reported similar double dissociations between peripersonal and far space. For example, Butler et al. (2004) used a visual search task in which letters and numbers had to be found on a large sheet amid numerous distracting symbols. Consistent with their performance on line bisection, neglect patients tended to miss targets on the left side of the sheet. At a group level among the seven patients tested, there were clear left-to-right gradients in target detection in both peripersonal and far space. At the individual patient level, however, these were clearly doubly dissociated. Based on lesion analyses with computed tomography (CT) scans, Butler and colleagues linked neglect of peripersonal and far space to damage to the dorsal and ventral visual pathways, respectively. This interpretation is consistent with evidence from healthy people that inactivation of regions of the dorsal and ventral visual pathways in the right hemisphere using TMS alters attention in peripersonal and far space, respectively (Bjoertomt et al., 2002).

While the studies with neglect provide evidence for distinct representations of peripersonal and far space in humans, they do not provide evidence for representations of the space immediately around the hand specifically. One line of evidence for such peri-hand space in humans comes from studies of patients with tactile extinction, discussed in more detail in chapter 13. In this condition, patients with unilateral brain damage fail to notice tactile, visual, or auditory stimuli in contralesional space when there are competing stimuli presented simultaneously to the intact side of space (Bender, 1952). Traditionally, extinction had been tested in situations where all the stimuli are from the same sensory modality. However, extinction can also be induced cross-modally, such that a stimulus of one modality (say, tactile) can be extinguished by the simultaneous occurrence of a stimulus of another modality (say, visual) on the opposite side of space (di Pellegrino et al., 1997; Làdavas et al., 1998; Mattingley et al., 1997). For example, di Pellegrino and colleagues described patient G.S., who had severe tactile extinction for stimuli on his left hand when simultaneous tactile stimuli were applied to his right hand. When visual stimuli were presented close to his right hand, he also failed to detect tactile stimuli on his left hand. When visual stimuli were presented at the same physical location but with the right hand behind his back, no extinction was present. This shows that what interfered with tactile detection on his left hand was not the presence of visual stimuli per se but rather the proximity of these stimuli to his hand. Similar extinction could also be induced by visual stimuli presented next to a prosthetic hand placed in front of the patient (Farnè et al., 2000). Other studies replicated this pattern and showed further that tactile extinction could be reduced by the presence of a visual stimulus near the impaired hand (di Pellegrino & Frassinetti, 2000; Làdavas et al., 1998).

Farnè et al. (2005) used similar logic to show that there are separate hand-centered and face-centered representations of peripersonal space. They tested seven patients with extinction and presented tactile stimuli either on the hand or the face. Visual distractors were presented close to (5 cm) or far from (35 cm) the contralateral hand or face. Visual stimuli disrupted tactile detection on the impaired side only when presented close to the body part that had been touched. No such impairment was seen when stimuli were presented close to the other body part or far from the same body part.

Peripersonal Space in Healthy Humans
Evidence for peri-hand space in healthy humans comes from a variety of paradigms. For example, in the crossmodal congruency effect (CCE; Spence et al., 2000), the participant holds a block of foam with two vibrotactile stimulators embedded inside it underneath the thumb and index finger. Their task on each trial is to make speeded judgments of whether the top or the bottom stimulator vibrates by lifting either the toe or heel of their

foot. In addition, two lights are attached to the foam block near each of the two fingers. On compatible trials, the light next to the stimulated finger lights up, whereas on incompatible trials, the light next to the other finger lights up. Although the lights are irrelevant to the participant's task, the CCE refers to the consistent finding that responses are faster and more accurate on compatible than on incompatible trials. Analogous effects of task-irrelevant tactile distractors are found when the participant's task is to respond based on the location of the lights.

These results show that visual stimuli near the hand are automatically integrated with tactile stimuli, even when this integration is detrimental to the task being performed. They also show a high level of spatial precision to these effects. It is not just that visual stimuli near the hand affect touch anywhere on the hand. Rather, visual stimuli near the index finger enhance touch on the index finger, while visual stimuli near the thumb enhance touch on the thumb. Numerous studies have shown spatial modulation of CCE effects. For example, when the hands are crossed so that the hand is located in the opposite visual hemifield, the CCE effects are determined by the current location of each hand, indicating that the effects follow the hands around in space (Kennett et al., 2002). When the lights are moved farther from the stimulated hand, even while remaining in the same visual hemifield, the magnitude of CCE is reduced (Spence et al., 2004). In healthy participants, clear CCEs were produced from visual stimuli near the hand, whichever visual hemifield the hand was placed in. In contrast, for split-brain patient J.W., who had undergone surgical section of the corpus callosum, no CCE was found for visual stimuli near his right hand when it was placed into the left hemifield (Spence et al., 2001).

Other studies have shown striking flexibility in terms of which visual stimuli are coded as being close to the hand. For example, CCEs can be found for lights appearing far from the actual hands but close to rubber gloves positioned in front of the participant (Pavani et al., 2000) or even close to the shadow of the hand (Pavani & Castiello, 2004). Finally, CCEs occur for visual stimuli appearing close to wielded tools (Holmes et al., 2004; Maravita et al., 2002)—an effect that will be discussed in more detail in chapter 10.

A series of studies by Brozzoli et al. (2009, 2010) showed that these multisensory interactions are linked to grasping actions. These authors modified the task so that the task-irrelevant visual stimuli were embedded at either end of a cylinder that participants were asked, in a secondary task, to reach and grasp with the thumb and index finger. This allowed them to present the visual and tactile stimuli at different phases of the action. The magnitude of the CCE effect increased progressively from before the action to the start of the action to the early phase of reaching (Brozzoli et al., 2009). Critically, however, no such changes were found when tactile stimuli were applied to the opposite hand. This shows that it is not merely that the grasping action draws attention to the object. In a subsequent study, Brozzoli et al. (2010) showed that this modulation differs depending on the specific action being performed. Similar changes to the CCE were found early in grasping and pointing movements, while these actions retained similar movement kinematics, but progressive increases later in the action were seen only for grasping. A recent study has shown that such increases can even be found during planning of grasping actions that have not yet begun (Patané et al., 2019).

One fascinating correlate of peri-hand space is the so-called magnetic touch illusion (Guterstam et al., 2016). In this effect, a prosthetic rubber hand is placed in front of the participant, while their own hand is hidden behind an occluding board. A paintbrush is used to apply tactile stimuli to the participant's hidden hand, while a second paintbrush is moved in mid-air above the rubber hand. This results in the experience of a "magnetic force" attracting the paintbrush and hand. By using motion tracking to determine the exact spatial relation between the hand and the paintbrush, Guterstam and colleagues mapped the spatial region producing magnetic sensations. They found a large change in the strength of magnetic sensations at a distance of around 40 cm between the paintbrush and hand, suggesting that it corresponds to a sort of bubble around the hand. Notably, the spatial extent of this effect is similar to that seen for the visual RFs of peripersonal neurons (Fogassi et al., 1996), leading Guterstam and colleagues to suggest that it is a perceptual correlate of visuo-tactile integration in peripersonal space.

A Parieto-Frontal Network for Peripersonal Space

In previous chapters, I have emphasized the importance of parallel parieto-frontal networks underlying grasping (AIP → vPMC) and reaching (PRR: V6a, MIP → dPMC). Peripersonal space appears to be linked to a third parieto-frontal network. As discussed above, neurophysiological studies in monkeys have identified neurons with congruent tactile and visual RFs in both the PPC (Duhamel et al., 1998; Leinonen et al., 1979) and the vPMC (Graziano et al., 1994, 1997; Rizzolatti et al., 1981). While both peripersonal space and grasping involve parieto-frontal links between nearby regions of the IPS and vPMC, these nevertheless appear to be functionally segregated (Luppino et al., 1999). The grasping network projects from a more anterior region of the IPS (AIP) to a region of the vPMC known as F5 in macaques. The peripersonal network, in contrast, projects from a more ventral region of the IPS (VIP) to a slightly more superior region of the vPMC known as F4. There thus appear to be three distinct, although certainly interacting, parieto-frontal networks involved in controlling arm and hand movements in space. In fact, there also appears to be a fourth network involved in the control of the eyes projecting from a more lateral region of the IPS (LIP) to the FEF (Passingham, 1993).

In humans, neuroimaging studies using fMRI have provided consistent evidence that peripersonal space relies on a parieto-frontal network. Makin et al. (2007) found that a region in the anterior IPS responded to a ball approaching the participant's hand but not to the ball approaching a prosthetic hand placed in the same location. Brozzoli et al. (2011) used an fMRI adaptation paradigm to identify brain areas that code for moving objects near the hand. The logic of this paradigm is that neurons coding for peri-hand space should show a reduced response when a second near-hand stimulus is presented shortly after a first. In contrast, no such adaptation should occur when the first stimulus is located far from the hand. Regions of the PPC and PMC showed such adaptation, including the anterior IPS and the vPMC. In a subsequent study, Brozzoli et al. (2012) showed that these adaptation responses generalized across different locations of the hand, showing clearly that they code for location in a hand-centered frame of reference. Consistent with these individual studies, a recent meta-analysis of fMRI studies investigating peripersonal space found consistent clusters of activation in both the IPS and vPMC (Grivaz et al., 2017).

Multisensory Modulations in Peri-Hand Space

Research on peripersonal space has revealed close links between sensory information from touch and from vision. The essential feature of peripersonal neurons is that they have spatially aligned RFs in both modalities. There is also evidence for a range of other multisensory modulations of sensory processing on and near the hand. The second half of this chapter will discuss these effects.

Hand-Assisted Blindsight

In chapter 4, we discussed the phenomenon of action blindsight, in which patients claim not to have visual experiences but are nevertheless able to perform reaching actions toward objects at greater than chance levels (e.g., Weiskrantz et al., 1974). Other studies have described different forms of blindsight in which the spatial proximity of the hand enhances patients' ability to see objects. Some studies in patients with hemispatial neglect have found that movements of the contralesional hand in the neglected hemispace can alleviate neglect (Eskes et al., 2003; Frassinetti et al., 2001; Robertson & North, 1993). While Robertson and North suggested that such effects were only effective with active and not passive movements of the hand, Frassinetti and colleagues showed that sufficiently complex passive movements can also alleviate neglect.

More striking is evidence that the mere physical presence of the hand in a region of space can enhance visual processing. Schendel and Robertson (2004) described patient

W.M., who was left with severe deficits in vision in the left visual field following a stroke that damaged his right primary visual cortex. Remarkably, when he extended his left hand into his blind visual field, his ability to detect stimuli was substantially improved. This improvement was found when the monitor on which stimuli were presented was relatively close to W.M.'s body (60 cm) but not when the monitor was placed farther away (180 cm). Notably, detection at the farther distance was enhanced when W.M. held a tennis racket in his left hand.

Brown et al. (2008) tested two patients who both had visual deficits affecting the left visual field. The patients were presented with six objects of different sizes and asked to judge object size either by matching perceived object size with the distance between the right thumb and index finger or by reaching out and grasping the object. Both patients reported not being able to see the presented objects in their left visual field. Nevertheless, there was a systematic relation between both measures of perceived size and actual object size when the left arm was placed next to the objects. In contrast, when the hand was farther away, resting on the patient's knee, there was no relation between judged and actual size. Brown and colleagues refer to this phenomenon as "hand-assisted blindsight."

The Hand as an Attentional Wand
The effects in blind individuals are also consistent with evidence from sighted humans and monkeys showing that visual processing is systematically altered in the space immediately surrounding the hand. Coslett and Lie (2004) have used the evocative term "attentional wand" to describe this idea. For example, Hari and Jousmäki (1996) asked participants to make rapid judgments of which of two lights projected onto the table came on by lifting the corresponding index finger. Responses were faster when the lights were projected directly onto the back of the participant's index fingers than when the fingers were positioned 5 or 15 cm to the sides of the lights.

Direct evidence that the proximity of the hand modulates the function of neurons in the visual cortex was reported by Perry et al. (2015), who recorded the tuning functions of neurons in early visual area V2 in monkeys. For each neuron recorded, Perry and colleagues first identified the location of the neuron's RF on a computer monitor and then measured its tuning function in two conditions. In the hand-near condition, the monkey held a bar positioned next to the RF; in the hand-away condition, the monkey's hand was not near the monitor. The proximity of the hand led to sharpening of the tuning functions, with relative enhancement of responses to preferred versus orthogonal orientations. Moreover, the hand also reduced the variability of neuronal responses.

This evidence in monkeys is consistent with research in humans showing that preparing a grasping action enhances perceptual sensitivity to detecting orientation changes

(Guo & Song, 2019; Gutteling et al., 2011). Indeed, using MVPA of fMRI data, Gutteling et al. (2015) showed that the type of movement being planned (grasping vs. pointing) could be decoded from the distributed pattern of neural responses in several regions of the early visual cortex, including both V1 and V2. This perceptual enhancement of orientation sensitivity during grasp preparation could be blocked by inactivation of AIP using repetitive transcranial magnetic stimulation (rTMS; Gutteling et al., 2013).

Brown et al. (2009) investigated the ability to localize the perceived location of visual stimuli projected onto the hand. On each trial, a target light was projected onto the participant's left hand, and they were instructed to rapidly point to the location of the stimulus with their right index finger. Shuttered glasses blocked vision as soon as the right hand started to move. Depending on the posture of the left hand, the stimuli were projected onto either the hairy skin of the hand dorsum or the glabrous skin of the palm, as shown in figure 7.3. The precision of localization was quantified by measuring the variability of localization between trials. As can be seen in figure 7.3, precision was substantially increased when stimuli were projected onto the glabrous skin than onto the hairy skin. No analogous effect was found for localization of visual stimuli projected onto the palm versus dorsum of a fake hand. Thus, the heightened spatial resolution of touch on the palm compared to the dorsum transferred to the localization of purely visual stimuli, even in a task in which no tactile stimuli were presented at all.

Other evidence was reported by Reed et al. (2006) who used the so-called Posner cueing paradigm, in which participants make rapid responses based on the left/right position of target stimuli, which are preceded by non-informative spatial cues. The presence of one of the participant's hands near a target location speeded reaction times to targets appearing at that location, regardless of the validity of the cue. In contrast, no such effects were observed when a broadly hand-shaped board was used, indicating that the effect is not simply due to the visual presence of an anchor stimulus near the target location.

Davoli and Brockmole (2012) used a flanker task, in which participants had to respond based on the identity of a central target letter ("H" or "S") while ignoring irrelevant distracting stimuli on either side. These distractors could be either congruent or incongruent with the target stimulus. When the participant's hands were placed in between the target and flankers, the congruency effect from the flanking stimuli was reduced. In contrast, when wooden blocks of broadly similar size and shape to the hand were placed between the target and flankers, no such reduction was observed. They suggested that the hands effectively shield visual attention from interference, focusing the spotlight of attention more sharply on the object of attention.

An interesting inversion of the effects of hand proximity comes from research investigating amputees. Makin et al. (2010) tested seven individuals with left-hand amputation

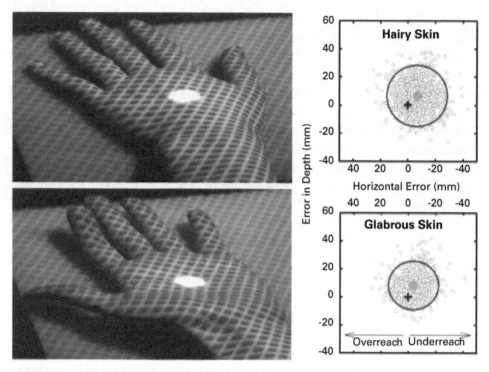

Figure 7.3
Visual localization of stimuli projected onto the hand dorsum (top) or palm (bottom). Although there was no tactile component to the task, localization was more precise on the palm than on the dorsum. Black crosses indicate actual stimulus location; empty gray circles indicate individual responses; filled gray circle indicates average judgment; a black ellipse indicates a 95 percent confidence interval. Adapted from Brown et al. (2009).

and five with right-hand amputation. Participants made judgments about the relative distance of visual targets in the left and right visual field from a central fixation cross, which allowed them to estimate biases to overestimate distance in one hemifield versus the other. When stimuli were presented far (150 cm) from the participants, no difference in bias was found between the two groups of amputees. In striking contrast, when stimuli were presented closer (50 cm), opposite biases were apparent in the two groups. Left-hand amputees tended to overestimate distances in the right visual hemifield, and right-hand amputees overestimated distances in the left visual hemifield. In both cases, it seemed that the amputees neglected the space around their missing hand.

Other research suggests that it is not the mere proximity of the hands that affects visual processing. Rather, these effects differ depending on the specific posture the hand is in. Thomas (2013) replicated the effect of Reed et al. (2006) for enhanced detection of near-hand stimuli when the hands were held in a flat, open posture. In contrast, when the hands were held as if making a precision grip with the thumb and index finger, no such facilitation was found. In a subsequent study, Thomas (2015) reported opposite effects of holding nearby hands in a power grip versus a precision grip posture. Thomas compared two tasks both involving perception of a large array of dots. In one task, participants had to judge whether the global motion of dots mixed with randomly moving distractors was to the right versus to the left. In the other task, participants had to judge whether the global form of a configuration of dots in noise was in a radial versus a concentric configuration. Performance on the motion task was enhanced by the proximity of the hand in a power grip posture, while performance on the form task was enhanced by the proximity of the hand in a precision grip posture. Further research has suggested that these effects can be modulated by specific training that participants receive in performing power and precision grips with different hand configurations (Thomas, 2017).

Hand Position and the Magnocellular System
Studies over the past fifteen years have reported an almost bewildering array of perceptual and cognitive effects from the presence of the hands. Hand proximity has been found to slow the disengagement from stimuli near the hands (Abrams et al., 2008), to reduce switching between global and local scopes of attention (Davoli et al., 2014), to bias figure-ground segmentation (Cosman & Vecera, 2010), and to increase the capacity of visual working memory (Tseng & Bridgeman, 2011). Hand proximity has even been found to modulate aspects of higher-level perception and cognition seemingly unrelated to hand function, such as perception of emotionally unpleasant scenes (Du et al., 2017) and reading (Davoli et al., 2010).

One intriguing hypothesis that potentially provides a principled basis for understanding such effects is that the presence of the hand biases vision toward specific processing streams. The visual system includes two largely segregated pathways, the parvocellular (P) and magnocellular (M) pathways, from ganglion cells in the retina, through the lateral geniculate nucleus of the thalamus, and which send information to largely segregated regions of early visual cortex (Livingstone & Hubel, 1987, 1988). The P pathway has high sensitivity to color and high spatial resolution but relatively low contrast sensitivity and temporal resolution. The M pathway, in contrast, is fast, with high sensitivity to motion and contrast but less sensitivity to color and spatial frequency. While the separation between these pathways becomes less strict at higher levels of cortical processing, the P and M pathways are thought to provide the dominant inputs to the ventral and dorsal visual pathways, respectively (Ungerleider & Mishkin, 1982). Milner and

Goodale (2006) have thus interpreted the P system as a processing stream specialized for conscious perception and the M system as a processing stream specialized for immediate control of action. Accordingly, some authors have argued that the various effects of hand proximity reflect a prioritization of signals from the M pathway and a corresponding deprioritization of signals from the P pathway (Goodhew et al., 2015; Gozli et al., 2012). This refocusing makes sense, given the M system's role in guidance of action.

The P and M pathways are known to have distinct psychophysical "signatures" (Livingstone & Hubel, 1987). Several studies probing these signatures have provided evidence that hand proximity upregulates processing of M versus P signals. Gozli et al. (2012) developed two tasks to reflect each of these systems. On each trial, a circle appeared on the screen either to the left or to the right of fixation. In the spatial-gap detection task, the participant's task was to judge whether the circle was complete or had a small gap in its contour. In the temporal-gap detection task, in contrast, their task was to judge whether the circle briefly disappeared or was continuously present. The spatial-gap detection task was designed to reflect the high spatial sensitivity of the P system, while the temporal-gap detection task was designed to reflect the high temporal sensitivity of the M system. Gozli and colleagues found an interaction between proximity of the hands and performance on these two tasks. Sensitivity to temporal gaps was enhanced by hand proximity, whereas sensitivity to spatial gaps was reduced.

Abrams and Weidler (2014) asked participants to make judgments of the orientation (upright or tilted) of Gabor patches that were either of low spatial frequency or high spatial frequency. When the hands were placed next to the monitor, sensitivity was higher for low than for high spatial-frequency stimuli. In contrast, when the hands remained in the participant's lap, the opposite pattern was found, with higher sensitivity for high spatial frequencies. Thus, the proximity of the hand did not simply enhance visual processing generally but also induced a trade-off, biasing the visual system to be more sensitive to lower spatial frequencies. Abrams and Weidler further investigated the relation between hand proximity effects and the M system by testing for the effects of presenting stimuli on a red background. A range of evidence has shown that activity of the M system is suppressed by diffuse red light (Livingstone & Hubel, 1987). Indeed, this effect of hand proximity on different spatial frequencies was eliminated when the screen background was made red.

Visual Enhancement of Touch

We have discussed ways in which the proximity of the hand alters visual processing. Other research has shown that vision of the hand also has effects on somatosensory processing. One line of evidence comes from studies showing that looking in the direction of the hand can affect touch, even if the hand is not actually seen. For example, one blind person claimed that "it helped him attend to touch if he 'looked' at his hands as he touched objects or patterns" (Heller et al., 2005, p. 166). Directing gaze toward the stimulated hand can increase the speed of responses to simple tactile stimuli (Honoré et al., 1989), reduce biases in tactile localization (Medina et al., 2018), enhance haptic recognition of raised outlines (Scocchia et al., 2009), and modulate early somatosensory evoked potentials (SEPs; Forster & Eimer, 2005).

Other research has identified specific effects of seeing the hand rather than merely looking in its general direction. Tipper et al. (1998) asked participants to make rapid responses with their feet when they felt a tactile vibration on their hand. In one condition, participants saw their hand on a live video image, while in another condition, the hand was covered by a cardboard box. Although the direction of gaze was constant across conditions, reaction times were faster when participants could see their hand. Kennett et al. (2001) extended this result to show that vision of the hand alters basic tactile acuity. They used a semi-silvered mirror to manipulate participants' vision of their hand while holding gaze direction constant. As shown in figure 7.4, participants placed their right arms into a box with a semi-silvered mirror embedded in one side. When a light inside the box was illuminated, the mirror became a window, and the participants had a direct view of their stimulated forearms (figure 7.4, left panel). When a light outside the box was illuminated, however, the mirror remained opaque, and the participants saw the reflection of a neutral object (figure 7.4, right panel). Tactile acuity was measured using the 2PDT. This threshold was reduced when participants could see their hand compared to the object. Intriguingly, when the stimulated skin surface was magnified, tactile acuity was enhanced even more (figure 7.4, center panel). Kennett and colleagues called this effect "visual enhancement of touch."

The 2PDT test used by Kennett et al. (2001) has been criticized as a measure of tactile spatial acuity (Craig & Johnson, 2000). Importantly, however, many subsequent studies have replicated the basic effect using other measures, notably orientation judgments of square-wave gratings (e.g., Cardini et al., 2011). Studies of stroke patients have also found that vision of the stimulated limb can ameliorate somatosensory impairment (Halligan et al., 1997; Serino et al., 2007).

Other studies have shown that the effects of seeing the body on touch are selective and differ depending on the task. Press et al. (2004) compared tasks in which participants merely had to detect the presence of a tactile stimulus on the forearm as quickly as possible and in which they had to discriminate which of two spatial locations was stimulated. Non-informative vision of the stimulated arm enhanced performance only for the spatial discrimination task. Indeed, Harris et al. (2007) and Zopf et al. (2011) found that vision of

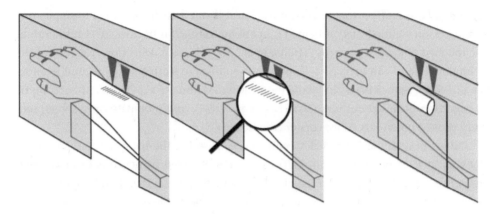

Figure 7.4
The paradigm of Kennett et al. (2001) to measure visual enhancement of touch.

the stimulated hand impaired performance on simple detection of a vibrotactile stimulus on the finger. In contrast, however, vision of the hand clearly improved performance when participants were asked to discriminate which of two sequentially presented tactile stimuli was more intense. Harris and colleagues argued that the effects of seeing the body are similar in nature to those induced by the presence of an adapting tactile stimulus presented immediately before the test stimuli. They suggest that both effects reflect a process of adaptive gain control of touch, driven by bimodal visual-tactile neurons in the PPC, as described above.

Visual Modulation of Somatosensory Intracortical Inhibition
One potential mechanism of visual enhancement of touch involves visual modulation of intracortical inhibition in the somatosensory cortex. Intracortical inhibition is largely driven by neurons with receptors sensitive to the neurotransmitter GABA (Jones, 1993). Lateral inhibition within somatotopic cortical maps is thought to sharpen the spatial resolution of touch (Brown et al., 2004). This is supported by data in humans using magnetic resonance spectroscopy to measure levels of GABA concentration in the primary somatosensory cortex, showing that this correlates across participants with the sensitivity of vibrotactile frequency discrimination (Puts et al., 2011). Mechanistically, GABA-mediated intracortical inhibition could enhance the discriminative sensitivity of touch by reducing the size of tactile RFs, so that each neuron shows a more spatially selective response. Numerous studies have shown that pharmacological modulation of GABAergic activity in the somatosensory cortex modulates RF size. For example, Dykes et al. (1984) injected the GABA antagonist bicuculline methiodide into the primary

somatosensory cortex of cats, finding clear increases in the size of tactile RFs. Similarly, enhancing versus reducing GABA activity in the somatosensory cortex of raccoons produced opposite effects on RF size (Chowdhury & Rasmussen, 2002). Whereas the GABA antagonists CGP 55845 and bicuculline methiodide produced rapid expansion of RF size, the GABA agonist baclofen shrunk RFs.

Haggard et al. (2007) tested the hypothesis that vision of the body shrinks the size of tactile RFs by measuring the spatial extent of tactile masking. Masking refers to the phenomenon that the perception of a tactile stimulus can be blocked by the presence of another nearby tactile stimulus (Laskin & Spencer, 1979b). From the perspective of an idealized neuron detecting the first stimulus, masking will only occur if both the stimuli fall within the neuron's RF. The spatial gradient of masking can therefore be interpreted as reflecting the RF size of populations of neurons involved in perceiving the stimuli. Haggard and colleagues asked participants to make judgments of the relative spatial location of taps along the proximodistal axis in the presence of masking stimuli that were either relatively near (1 cm) or far (2 cm) to the left or right of the target stimuli. Vision of the stimulated arm, compared to vision of a neutral object, sharpened the spatial gradient of masking. The amount of masking from the far stimuli was substantially reduced by vision of the arm—an effect that was not seen for the near maskers. Haggard and colleagues interpreted these results as evidence that vision of the body shrinks the size of tactile RFs in early somatotopic maps.

Consistent evidence comes from studies showing that vision of the body modulates SEPs following mechanical tactile stimuli (Sambo et al., 2009; Taylor-Clarke et al., 2002), electro-cutaneous stimulation of the skin (Cardini et al., 2011), and direct electrical stimulation of the median nerve (Longo et al., 2011). Each of these studies has reported modulation of SEP components consistent with visual modulation of processing in the primary somatosensory cortex. Allison et al. (1991) have distinguished between short-latency (before ~40 ms) and long-latency SEPs, the former of which appear to arise exclusively from processing in areas 3b and 1 of contralateral S1. While mechanical and electro-cutaneous stimuli do not produce short-latency SEPs, the one study that used electrical nerve stimulation (Longo et al., 2011) found that seeing the stimulated hand reduced the amplitude of the P27 component, clearly implicating S1. Interestingly, P27 amplitude appears to be inversely related to intracortical inhibition in S1. For example, administration of GABA-agonist benzodiazepines, which enhances intracortical inhibition, has been found to reduce P27 amplitude (Sloan et al., 1990). Conversely, P27 amplitude is increased in conditions featuring disrupted somatosensory intracortical inhibition, including carpal tunnel syndrome (Tinazzi et al., 1998) and dystonia (Tinazzi et al., 2000).

Further evidence for an involvement of intracortical inhibition in the visual enhancement of touch effect was provided by Cardini et al. (2011), who measured the suppression of SEPs when adjacent fingers were stimulated simultaneously. The SEPs elicited by simultaneous stimulation of adjacent fingers of the same hand is smaller than the sum of SEPs evoked by stimulation of each finger individually (Gandevia et al., 1983)—a reduction linked to intracortical inhibition in S1 (Hsieh et al., 1995). Cardini and colleagues applied electro-cutaneous stimuli to participants' right index finger, middle finger, or to both fingers simultaneously while participants viewed their hand directly or a wooden box placed above their hand. At random intervals between electrical stimuli, square-wave gratings were applied to either the index or middle finger, and participants judged their orientation. Cardini and colleagues found that the magnitude of suppression of the P50 SEP component—that is, the amount by which SEP amplitude during combined stimulation is smaller than the sum of amplitudes in the two individual finger stimulation conditions—was increased when participants saw their hand. Moreover, the magnitude of this change in suppression was correlated across participants with the behavioral visual enhancement of touch measured with the grating-orientation judgment task.

There is also evidence that seeing the stimulated body part can reduce the intensity of both acute pain (Longo et al., 2009; Mancini et al., 2011) and chronic pain (Diers et al., 2013). This visually induced analgesia may also result from changes in intracortical inhibition in the somatosensory cortex. Chronic pain has been found to be linked to reduced intracortical inhibition in the sensorimotor cortex (Schwenkreis et al., 2003), while GABA-agonist benzodiazepines are effective at controlling such pain (Canavero & Bonicalzi, 1998). An fMRI study found that seeing the hand while experiencing acute pain was linked to increased functional connectivity between the PPC and several regions of the cortical pain network, including both S1 and S2 (Longo, Iannetti, et al., 2012). This suggests that visually induced analgesia may result from top-down modulation from the PPC, which increases intracortical inhibition in the somatosensory cortex. A similar mechanism could account for visual enhancement of touch. Indeed, visual enhancement of touch can be blocked by TMS applied either to the PPC (Fiorio & Haggard, 2005) or to S1 (Konen & Haggard, 2014).

8 The Expert Hand

In Greek mythology, Orpheus was the greatest of all mortal musicians. In one tale, Orpheus sails with Jason and the Argonauts, where the beauty of his playing breaks the spell of the Sirens' song. In another, he travels to Hades to rescue his wife Eurydice from death itself, where his lyre calms even the hellhound Cerberus (Hamilton, 1942). It is uncertain whether there existed a historical musician whose talents inspired the Orphic myths. On some accounts, Orpheus, although mortal, was only half human, his mother being the muse Calliope. In contrast, Italian violinist Niccolò Paganini (1782–1840) most certainly existed and appears—on balance of considerations—to have been entirely human.

Paganini's virtuosity is legendary. He is widely rated among the greatest musicians who ever lived (Schwarz, 1983). In the words of one biographer, "Paganini took a wooden box with four strings and demonstrated on it the possibilities of the entire orchestra" (Day, 1929, p. 131). To German composer Giacomo Meyerbeer, "Paganini begins where our reason stops" (quoted in Schwarz, 1983, p. 175). His playing bewitched all of Europe, and he was frequently dubbed the "modern Orpheus" (figure 8.1). Such was the awe in which Paganini's abilities were held that it was frequently claimed (and perhaps, occasionally, believed) that there was something diabolical about his virtuosity. "For a digital dexterity which was not humanly possible, he had given Satan the majority of shares in his soul" (Day, 1929, p. 4).

Paganini's talents were certainly a result of innate musical genius combined with ceaseless training (reportedly as much as seven hours per day since childhood). Those factors alone, however, may not have been sufficient. Paganini's hands themselves may have had unusual characteristics that enhanced his playing. His personal physician, Dr. Francesco Bennati (1831), described Paganini's hands:

> Paganini's hand is not larger than normal; but because all its parts are so stretchable, it can double its reach. For example, without changing the position of the hand, he is able to bend the first joints of the left fingers—which touch the strings—sideways, at a right angle to the natural

Figure 8.1
Paganini billed as "The Modern Orpheus" during his 1831 visit to England. From the digital collection of the New York Public Library.

motion of the joint, and he can do it with effortless ease, assurance, and speed. Essentially, Paganini's art is based on physical endowment, increased and developed by ceaseless practicing. (Schwarz, 1983, p. 198)

Modern analyses of contemporaneous drawings of Paganini's hands, and of a plaster cast of his hand made while he was alive, suggest a condition known as "arachnodactyly" or "spider hand," in which the fingers are abnormally long and slender (Pedrazzini et al., 2015). This characteristic is typical of the genetic condition Marfan syndrome, which affects connective tissues across the body, resulting in hyper-flexible joints, elongated limbs and fingers, as well as heart problems. In an intriguing paper, Schoenfeld (1978) argued that Paganini suffered (or, perhaps, benefited) from Marfan syndrome, although the condition was not described medically until half a century after his death. In this sense, the things Paganini did with his hands may indeed be impossible for most humans, no matter how much they train. The credit for this, however, goes not to Satan, but to a mutation on the *FBN1* gene.

Paganini shows us the outer limits of human manual skill. His status as a paragon of virtuosity makes him obviously unrepresentative. Nevertheless, to understand human hand function, we need to consider the full range of abilities in human populations, not merely those that are universally shared. Moreover, although professional musicians such as Paganini have specific expertise that most of us do not have, all of us are expert users of our hands for a wide range of everyday activities. In this chapter, I will discuss the nature of hand expertise, discussing general principles of skill acquisition before turning to specific examples of expert performance, both virtuosic and everyday.

Principles of Motor Skill Acquisition

We all have a wide variety of manual skills and the ability to learn new ones. In this section, I will review evidence on the process of skill formation. The same general mechanisms of skill acquisition are likely to underlie both specialized forms of expertise, such as musicianship, and everyday forms, such as using a knife and fork.

Neural Recruitment versus Increased Efficiency

A major challenge in studying motor skill learning, especially using neuroimaging methods, is that learning involves complex combinations of increases and decreases of neural activity (Diedrichsen & Kornysheva, 2015). On one side, the acquisition of expertise commonly recruits additional neurons that were not previously involved in the task. For example, in an influential study, Elbert et al. (1995) used MEG to measure the size of finger maps in skilled string musicians, such as violinists. Compared to nonmusician controls, musicians had stronger responses in S1 to tactile stimulations of the fingers of the left hand, as well as shifts in the location of fingers. Together, these results indicate that musical expertise is associated with an expansion of finger maps in S1. Intriguingly, these changes appeared limited to the nondominant left hand and were not found for the dominant right hand. This asymmetry is likely due to the use of the left hand for fingering on the strings, which requires an extreme level of individuated control over each finger. While a violinist's use of the right hand to control the bow is equally skilled, it does not require the same degree of fractionated control over each finger. Similar expansion of S1 finger representations have been reported for people who read braille, which are also specific to the fingers actually used for reading (Pascual-Leone & Torres, 1993).

On the other side, a common finding in fMRI studies is that expert performers, such as professional musicians, show reduced or more focal activations in M1 and the premotor cortices compared to less skilled performers (Haslinger et al., 2004). Such results are frequently interpreted as evidence that fewer neurons need to be active in expert performers on account of increased efficiency and reduced task difficulty. Interpretation of reduced fMRI responses, however, is complicated by the fact that the BOLD signal does not reflect neural activity directly but rather the metabolic consequences of such activity. An elegant study by Picard et al. (2013) measured both neural activation using microelectrode recording and metabolic uptake of 2-deoxyglucose, showing that this distinction is critically important in assessing the effects of expertise. Monkeys with extended experience in producing specific sequences of reaching actions showed reductions in metabolic activity in M1, but no reduction in the number of neurons activated. In fMRI, this would show up as a reduced BOLD signal, which would likely be interpreted as reduced neural activity. In fact, however, the results of Picard and colleagues suggest that expertise may produce more metabolically efficient generation of neural activity without any reduction in the total amount of neural activity.

This increased efficiency suggests that skill learning involves progressive stabilization of neural patterns underlying a trained action with a corresponding reduction in neural variability. Wiestler and Diedrichsen (2013) investigated this process using MVPA to compare the neural representations of trained and untrained finger movement sequences. Participants practiced a set of sequences for four days and were then tested in the MRI scanner on both the trained sequences and another set of sequences they had not performed before. Univariate fMRI analyses showed broadly similar levels of activation during both types of sequences in M1, the PMC, and the SMA. MVPA, however, showed clearly that training had produced clear changes in the underlying representational space. While the distributed patterns of activity could be successfully discriminated even for untrained sequences, classification accuracy was substantially higher for trained sequences. They then used a multivariate method called "pattern component modeling" to investigate how training alters the underlying representational space in which actions are coded. The trained actions were farther apart in pattern component space than the untrained actions. Thus, training involves a reconfiguration of the abstract representational space of the motor cortices, maximizing the dissimilarity of trained actions.

Multiple Processes Operate at Different Timescales

Skill learning also appears to depend on different mechanisms operating over different timescales (Karni et al., 1998). Performance improves rapidly within a single training session (fast learning) but also shows slower continual improvement over successive training sessions (slow learning). Fast and slow learning appear to depend on distinct neural networks (Floyer-Lea & Matthews, 2005; Karni et al., 1995). Floyer-Lea and Matthews used a task in which participants held an MRI-compatible pressure sensor and had to learn an eight-second-long sequence of varying forces to try to keep to bars shown on a monitor at the same height. One group of participants were tested in a single session in which they gradually got better at the task. By comparing brain activations during the first versus second half of the session, Floyer-Lea and Matthews identified brain areas involved in short-term motor learning. Such short-term within-session learning was associated with a reduction of cortical activation and increases in subcortical areas, including the thalamus, basal ganglia, and cerebellum. This shift likely reflects increasing automatization of the learned action. Another group of participants completed three weeks of training on the sequence, practicing for fifteen minutes five days per week. By comparing brain activations from before to after this training, Floyer-Lea and Matthews identified areas involved in long-term motor learning. Such slow learning across sessions relied on different brain networks, notably including increases in activation of contralateral S1 and M1. Similarly, Karni et al. (1995) had participants train finger movement sequences for several weeks, finding that, after four weeks, activations in M1 were enhanced for practiced sequences compared to novel sequences.

Another study used MRI repetition suppression to investigate sequence-specific representations across six weeks of training (Wymbs & Grafton, 2015). By measuring the reduction of responses when the same sequences were presented twice in a row (i.e., repetition suppression), the formation of representations of specific movement sequences could be distinguished from more general representations of the task. In several areas of the cortical motor system, Wymbs and Grafton identified three distinct stages of motor learning. In the first stage, occurring over the initial training session, a rapid increase of repetition suppression was found in regions such as M1 and the PMC, indicating the formation of distinct representations of the particular sequences being practiced. Over a longer timescale of a few weeks, this pattern changed, and levels of repetition suppression declined, which Wymbs and Grafton interpret as reflecting generalization across specific sequences, indicating increased efficiency of the more general skill. Finally, after several weeks, repetition suppression increased again in a subset of regions such as M1, the PMC, and the SMA, likely reflecting the emergence of expert representations of specific sequences nested inside more general skill in the task more widely.

Such differences in the timescale of learning likely also correspond to differences in the level in the motor hierarchy at which learning is occurring. The development of motor skills certainly involves the formation of new motor synergies and automatization, but also influences higher-level, more abstract motor representations and their interfaces with perceptual systems and semantic knowledge (Diedrichsen & Kornysheva, 2015).

Consider the example of a child learning to play the piano. Her initial efforts may involve learning the specific sequence of finger movements required to play a single song, such as *Twinkle Twinkle Little Star*. As she plays a range of pieces, however, she starts to build more abstract skill in piano playing—skill that will generalize to completely new songs she tries to play. Later, as an experienced pianist, she may practice and build fluency and expression in a specific piece (perhaps one of Chopin's *Etudes*). The skill in playing this specific piece, however, builds on and is embedded within her more general piano-playing expertise, spanning several levels of the motor hierarchy, as well as perceptual processes from audition, touch, and vision, memory of familiar compositions, and conceptual knowledge of music theory.

Motor Chunking and Intermediate-Level Representations

Critical to this process is the formation of intermediate-level representations of skill. These may be specific to a particular activity but abstracted above the level of specific muscular contractions or even of motor synergies. For example, a pianist will form representations of the finger patterns needed to produce chords, progressions, and scales, but which generalize across different songs. For a novice, playing a piano chord such as C major will involve considering three distinct finger movements to press the C, E, and G keys. With practice, these three movements will become "chunked" together, allowing the chord to be processed at a high-level as a single unit. With increasing expertise, larger and more complex chunks will be formed, and these chunks may be applied in more sophisticated ways and in a wider range of contexts.

Figure 8.2 depicts a simplified model of the formation of such intermediate-level representations through a process of motor chunking, following that proposed by Diedrichsen and Kornysheva (2015). Before learning, implementing a complex action at a high level requires explicit processing of each of the component motor synergies. This level

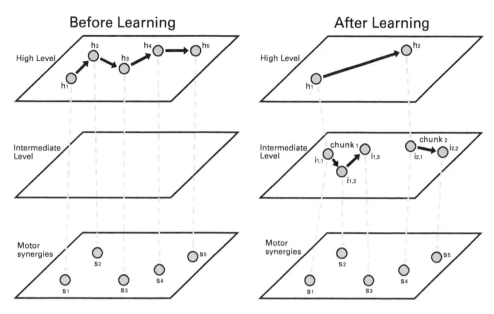

Figure 8.2
The process of motor "chunking" during skill learning. With practice, an intermediate level of representation is formed between high-level representations accessible to conscious control and motor synergies. Motor "chunks" at this intermediate level, consisting of sequences or combinations of synergies, allow the action to be represented in a simpler form at the highest level. Following Diedrichsen and Kornysheva (2015) and Sadnicka et al. (2017).

of conscious control is effortful, time-consuming, and exerts a high cognitive load on performance. With practice, an intermediate level of representation is formed between high-level motor representations available to consciousness and motor synergies. Series of consecutive movements, or combinations of simultaneous ones, that had previously been represented as distinct now become chunked into larger units. This allows the action to be represented in a simpler way at the highest level, reducing the cognitive load involved in implementing it. In the example in the figure, the action is composed of five synergies, which, at the start of learning, must each be processed individually. Eventually, two chunks are formed: $chunk_1$, consisting of three synergies, and $chunk_2$, consisting of two synergies. This allows the same five synergies to be produced while substantially simplifying the high-level representation of the action. This simplification produces a corresponding reduction in the amount of attention required to perform the action successfully.

Studies of motor sequence learning in humans and animals have implicated both parieto-frontal networks (Koechlin & Jubault, 2006; Pammi et al., 2012) and the basal ganglia (Graybiel, 1998) in motor chunking. There is also evidence, however, that these brain regions may produce chunks in different ways. Wymbs et al. (2012) argued that two quite different processes could be involved in chunk formation. One bottom-up process, which they call "concatenation," involves putting together adjacent movements that tend to co-occur. For example, if, in a sequence of finger movements, ring-finger movements very frequently follow middle-finger movements, the sequence "middle → ring" may become concatenated into a discrete chunk. Another, more top-down process, which they call "segmentation," works from the opposite direction by identifying meaningful subsections of longer sequences and breaking them down into smaller chunks. Using a sequential finger movement paradigm with fMRI, Wymbs and colleagues argue that the basal ganglia is involved in chunking via concatenation, while frontoparietal cortical regions, such as the IPS, and PMC regions, such as Broca's area, are involved in chunking via segmentation.

Automatic versus Effortful Processing
Another important feature of skill learning is that, early in learning, people need to think consciously about what they are doing, but this becomes increasingly automatized as learning progresses. This is typically linked to a distinction between declarative and procedural knowledge. We all have the experience of getting into a car and finding ourselves at home without having any conscious recollection of the actions we performed to have driven the car safely. Dramatic evidence for this distinction comes from studies of individuals with anterograde amnesia, who can learn new motoric skills despite being unable to remember having performed the task at all. For example, Milner (1962) trained the amnesic patient H.M. on a mirror drawing task in which he traced the outline of a figure while seeing his hand only through a mirror image. Like other people, H.M. showed continuous improvement during an initial session of this task. More remarkably, this improvement remained when he was tested days later, although he denied that he had ever performed the task before. H.M.'s intact ability to learn new manual skills was in striking contrast to his profound inability to remember events and people.

Automatization is widely considered to be central to the acquisition of new motor skills (Logan, 1988). Nevertheless, conscious control can be used even for well-practiced skills. In some cases, however, conscious control may interfere with performance. For example, Charles Darwin commented in 1838 in his notebook (Barrett, 1974) on his wife's piano playing: "Emma W. says that when in playing by memory she does not think at all, whether she can or cannot play the piece, she plays better than when she tries is this not precisely the same, as double consciousness kept playing so well" (p. 342).

This idea of "double consciousness" was developed by Willingham (1998), who argues that any action can be performed in either a conscious or an unconscious mode—what he calls the "dual mode principle." Everyday manual actions, from writing to tying our shoelaces to brushing our teeth, can be performed rapidly and with little explicit attention paid. However, we can also focus our attention on these actions to perform them more precisely. For example, when making a shopping list, we may be content for writing to operate in unconscious mode, focusing our attention on what's in the pantry and what meals we plan to cook in the upcoming days. In contrast, when writing a message in a birthday card, we are likely to switch to conscious mode, ensuring that our writing is neat and tidy. The dual mode principle thus provides us with a trade-off where we can balance the amount of precision needed for a task with the level of attention required for successful execution and monitoring.

In the initial stages of learning, before automatization, no such trade-off exists, as attention is required even to perform a task at all. Jueptner et al. (1997) used PET to investigate brain areas active during learning and performance of sequences of finger movements. Areas of the dorsal prefrontal and anterior cingulate cortices were activated during initial learning, but they stopped responding after the task had been practiced. Similarly, Bassett et al. (2015) used functional-connectivity analyses of fMRI data to investigate how functional brain networks change during manual sequence learning. With training, the sensorimotor cortex showed increasing modularization, become progressively more autonomous from frontal and cingulate regions involved in cognitive control.

There is also evidence that adopting conscious control over a learned action may involve similar processes of attentional monitoring as are involved in initial learning. When Jueptner et al. (1997) asked participants to pay attention to the task and think explicitly about the upcoming movement, the same prefrontal and cingulate regions were reactivated.

M1 is also critical for motor learning. Pascual-Leone et al. (1994) mapped the area of scalp from which MEPs at hand muscles could be evoked by TMS, finding an increase in area while learning a sequence of finger movements. This area increased progressively, in concert with increases in the speed of responding and hence of learning. Eventually, however, motor maps returned to their baseline topography. Intriguingly, this return to baseline occurred just as participants gained conscious, introspective insight into the sequence. Similar findings have been obtained using structural MRI. Wenger et al. (2017) asked right-handed participants to practice writing using their left hand and used voxel-based morphometry (VBM) to measure structural changes in the brain over seven weeks. After four weeks, there were increases in gray-matter density in M1 in both hemispheres.

Subsequently, however, these increases returned to near-baseline levels by seven weeks, although behavioral performance remained high.

Research in rodents has provided consistent evidence for a critical role for the primary motor cortex in motor learning. Kawai et al. (2015) trained rats to perform sequences of lever presses and investigated how lesions of the motor cortex disrupted learning. When the motor cortex was lesioned before training, rats were unable to learn the sequences. In striking contrast, however, when lesions were performed after training was complete, no impairments on the task were seen. Thus, the motor cortex was necessary for the process of learning, but the outcome of learning was implemented in other areas. Kawai and colleagues suggest that the motor cortex has the role of "tutoring" subcortical regions, which progressively take over task performance as expertise develops.

Another striking example of dissociation between conscious and unconscious learning processes comes from a study by Mazzoni and Krakauer (2006). They asked participants to reach targets with a cursor that followed their hand movements. After a few trials, they introduced a visuomotor rotation, such that the direction of cursor movements was rotated by forty-five degrees anticlockwise. After two trials in which participants made large errors, they were encouraged to adopt an explicit strategy of moving to a location forty-five degrees clockwise of the desired target in order to offset the rotation. This cognitive strategy was initially effective, and errors were reduced to near zero. Gradually, however, participants started to make larger and larger errors in the *opposite* direction from the rotation. This likely reflects an implicit learning process that gradually corrects for the visuomotor distortion. What is striking in this case, however, is that this implicit process emerges even in a situation where an explicit strategy has already corrected the error and, in fact, has the effect of increasing the error produced on each trial.

Specialized Hand Expertise

Like Paganini, many people have extraordinary skill in using their hands for specific activities. These people are engaged in all fields of human endeavor, from potters to chefs, tailors to surgeons, blacksmiths to circus performers. I will briefly mention three specific examples that have recently struck me:

1. A video shared on social media[1] showed a Mexican *taquero* (taco chef) chopping an onion with a long kitchen knife. His right hand controls the knife with elegant and minimal chopping motions, while his left hand spins and flips the onion into place. The precision and efficiency of his movements are extraordinary, as is the calm confidence with which the operation is performed.

2. At the *Museu das Rendas de Bilros* (Bobbin Lace Museum) in Vila do Conde, Portugal, I was hypnotized by the movements of a woman making lace. Spools of thread were connected to a dizzying profusion of bobbins, which she whipped back and forth over each other, creating delicate webs of filigreed lace.

3. In December 2020, thirteen-year-old Michael Artiaga (screenname "Dog") of Fort Worth, Texas, became the youngest winner of the Classic *Tetris* World Championship, defeating his brother Andy 3-2 in the finals.[2] Dog employed a recently developed playstyle called "hypertapping" in which the controller is held sideways, with one hand spread across its entire length, allowing rapid control of block placement. Using hypertapping, Dog successfully defended his title in 2021. In 2022, however, a new champion emerged, Eric Tolt (screenname "EricICX"), using a new grip style called "rolling." That new grip styles continue to roil the world of competitive *Tetris* decades after the game's appearance shows that *Tetris* skill is determined not only by abstract spatial cognition but also by how this spatial reasoning interacts with expert control of the hand.

These people all have extraordinary manual skills, whether for chopping vegetables, making lace, or playing video games. I am sure that you have examples from your own experience that you might add to this list. I would encourage you to pause briefly and consider a couple of these. In the remainder of this section, I will discuss in more detail one specific field of expert hand performance that has been widely studied: that of musical performance.

Musical Expertise

The demands that high-level musical performance places on the sensorimotor system are extreme. As one example, figure 8.3 shows a professional string musician applying the double-stopping technique on a viola in the opening solo of Paul Hindemith's 1935 concerto *Der Schwanendreher* (the swan turner). The index and little fingers both hold E notes an octave apart on different strings that are played simultaneously. Although the middle and ring fingers are not involved in playing notes, highly dexterous control is nevertheless required to prevent them from interfering with the two strings being played. This demonstrates the highly individuated movements of each finger required for skilled performance. At the same time, it also highlights the coordination across all fingers to implement the desired musical outcome. These simultaneous demands for individuation and coordination highlight the extreme demands that musical performance places on the motor system.

Figure 8.3
Manual dexterity in expert musicianship. A professional string musician demonstrates the double-stopping technique on a viola.

Increased Efficiency of Movement

Many studies have investigated the kinematics of playing in expert pianists compared to more recreational or novice players. For example, expert pianists apply less force than novices to produce a key press of a given duration (Parlitz et al., 1998). This not only produces an obvious reduction in peak force exerted but also allows the experts to relax the playing finger much sooner.

Another important feature of skilled piano playing is that while the keys themselves are pressed by the fingers, the movements themselves are produced with contributions from the more proximal musculature controlling the upper arm and forearm, the extrinsic hand musculature. This is especially important, as the intrinsic hand musculature is highly prone to fatigue and cramping. In a 1930 study, Bernstein and Popova (English translation in Kay et al., 2003) used high-speed movies of concert pianists to investigate how forces are produced by different arm segments. Pianists made rhythmic presses of two keys an octave apart using the thumb and little finger of one hand at several different tempos. At all tempos, movements of both the forearm and hand were involved. At slow tempos, these movements were clearly segmented, with movements of the forearm occurring first and being separated by a short pause before movements of the hand. At faster tempos, more than six-and-a-half strikes per second, however, the movements merged into a single continuous movement. The movement of the wrist follows naturally from, and indeed is "forced" by, the momentum produced by the forearm. Thus, through coupling between the biomechanics of different segments of the arm, hand movements and finger presses were produced without any involvement of the intrinsic hand musculature.

Many more recent studies have identified a characteristic proximal-to-distal sequence of movements in expert pianists. The total force produced onto a key by the fingertip is the sum of torques from several sources—from muscles certainly but also from gravity and from intersegmental dynamics. For example, deceleration of downward movement at the elbow can cause the forearm and hand to swing downward, producing force at the fingertip without exertion of the hand muscles. Furuya and Kinoshita (2008) showed that expert pianists take greater advantage of these nonmuscular forces than novice players do, producing keystrokes with less muscle torque at both the elbow and wrist. Even for a given muscle torque, expert players are more efficient. The same total muscle torque can be produced by different combinations of agonist and antagonist muscular activations. Whereas novice players produce downward extension of the elbow joint on a keystroke by *increasing* activation of extensor muscles, experts produce the same extension by *decreasing* activation of flexor muscles (Furuya et al., 2009). Thus, the experts effectively allow gravity to do much of the work for them. This increased efficiency allows experts to play longer without inducing fatigue. For example, following thirty minutes of producing forceful and repetitive keystrokes, novice players showed clear muscle fatigue, whereas experts did not.

Formation of New Motor Synergies

Musical expertise involves not only increased efficiency of movement but also the formation of qualitatively new synergies. Recall from chapter 3 that a typical approach to measuring motor synergies is to apply statistical factorization methods such as principal component analysis to kinematic or EMG data to show that a comparatively small number of latent variables account for a large proportion of the variance in the data. Furuya et al. (2011) recorded motion at finger joints while five expert pianists played thirty different musical pieces. They then used principal component analysis to decompose these joint motions into distinct patterns of covariation. During thumb key presses, there were two distinct patterns of covariation in movement of the other fingers. For movements of the other fingers, there were three distinct patterns of joint rotation of the striking finger. Remarkably, however, these components did not involve movements of the other fingers, showing a high level of independent motion. In each case, complex piano-playing movements could be reconstructed as linear combinations of a relatively small number of components. Thus, despite the variability and complexity of the musical pieces played, a modest number of fundamental motor synergies may be involved. Thus, in much the same way that a vast number of songs can be implemented by the spatiotemporal combination of a modest number of discrete piano keys, so too are the dexterous movements of the playing hand composed of the spatiotemporal combination of a modest number of motor synergies.

A similar conclusion was reached by Gentner et al. (2010), who used principal component analysis to reconstruct patterns of finger movements evoked by TMS applied to M1. A TMS pulse to M1 elicits movements not in just one single muscle but in several muscles controlling different fingers of the hand. By investigating patterns of covariation across joints in TMS-evoked movements, the authors reconstructed the space of motor synergies controlling hand movements in violinists, pianists, and nonmusicians. Remarkably, in all groups, just four components were able to account for more than 90 percent of the variance, again showing that a comparatively small set of synergies is involved in the control of hand actions. Critically, however, the action space spanned by these synergies was systematically different in the musicians than in the nonmusicians. Gentner and colleagues used linear combinations of synergies in one group of participants to reconstruct naturalistic hand movements of other participants. Naturalistic violin and piano-playing movements could be reconstructed more effectively from linear combinations of TMS-evoked principal components from violinists and pianists, respectively, than from nonmusicians. In contrast, naturalistic grasping movements could be reconstructed equally well from TMS-evoked components of all groups. These results show that the action space of motor synergies is systematically altered by musical training.

A recent study by Ogawa et al. (2019) used representational similarity analysis of fMRI data to investigate patterns of overlap between M1 finger representations in expert pianists and matched participants without musical training. Participants made tapping movements, and Ogawa and colleagues measured the distributed pattern of activations produced in M1 for each finger separately. They then used statistical classification methods to assess how well the representations of each finger could be distinguished from these patterns. Clear above-chance classification was found for both groups, but this was significantly lower in the pianists than in the control participants. These results provide further evidence that long-term training produces changes in hand representations in M1. However, these changes do not simply reflect increased individuation of the representation of each finger. As important as fractionated control of individual fingers is for skilled piano performance, the changes seen in musicians appear to reflect the emergence of coordinated movement representations involving multiple fingers.

Changes in Somatosensory Function

Musical expertise involves changes in the somatosensory system as well as the motor system. Professional pianists show increased tactile spatial acuity at their fingertips compared to novice players (Ragert et al., 2004). A recent study by Hirano, Kimoto, and Furuya (2020) showed further that expert pianists show specialized integration between somatosensory and motor signals. They measured MEPs following TMS to M1 that was preceded by either a tactile (electric shock) or proprioceptive (passive finger displacement) stimulus. In each case, the latency between the somatosensory stimulus and TMS was set for each participant on the basis of the peak latency of SEPs recorded in an initial session for each type of stimulus (~20 ms for tactile stimuli and ~50, 100, and 200 ms for proprioceptive stimuli). Somatosensory stimuli inhibited motor cortex excitability in all participants. However, in expert pianists compared to novices, the magnitude of this inhibition was reduced for tactile stimuli but increased for proprioceptive stimuli. The reduction of tactile influences on the motor cortex may be related to suppression of cutaneous reflexes, which allows the experts to suppress unwanted movements and enhance fine control of fast finger movements. Conversely, the increased proprioceptive inhibition of the motor cortex may decrease joint stiffness, resulting in smoother, more fluid movements.

This interaction between the somatosensory and motor systems may also have important implications for advanced training and learning. There is evidence that expert musicians may hit a ceiling effect in terms of the advantages they can gain from further purely motor training. Hirano, Sakurada, and Furuya (2020) compared the effects of different types of training on expert pianists and nonmusicians. In a purely motor training task, participants tried to press a key with a target force and received feedback about their performance after each trial. This training led to clear improvements in performance for the nonmusicians but not for the experts, consistent with a ceiling effect in their performance. In another type of training, a robotic device was attached to a piano key, pulling upward so that the effective weight of the keys could be manipulated. Training involved pressing two keys in sequence and judging which one was heavier. In contrast to the previous task, performance on this task requires that information about the force produced on the key be combined with tactile and proprioceptive feedback. In this latter task, improvements were seen, even for the experts.

One recent study has shown intriguing changes in motor skills in expert pianists from passive somatosensory experience. Furuya et al. (2023) applied a robotic exoskeleton to the hand, which allowed complex combinations of finger movements to be produced without any motor contribution from the participant. The tactile and proprioceptive feedback from these movements, however, produced clear changes to subsequent motor performance, increasing the maximum speed with which repetitive piano key presses were made.

Everyday Hand Expertise

When we think of hand expertise, we are apt to think of specific individuals who have expert skills that most of us lack. This focus, however, can obscure the extraordinary dexterity involved in everyday activities. Skills are no less expert simply because they are shared by nearly all members of a population. In modern societies, virtually all adults share a wide range of expert manual skills, such as those used in writing and typing (as will be discussed further below) as well as skills such as tying shoelaces and using cutlery. In many East Asian countries, virtually everyone has highly dexterous skills in using chopsticks. Across the world, skills involved in activities such as sewing, weaving, and pottery are widespread, if perhaps gender specific. In early hominin and Neanderthal populations, skills in knapping stone tools and their use for activities such as butchery may have been similarly universal, as will be discussed in chapter 10.

Writing

Writing is a good example of a skill that we tend to underestimate simply due to its ubiquity. Writing with a pencil involves forming a large number of highly precise symbols, which may be substantially smaller than our fingertips themselves. Moreover, we must control the movement not of the hand itself but rather of the tip of an object (i.e., a pencil or pen) from which we obviously receive no direct sensory feedback. In literate cultures, children generally start writing around the age of six and may not achieve adult writing speeds until around the age of fifteen (van Galen, 1991).

Handwriting requires not only dexterous manual motor skills, but also that this ability interface with linguistic knowledge about orthography, grammar, and the lexicon. Neuroimaging studies of writing in healthy human adults have identified a broad left-lateralized network of regions involved in hand control, including the IPS, PMC, and SMA, as well as regions involved in language, such as the posterior temporal lobe (Planton et al., 2013).

One important line of evidence about the neural bases of writing comes from deficits in writing ability following brain damage—a condition known as "agraphia." The study of agraphia has a remarkable history, dating to the mid-1800s, preceding even Broca's famous studies of aphasia (Barrière & Lorch, 2004). In many cases, problems in spelling seen in writing are paralleled by corresponding problems when patients are asked to spell words verbally. This indicates a problem not in motor control of writing but rather in a more abstract linguistic representation. Neurocognitive models of writing have therefore included an "orthographic buffer" in which sequences of letters are organized and which feeds into both written and verbal spelling (Margolin, 1984) and presumably typing as well.

Shared problems in both written and verbal spelling could thus result from damage not only to the orthographic buffer itself but also to the pathways providing input to this buffer. Dissociations in agraphic patients have provided evidence for at least two such pathways. Patient P.R. showed agraphia following a left middle cerebral artery stroke (Shallice, 1981). Although he could repeat words he heard, he was highly impaired at spelling nonsense words. He spelled 94 percent of actual English words correctly but only 18 percent of nonwords. For example, given the pseudoword *sare* he wrote *sev*. P.R. thus retained the ability to spell based on stored lexical knowledge about existing words but was impaired at spelling based on phonological inputs.

The converse pattern was seen in patient R.G., who showed agraphia as a result of a tumor in his left parieto-occipital cortex (Beauvois & Dérouesné, 1981). R.G. appeared to spell words based on a purely phonological analysis of their sounds. This led to problems specifically when (French) words were spelled in nonstandard or ambiguous ways. For example, shown a picture of a milk bottle, he wrote *lai*, leaving off the silent "t" in the French word *lait*. Critically, however, he showed no impairment in reading words with irregular spelling, showing that the relevant knowledge was not lost but simply inaccessible for writing. This double dissociation between phonological (Shallice, 1981) and lexical (Beauvois & Dérouesné, 1981) agraphia thus provides evidence for two distinct input pathways to the orthographic buffer: one based on a phonological assessment of the sounds of words, and another based on a lexical knowledge of known words.

In other patients, problems occur for written but not verbal spelling, indicating a disruption of processes more directly related to the manual motor act of writing (Kinsbourne & Rosenfield, 1974). An especially pure case of this is seen in patient I.D.T., who developed writing problems as a result of a tumor in his left parieto-occipital cortex (Baxter & Warrington, 1986). Remarkably, I.D.T. was able not only to spell words verbally but also to write them out when asked to copy seen text (figure 8.4a). In striking contrast, when given words or phrases by dictation, he was completely unable to write them out (figure 8.4b). I.D.T.'s preserved verbal language skills show that his difficulty is not with the linguistic aspect of writing, while his preserved copying and drawing abilities show that he is also unimpaired in basic sensorimotor control of the hand. His preserved copying ability likely uses the same mechanisms he would use to copy nonlinguistic stimuli, such as a drawing of a star or cube or, for that matter, reproducing letters in an alphabet that was unfamiliar to him. Baxter and Warrington describe this condition as "ideational

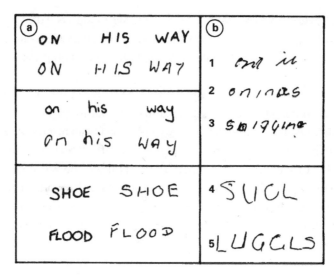

Figure 8.4
Ideational agraphia in patient I.D.T. (a) When asked to copy text he was shown visually, I.D.T. performed well. (b) In striking contrast, when asked to write words and phrases from dictation, he was severely impaired. From Baxter and Warrington (1986).

agraphia," and it should probably be best thought of as a task-specific form of apraxia—a condition that I will discuss further in the next chapter.

Agraphia can be strikingly specific. Patient D.K. was more impaired at writing lowercase than uppercase letters (Patterson & Wing, 1989), while patient M.N. was unable to write in uppercase print and, when asked to do so, would revert to lowercase cursive (Menichelli et al., 2008). Another patient was unable to write in cursive but could print neatly and read writing in either style (Matiello et al., 2015). When her brain tumor was surgically removed, her ability to write in cursive returned. Patient G.O.S. was able to write nouns but was impaired for adjectives and verbs (Baxter & Warrington, 1985). Another two stroke patients were able to write consonants but not vowels (Cubelli, 1991).

The specific nature of the errors such patients make can provide important insight into the underlying cognitive processes involved. Rapp and Caramazza (1997) investigated whether spelling errors involved substituting letters that are perceptually similar or ones that involve similar stroke movements to produce. For example, "F" and "P" are similar perceptually but involve very different movements to produce (e.g., "P" requires a curved stroke, whereas "F" does not). In contrast, "F" and "L" involve similar strokes to write but are perceptually quite different. Two patients (H.L. and J.G.E.) who had strokes in the occipital-temporal region made many more substitutions involving stroke similarity than perceptual similarity.

The prevalence of such errors suggests the operation of a processing stage in which letters to be written are coded in terms of the abstract movement features required to produce them. Indeed, such substitutions have also been described in analyses of writing slips by healthy individuals (Ellis, 1979). These movement features appear to be represented in an effector-independent manner, consistent with the principle of motor equivalence. Although H.L. could spell correctly verbally, he made similar patterns of errors when asked to write using his (dominant) right hand, his left hand, and his right foot (Rapp & Caramazza, 1997). These abstract, effector-independent representations are likely based in the PPC at the top of the motor hierarchy. This is consistent with neuroimaging studies of healthy adults discussed in chapter 3, which have found common PPC activations when participants write at different spatial scales (Kadmon Harpaz et al., 2013) and with different effectors (Rijntjes et al., 1999).

Studies in healthy adults have provided further insight into the nature of these spatial representations of written forms. For example, Wing (1980) investigated the control of letter height in writing. He noted that letter height may differ within a word, such as "l" being taller than "e," or because the overall scale of writing has changed. Wing measured changes in horizontal and vertical displacement as a function of time while participants wrote words such as "elegy" at different spatial scales. On average, the height of the letter "l" was 82 percent larger than that of the letter "e," although the duration of the associated movement was only 28 percent longer. When participants were asked to write larger than normal, in contrast, the increase in movement duration (24 percent) was much closer to the change in letter height (27 percent). This suggests that distinct processes are involved in scaling the height of different letters within a word and controlling the overall scale of writing.

There is also evidence that regions of the PMC are important for writing. Based on studies of brain-damaged patients in the late nineteenth century, the German neurologist Sigmund Exner proposed that there was a "handwriting center" in the middle frontal gyrus. While few patient studies have provided clear evidence for this, evidence consistent with the existence of "Exner's area" was provided by Roux et al. (2009), who studied patients undergoing brain surgery for the removal of tumors. In six of twelve patients, direct electrical stimulation of specific locations within the PMC produced rapid and specific impairments in writing, without corresponding difficulties in speaking, reading, or other hand functions. In two of these patients, the location of the tumor meant that this area needed to be surgically removed, and in both, handwriting difficulties continued afterward. In five other patients, similar handwriting problems could be produced but were associated with other language problems, such as with reading or speaking. Intriguingly, in some of these patients the locations linked to agraphia were in Brodmann areas 44 and 45—that is, Broca's area. These results suggest that areas of the PMC involved in writing may be partly distinct from areas involved in speech production but that these two aspects of language production are also highly overlapping.

Typing

Typing is another highly skilled manual behavior, which is nearly ubiquitous among people living in modern societies. While most of us cannot match the two hundred words per minute of champion touch typists (Rumelhart & Norman, 1982), even more everyday typing ability is highly impressive. In one sample of 356 American university students, average performance was sixty-seven words per minute or around five to six keystrokes every second (Crump & Logan, 2010).

Logan and Crump (2009, 2010) proposed a theory of skilled typing that involves two distinct, hierarchically nested control loops. The outer loop deals with linguistic knowledge about word order and interfaces with more abstract semantic knowledge and high-level behavioral goals, such as my goal to write this chapter. The inner loop, in contrast, receives individual words from the outer loop and implements the finger movements required to type each word.

Several lines of evidence support this distinction between outer and inner loops. For example, in one study, participants were given words and asked to type only the letters that would typically be typed with one hand (Logan & Crump, 2009). This dramatically impaired typing performance, leading to a 47 percent increase in the time taken to type words. It is worth trying this yourself, as the difficulty is immediately obvious. Critically, this disruption was specific to limiting typing to one hand. In a control experiment, letters were presented in red or green and participants were asked to type only the letters in one color. In contrast to the previous results, typing only letters of one color had virtually no effect on response time. These results suggest information encapsulation between the two loops. While the outer loop may cause several accurate finger presses each second by sending instructions to the inner loop, it lacks any knowledge about which hand is used to type each letter.

In another study, Logan and Crump (2010) investigated error detection processes in the two loops by providing false visual feedback about typing performance. They corrected errors made by the typists and introduced false errors, such that the letters that appeared on the screen did not match the actual keys pressed. When asked to assess the accuracy of their performance, participants took credit for corrected errors and accepted blame for false errors. This suggests that the outer loop relies on visual feedback for determining the accuracy of performance. Logan and Crump also investigated error monitoring by measuring post-error slowing—a phenomenon in which performance slows immediately after an error is made. This slowing was found when participants had actually made an error and was not affected by false visual feedback about errors. Thus, the inner loop is aware when an error has actually been made, even if the participant is subjectively unaware of this.

The speed with which key presses are made by skilled typists rules out the use of detailed sensory feedback from the previous keystroke in the planning of the next one (Lashley, 1951; Rumelhart & Norman, 1982). Indeed, when successive keystrokes are made by different fingers, there is evidence that the movement of the second finger can start 20–50 ms before the first keystroke is even made (Soechting & Flanders, 1992). Nevertheless, there is also evidence that sensory feedback from touch and proprioception is important for skilled typing. Cutaneous anesthesia of the index fingertip produced a fivefold increase in the frequency of key press errors with that finger (Rabin & Gordon, 2004).

9   The Diseased Hand

In April 1875, two anonymous reports appeared almost simultaneously in the *British Medical Journal* (1875b) and *The Lancet* (1875a) describing a completely novel disorder of the hand. At a meeting of the Paris Société de Biology, Dr. Ernest Onimus reported two cases of people working as telegraph clerks who suffered from muscle cramps in the hands when making signals. As the *Lancet* report described one telegraphist: "First his thumb failed, then the first and second fingers, and when he had recourse to his wrist as a substitute for the hand, this became disabled too" (p. 585).

Over the next few decades, telegraphists' cramp became an increasing problem. A 1911 report by His Majesty's Stationery Office in the UK found that as many as 60 percent of operators reported manual difficulties when sending signals using the Morse key telegraph device. By the 1930s, however, the incidence of telegraphists' cramp declined, likely due to increasing automation of technology and the development of more ergonomic alternatives to the Morse key. Following a brief resurgence during World War II, telegraphists' cramp disappeared from history (Haward, 2019).

The curious history of telegraphists' cramp reminds us that despite—and sometimes because of—our expertise in using our hands in innumerable skilled ways in our daily lives, our ability to use our hands can be impaired by disease. Throughout this book, I have described a wide range of clinical disorders that involve disruptions or alterations of hand function. In general, my focus in these descriptions has been on the implications of these disorders for understanding the underlying mechanisms of hand function more widely. In this chapter, I will focus on diseases of hand function themselves. I will first discuss focal hand dystonias, in which expertise itself has gone awry. I will then discuss Parkinson's disease, a degenerative disease predominantly affecting subcortical function. Finally, I will discuss apraxias—deficits in higher-level hand function resulting from cortical damage from a stroke or tumor.

Task-Specific Focal Hand Dystonia

Telegraphists' cramp is an example of a focal hand dystonia. Dystonia is a motor disorder characterized by involuntary muscle contractions, which result in abnormal postures and movements. This can make it impossible to perform familiar actions. The motor problems in dystonia can, however, be strikingly task specific, with patients unimpaired at performing other actions, even ones involving the same fingers and muscles. Moreover, the affected behaviors are often actions for which the patient is particularly expert. Indeed, many task-specific dystonias appear closely related to the development of expert performance itself. I will discuss three specific forms of task-specific dystonia—affecting writing, musicianship, and athletic performance—before discussing the mechanisms that underlie them.

Writer's Cramp

The earliest described task-specific dystonia was writer's cramp or, as it was originally known, scrivener's palsy. Although described as early as the eighteenth century, the condition was first systematically discussed in the mid-nineteenth century and, as in the case of telegraphist's cramp, was linked to the emergence of new technology (Pearce, 2005). In the 1830s, the British Civil Service switched from the traditional use of feather quills for writing to pens with steel nibs. Despite the clear advantages of the new pens, there was a dramatic rise in clerks having motor difficulties in the act of writing. In part, this was likely due to the way in which the new pens altered the dynamics of writing movements. Part of the problem, however, may have come specifically from the advantages of steel nibs. Whereas quills needed to be periodically sharpened, requiring the writer to take a brief break, steel pens could be used continuously, which increased efficiency but may also have had deleterious consequences (Sadnicka et al., 2017).

Writer's cramp can produce devastating impairments in the ability to write. Notably, however, it does not reflect a general paralysis, weakness, or problem with the peripheral motor apparatus. As described in an early report by Samuel Solly (1864): "The loss of powers is not sudden, as in a paralytic stroke; nor is it a complete paralysis of any group of muscles. The paralysed scrivener, though he cannot write, can amuse himself in his garden, can shoot, and cut his meat like a Christian at the dinner-table; indeed he can do almost anything he likes, except earn his daily bread as a scribbler" (p. 709).

Those suffering from writer's cramp often experience muscle spasms as soon as they pick up a pen to start writing. A progressive increase in muscle tone of the affected hand, as well as sudden jerks and tremors that may occur periodically, disrupts the process. The pen is often held with excessive force but at the same time insecurely so that it may fall

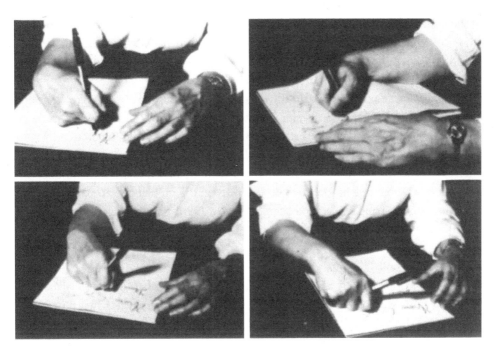

Figure 9.1
Consecutive photographs of a woman with writer's cramp. Note the unusual way she holds the pen and the hyperflexion of the wrist, leading the wrist to lift from the table, until writing eventually breaks down entirely. From Sheehy and Marsden (1982).

through the fingers. The fingers and wrist frequently flex too much, making the act of writing awkward and inefficient. This latter sign is clearly shown by the woman in figure 9.1. The hyperflexion in her wrist leads her entire forearm to rotate, lifting the wrist off the table, producing a complete breakdown of her attempt at writing. Patients often complain of extreme fatigue in the hand and wrist, which emerges rapidly when they start to write. Studies measuring EMG show that in contrast to the typical alternation of contraction of agonist and antagonist muscles, these contract simultaneously (von Reis, 1954).

Despite the high level of impairment patients experience in the act of writing, their impairment is often strikingly specific (Marsden & Sheehy, 1990). Individuals who are unable to write prose may still be able to write nearly normally using shorthand notation. Similarly, patients unable to write on a sheet of paper may be able to write on a blackboard. These examples provide important clues about the nature of the underlying problem. The former example shows that the problem is not with the muscles or low-level motor control, since longhand and shorthand writing use the same hand musculature.

At the same time, the latter example shows that the problem isn't with abstract representations of writing, which are shared between writing on paper and on a blackboard (i.e., the principle of motor equivalence discussed in chapter 3). Rather, the issue appears to be with an intermediate level of motor representation, as discussed below.

In some cases, especially early in the condition, problems may be limited to specific letters or even single letters in specific contexts. One patient who had to sign his name several hundred times per day in a stressful work environment had difficulties only with the letter "J" at the start of his signature (Shamim et al., 2011). Writer's cramp is also often specific to the dominant hand, and many patients switch to using their nondominant hand. In about one in four such cases, however, the problems eventually spread to the nondominant hand as well (Marsden & Sheehy, 1990). Such linkages between representations of the two hands will be discussed further in chapter 13.

Musician's Dystonia

In 2004, the world premiere of Paul Hindemith's piano concerto *Klaviermusik mit Orchester* (piano music with orchestra) was performed, more than eighty years after it had been composed. With Sir Simon Rattle conducting, the Berlin Philharmonic accompanied American pianist Leon Fleisher who—unusually—played using only his left hand. Hindemith wrote *Klaviermusik* in 1923 for the left hand alone, commissioned by Austrian pianist Paul Wittgenstein (brother of philosopher Ludwig), who lost his right arm during World War I. Although Fleisher had both his arms, he had suddenly lost the ability to play piano at a high level at the age of twenty-six while at the very height of his career (Fleisher & Midgette, 2010). The ring and little fingers of his right hand were in a constant state of flexion, making it impossible for him to play effectively. While Fleisher's dystonia derailed his career as a concert pianist, he did continue to perform by taking advantage of the repertoire composed specifically for the left hand.

As many as 1–2 percent of professional musicians are affected by dystonia (Furuya & Altenmüller, 2013). Like in writer's cramp, those suffering show involuntary and maladaptive flexions of their fingers and wrist as well as simultaneous contraction of antagonistic muscle pairs. This results in a loss of accuracy in terms of both finger placement and timing, which seriously impairs musical performance. Indeed, in many cases, dystonia is career ending for musicians. Like writer's cramp, however, it is also highly specific to the task of playing music. Hofmann et al. (2015) studied groups of musicians with and without dystonia on a range of tasks of hand dexterity, finding little evidence for differences in hand function beyond the musical context.

One famous case is the nineteenth-century German composer Robert Schumann, who developed a task-specific problem in controlling his right middle finger while playing the piano (Altenmüller, 2005). Consequently, he had to give up his ambitions as a concert pianist. Happily, however, he remained able to write with his right hand, allowing him to become a successful composer and musical critic. As a young pianist, Schumann was notorious for the intensity of his training and focus on virtuosic displays of finger dexterity. His *Toccata in C Major (Op. 7)*, written during the period when his dystonia first developed, is renowned as one of the most technically difficult pieces in the piano repertoire. It is likely that Schumann's focus on this demanding type of composition, designed to showcase his virtuosity, contributed to his dystonia. Intriguingly, dystonia appears, in turn, to have shaped his composition. For all its fiendish difficulty, the final version of *Toccata* is surprisingly playable without the use of the middle finger of the right hand.

Musician's dystonia appears to be closely linked to high demands for spatial and temporal precision. This can be seen in terms of which hand is affected. In piano players, the right hand is more likely to be affected, whereas in string musicians, dystonia is more common for the left hand, which is responsible for the fingerings on the strings (Altenmüller et al., 2012). Moreover, among string musicians, the incidence of dystonia increases as the size of the instrument decreases, being higher, for example, among violinists than among cellists.

Sports-Related Dystonia and the "Yips"

A final form of task-specific dystonia affects athletes and is commonly known as the "yips" (McDaniel et al., 1989). In June 1927, Scottish golfer Tommy Armour competed in the Shawnee Open golf tournament in Pennsylvania. Armour had won the US Open tournament less than one week before and was at the top of the golf world. At Shawnee, however, Armour suffered a devastating loss of form. At the par-5 seventeenth hole, he scored a remarkable twenty-three strokes—at the time, the worst hole in PGA history.[1] Armour described motor problems during putting, which led to his premature retirement from the game, coining the term "yips," which he described as "a brain spasm that impairs the short game" (Owen, 2014). Studies of the prevalence of the yips have produced varying estimates of between a quarter and a half of professional-level golfers (Lenka & Jankovic, 2021), including many of the best golfers in history, such as Ben Hogan, Tom Watson, and Tiger Woods. Golfers suffering from the yips experience tremor, spasms, and jerks of their hands and forearms, which tend to occur most strongly when putting, especially short downhill putts, which require the most subtle hand movements. As in other dystonias, agonist and antagonist muscles contract simultaneously rather than in alternation (Adler et al., 2005).

While the yips has been most widely discussed in golf, similar conditions have been described in many sports, ranging from trap shooting (Ajax, 1982) to cricket (Bawden &

Maynard, 2001). In baseball, the condition is known as "Steve Blass Disease," after the pitcher who, despite leading the Pittsburgh Pirates to the 1971 World Series Championship, subsequently suffered a mystifying loss of his ability to throw strikes (Angell, 1975). Such difficulties can be strikingly specific. In 2000, New York Yankees second baseman Chuck Knoblauch developed problems throwing the ball to first base, at one point sailing a ball into the stands where it hit journalist Keith Olbermann's mother (Owen, 2014). However, Knoblauch had no problems throwing from the outfield and successfully transitioned to play left field for the final two years of his career.

Mechanisms Underlying Task-Specific Dystonia
In the nineteenth century, conditions such as scrivener's palsy and telegraphists' cramp were interpreted as manifestations of diseases of "modernity," such as neurasthenia and hysteria (Haward, 2019). More recently, conditions such as the yips have often been interpreted as "choking" under pressure. There is evidence that factors such as personality and anxiety may have important effects on the emergence and expression of dystonia. One study found that musicians suffering from dystonia were six times more likely than similar musicians without dystonia to show elevated levels of trait anxiety and perfectionistic traits (Ioannou & Altenmüller, 2014). Similarly, both anxiety and perfectionism are strongly associated with the emergence of yips in professional athletes (Clarke et al., 2020). Such findings have increasingly led to a broadening of the concept of dystonia to include a range of non-motor symptoms (Stamelou et al., 2012), reflecting that indisputable fact that all our motor activities are embedded within contexts full of personal, social, and ecological significance. Despite this, there remains a clear neurological core of dystonia, which cannot be dismissed or explained away as a personality defect or inability to deal with pressure (Sadnicka et al., 2017).

One influential animal model of dystonia was developed by Byl et al. (1996). Two owl monkeys were trained in a simple grasping task, which they performed repetitively hundreds of times per day to obtain food rewards. After several months of this training, both monkeys developed difficulties in hand movements. They had trouble removing their hand from the device on which the food rested, reductions in the force applied at the fingertips, and eventually difficulty making contact with the target at all. The monkeys refused to perform the same number of trials per day that they had previously, and they sometimes switched to use their other hand. These results thus demonstrate that extreme levels of practice with a highly specific motor task can cause motor problems similar to those of humans with hand dystonia.

Alongside these behavioral problems, Byl et al. (1996) also used single-unit recordings to investigate plastic changes in the representations of the fingers in S1, as shown

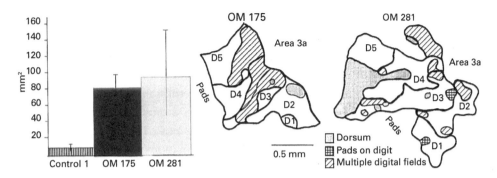

Figure 9.2
Changes in somatotopic finger maps in area 3b in owl monkeys overtrained in one specific grasping action (Byl et al., 1996). *Left panel*: The two trained monkeys (OM175 and OM281) showed large increases in the area of skin encompassed by RFs in area 3b compared to a control monkey. *Center and right panels*: Somatotopic maps in area 3b in monkey OM175 (center) and OM281 (right). In both monkeys, there are many areas with multifinger RFs, which in other monkeys are not seen in area 3b. Compare with the map in figure 2.3. From Byl et al. (1996).

in figure 9.2. Compared to a control monkey, the two trained monkeys (OM175 and OM281) showed a dramatic enlargement of the size of RFs on the fingers in area 3b, with the average area of RFs increasing by nearly a factor of ten (figure 9.2, left panel). Similarly, the orderly somatotopic arrangement of the digits was disrupted, with the appearance of multifinger RFs in area 3b (figure 9.2, center and right panels). Thus, the extreme training on a single-finger task produced a blurring of the representations of individual digits in area 3b.

Consistent with this animal model, studies in humans have provided evidence for disorganization of somatotopic maps of the hand in patients with dystonia. Studies using EEG (Bara-Jimenez et al., 1998), MEG (Elbert et al., 1998), and fMRI (Butterworth et al., 2003) have found evidence for reduced individuation of the fingers in S1 in these patients. Similarly, Tamburin et al. (2002) showed a reduction in the somatotopic specificity of inhibitory tactile-motor interactions. In control participants, electrical stimulation of one finger produces a suppression of TMS-induced MEPs in muscles controlling fingers adjacent to the stimulated finger but not in muscles on the other side of the hand. Patients with dystonia showed clear suppression of MEPs, but this did not show the spatial specificity seen in controls. Evidence from structural MRI has found increased gray-matter density in the hand area of S1 in patients with focal hand dystonia compared to controls (Garraux et al., 2004). All of these studies point to abnormal organization of somatotopic maps in patients with dystonia and have led to a widespread view of dystonia as a result of maladaptive consequences of excessive amounts of neural plasticity (Quartarone et al., 2006).

Closely related to the maladaptive plasticity model, many studies have found disruption of intracortical inhibition within the sensorimotor system in dystonia. For example, Tinazzi et al. (2000) measured SEPs over S1 following electrical stimulation of the median and ulnar nerves. In healthy participants, the SEPs evoked following simultaneous stimulation of both nerves were smaller than the sum of SEPs evoked by separate stimulation of each nerve, indicating inhibition between the representations of each side of the hand. In patients with dystonia, the magnitude of SEPs evoked by separate stimulation of each nerve was similar to controls, indicating that the basic somatosensory representations are intact. Critically, however, the patients shower *larger* responses to simultaneous stimulation of both nerves, showing a lack of the suppression seen in healthy participants.

Similarly impaired inhibition has also been described in the motor system. Sohn and Hallett (2004) applied TMS to M1 either when the hand was at rest or time locked to the onset of voluntary movements of the index finger. In healthy participants, MEPs in muscles controlling the little finger were suppressed when stimulation was locked to index finger movements compared to the resting baseline. In striking contrast, in patients with focal hand dystonia, responses were not only not suppressed but also dramatically larger than at rest. The inhibition in the motor system of muscles controlling adjacent fingers when we move one finger helps produce precise and highly individuated movements. The loss of such inhibition in dystonia, and indeed its switch to facilitation, thus provides a potential neurophysiological mechanism for the loss of dexterous hand control and for the inappropriate co-contraction of antagonistic muscles.

Furuya et al. (2018) provided evidence linking such deficits in cortical inhibition to motor deficits in musician's dystonia. Expert pianists with and without dystonia were tested both on paired-pulse TMS measures of cortical inhibition as well as on measures of dexterity when playing sequences on an electric piano that allowed detailed recording of the timing of key presses. At the group level, dystonic patients showed reduced inhibition in M1, consistent with previous findings. Moreover, regression analyses showed that the magnitude of M1 disinhibition was related to the severity of dystonic symptoms when playing piano sequences. For example, increased M1 disinhibition was linked to higher variability in the duration between successive keystrokes, suggesting that it produced lower precision in the timing of finger movements while playing piano.

Recent evidence using more sensitive neuroimaging measures, however, has called the maladaptive plasticity model into question. Sadnicka et al. (2023) used representational similarity analysis of fMRI data to map the overlap between fingers in S1 during finger movements in musicians with dystonia and matched control musicians. In contrast to the maladaptive plasticity model, there was no evidence for abnormal somatotopic maps of the fingers. Another study by Mainka et al. (2021) used the paradigm developed by Fiori and Longo (2018) described in chapter 2 to investigate the organization of tactile space in dystonia patients. Both patients and controls showed the characteristic distortions of tactile space, overestimating the distance of touches oriented across the width of the hand, with no apparent difference between groups. Thus, both of these recent studies have found broadly intact tactile representations in dystonia.

Other recent studies have found evidence that functional connections among larger-scale brain networks may play a central role in dystonia. Simonyan et al. (2017) used high-resolution tomography with a radioactive tracer that allowed them to measure functioning of dopamine receptors in the basal ganglia. Patients with writer's cramp showed increased and decreased function of two different classes of dopamine receptor (D1 and D2, respectively) in the striatum compared to controls. This suggests dysfunction in the motor circuit between the basal ganglia, thalamus, and sensorimotor cortex, as will be discussed in the next section where this pattern will be contrasted with that involved in Parkinson's disease.

Another study by Merchant et al. (2020) found that resting state fMRI functional connectivity between the inferior parietal lobe and PMC was increased in patients with writer's cramp compared to matched controls. They also used double-pulse TMS to investigate how stimulation of the inferior parietal lobe modulated the magnitude of MEPs in peripheral muscles induced by M1 TMS. In controls, opposite effects were found following conditioning TMS pulses delivered to more dorsal and more anterior inferior parietal regions, the former producing facilitation in M1 and the latter inhibition. In the patients with writer's cramp, this pattern was reversed. The functional implications of these specific patterns of facilitation and inhibition within frontoparietal circuits remain poorly understood. These results, however, do suggest changes in the functional interactions of large-scale brain networks.

Finally, a recent study has also implicated functional connectivity between the cerebellum and cortex in dystonia. Kita et al. (2021) tested expert pianists with and without dystonia on a task in which they made sequential finger movements on an MRI-compatible piano keyboard. Comparing simple levels of brain activation in response to key presses showed no differences at the cortical level distinguishing patients and controls. Similarly, multivariate pattern analyses of activity in the sensorimotor cortex also failed to distinguish between these groups. In contrast, however, the presence of dystonia could be predicted from the distributed patterns of effective connectivity between the cerebellum and sensorimotor cortices in response to button presses. Critically, it is not simply that there was more or less effective connectivity between the cerebellum and sensorimotor cortex in dystonia but rather a different distributed pattern of such connectivity. Together, these recent studies suggest that dystonia is linked to subtle changes in cross-region interactions within large-scale brain networks involved in somatosensory and motor function.

Sadnicka et al. (2017) have recently proposed a model of task-specific focal dystonia based on the model of chunking in motor learning described in the previous chapter. With increasing practice and expertise, longer and longer motor chunks are formed. These intermediate-level representations facilitate highly skilled performance, allowing automaticity and reducing attentional demands. However, they can also be inflexible. While intermediate-level chunks formed during practice of a broad range of related skills may allow a high level of generality, those formed by intense periods of practice of highly specific activities may have the opposite consequence, becoming increasingly rigidly tied to the specific activity and context.

Consider a pianist intensely practicing a single piece before a major public performance. The motor chunks that had previously implemented generalizable musical elements such as chords and scales may become increasingly linked to the specific piece being practiced. This may leave her skills more rigid and inflexible than they were previously and less able to accommodate different contexts and situations.

In many cases, dystonia also appears to be triggered by seemingly innocuous issues, such as fatigue or minor injury to the hand. For example, the onset of Leon Fleisher's dystonia was preceded by a small cut to his right hand. Although the cut soon healed, with no apparent long-term consequences, it may nevertheless have had an important role in causing his dystonia. Attempting to play piano while making accommodations for the injury may have pushed Fleisher's intermediate-level representations outside the boundaries within which they could generalize. Similarly, postural adjustments he made in response to playing while injured may have themselves become encoded in motor chunks that persisted even after the hand healed. In both ways, the motor chunks that had supported Fleisher's virtuosity may have become garbled. Accommodations to a temporary bodily state, such as injury or fatigue, may become crystallized in motor chunks and persist even after this temporary state has gone.

One important commonality of all the task-specific dystonias described above is that they involve activities (writing, musical instruments, sports) that are comparatively recent cultural developments. These are not activities for which the human hand evolved to perform. Nevertheless, the motor learning mechanisms in the brain allow a high level of generality, allowing us to develop skills that did not exist when the mechanisms themselves evolved. When we develop high levels of expertise in a specific skill, such as playing the piano or writing, however, these general-purpose learning mechanisms are pushed to develop highly specialized, task-specific motor representations. In extreme cases, these representations become so highly specialized that become inflexible and rigid, maladaptively narrowing in their ability to accommodate any alteration in task requirements.

On this view, the cause of task-specific dystonia is expertise itself. The very mechanisms that support the development of a task-specific skill such as playing the piano become hyper-specific. For example, they may become specific to the task not of playing the piano generally but rather of playing *Toccata in C Major* on one specific Steinway grand piano in one particular state of muscle fatigue. In this case, the architecture of intermediate-level motor representations that had supported piano playing more generally may be lost or distorted.

Parkinson's Disease

On September 28, 1999, Canadian actor Michael J. Fox appeared at a hearing of the United States Senate Appropriations Subcommittee in Washington, DC. Fox, known for his roles as Alex P. Keaton in the sitcom *Family Ties* and Marty McFly in the *Back to the Future* trilogy, had announced that he suffers from Parkinson's disease less than one year before. Advocating for increased governmental funding for Parkinson's research, he deliberately appeared without taking the medications that can help suppress the visible symptoms of the disease. For many senators, as well as viewers around the world, Fox's appearance was eye-opening on the effects of Parkinson's. As Fox (2002) describes:

> I looked as though an invisible bully were harassing me while I read my statement. My head jerked, skewing my reading glasses as if the back of my skull were being slapped. I was fighting to control the pages of my speech, my arms bouncing as if someone were trying to knock the paper out of my hands. But through it all, I never wavered. I saw in my eyes an even, controlled sense of purpose I had never seen in myself before. There was, ironically enough, a steadiness in me, even as I was shaking like a leaf. (p. 297)

General Clinical Characteristics

Parkinson's disease is a progressive neurodegenerative disease that results from a loss of dopamine-containing neurons in the basal ganglia. This produces a range of motor problems, which prominently include difficulties in controlling the hand. Indeed, James Parkinson (1817), who first described what he called "the shaking palsy," identified problems in controlling the hand as central to the condition and among the first symptoms to appear:

> Hitherto the patient will have experienced but little inconvenience; and befriended by the strong influence of habitual endurance, would perhaps seldom think of his being the subject of disease, except when reminded of it by the unsteadiness of his hand, whilst writing or employing himself

in any nicer kind of manipulation. But as the disease proceeds, similar employments are accomplished with considerable difficulty, the hand failing to answer with exactness to the dictates of the will. (p. 4)

One hundred and seventy-three years later, a tremor in his left hand was Michael J. Fox's first sign that something was wrong (Fox, 2002): "That's when I noticed my pinkie. It was trembling, auto-animated. How long this had been going on I wasn't exactly sure. But now that I noticed it, I was surprised to discover that I couldn't stop it" (p. 2).

The most obvious motor symptom of Parkinson's disease is tremor or involuntary rhythmic shaking of the hands and other body parts. In many cases, the tremor occurs predominantly when the hand is at rest, showing a regular, rhythmic shaking at around 5 Hz. Many patients show a characteristic pattern of movements of the thumb and index finger combined with a rotation of the forearm that has been called the "pill-rolling" tremor. Such resting tremor generally stops when the patient performs a voluntary action or even when the arm swings during walking. In other cases, the act of grasping an object may prevent the tremor, and some patients will carry objects specifically for this reason. Michael J. Fox (2002), for example, describes using this technique to hide his tremor when acting on set before making his diagnosis public: "Especially in the early stages, I was able to mask trembling through the most basic of manipulations: by picking up and putting down a coffee cup, twiddling a pencil or threading a coin through the fingers of my left hand" (p. 230). Some patients, however, also suffer from tremor that continues during skilled actions of the hand. Such action tremor is substantially more debilitating and may make everyday activities such as writing, eating, and typing difficult or impossible.

While tremor is a visually obvious symptom, the main deficits of Parkinson's concern not an excess of movements but rather their absence. Parkinson's patients are stiff and rigid, in extreme cases like statues—what Fox (2002) describes as being "a prisoner in my own body" (p. 258). This can be seen in three related motor signs (Berardelli et al., 2001): bradykinesia, which refers to slowness of movement, hypokinesia, which refers to reduced amplitude of movement, and akinesia, which refers to difficulties in initiating movements in the first place or in the absence of spontaneous movements. While these three signs are potential distinct and sometimes dissociate in individual patients, they are often collectively labeled as "bradykinesia." A patient asked to perform a repetitive action, such as moving the hand from side to side will be slow to initiate the movement in the first place and will typically show a progressive reduction in movement amplitude over time. Bradykinesia appears not only for actual hand movements but also for imagined movements, demonstrating that it is a result of central processes in the brain and not peripheral weakness (Dominey et al., 1995).

Parkinson's Disease and the Basal Ganglia

As already mentioned, Parkinson's disease results from the loss of dopamine-producing neurons in the basal ganglia. The basal ganglia have bilateral connections with several areas of the cortical motor system, which have already been introduced, including AIP, SMA, PMC, and M1 (Clower et al., 2005; Hoover & Strick, 1993). They are thus strategically connected into areas involved in hand control, giving them an important role in coordinating and regulating the invisible hand.

Classical models of basal ganglia function have emphasized the role of two pathways by which they modulate activity in the thalamus and thus the cortex (Alexander & Crutcher, 1990; Graybiel, 2000), as shown in figure 9.3. Both pathways start in a region of the basal ganglia called the "striatum" and terminate in output nuclei known called the "globus pallidus internal segment" (GP_i) and "substantia nigra pars reticulata" (SN_r), which in turn send inhibitory signals to the thalamus. The direct pathway sends inhibitory signals directly from the striatum to the output nuclei. The indirect pathway, in contrast, sends excitatory signals to the output nuIlei indirectly by way of the globus pallidus external segment (GP_e) and subthalamic nucleus (STN). These output nuclei themselves have an inhibitory input on the thalamus, meaning that the "sign" of the signal in terms of excitation/inhibition is reversed once more. Thus, the overall effect on the cortex is excitatory for the direct pathway and inhibitory for the indirect pathway. The direct and indirect pathways thus have opposite effects on cortex, producing cortical excitation and inhibition, respectively, and have therefore been described as operating like an accelerator and brake (Graybiel, 2000). Activity within these two pathways is controlled by distinct populations of dopamine receptors in the striatum, known as D1 and D2.

In the healthy state, these braking and accelerating mechanisms are balanced, allowing behavior to be controlled and regulated effectively. When the balance between these pathways is disrupted, however, motor behavior can be severely disrupted. A prominent model of Parkinson's disease is that a deficit in dopamine receptors results in excess activation in the indirect pathway and reduced activation in the direct pathway (figure 9.3, middle panel). The brake dominates over the accelerator, resulting in the difficulty initiating actions and reductions in movement amplitude and speed seen in Parkinson's patients. In contrast, other conditions may reflect the opposite problem, with the direct pathway accelerator dominating over the indirect pathway brake. For example, as mentioned above, dystonia appears to be associated with increased activity of the direct pathway (Simonyan et al., 2017), which may be related to the excess movements seen in dystonia and the inability to appropriately alternate between contractions of antagonistic muscles. Other conditions involving excessive movement, such as Huntington's disease, may similarly involve excess activity in the direct pathway.

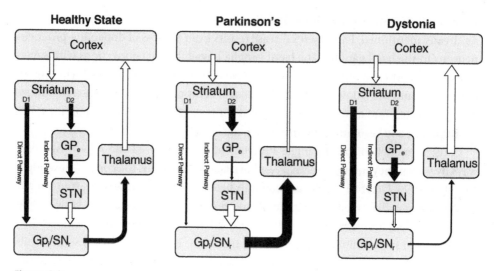

Figure 9.3
The brake–accelerator model of basal ganglia function. The basal ganglia comprise two pathways that modulate thalamic, and indirectly cortical, activity. The direct pathway operates as an accelerator, stimulating thalamocortical activity. The indirect pathway operates as a brake, inhibiting thalamocortical activity. White arrows indicate excitatory projections. Black arrows indicate inhibitory connections. The width of the arrows indicates projection strength. GP_e, globus pallidus external segment; GP_i, globus pallidus internal segment; STN, subthalamic nucleus; SN_r, substantia nigra pars reticulata.

This model is certainly an oversimplification but has nonetheless been highly fruitful in understanding basal ganglia function and Parkinson's disease as well as in guiding the development of therapies such as dopamine replacement drugs and deep-brain stimulation. The model has also received support from studies of experimental animal models of Parkinson's disease. For example, the drug MPTP rapidly results in the destruction of dopaminergic neurons in the substantia nigra pars compacta (SN_c) of the basal ganglia and induces Parkinson's-like symptoms. In monkeys rendered parkinsonian by administration of MPTP, symptoms were eliminated by lesioning of the STN (Bergman et al., 1990). As can be seen in figure 9.3, the STN is the final stop of the indirect pathway. Lesions of the STN thus remove the excessive "brake" producing parkinsonian symptoms. Another study used optogenetics in mice to experimentally activate either the direct or indirect pathways (A. V. Kravitz et al., 2010). Remarkably, stimulation of the indirect pathway rapidly elicited parkinsonian symptoms such as bradykinesia and increases in "freezing" behaviors. In contrast, in mice with parkinsonism induced by a drug known as 6-OHDA, optogenetic stimulation of the direct pathway produced an immediate reduction of symptoms.

Deficits in Hand Function

One characteristic manifestation of bradykinesia is the phenomenon of micrographia or writing in small letters. McLennan et al. (1972) investigated handwriting longitudinally in nearly eight hundred Parkinson's patients. By identifying standardized forms of writing, such as signatures on canceled cheques and entries in logbooks, they were able to compare premorbid handwriting to that following onset of the disease. For example, the signatures in figure 9.4 show annual signatures at the same scale from a woman, five before and four after the onset of Parkinson's symptoms. There is a clear progressive reduction in the size of her writing across the years. Notably, however, the style of her signature remains strikingly similar. Her signature remains well formed and legible—even neat—even as it gets smaller. This patient shows what is called "consistent micrographia," in that at any moment in time, her writing is of a consistent (even if small) size. Neuroimaging studies have found the consistent micrographia is associated with reduced activity in the basal ganglia and medial cortical regions such as the SMA as well as reductions in functional connectivity between these areas (Wu et al., 2016). Other patients show a difference pattern known as "progressive micrographia" in which the size of writing decreasing across the writing of a single word or sentence.

Another interesting feature of the example in figure 9.4 is that the onset of micrographia appears in the two years before the onset of other Parkinsonian symptoms. It is common that micrographia is an early symptom and indeed may not be recognized as problematic at all. Nevertheless, the size of writing and related variables such as stroke velocity and pressure appear to be highly diagnostic of the presence of Parkinson's (Rosenblum et al., 2013). This makes assessment of writing promising as a fast, easy, inexpensive, and noninvasive means for the screening and identification of early Parkinson's, which is an area of ongoing research (Cascarano et al., 2019).

Parkinson's is also associated with a range of other problems in reaching and grasping. In a classic study, Flowers (1976) asked patients to make reaches to visual targets using a joystick. Compared to control participants, Parkinson's patients took longer to initiate reaches. Most interestingly, while controls generally made efficient ballistic movements to the targets with rapid acceleration and deceleration, patients made much slower reaches with largely constant velocity. Thus, rather than moving directly to the target location, the patients appeared to rely on close visual monitoring of the hand's trajectory throughout the reach.

Studies measuring EMG activity in muscles provided similar evidence (Hallett & Khoshbin, 1980). Smooth ballistic reaches in healthy individuals involve a triphasic pattern of activation of agonist and antagonist muscles (agonist → antagonist → agonist). Parkinson's patients also show this pattern, but the intensity of each burst is too small.

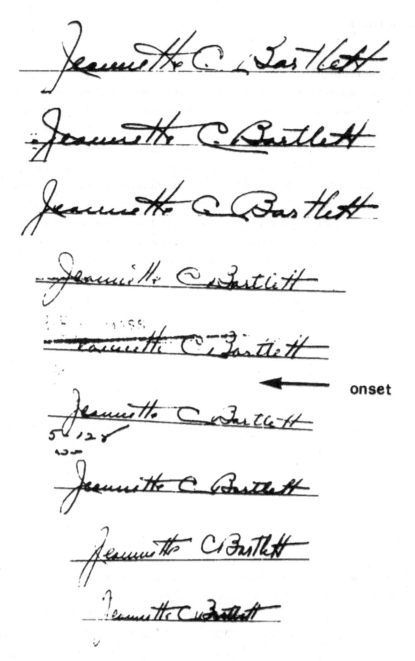

Figure 9.4
Micrographia in Parkinson's disease. The figure shows nine consecutive signatures at annual intervals from before and after the emergence of symptoms of Parkinson's (indicated by arrow). Despite the reduced size, the signature remains otherwise well formed and recognizable. Intriguingly, the emergence of micrographia is apparent in the two years preceding the onset of Parkinson's symptoms. From McLennan et al. (1972).

This means that multiple cycles of muscular contraction are needed to reach the target. This results in a slow, jerky movement, in dramatic contrast to the smooth ballistic reaching seen in controls. On this interpretation, the slowness of movement (bradykinesia) is essentially a consequence of reduced movement amplitude (hypokinesia; Berardelli et al., 2001).

A different conception comes from an elegant study by Mazzoni et al. (2007), who asked participants to make a number of reaching movements within a specified speed range. They found that Parkinson's patients took more attempts to reach this criterion, having a tendency to make movements that were too slow. Critically, however, on the trials in which patients did make reaches at the required speed, the accuracy of their reaches was just as high as for controls. This result suggests that the patients are not actually less accurate. Rather, Mazzoni and colleagues suggest that the slow movements reflect a shift in the underlying cost/benefit assessment mediating action planning. Subsequent studies have provided consistent evidence that alterations in reward sensitivity are linked to features of bradykinesia in Parkinson's disease (Manohar et al., 2015) as well as to broader features of apathy and impulsivity that are seen in many patients (Sinha et al., 2013).

Deficits in Internal Generation of Action

Patients with Parkinson's disease often have trouble initiating movements, especially when the movement is not cued by an immediately present stimulus (Bloxham et al., 1984). Once a movement is started, however, patients commonly have less difficulty in continuing it. Many studies have shown that Parkinson's patients are unusually dependent on visual feedback for guiding actions. For example, Flash et al. (1992) measured kinematic trajectories of reaches to visual targets while manipulating the availability of continuous visual feedback about arm location. While the absence of visual feedback led to an increase in errors in all participants, this effect was dramatically larger in patients with Parkinson's than in controls.

There is also evidence that patients have difficulties producing actions when they are not driven by external stimuli. Jackson et al. (1995) asked participants to reach out to grasp a wooden block when an auditory tone occurred. On some trials, the tone sounded while the participant could see the block; on other trials, the participant was asked to close their eyes, and the tone started two seconds later. Parkinson's patients performed similar to controls on visually guided reaches but showed substantial slowing on memory-guided reaches, with longer movement times and lower peak velocities.

Such results suggest that Parkinson's disease specifically impairs volitional actions produced based on internal states, generally sparing actions triggered by external stimuli.

Recall from chapter 4 that these two types of actions are broadly dependent on separate medial and lateral premotor systems (Goldberg, 1985; Passingham, 1993). Consistent evidence for this interpretation comes from neuroimaging studies showing that Parkinson's is associated with reduced activations in medial motor areas such as the SMA. As discussed in chapter 4, Playford et al. (1992) found increased activations of the SMA when participants moved a joystick in a freely chosen direction on each trial compared to a constant direction. In patients with Parkinson's, in contrast, there were activations in M1 and the lateral PMC but no clear activations in more medial regions such as the SMA. These results suggest that the medial premotor system underlying internally generated actions is specifically impaired in Parkinson's, and the lateral system involved in externally triggered actions is co-opted to compensate. Intriguingly, another study from the same group showed that when patients took the dopamine agonist drug apomorphine, SMA activations normalized in parallel with clinical improvements in motor control (Jenkins et al., 1992).

This reliance on external stimulation can produce paradoxical improvements in performance when tasks are made harder in ways that require reaction to events in the environment. In one study, Majsak et al. (1998) asked patients to reach for balls that were either stationary or rolling down a ramp. Kinematic analysis of wrist velocity showed clear bradykinesia when patients reached for the stationary ball. In striking contrast, however, their reaches to the moving balls were faster and smoother, nearly identical to those of control participants. Thus, bradykinesia was present only when the timing requirements for the reach needed to be internally generated by the participant and disappeared when they were perceptually specified by events in the environment.

Parkinson's patients also show impairments in prespecifying features of grasping and coordinating them with reaching (Fasano et al., 2022). The onset of pre-shaping of the fingers for grasping is delayed from the start of reaching, and the maximum grip aperture happens later in the reach (Schettino et al., 2003). The maximum grip aperture itself is smaller than in controls and shows poor correspondence with the actual size of the object to be grasped (Negrotti et al., 2005). Moreover, the placing of fingertips onto the objects is suboptimal, showing poor planning of grasps (Lukos et al., 2010).

Sensory Deficits
While Parkinson's is typically considered a motor disorder, a wide range of evidence has found a variety of somatosensory deficits as well (Conte et al., 2013). For example, Sathian et al. (1997) found that the spacing between tactile gratings needed to determine their orientation on the index fingertip was more than twice as large in Parkinson's patients than in age-matched control participants. Several other studies have found

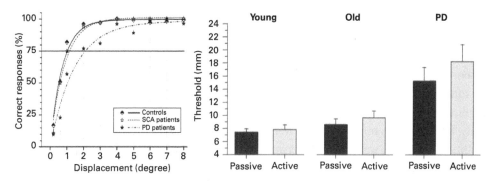

Figure 9.5
Proprioceptive impairments in Parkinson's disease. *Left panel*: Accuracy at detecting passive elbow rotations in Parkinson's patients compared with healthy controls and patients with spinocerebellar ataxia (SCA). From Maschke et al. (2003). *Right panel*: Thresholds for detecting haptic curvature during active and passive arm movements in Parkinson's patients, age-matched adults, and young adults. From Konczak et al. (2012).

deficits in the temporal discrimination of two touches. For example, Artieda et al. (1992) measured the delay between two sequential touches for patients to perceive them as two distinct events. These thresholds were substantially larger for Parkinson's patients than for matched controls and, within patients, were closely related to the clinical severity of the disease. Temporal discrimination thresholds were normalized for approximately two hours following oral administration of levodopa.

Similar results have been found for haptic judgments of the shape of two-dimensional objects (Weder et al., 1999). This latter effect was systematically related to levels of dopamine uptake in the basal ganglia measured using PET. Deficits in active haptic perception may relate to basic tactile processing but may also involve impairments in proprioception. Maschke et al. (2003) measured thresholds for detecting passive rotations of the elbow, finding that whereas control participants could detect rotations of one degree with 75 percent accuracy, Parkinson's patients were only able to detect rotations of more than two degrees, as shown in the left panel of figure 9.5. These proprioceptive impairments correlated strongly with the clinical severity of the disease. In another study from the same group, Konczak et al. (2012) measured the ability to detect curvature haptically. Participants judged which of two felt surfaces was more curved following either active exploratory movements of their hand or passive movements produced by a robotic manipulandum. As shown in the right panel of figure 9.5, Parkinson's patients showed substantially higher thresholds than both young adults and age-matched controls for both active and passive tasks.

Similar results have been found in experimentally induced animal models of Parkinson's. Boraud et al. (2000) measured responses of basal ganglia neurons in macaque monkeys to passive rotations of limb joints, finding that most neurons in the globus pallidus responded to rotations of a single joint. Indeed, many neurons were actively inhibited by rotations of non-preferred joints, showing lower-than-baseline firing rates. Following administration of MPTP, however, the selectivity of these responses reduced dramatically. Most neurons now responded in response to rotations of several joints. Responses to limb movements were thus noisier and less differentiated, consistent with the pattern seen in human patients.

Motor Deficits Following Cortical Damage: Hemiparesis and Apraxia

Deficits in hand function can also be caused by acute damage to the brain, whether caused by stroke, a tumor, or head trauma. Damage to basic sensorimotor cortices, such as M1 or S1, frequently causes motor and sensory problems on the contralateral side of the body, including weakness (hemiplegia), paralysis (hemiparesis), and somatosensory loss (hemianesthesia). Experimental lesions of M1 and S1 can also produce these conditions in animals, as shown in the nineteenth century by David Ferrier (Ferrier & Yeo, 1884) and in a distinguished tradition in the twentieth century, including Charles Sherrington (Leyton & Sherrington, 1917), Karl Lashley (1924), and Margaret Kennard (1936). In humans, reversable hemiparesis and hemianesthesia can also be induced via the "Wada test" (Wada & Rasmussen, 1960). As discussed in chapter 6, these conditions are commonly associated with bodily delusions such as anosognosia, somatoparaphrenia, and supernumerary phantom limbs.

In other stroke patients, problems in motor control are more subtle. Patients may make smooth and rapid movements, but these may be uncoordinated, clumsy, and inappropriately structured. The patient may try to pick up an object with an inappropriate grip type or choose the wrong object for the desired function. "Apraxia" is a label given by exclusion to problems in skilled motor behavior that cannot be accounted for by basic deficits in sensorimotor function (Geschwind, 1965; Leiguarda & Marsden, 2000).

Apraxia as a Deficit in Higher-Level Motor Representations

The conception of apraxia as a deficit of higher-order control of action distinct from problems in basic sensorimotor function is closely linked with German neurologist Hugo Liepmann. In 1899, Liepmann returned to his home city of Berlin from Breslau, where he had worked under Karl Wernicke. Much influenced by Wernicke's work on different forms of aphasia—neurological deficits of language—Liepmann interpreted motor deficits within a similar framework, focusing on the brain regions involved in conceptual understanding of actions, regions involved in implementing actions in patterns of muscle contraction, and the white-matter pathways connecting these regions.

Liepmann's (1900, 1977) first case of apraxia was the famous case of the *Regierungsrat*, a high-level imperial counselor in the German government. As Liepmann (1977) describes:

> He was asked to point at certain objects placed before him and to carry out certain hand movements. He failed in almost everything, handling objects quite absurdly. At first sight it appeared as if the patient did not understand—that he was cerebrally deaf, possibly also cerebrally blind. However, I noticed certain bizarre and distorted movements which he made during the course of the examination; they were confined to the right upper extremity, which the patient used exclusively during the period of observation. This peculiar motor behavior made me wonder if his incorrect responses reflected a basic lack of comprehension (of auditory-verbal or visual-object impressions), or whether these responses related, rather, to faulty motor execution. . . . To resolve the question I held on to the patient's right hand and forced him to use his left hand. Now, all of a sudden, the picture changed. With his left hand he immediately selected the card that was asked for from among five cards set out before him. (p. 159)

A critical feature of the *Regierungsrat*'s case was that he did not suffer from hemiparesis or hemiplegia of his right hand. Although semantically and situationally absurd, movements of the right hand were performed dexterously, with normal speed and smoothness. The low-level mechanisms controlling the hand appeared to be intact. At the same time, his ability to perform actions with his left hand showed that high-level conceptual knowledge about actions and their meaning was also intact. The problem appeared to be in communicating this high-level knowledge specifically to the right hand. As Liepmann concludes:

> So far as I know, it has not been observed—or at least not yet reported—that a human being might act with his right extremities as if he were a total imbecile, as if he understood neither questions nor commands, as if he could neither understand the value of objects not the sense of printed or written words, yet prove by an intelligent use of his left extremities that all of those seemingly absent abilities were in reality present. (p. 155)

A Taxonomy of Apraxia

Over the next twenty years, Liepmann described numerous patients with a variety of complex motor deficits following brain damage. In characterizing these deficits, he argued that different patterns of errors by apraxic patients indicated deficits at different levels of action representation. Lesions in the parietal lobe, especially centered on the left supramarginal gyrus, would affect more semantic, conceptual knowledge about actions. More anterior lesions in the PMC, or lesions to the white matter connecting the parietal and frontal lobes, would, in contrast, leave understanding of the actions unimpaired but impair the ability to translate these engrams into overt behavior. Finally, damage to motor synergies might disrupt the coordination of actions without producing the paralysis or weakness seen following M1 damage.

It is no accident that Liepmann had trained under Karl Wernicke, as this basic model of praxis has more than a family resemblance to Wernicke's model of language in which posterior lesions produced deficits in language semantics (Wernicke's aphasia) while frontal lesions produced deficits in language production (Broca's aphasia). This analogy between language and action would be developed more explicitly by Norman Geschwind (1965) in his seminal work on disconnection syndromes. While much has been learned about brain representations of action in the century since Liepmann, his basic model remains broadly consistent with more recent models of apraxia (Buxbaum, 2001; Rothi et al., 1991) as well as with the hierarchical approach to motor representation described in chapter 3.

In limb-kinetic apraxia, patients show clumsiness or a lack of dexterity in their actions without the weakness or paralysis that characterizes hemiplegia and hemiparesis. Some authors have referred to this as a "loss of deftness" (Hanna-Pladdy et al., 2002). Limb-kinetic apraxia appears linked to damage to the PMC, although, in pure form, appears to be quite rare (Leiguarda & Marsden, 2000). This rarity is likely due to the fact that strokes or tumors that affect the PMC are likely also to affect M1 and thus produce hemiparesis. The presence of hemiparesis makes it impossible to assess apraxia, since there is no way to measure the deftness of actions that do not occur at all.

In ideomotor apraxia, the deficit is at a higher level. Movements may be executed smoothly and rapidly, showing that low-level kinematic representations and muscle synergies are intact. But the action may be inadequate in other ways. It may show irregular timing, or the hand may show improper spatial orientation or configuration. Critically, however, while the movement is incorrectly produced, the goal of the action can nevertheless usually be recognized. Or, as in the case of the *Regierungsrat*, the action may be performed correctly using the other hand. In either case, the patient shows by their behavior that their conceptual understanding of what the action should consist of is intact, but they are unable to implement this knowledge into correct actions.

One striking behavior seen in ideomotor apraxia is the use of body parts as objects (Goodglass & Kaplan, 1963). Asked to pantomime how they would brush their teeth, the patient might extend their finger and move it along their teeth as if the finger was the toothbrush rather than correctly modeling how they would hold the toothbrush. This action shows that low-level motor control over the musculature is intact, since the movement is made smoothly and deftly. Moreover, high-level representations of the nature of the action are intact in the sense that the patient clearly understands what the function of a toothbrush is and broadly how it is used. The conversion of this knowledge into appropriate movements of the hand, however, has become garbled.

Another important feature of ideomotor apraxia is that it may depend on the manner in which the action is cued. De Renzi et al. (1982) compared the ability of 150 patients with left-hemisphere lesions to pantomime the use of objects cued either verbally (e.g., "pretend to hold a hammer and show me how you would use it."), visually (placed on the table in front of the patient), or tactually (the blindfolded patient was given the object to hold). In most cases of apraxia, similar deficits were seen across modalities. Such cases of supramodal apraxia can be interpreted in terms of damage to the representations of action themselves. Critically, however, there were numerous cases of selective deficits for each modality individually. Collectively, these cases constitute a triple dissociation between visual, auditory, and tactile apraxia. Such modality-specific apraxia has the hallmarks of classic disconnection syndromes (Geschwind, 1965) and may result from damage to white-matter pathways connecting primary sensory regions for one modality from more abstract action representations in the PPC.

Finally, patients may have impairments in conceptual knowledge of action itself—a condition known as ideational apraxia. Asked to pantomime the use of a toothbrush, the patient may hold their hand up to their ear as if they were holding a telephone. Such content errors indicate a problem at a higher level of representation. The deficit here is not in dexterity, nor in converting an abstract action plan to more basic movements, but rather in the conceptual understanding of action itself.

Some authors have interpreted this as a form of agnosia for object use (Denny-Brown, 1958), implying loss of semantic knowledge of learned associations of familiar tools. This is almost certainly true in many cases. However, many patients also show deficits in their ability to reason mechanically about the properties of objects beyond conventional uses. For example, Heilman et al. (1997) presented patients with mechanical problems for which the conventional tool was not available. Presented with a nail that needed to be driven into a board, the patient was shown tools such as a pair of pliers and a needle but *not* a hammer. While neither conventional nor ideal, the pliers have physical characteristics that can be used to drive the nail, while the needle does not. Patients with conceptual apraxia performed poorly on this task, demonstrating a problem with reasoning about novel functions of objects and not merely with stored semantic associations of familiar tools.

Similarly, Goldenberg and Hagmann (1998) developed the Novel Tool Test, shown in figure 9.6. Patients had to select from three novel tools which one could be used to lift a block. Patients with right-hemisphere damage frequently had difficulty in actual tool use but performed nearly perfectly on this task, making errors on around 3 percent of trials.

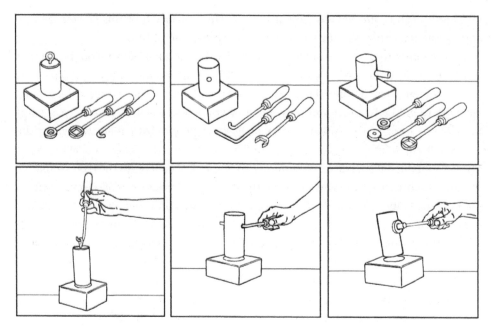

Figure 9.6
The Novel Tool Test. The top row shows three objects to be lifted using one of the three novel tools depicted. The bottom row shows the correct answer. From Goldenberg and Hagmann (1998).

In striking contrast, patients with left-hemisphere damage made errors on more than 20 percent of trials.

Such results indicate that apraxia not only affects the ability to use tools, or even semantic knowledge about familiar tools, but also can impair the ability to reason about the physical or mechanical properties of novel objects. Osiurak et al. (2020, 2021) have argued that such deficits reflect a problem not in abstract motor representations, as argued by Liepmann, but rather in a distinct cognitive domain of technical reasoning—what they label "technition." They link this ability with a region in the left inferior parietal lobe, as will be discussed in more detail in chapter 10. A lesion mapping study of apraxic patients implicated the inferior parietal lobe in reasoning about novel tools, whereas semantic knowledge about known tools was linked to the temporal lobe (Goldenberg & Spatt, 2009).

Another approach to distinguishing between loss of conceptual knowledge and the ability to convert this knowledge into action is to investigate the recognition of gestures that are passively observed. Heilman et al. (1982) tested this by dividing stroke patients with apraxia into a "posterior" group with predominantly parietal lobe lesions and an "anterior" group with predominantly frontal lobe lesions. Patients were first tested on a motor task in which they had to pantomime actions from prompts such as "Show me how you would open a door with a key." Both groups struggled with this task, making errors on around 60 percent of trials. Patients were then tested on a perceptual task in which they judged which of three videoed hand actions matched a specified action—for example, "Which of these shows someone using a key?" In striking contrast to the similar performance on the motor task, the two groups performed very differently on this perceptual task. Patients with posterior lesions made errors on nearly a third of trials, whereas those with anterior lesions did so on less than 10 percent of trials.

Both groups of patients were unable to produce manual pantomimes. This similarity, however, masks important differences in the reasons they were unable to do so. Patients with frontal apraxia understand what the action should consist of but are unable to translate this knowledge into a dexterous action. Patients with parietal apraxia, in contrast, appear to have lost their conceptual understanding of the action entirely.

Motor Knowledge and the Left Parietal Lobe
Liepmann argued that conceptual knowledge about actions is a specific function of the left hemisphere. Whereas low-level control of movements of each hand is implemented in the contralateral motor cortex, the more abstract ideas of actions appear to be stored in the left hemisphere, regardless of which hand is used. In an early study, Liepmann (1905) tested a large group of stroke patients with unilateral hemiparesis to investigate whether they showed deficits in their non-paralyzed hand. He asked them to pantomime the use of familiar objects, make expressive gestures such as military salutes, and to imitate movements of the experimenter. Among patients with right hemiplegia following left-hemisphere damage, nearly half (twenty of forty-one) showed apraxia for their (supposedly unimpaired) left hand. In dramatic contrast, no apraxia was seen in the right hand of patients with left hemiparesis after a right-hemisphere stroke. Critically, it was not that left-hemisphere damage produced symmetrical deficits in the two hands. No weakness or paralysis was seen at all in the ipsilesional left hand. Basic sensorimotor function in the left hand was intact. Liepmann argued that this left-hand apraxia following left-hemisphere lesions suggested that the conceptual knowledge of familiar actions and tools was stored in the left parietal lobe.[2] Strokes affecting the corpus callosum could thus leave the right hemisphere cut off from higher-level knowledge about actions but still retaining its basic sensorimotor function.

Similar, although somewhat milder, apraxia has been described in the left hand of "split-brain" patients with surgical removal of the corpus callosum (Gazzaniga et al., 1962, 1967). For example, patient W.J. had his corpus callosum removed to treat severe epilepsy as a result of wounds he suffered in World War II (Gazzaniga et al., 1962). Although basic sensorimotor function in his left hand was intact, he showed strong apraxia for more complex and difficult motor tasks with the left hand, especially for those cued by verbal instructions. None of these difficulties was seen for the right hand. That the left hand—but not the right hand—was made apraxic by disrupting communication between the hemispheres indicates a special role of the left hemisphere in controlling action for both sides of the body.

More recent studies have broadly confirmed Liepmann's finding that left-hemisphere lesions produce apraxia in both hands, while right-hemisphere lesions largely affect only the left hand (Geschwind, 1965; Hanna-Pladdy et al., 2002). Lesion overlap studies of large, unselected samples of stroke patients have confirmed this pattern, finding apraxia to be linked to damage in broad left-hemisphere frontoparietal networks (Haaland et al., 2000; Stoll et al., 2022) as well as white-matter tracts linking these areas (Rosenzopf et al., 2022). This clinical research is consistent with research using intracranial electrical recordings (Merrick et al., 2022) and fMRI (Verstynen et al., 2005), which has shown that the left hemisphere has rich bilateral movement representations, whereas the right hemisphere has more purely contralateral representations. This issue will be discussed in more detail in chapter 13.

Conversely, there is also evidence that neural stimulation over the left PPC can alleviate ideomotor apraxia. Bolognini et al. (2015) used transcranial direct current stimulation (tDCS)—a method in which electrical currents from scalp electrodes are used to stimulate the brain. tDCS was applied to the left PPC and right M1 of patients with left-hand ideomotor apraxia while asking them to reproduce gestures they saw performed by an experimenter. Bolognini and colleagues investigated different stages of processing by dividing the total elapsed time between the start of the experimenter's gesture and the completion of the patient's response into two phases. Planning time was quantified as the duration from the start of the demonstration to the onset of movement by the patients, while execution time was quantified as the duration from movement onset to movement completion. There was a double dissociation in the effects of tDCS applied to the left PPC and right M1. Stimulation of the left PPC reduced planning time but had no effect on execution time, while stimulation of right M1 showed the converse pattern.

10 The Extended Hand

In July 1960, Jane Goodall arrived at the Gombe Stream Chimpanzee Reserve in Tanzania. Goodall had been recruited by the anthropologist Louis Leakey to study chimpanzee behavior, alongside Dian Fossey, who went to study gorillas, and Biruté Galdikas, who went to study orangutans, in the belief that the behavior of great apes would provide insight into human evolution. Less than four months later, Goodall (1971) made an extraordinary observation of the behavior of the chimpanzee she had named David Greybeard:

> He was squatting beside the red earth mound of a termite nest, and as I watched I saw him carefully push a long grass stem down into a hole in the mound. After a moment he withdrew it and picked something from the end with his mouth. I was too far away to make out what he was eating, but it was obvious that he was actually using a grass stem as a tool. (p. 35)

Through the middle of the twentieth century, the use and making of tools was thought to be a uniquely human characteristic. Goodall's (1964) first reports of tool use in the chimpanzees of Gombe electrified the field. She reported several types of tool use. Chimpanzees would choose sticks to push into the ground to catch the ants, which would crawl onto it. They would fashion clumps of grass or stalks of plants, which they used to "fish" for termites by sticking it down the termite nest. Goodall also reported the use of leaves to fashion simple drinking bowls and to wipe the body clean of mud or sticky foods. Louis Leakey famously sent Goodall a telegram in response to hearing about this phenomenon, reading: "Now we must redefine tool, redefine Man, or accept chimpanzees as humans" (Goodall, 1998, p. 2184).

The Emergence of Technology

Tool Use in Animals

Over the decades since Goodall's report, extensive research has shown widespread use and manufacture of tools in wild chimpanzees. Most chimpanzee populations have a "tool kit" of around twenty tools (McGrew, 2010). These behaviors show some level of sophistication and complexity. Chimpanzees consider a range of material and contextual features when selecting tools (Sirianni et al., 2015) and tend to use the same combination of tools on multiple occasions (Carvalho et al., 2009). They can also use combinations of tools in complex sequences to achieve a single goal (Boesch et al., 2009; Carvalho et al., 2008). For example, chimpanzees at the Loango National Park in Gabon use up to five distinct tools in order to extract honey (Boesch et al., 2009). Most strikingly, different populations of chimpanzees use different sets of tools, and in different ways, with greater variability between than within communities (Boesch et al., 2020; Whiten et al., 1999). These group differences suggest continuous cultural traditions, with technological knowledge passed on across generations.

Tool use can also be found among other primates—both great apes, such as orangutans (Galdikas, 1982) and gorillas (Breuer et al., 2005), and species of New World (Mannu & Ottoni, 2009) and Old World (Gumert & Malaivijitnond, 2013) monkeys. However, the use, and even manufacture, of tools is not unique to primates but more widespread in the animal kingdom than had been previously believed (Shumaker et al., 2011). For instance, one animal that has received substantial recent attention is the New Caledonian crow, which uses tools made from twigs or leaves to obtain food such as insects and spiders from holes in trees (Hunt, 1996; Kenward et al., 2005). This behavior has many characteristics interpreted as showing a high degree of sophistication. For example, these crows show sensitivity to multiple functional properties of potential tools (St. Clair & Rutz, 2013) and construct composite tools from multiple distinct parts (von Bayern et al., 2018). They shape hooks in different ways to achieve specific goals (Weir et al., 2002) and show advance planning for specific future uses (Boeckle et al., 2020). Some authors have even suggested that different populations of crows show differentiation of tool use indicative of cultural traditions and cumulative evolution over historical time (Hunt & Gray, 2003).

At the same time, the frequency and sophistication of tool use and manufacture in the animal kingdom should not be overstated. As McGrew (2013) notes, the large majority of primate species do not use tools at all, whether in the wild or in captivity. Even when tool use has been reported in a species, it is also important to consider how frequent and widespread the behavior is. Whiten et al. (2001) distinguished between patterns of tool use that are *customary* (i.e., occurring in most or all members of a group) and those that are *habitual* (i.e., occurring repeatedly in multiple individuals) or merely *present*. When, after decades of study, two isolated examples of tool-using behaviors by gorillas are at last observed (Breuer et al., 2005), it is notable, but it is also, to some extent, the exception that proves the rule. That Giotto once drew a perfect circle freehand (Vasari, 1568) should not lead us to conclude that this is a common occurrence, or representative of human manual ability. As Mithen (1996) notes, the very fact that some groups of termite-eating chimpanzees have not figured out that they could get these insects more easily using a stick isn't evidence of culture so much as "simply not being very good at thinking about making and using physical objects" (p. 84).

Paleolithic Tool Manufacture

While the use of tools may not be exclusive to humans, it has certainly reached a state of development in our species that differs in kind from that seen in other animals. However, it is also true that, historically, the emergence of the toolmaking and using tradition we have inherited began before our species evolved. There is clear evidence of widespread and customary tool use in most species within the genus *Homo*, including Neanderthals (Hoffecker, 2018), Denisovians (Brown et al., 2022), *Homo heidelbergensis* (Thieme, 1997), *Homo erectus* (Diez-Martín et al., 2015), and *Homo habilis* (Leakey et al., 1964). Moreover, there is increasing evidence that pre-*Homo* australopithecines also habitually used stone tools, including *Australopithecus afarensis* (McPherron et al., 2010), *Kenyanthropus platyops* (Harmand et al., 2015), and *Paranthropus* (Backwell & d'Errico, 2001; Plummer et al., 2023). The earliest known stone tools, from Lomekwi 3 in Kenya, date to as much as 3.3 million years ago (Harmand et al., 2015), before the emergence of the genus *Homo*.

In the first part of this chapter, I will review the evidence for the evolution of Paleolithic technologies. However interesting it may be, this extended discussion of archaeological evidence may seem out of place in a book on the psychology and neuroscience of the hand. The manufacture and use of stone tools, however, is highly demanding and requires rich cognitive and sensorimotor skills of the sort that I have discussed throughout this book. Understanding these tools and how they were made and used can thus provide rich insight into the evolution of the invisible hand. For example, the authors of the paper reporting the Lomekwi 3 tools interpret these artifacts as evidence that "their makers' hand motor control must have been substantial and thus that reorganization and/or expansion of several regions of the cerebral cortex (for example, somatosensory, visual, premotor and motor cortex), cerebellum, and of the spinal tract could have occurred before 3.3 Ma" (Harmand et al., 2015, p. 314).

The most widely used system for classifying stone tools was developed by the Cambridge archaeologist Sir Grahame Clark (1977), who distinguished five *modes* of lithic technology, distinguished by increasing complexity in their methods of production. Clark argued that these modes follow in a fixed sequence with progressive development across time, although populations at different locations in the world at any moment might be at different stages. These five modes are:

- Mode 1: The Oldowan industry, unifacial flaking of simple stone cores
- Mode 2: The Acheulean industry, bifacial flaking of symmetrical hand axes

- Mode 3: The Mousterian industry, including the Levallois technique
- Mode 4: The Aurignacian industry
- Mode 5: The Microlithic industries, including the Magdalenian culture

Mode 1 Stone Technology: The Oldowan Industry

Oldowan tools were first identified by Louis Leakey et al. (1931) in Bed I at Olduvai Gorge, Tanzania, and subsequently described in detail by Mary Leakey (1966, 1971). Oldowan tools are based on a single rock core from which flakes have been removed by striking the core with (or against) other rocks. As the final tool retains the overall shape of the initial core, these are sometimes called "pebble tools," as pebbles smoothed by flowing river water make ideal cores. By striking the core with a hammerstone, flakes can be removed from the core, producing characteristic patterns of what is called "conchoidal" fracture.

Oldowan tools have been found from as early as 2.6 million years ago at Gona (Semaw et al., 1997, 2003) and Ledi-Geraru (Braun et al., 2019), both in Ethiopia. Recently, stone tools dated to as early as 3.3 million years ago have been described at Lomekwi 3 in Kenya (Harmand et al., 2015), which have been attributed to a pre-Oldowan, "Lomekwian" culture (Lewis & Harmand, 2016). Oldowan tools have been found across huge stretches of the world, not only across the entirety of Africa from South Africa (Brain, 1970) to Algeria (Sahnouni et al., 2018) but also across the Eurasian landmass, from China (Huang et al., 1995) to Spain (Carbonell et al., 2008).

An important approach to understanding ancient hominin tool manufacture is the experimental production of such tools by archaeologists (Schick & Toth, 1993). Such studies have revealed four major techniques by which Oldowan tools were likely made:

1. Hard-hammer percussion: This is a simple method in which the core is held in the nondominant hand and struck with a hammerstone held in the dominant hand. By striking the core with a glancing blow at a sharp angle, flakes can be successively removed from the core. This appears to be the most frequently used method at the earliest Oldowan sites.
2. Bipolar technique: In this method, the core is held resting on a larger stone and struck with a hammerstone held in the dominant hand. This method can produce distinctive shapes resulting from forces affecting the core from both above and below. It is effective for making very small flakes and may be safer than hard-hammer percussion when working with cores that may fracture unpredictably.
3. Anvil technique: In this method, the core is struck against a larger stone anvil rather than with a hand-held hammerstone. It may be useful when working with a large core, which is difficult to hold in one hand.

4. Throwing: In this simple method, the core is thrown from a distance against a large stone anvil. While the results are unpredictable, this can produce useful fracture patterns and has obvious advantages in terms of safety.

What were Oldowan tools used for? A wide range of potential functions have been proposed (Schick & Toth, 1993). At the simplest level, the tools could have been used to crack open nuts, in much the same way as wild chimpanzees do (McGrew, 2010). They may also have been used for woodworking. Oldowan "choppers" can be used effectively to remove samplings or branches from trees, while "scrapers" can be used to form, smooth, and polish these sticks into sharp-tipped tools, which can serve as a spear or digging stick. Such digging implements, in turn, can open new sources of water, food such as tubers or burrowing animals, and raw materials such as stones or clay. Stone tools can also be used to hollow out pieces of wood to use as containers to transport and store water, food, or other items. They can be used to scrape flesh from animal hides, which, when dried or cured, can be used to make carrying containers, shelter such as tents or roofing for huts, and clothing. Microscopic analyses of patterns of wear on Oldowan tools have provided direct evidence that they were used for many of these functions (Keeley & Toth, 1981).

Two uses of Oldowan tools, however, have received the most attention: cutting and percussion. Evidence for the use of these tools for cutting comes from modern experimental studies showing that Oldowan-style flakes can be used effectively for the butchering of animal carcasses (Jones, 1980; Schick & Toth, 1993). A variety of African animals, typical of those the Oldowan makers likely encountered, have been successfully butchered using only these flakes, including zebras, wildebeests, sheep, goats, horses, pigs, and deer (note that these studies were conducted on animals that had died of natural causes). It is even possible to cut through the thick skin of an elephant in just a few minutes using only stone flakes (Schick & Toth, 1993; Starkovich et al., 2020). Studies of cut marks on bones found at tool sites have provided more direct evidence for the use of stone tools in butchery (Bunn, 1981; Diez-Martín et al., 2009; Potts & Shipman, 1981). For example, analysis of two-million-year-old animal bones found alongside Oldowan tools at Gona in Ethiopia have shown a range of marks and damage indicating that they had been butchered using stone tools (Domínguez-Rodrigo et al., 2005).

Another important use of Oldowan tools is for pounding and percussion. Obtaining nuts from within hard shells has already been mentioned. But modern experimental studies have shown that Oldowan choppers can smash marrow-containing bones from many African animals and smash skulls to obtain the brain (Schick & Toth, 1993). Direct evidence for the use of Oldowan tools to obtain bone marrow has come from analyses of damage to bones (Bunn, 1981), patterns of debris at tool-using sites (Mora & de la Torre, 2005), and analyses of wear patterns on tools themselves (Arroyo & de la Torre, 2016). In addition, pounding could be used to tenderize meat (Arroyo & de la Torre, 2016) or soften plants (de Beaune, 2004), both of which would have made foods easier to chew and eat—an important consideration in the absence of regular fire to cook food.

Mode 2 Stone Technology: The Acheulean Industry

The emergence of the Acheulean industry represents a major transition in human evolution and the historical development of technology (de la Torre, 2016; Gowlett, 2015; Shea, 2017). Named after the nineteenth-century discovery of stone tools in the village of St. Acheul in France, similar artifacts have been found at widespread locations across Africa and Eurasia. Its initial spread is usually linked with the expansion of *H. erectus* populations out of Africa and across Eurasia, but the technology is also linked with both Neanderthals and modern humans. Acheulean artifacts have been identified in sites more than 1.7 million years old in Kenya (Lepre et al., 2011), Ethiopia (Beyene et al., 2013), and Tanzania (Diez-Martín et al., 2015). In contrast, recent estimates suggest that the Acheulean industry ended at some places as recently as twenty-nine thousand years ago (Key et al., 2021). The use of Acheulean technology thus lasted for more than 1.5 million years, making it the longest-lasting cultural tradition ever practiced by humans. Indeed, some features of Acheulean technology continue to be used in the present day by people living in the remote highlands of New Guinea (Toth et al., 1992).

Whereas Oldowan tools were flaked on a single surface, Acheulean tools feature bifacial flaking, which allows the creation of sharper, longer, and more controlled cutting surfaces. While Acheulean tools come in a variety of forms, such as picks and cleavers, the outstanding example is the hand axe, as shown in figure 10.1. Hand axes are bifacially flaked tools, which often show a symmetrical teardrop shape. They are large tools, typically between 10 and 20 cm in length, fitting snugly into the palm of the hand, in contrast to Oldowan flakes, which would be held between two fingers. This allows dramatically more force to be applied. Figure 10.1 shows a range of hand axes from different layers of a single site at Konso in Ethiopia across almost one million years (Beyene et al., 2013). The refinement of the tools across time is clear, as is the basic continuity of the technology.

Insight into the functions of the hand axe comes from studies using manual pressure devices to measure the forces acting on the hand while participants engaged in stone tool use and manufacture. This research has shown very high biomechanical stresses on the hand when using an Oldowan flake (Rolian et al., 2011). Indeed, Oldowan-style activities such as marrow extraction using a hammerstone and hard-hammer

Figure 10.1
The development of Acheulean hand axes across time at Konso, Ethiopia. From left to right, two hand axes are shown from layers dated to 1.75, 1.6, 1.25, and 0.85 million years ago. The top and bottom rows show both sides of the same tools. Clear development is seen across time, with later tools being smaller, more precisely flaked, and more symmetrical and having smoother cutting edges. Adapted from Beyene et al. (2013).

flake production produced substantially larger forces on the hand than hand axe use (Williams-Hatala et al., 2018). Such considerations suggest that ergonomic improvements in hand axes over previous tools may have been of critical importance (Wynn & Gowlett, 2018), and which may have developed in parallel with derived features of the modern human hand (Marzke, 2013).

Experimental studies show that hand axes are highly efficient for the butchery of large animals, much more so than Oldowan flakes (Jones, 1980; Schick & Toth, 1993). Nevertheless, numerous other uses for Acheulean tools have been proposed, including digging to obtain plants, burrowing animals, or water; as stylized cores for obtaining further flakes; for woodworking, such as chopping, smoothing, and bark-stripping; and as projectile weapons. Studies of patterns of wear have provided direct evidence for many of these activities. For example, a study of early Acheulean artifacts from Olduvai Gorge showed use for scraping tuberous plants and for sawing and scraping of wood (Bello-Alonso et al., 2021). Similarly, analysis of tools from the late Acheulean site of Revadim in Israel shows that they were used for cutting, scraping, flaking, and marrow extraction (Venditti et al., 2021).

Mode 3 Stone Technology: The Middle Paleolithic and the Mousterian Industry

Collectively, Oldowan and Acheulean tools are referred to as Lower Paleolithic technology. The advance to the Middle Paleolithic (or, in Africa, Middle Stone Age) represents a dramatic development in technology. Middle Paleolithic technologies are sometimes referred to as the "Mousterian industry," after the site of Le Moustier in Southwest France, where such tools were found alongside a complete skeleton of a Neanderthal youth in the 1860s. In Europe, Middle Paleolithic technology is linked to the emergence of Neanderthals, while in Africa, it is linked to the emergence of anatomically modern humans (Foley & Lahr, 1997; Klein, 2000). But this technology has been described across Africa and Eurasia, from South Africa (Schmidt et al., 2022) to Armenia (Adler et al., 2014), from China (Hu et al., 2019) to the Netherlands (Niekus et al., 2019).

A central feature of Middle Paleolithic tools is their use of prepared core methods (the Levallois technique). Rather than using a hammerstone to strike a flake directly off of a core, a more sophisticated approach is used. A striking platform is first created on one surface of the core, and the edges of the intended flake are trimmed. This produces a domed "tortoise core" shape on the surface of the core. When the flake is finally struck from the core, the entire prepared contour has an edge much sharper than in Lower Paleolithic methods. Levallois methods thus represent a substantial increase in the level of planning and forethought involved in tool manufacture.

Another key feature of Middle Paleolithic tools is their integration into composite, hafted tools (Ambrose, 2001). A Levallois-style flake could be attached to a wooden shaft to make a spear, arrow, or axe. This produced dramatic improvements in the tool's range and the force with which it could be used. As will be discussed below, hafted tools are commonly thought to be a key indicator of complex, "modern" cognitive abilities due to the requirement of combining multiple elements in a hierarchical way to produce a single artifact.

Several lines of evidence provide strong indications that at least some Middle Paleolithic tools were hafted. First, many points show characteristic changes in shape, such as progressive thinning toward the base, which suggest that they were shaped to be inserted into slots in a handle. Evidence from the Kathu Pan site in South Africa suggests that hafted tools were being used by Middle Stone Age residents as much as five hundred thousand years ago (Wilkins et al., 2012). Second, experimental studies show that modern points made in similar shapes can be used effectively as hafted spears or arrows (Lombard & Pargeter, 2008; Shea et al., 2001). Third, microwear analyses of tools show patterns of fractures on points, which are characteristic of the stresses imposed on hafted spear points (Lombard & Pargeter, 2008; Shea, 1988). Finally, chemical analyses of residue found on tools show evidence of the use of adhesives to connect sharpened flakes to wooden shafts to make spears or arrows. There is evidence for the use of tar made from tree bark (Mazza et al., 2006; Niekus et al., 2019), bitumen from petroleum (Boëda et al., 1996; Hauck et al., 2013), and distilled leaves (Schmidt et al., 2022) as adhesives used for hafting of spear- and arrowheads. Even more striking is the use of compound adhesives made from combinations of raw ingredients mixed in complex ways to produce properties absent in any of the original materials alone. The toolmakers at Sibudu Cave in South Africa seventy thousand years ago hafted tools using a mixture of plant gum and red ochre (Wadley et al., 2009). Modern experimental replications show that while the plant gum on its own works as a hafting adhesive, the addition of red ochre makes the spear less brittle and also prevents the adhesive from dissolving when wet.

Modes 4 and 5 Stone Technology: The Aurignacian and Magdalenian Industries
The final developments in stone tool technology that I will discuss involve the transition to the Upper Paleolithic (or, in Africa, the Later Stone Age). Mode 4 tools are linked to further refinements of Levallois methods, allowing the production of long thin blades, with long, continuous, and sharp edges. Such blades allow much more efficient reduction of cores and were used for a wide range of activities. They could be hafted to create spears and knives and used as scrapers for working hides or wood and as burins for engraving. Mode 4 technology is often referred to as the "Aurignacian industry," after the Cave of Aurignac in France, where stone artifacts were discovered in the 1860s. Mode 5 tools consist of microliths—small pieces (1–2 cm in length) that can be used to make extremely sharp tips for hafted spears and knives. Mode 5 technology is often called "Magdalenian," after the La Madeleine rock shelter in France where such tools were discovered in the nineteenth century.

Upper Paleolithic technology is generally associated with the emergence of fully modern humans. In Africa, the earliest sites showing the transition from Middle to Later Stone Age tools have been dated to around sixty thousand years ago (Ranhorn & Tryon, 2018). In Europe, the appearance of such tools is linked to the displacement of Neanderthal populations by anatomically modern humans starting around forty-six thousand years ago (Hublin et al., 2020).

The appearance of Upper Paleolithic technologies is also linked to a variety of other behavior patterns, which have been described as a "revolution" (Bar-Yosef, 2002) and as indicative of behavioral modernity (Klein, 2000). These include the use of bone and antler as raw materials to create tools and ritual artifacts; long-distance transport and exchange of raw materials over distances of hundreds of kilometers; complex hunting tools such as spear throwers and bows; representational art, including carved figurines and cave paintings and engravings; use of storage facilities such as underground "freezers," allowing long-term preservation of food; organization of habitations into distinct locations for activities such as sleeping, butchering, and cooking; and burial of the dead.

The Manual Requirements of Paleolithic Tools

The opposable human thumb and its ability to produce perfect pad-to-pad precision grips is commonly lauded as the pinnacle of manual dexterity. But this two-fingered precision grip is not especially helpful in most cases of tool use. Most tools are too big or heavy to be held this way and could not be wielded with sufficient force for most tasks. As discussed in chapter 1, in an elegant series of studies, Mary Marzke and her colleagues showed that the grips most important for making and using tools blur the distinction between precision and power grips. These grips allow us to use objects *both* precisely and forcefully and underlie the extraordinary power and flexibility of human tool use.

Marzke and Shackley (1986) investigated the grips used in the experimental production by an archaeologist of tools modeled on those found at Olduvai Gorge and Koobi Fora using stones common at those places. Figure 10.2 compares a standard pulp-to-pulp precision grip (top left) to three other grips, which blur the distinction between precision and power grips. A cradle grip (top right) can be used to hold a stone core securely in the palm of the nondominant hand while using the fingers to rotate it. This allows the core to be easily repositioned between strikes from the hammerstone held in the dominant hand. This can be quite a strong grip, since the object is held securely by all five fingers and by the palm of the hand. At the same time, the fingers can spin and flip the object quickly and precisely.

The hammerstone itself can be held using the thumb, index, and middle fingers as a tripod—the three-jaw chuck grip (figure 10.2, bottom left). This variant of the precision grip also increases the strength and force that can be applied. It is also the most natural grip for overhand throwing of, for example, a baseball. Finally, objects can be cut with stone flakes using the two-jaw pad-to-side grip (bottom right), in which the pad of the thumb holds an object tightly against the side of the index finger. This is the grip we use to unlock a door with a key, where we need to precisely insert the key into the keyhole but also have sufficient force to turn the mechanism in the lock.

Marzke and her colleagues (Marzke, 1997; Marzke & Shackley, 1986; Marzke & Wullstein 1996) have thus argued that a relatively small number of grips are critical to the manufacture and use of stone tools. Notably, none of these grips involves the precise pad-to-pad precision grip. A recent study by Key et al. (2018) examined the diversity of hand grips used in a much larger sample of individuals ($N = 123$) using Lower Paleolithic stone tools to cut a range of objects. As many as twenty-nine distinct grip types were observed: twenty-six forms of precision grip and three types of power grip. However, Key

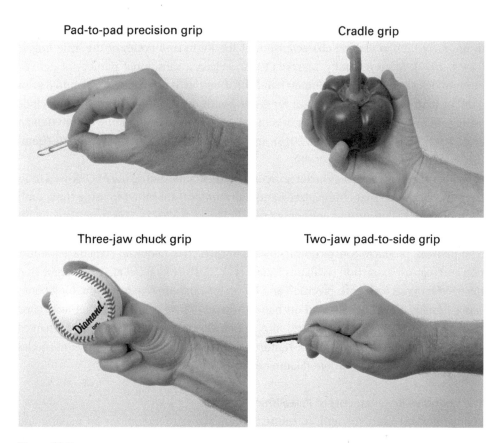

Figure 10.2
Comparison of the pad-to-pad precision grip (top left) with grips described by Marzke and colleagues in the experimental manufacture and use of Oldowan tools.

and colleagues showed that these fell into five fundamental grip types, which again did not include the pad-to-pad precision grip.

This pattern has been confirmed by subsequent studies using EMG to measure muscle involvement (Marzke et al., 1998) and pressure sensors to measure forces on the hand (Key et al., 2019). These studies have revealed intriguing links between the grips and forces involved in early stone tool manufacture and derived features of the human hand. For example, the little finger has been found to be unexpectedly important in tool production. When a core is held using a cradle grip, the little finger is frequently used to rotate the core into the desired position for an upcoming hammer strike and also to stabilize the core during impact (Marzke et al., 1998; Marzke & Shackley, 1986). Indeed, the pressures recorded on the little finger were frequently as large as those on the thumb and index finger (Key et al., 2019). This importance of the little finger appears to correspond to uniquely human derived characteristics of the joints and bones of the little fingers (Lewis, 1977). More generally, Marzke (2013) reviews a substantial number of derived morphological features of the human hand consistent with producing the required grips used for hard hammer percussion or for dealing with the resulting forces. These include features of bones, such as the robustness of the pollical metacarpal of the thumb (Rolian et al., 2011; Williams-Hatala et al., 2018), and the independence of the FPL muscle (Hamrick et al., 1998; Marzke et al., 1998).

Studies that have assessed the forces acting on the hands during use of Oldowan tools have also suggested that chimpanzee hand morphology is ill-suited to using these tools with full force in cutting and pounding tasks (Rolian et al., 2011). This is consistent with research showing that stone percussion to crack open nuts, which is well documented in chimpanzees, produces conspicuously lower pressures than Oldowan-style flake production and marrow extraction (Williams-Hatala et al., 2018). Marzke et al. (1998) argue that the need for grips with both precision and power for extracting marrow from long bones was a primary cause for selection on hand structure. There is evidence that associated changes in bone (Susman, 1994) and muscle (Karakostis et al., 2021) structure emerged about two million years ago in some hominin lineages and possibly a bit earlier in species such as *Australopithecus africanus* (Skinner et al., 2015).

The Cognitive Requirements of Paleolithic Tools

Manual dexterity and strength, of course, are not the only limitations on making and using tools. Often, the hardest part is simply understanding that an object can be used in a particular way. This can be clearly seen in Wolfgang Köhler's (1925) famous studies of insight learning in chimpanzees. Köhler confronted the chimpanzees with problems in which objects needed to be used in unfamiliar ways to obtain food rewards. For example, two short sticks might need to be connected to pull in a distant reward, or boxes might need to be stacked to reach a reward on a high shelf. Köhler found that chimpanzees were often unsuccessful in their initial attempts, in which they approached the task in an ineffective way. After a pause, the animals would often appear to have an aha moment before coming back to use the tools effectively to obtain the food. The initial failure, of course, had nothing to do with limited hand function but rather resulted from the cognitive conceptualization of the problem.

In terms of manufacturing tools, Lewis and Harmand (2016) argue that there are three main cognitive requirements for the knapping of stone tools. First, the knapper must have an understanding of the fracture mechanics of the raw stone materials being used. Second, it requires precise sensorimotor control over force and accuracy of the percussive hand movements used. Finally, it requires a visuospatial understanding of locations and angles at which to strike the core and detach the flakes.

Wynn and McGrew (1989; Wynn et al., 2011) argue that the manufacture and use of Oldowan tools is broadly compatible with tool behaviors found in chimpanzees. They suggest that there are only two aspects of Oldowan behavior distinct from that seen in other primates: (1) the transport of stone across long distances (e.g., several kilometers) from its geological source, and (2) competition with other carnivores for animal carcasses. The link between Oldowan-style tools and primate tool use is supported by work that has taken an archaeological approach to describing 4,300-year-old sites of chimpanzee nut-cracking activities in the Ivory Coast (Mercader et al., 2002, 2007). The stone tools and debris at this site were interpreted as being highly similar to Oldowan sites and as evidence for a "Chimpanzee Stone Age" (Mercader et al., 2007). Research on living chimpanzees showing complex sequences of tool use has also been interpreted as similar to the Oldowan methods (Carvalho et al., 2008). One recent study has reported that stone flakes, remarkably similar to those found at Oldowan sites, are created as a byproduct of stone pounding by bearded capuchin monkeys in Brazil (Proffitt et al., 2016). These monkeys lick the cracked stones (possibly as a source of silicon). Critically, however, the monkeys ignore the flakes completely, indicating that their production is non-intentional.

It is important to note, however, that detailed studies of the creation and use of Oldowan artifacts are not consistent with these tools being accidental byproducts of other activities. On the contrary, this research indicates that the individuals who made these tools had good abilities in terms of their manual dexterity, hand-eye coordination, ability to select appropriate raw materials, understanding of fracture dynamics, and ability to plan several steps ahead (de la Torre, 2004; Roche et al., 1999). For example, Roche and colleagues (Roche et al., 1999; Delagnes & Roche, 2005) conducted a detailed analysis of the more than three thousand Oldowan artifacts recovered from the 2.34-million-year-old Lokalalei 2C site in Kenya. They were able to refit many of the individual flakes to the cores from which they had been struck, in some cases piecing together as many as thirty-nine separate elements. This allowed them to reconstruct the process by which the knappers had struck successive flakes from each core. This analysis revealed a sophisticated and precise strategy, which allowed a long series of flakes to be obtained from a single pebble core. Such abilities are only seen in modern stone knappers following extensive experience and the development of a high level of expertise (Nonaka et al., 2010). Similarly, comparisons of the specific rocks used by knappers at the 2.5- to 2.6-million-year-old site at Gona in Ethiopia with their geological abundance in the area show that even the earliest Oldowan toolmakers were highly selective for the most appropriate stones for tool manufacture (Stout et al., 2005).

Some researchers have tried to teach great apes to make Oldowan-style tools. Wright (1972) showed that an orangutan could be trained to produce flakes to cut a cord to access the food inside. A more extensive approach to training was used with Kanzi, an enculturated bonobo, who learned to make stone tools over a period of several years (Roffman et al., 2012; Schick et al., 1999; Toth et al., 1993). Kanzi was able to successfully produce stone flakes and even spontaneously developed his own method that he had not been trained on, in which he would throw a stone against a hard tile floor or another rock to produce cracks. Nevertheless, Kanzi's toolmaking did not show the skill of even the earliest Oldowan toolmakers. Toth et al. (2006) made a systematic comparison of tools produced by Kanzi and his half-sister Pan-Banisha, those produced by modern human knappers, and the 2.6-million-year-old tools found at Gona, Ethiopia (Semaw et al., 1997, 2003). Across many measures of skill, the bonobos performed worse than both modern humans and the Gona toolmakers. For example, the Gona cores have, on average, more than twice as many flakes successfully removed as those of the bonobos. Similarly, the cores of Kanzi and Pan-Banisha show substantially more battering from unsuccessful hammer strikes. Toth and colleagues suggest that the performance of the bonobos is limited by both biomechanical constraints related to lower impact velocity with the hammer stone and cognitive constraints in terms of understanding where to place blows and getting the hammer stone there precisely.

It is also important to keep in mind that if Kanzi and Pan-Banisha make stone tools, it is only because they have been taught to do so by humans. The physical movements involved in using one stone to shape a tool from another stone may be quite similar to those involved in using a stone to crack open a nut. But there is, nevertheless, a "cognitive leap" (de Beaune, 2004) in the recursive application of stone percussion to another stone rather than to an immediate object of desire such as a nut. Whether or not some chimpanzees are capable of making this leap, its significance should not be understated. For this reason, Rogers and Semaw (2009) refer to the appearance of the earliest stone tools as going "from nothing to something."

However, it is not only going from nothing to something that involves a leap. The progressions between successive modes of Paleolithic toolmaking reflect equally significant advances. For example, Mode 2 (Acheulean) tools represent a dramatic advance over Oldowan tools (de la Torre, 2016). In contrast to the "pebble" Oldowan tools, in which the final tool retains the basic overall shape of the raw stone chosen, the hand axe was created from a much larger piece of stone. Thus, rather than opportunistically taking advantage of the found shape of stones, the hand-axe makers show evidence of using a preconceived mental template of what they wanted to produce. Similarly, the transition to Mode 3 (Mousterian) tools again shows substantial cognitive advances. The use of "prepared core" methods such as the Levallois technique require extended periods of preparatory work on the stone core, which is not transparently related to the final flakes eventually produced. The emergence of hafted tools in Mode 3 is also evidence of important cognitive advances. Whereas Oldowan and Acheulean tools are made from a single rock core, hafted tools involve the combination of at least three distinct components: the stone head, a (usually wooden) shaft, and materials for binding the head to the shaft. As mentioned above, in some cases, the materials used for binding hafted tools were themselves composites of multiple materials, requiring complex initial processing (Wadley et al., 2009). Ambrose (2001) argues that the hierarchical nature of hafted tools in which multiple components are integrated to produce a single composite tool are analogous to the hierarchical organization of grammar in spoken language, suggesting that both abilities coevolved as a result of changes in Broca's area in the PMC.

As mentioned in the previous chapter, Osiurak et al. (2020, 2021) have argued that there is a distinct cognitive domain of technical reasoning, which they label "technition" and relate to the left inferior parietal lobe on the basis of impaired tool use in stroke patients with apraxia. On this view, the emergence of successive stages of Paleolithic technology may have been linked to changes in specifically technical reasoning abilities.

Another approach has used neuroimaging methods to investigate neural networks recruited during production of stone tools. This literature has suffered from very small sample sizes, and the results are not easy to interpret. But in general, the results suggest that Oldowan-style toolmaking recruits bilateral regions of the PPC and vPMC (Putt et al., 2017; Stout et al., 2008, 2015; Stout & Chaminade, 2007)—areas involved in basic sensorimotor control, including the reaching and grasping networks. Acheulean toolmaking recruits these same regions, plus additional regions, such as the inferior frontal gyrus (Putt et al., 2017; Stout et al., 2008), the temporal pole (Putt et al., 2017), and the dorsal prefrontal cortex (Stout et al., 2015). The activations in the inferior frontal gyrus are particularly notable, as they are broadly consistent with the location of Broca's area (and its right-hemisphere homologue), suggesting a potential link between stone toolmaking and language—a link that will be discussed in more detail in chapter 15. Another study used structural MRI methods, including diffusion tensor imaging (DTI) and VBM, to investigate structural changes in the brain linked to the acquisition of skills in making stone tools (Oldowan, Acheulean, and Levallois). Participants were scanned both before and after learning how to make tools, allowing structural changes to the physical structure of the brain to be measured. Increases in both gray matter and white matter were found in parieto-frontal regions involved in hand control, including the inferior parietal lobe and inferior frontal gyrus (Hecht et al., 2015).

A common theme across the last three paragraphs has been the link between the emergence of stone tools and regions of the PPC and PMC, which are familiar throughout this book as being central to skilled control of the hand. This should remind us that beyond high-level cognitive skills, making stone tools is an extraordinarily complex sensorimotor task, requiring precise coordination of vision, touch, proprioception, and motor control of both hands. One recent study of the experimental manufacture of late-Acheulean hand axes (Pargeter et al., 2020) found that the development of expertise was more closely linked to changes in sensorimotor function than to a conceptual understanding of flake dynamics. Such results suggest that the most critical aspect of tool manufacture and use may be the ability to integrate external objects efficiently into the sensorimotor processes involved in controlling the hand—that is, with the invisible hand.

Integration of Tools with the Invisible Hand

We now turn from thinking about the historical emergence of tool technology to research in psychology and neuroscience investigating the neural mechanisms involved in using tools. As we will see, tools appear to be tightly integrated with many of the domains of hand function we have discussed in previous chapters, including peripersonal space, somatosensation, motor control, and higher-level mental body representations.

Brain Regions Involved in Tool Use

One set of studies has investigated the effects of training monkeys to use tools on brain activation and structure. One study using PET compared brain activation in macaque monkeys when they used a tool to recover food pellets compared to when they merely manipulated the tool (Obayashi et al., 2001). Active tool use produced stronger activations in several brain areas, including areas of the inferior parietal cortex and PMC. Another study used structural MRI to quantify gray-matter density longitudinally in macaques as they were trained to use a rake to retrieve food items (Quallo et al., 2009), finding increased gray-matter density in several areas, notably including the inferior parietal cortex and S2. A follow-up study by the same group showed that tool use involved activation of neurons in M1 with direct CM projections (Quallo et al., 2012). Thus, in monkeys, tool use appears linked to many of the same underlying processes involved in other hand functions.

In humans, there is also evidence for a left-lateralized tool-related network. Neuroimaging studies in which participants are shown images of tools, compared to other graspable or non-graspable objects, have consistently found activations in a left-lateralized network of areas, including the IPS, vPMC, and regions of the occipitotemporal cortex (Bracci et al., 2012; Chao & Martin, 2000; Valyear et al., 2007). A meta-analysis by Lewis (2006) found broadly comparable activations of this network when participants saw tools, actively used them, planned to use them, pantomimed their use, or even heard sounds associated with their use.

These regions have clear links with the grasping and reaching networks described in chapter 3. There is also evidence, however, that there may be regions more specifically linked with tool use. Orban and colleagues (Orban & Caruana, 2014; Peeters et al., 2009, 2013) identified a putatively tool-specific region in the left inferior parietal lobe (the anterior supramarginal gyrus; aSMG). Peeters et al. (2009), for example, recorded fMRI responses while humans or macaque monkeys watched videos of actions performed using either simple tools or hands. For both humans and monkeys, observation of both hand actions and tool actions produced activations in the grasping network, including the anterior IPS and the vPMC. Critically, however, in humans, there was an additional activation in the inferior parietal lobe, which was selective for actions with a tool. No such activation was found for monkeys, even those that had been trained to use tools themselves.

Kastner and colleagues (Kastner et al., 2017; Mruczek et al., 2013) have also argued that there is a tool-specific region in the left parietal cortex, although they place it in the anterior IPS rather than in the supramarginal gyrus. Mruczek et al. (2013) showed human participants images of tools, animals, or graspable (non-tool) objects while in the MRI scanner. They found a region of the anterior IPS that showed selective responses to tools but not to other graspable objects. This region was distinct from the parietal grasp region, which they identified using an active-grasping localizer scan. Moreover, there was a gradient of responsiveness across the entire IPS, with more posterior regions showing selective responses for any graspable object (tool or not) compared to animals and with more anterior regions becoming progressively more tool selective.

Some other studies that have compared images of tools to graspable objects have reported similar left parietal tool-specific activations (Buxbaum et al., 2006; Valyear et al., 2007). It is not entirely clear what the relation is between these putatively tool-specific regions. The contrasts used to identify these regions are quite different, involving seeing video stimuli of actions in some studies (Peeters et al., 2009, 2013) and still images of objects in others (Buxbaum et al., 2006; Mruczek et al., 2013; Valyear et al., 2007). Notably, however, these studies agree that there is a left-lateralized tool-specific region of the anterior PPC distinct from grasp-specific regions such as AIP. Figure 10.3 shows the peak coordinates of tool-specific regions from each of these studies, showing that they are clearly not merely different descriptions of the same region. The two studies by Peeters

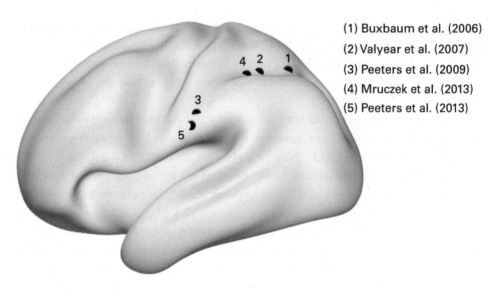

Figure 10.3
Peak MNI coordinates of tool-specific activations in neuroimaging studies involving visual perception of tools.

and colleagues involved video images that activated a region more anterior and inferior to that seen in the other studies that involved static images.

Another study applied a similar logic but used active reaches by participants rather than just showing them images of tools or tool actions (Gallivan, McLean, Valyear, et al., 2013). Participants made either grasping (i.e., picking up) or reaching (i.e., touching) movements toward objects using either their hand or a pair of tongs. Gallivan and colleagues used MVPA to identify brain areas in which they could decode from the pattern of brain activity whether the action had involved grasping or reaching. Of particular interest was whether such decoding was based on effector-specific or effector-independent representations. To test this, Gallivan and colleagues tested whether classifiers trained on data from one effector (e.g., the hand) would correctly classify reaches performed by the other effector (e.g., the tongs). Tool-specific representations, which did not support cross-decoding with hands, were found in both the supramarginal gyrus and anterior IPS, consistent with the results described above for visual stimuli. There was also evidence for effector-independent representations that generalized across hands and tools in more posterior regions of the IPS and PMC.

One recent study directly investigated the relation between brain regions involved in visual perception of tools and hands and actual tool use. Knights et al. (2021) first used visual localizers to identify regions in the occipitotemporal and intraparietal cortex which showed selective responses to hands or to tools. They then asked participants to grasp 3-D printed tools using either typical grasps (e.g., grasping a knife by the handle) or atypical grasps (e.g., grasping a knife by the blade). Using MVPA, they asked which regions contained information differentiating whether tools were grasped in typical or atypical ways. Decoding was above chance in regions that showed hand-selective visual responses both in occipitotemporal cortex and the IPS, but surprisingly not in regions that were tool-selective.

Studies using TMS to disrupt neural processing in the inferior parietal lobule have provided complementary evidence. Pelgrims et al. (2011) showed participants images of pairs of objects or of an object and a hand posture. They were asked to make judgments either about manipulation (e.g., "Is the same hand posture normally adopted to use these two objects?") or about conceptual knowledge (e.g., "Are the two objects used together to achieve a common goal?"). Repetitive TMS applied to the left supramarginal gyrus impaired performance on manipulation judgments but not on conceptual judgments. Similarly, Andres et al. (2017) found that rTMS to the left supramarginal gyrus impaired judgments of where fingers would be placed on an object to grasp it but not judgments about the spatial location of features.

As discussed in the previous chapter, lesions of the left parietal lobe have been shown to produce deficits in a variety of tool-related judgments, including appropriate use of common tools (De Renzi et al., 1982), pantomiming the use of a shown object (Barbieri & De Renzi, 1988; Goldenberg et al., 2003), matching objects based on common function (De Renzi et al., 1969), matching objects to demonstrated pantomimes of use (Buxbaum et al., 2005; Vaina et al., 1995), and inferring functions of novel objects (Goldenberg & Hagmann, 1998; Heilman et al., 1997). Osiurak et al. (2020, 2021) have linked this to a specific region in the left inferior parietal cortex, which they claim is specific for technical reasoning about objects.

Integration of Tools with Peripersonal Space

As discussed in chapter 7, there is an integrated multisensory representation of the peripersonal space surrounding the body. Recall that Iriki et al. (1996) described neurons in the posterior bank of the postcentral gyrus of macaque monkeys that had both tactile RFs and visual RFs, including either the space immediately adjacent to the tactile RF (distal type neurons) or the space reachable by the monkey's arm (proximal type neurons). More strikingly, Iriki and colleagues trained these monkeys to use a long rake to retrieve food from distant locations. After this training, they measured the visual RFs of neurons immediately before and five minutes after continuously using the rake. Remarkably, after tool use, these RFs had extended to reflect the new spatial limits of effective actions. For

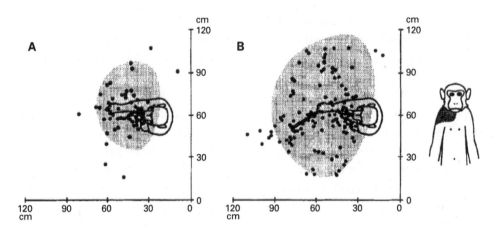

Figure 10.4
The visual RF of a proximal-type neuron (A) immediately before and (B) after a period of tool use. The image on the right shows the tactile RF of the neuron on the monkey's right shoulder. From Iriki et al. (1996).

distal-type neurons, visual RFs expanded to include the space immediately surrounding the rake, while for proximal-type neurons, visual RFs expanded to include the space reachable by the tool (figure 10.4).

One line of evidence concerning peripersonal space comes from studies of cross-modal extinction, in which perception of touch on one hand is blocked when a visual stimulus is presented close to the contralateral hand (di Pellegrino et al., 1997; Farnè et al., 2000). Similar cross-modal extinction can be produced when a visual stimulus appears near to the tip of a long rake that the patient holds in the contralateral hand but which is not near the hand itself (Farnè & Làdavas, 2000; Maravita et al., 2001). Farnè, Bonifazi, and Làdavas (2005) showed that active tool use is necessary to produce this effect, as no extinction is found when the patient holds the tool without actively wielding it. Similarly, pointing toward the location where the visual stimuli was located did not produce extinction, showing that the effect of tool use is not only due to the fact of making directional motor responses toward distal locations.

In another study, Farnè, Iriki, and Làdavas (2005) showed that the extension of peri-hand space appears to be systematically related to the length of the tool. A visual stimulus was presented 60 cm from the hand while participants held a rake that was either 60 or 30 cm long. Larger cross-modal extinction was found with the longer tool. These authors also dissociated the physical length of the tool from its functional length by attaching the rake head halfway along the length of a 60 cm handle. This tool produced extinction similar to the 30 cm rake, indicating that it is the functionally effective length

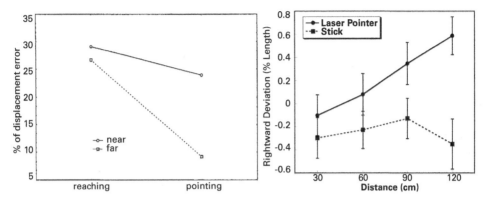

Figure 10.5
Extension of peripersonal space through tool use. *Left panel*: Line bisection results from patient P.P. (Berti & Frassinetti, 2000) showing rightward displacement error as a percentage of line length at near (50 cm) and far (100 cm) distances when using a laser pointer ("pointing") or a stick ("reaching"). *Right panel*: Line bisection in healthy participants using a laser pointer or stick. From Longo and Lourenco (2006).

of the tool that matters. These results are consistent with the interpretation that the peri-hand space has plastically extended to include the entire functional length of the tool. But they are also consistent with the interpretation that peri-hand space has simply shifted or translated to the functionally relevant extremity of the tool. To distinguish between these possibilities, Farnè et al. (2007) applied visual stimuli both at the extremity of a wielded tool and at an intermediate location along the shaft of the tool with no special functional relevance. Clear extinction was elicited by visual stimuli presented at both locations, providing evidence for a continuous extension of the size of peri-hand space.

Tool use can also remap hemispatial neglect. Berti and Frassinetti (2000) described patient P.P., who showed neglect of the left side of space following a stroke that damaged extensive regions of her right hemisphere. When asked to bisect lines using a laser pointer in both peripersonal space (50 cm) and far space (100 cm), she showed the rightward bias characteristic of neglect only in peripersonal space, as shown in the left panel of figure 10.5. In striking contrast, when P.P. was asked to bisect the distant lines using a long stick, this rightward bias was clearly apparent at the farther distance. The use of a tool to respond thus projected her neglect into more distant regions of space.

Similar results have also been found when measuring the more modest attentional biases that healthy participants show in line bisection tests. On paper-and-pencil line bisection tasks, healthy adults show a small leftward bias known as "pseudoneglect." When stimuli are presented at progressively farther distances, this small leftward bias gradually shifts into a somewhat larger rightward bias. Longo and Lourenco (2006) compared line bisection performance at four distances when participants responded using either a laser pointer attached to a tripod or a stick. As shown in the right panel of figure 10.5, when participants responded with the laser pointer, there was a gradual rightward shift with increasing distance. In striking contrast, however, when participants used a stick to bisect the lines, pseudoneglect was consistently found at all distances.

Integration of Tools with Somatosensory Hand Representations
We can also use tools as active sensing devices. This is seen most obviously in the use of canes by blind people, by which they are able to navigate and avoid obstacles. But this ability is not unique to the blind, and it is surprisingly easy to perceive quite detailed features of the world around us via tools, such as using a wooden spoon to judge whether a sauce is sticking to the bottom of a pan (Katz, 1925). Take a moment to try this yourself. Put some objects on the table in front of you, close your eyes, and use a pencil or a spoon to explore the objects. You will feel the contours of the objects, how easily the probe slips across their surfaces, the vibrations as you draw the probe across bumps, and the size of the objects. The experience of objects explored in this way may lack some of the vividness and richness of haptic exploration with the hand itself, as described in chapter 5, but it is not dramatically different in nature. Indeed, experimental studies have provided evidence that people are easily able to perceive a range of stimulus features by exploration with a probe, including texture (Klatzky et al., 2003; Klatzky & Lederman, 2002), size (Barac-Cikoja & Turvey, 1993), shape (Vaught et al., 1968), distance (Chan & Turvey, 1991), vibration (Brisben et al., 1999), and softness (LaMotte, 2000) as well as the length of the probe itself (Carello et al., 1992; Turvey et al., 1998).

A particularly interesting feature of this type of perception is that there are obviously no mechanoreceptors in the probe itself. The sensory processes are thus based on activation of the same populations of receptors in the skin that are used for exploration with the hand. Despite this, people generally experience the sensations subjectively as located at the tip of the probe. Lotze (1885) famously described this projection of tactile sensations to the tips of tools: "We fancy that we feel the contact of the rod with the object at a distance from us as directly through sense as we do its contact with the surface of our hand" (p. 588). Gibson (1966), similarly, wrote that "when a man touches something with a stick he feels it at the end of the stick, not in the hand" (pp. 100–101). William James (1890) made the same point even more memorably: "The draughtsman's immediate perception seems to be of the point of his pencil, the surgeon's of the end of his knife, the duellist's of the tip of his rapier as it plunges through his enemy's skin" (pp. 37–38). Yamamoto and Kitazawa (2001b) investigated this phenomenon using the crossed-hands deficit they discovered, which I discussed in chapter 2. They showed that a similar deficit in determining which of two touches came first occurred when touches were applied at the tips of crossed tools, even when the hands themselves remained uncrossed. Remarkably, when the hands were crossed and then the tools were crossed again, performance returned nearly to the level seen in the uncrossed condition.

Another approach to investigating sensory integration of the tool is to investigate localization of tactile stimuli on the tool. Miller et al. (2018) applied taps to seven locations along the length of a long rod and asked participants to indicate the perceived location of the touch by positioning a cursor onto the corresponding location on a picture of the tool shown on a screen. Localization performance was quantified by comparing the relation between actual and judged location using linear regression. Perfect performance would produce a regression slope of exactly one—that is, a 1 cm distal shift of the actual location of touch should produce a 1 cm shift of judged location. In contrast, if people are unable to do the task at all, there should be no relation between actual and judged location, and the regression slope should be zero. When touches were passively applied to the rod, Miller and colleagues found a mean regression slope of 0.57, showing quite precise and systematic localization of touch along the rod. Strikingly, when touch was applied by the participants actively lowering the rod, performance was even better, with a mean regression slope of 0.93. Notably, participants did not just perceive touch as being located at the *tip* of the tool but were able to perceive location in a graded and continuous way. These results show that the tool functions as a metrically precise sensory extension of the hand.

Miller et al. (2018) suggest that the ability to localize touches on the rod depends on the detection of vibrations. By recording mechanical vibrations from the handle of the rod and from the participant's index finger, they showed that the pattern of vibration frequencies and amplitudes differed systematically, depending on where touch is applied. They then used the TouchSim computational model, which simulates the patterns of activations of the population of receptors in the glabrous skin of the hand to stimulus inputs (Saal et al., 2017). Miller and colleagues found that the responses of populations of Pacinian mechanoreceptors to these vibrations contained sufficient information to decode the location of touch on the rod.

In a subsequent study, Miller et al. (2019) investigated the brain processes involved in localization of touch on a tool. They recorded EEG and used a repetition-suppression paradigm in which two successive touches were applied at either the same or at different locations on a tool. Repetition suppression was identified in a contralateral cluster of electrodes over sensorimotor cortices between around 50 and 110 ms following stimulation. Source localization of this suppression effect showed that it first involved processing within the hand region of S1 and M1 and subsequently spread to the PPC. Miller and colleagues then compared these responses to those produced during localization of touch on the actual forearm, finding very similar patterns. They then compared the similarity of processing for localization on the tool and arm using a decoding approach. There was clear evidence for cross-surface decoding, such that a classifier trained to discriminate trials in which location matched or mismatched for the forearm could successfully classify trials on the tool, and vice versa. These results provide clear evidence that the same somatosensory processes involved in tactile localization on the body itself are used for localization on a wielded tool.

Integration of Tools with Hand Motor Control Programs
Umiltà et al. (2008) recorded from individual neurons in area F5 of vPMC in macaque monkeys that had learned to use pliers to pick up objects. They first identified neurons that fired when the monkey performed hand grasping movements without the pliers. These same neurons also responded when the monkey picked up objects using the pliers, suggesting that the same motor programs involved in controlling hand grasping also controlled analogous actions performed using a tool. Interestingly, Umiltà and colleagues trained the monkeys to use two types of pliers. With the "normal" pliers, the tips of the pliers closed when the monkey squeezed the handle, showing a match between the proximal action produced and the distal effect at the tips of the tool. With the "reverse" pliers, in contrast, squeezing the handle spread the tips of the pliers apart, so that grasping an object involved releasing hand pressure—a mismatch between the proximal action and distal effect. In area F5, neurons coding for hand grasps responded to grasps produced with both types of tools, indicating that they were coding for distal goals and not proximal movements. In contrast, when Umiltà and colleagues recorded from the primary motor cortex, they identified two groups of neurons: one that responded to distal goals (as in F5) and another that responded based on the specific proximal movements. This pattern, with premotor neurons coding for goals and primary motor cortex neurons coding for a mix of goals and movements, is very similar to that found by Kakei et al. (1999, 2001) described in chapter 3.

A similar study was conducted in humans by Jacobs et al. (2010) using a task in which participants reached and grasped objects in different orientations. Depending on the orientation, the object could be more comfortably grasped using an overhand or an underhand grasp. Participants were also trained to perform this task using a mechanical grabber that was 56 cm long. Critically, the tips of the grabber were rotated, so that the distal opposition space at the tip of the tool was rotated ninety degrees from the opposition space of the proximal hand movements required to move the tool. Jacobs and colleagues measured the proportion of trials for which participants chose to make overhand versus underhand responses, showing that this choice differed systematically between reaches with the hand and with the tool. Despite the evidence for distinct motor programs being used, fMRI data showed that the grasp network, including the anterior IPS and the vPMC, was activated in both cases. This suggests that the parieto-frontal network involved in object-directed actions with the hands also controls object-directed actions using a tool.

Another study investigated whether there is a precise level of motor equivalence between precision grips performed with the fingers and with a grabber (Tang et al., 2016). This approach takes advantage of the fact that the specification of motor-control parameters on a given trial is influenced by those on the previous trial, but only when the same act is being performed on both trials. For example, in an earlier study, these authors showed that the details of the previous trial influenced the next trial when both trials were grasps or when both were points but critically not when one was a grasp and the other a point (Tang et al., 2015). In their 2016 study, Tang and colleagues found that there was no transfer between precision grips performed with the fingers and with a tool, indicating that they rely on distinct motor-control programs. In contrast, transfer was found between precision grips performed in a typical manner (e.g., opposition between thumb and index finger of a single hand) and in an unusual manner (e.g., opposition between the middle fingers of the two hands).

Together, these results indicate that skilled use of tools involves the same overall brain networks as hand actions and the same general type of motor program, but not exactly the same program. One model for how such distinct motor programs might work comes from Ingram et al. (2010), who used a robotic force-feedback device to give participants experience using a series of "virtual" hammers that they saw projected over their hand and could control by moving a handle. The researchers introduced force fields, such that movements of the hammer in different directions required different amounts of force. Across training trials, participants adapted to the force field, adjusting the force produced for movements in different directions to produce smooth movements. The key question was what would then happen when they were presented with a different virtual hammer, which extended from the hand at a different orientation. Ingram and colleagues found some generalization of adaptation to different virtual tools, but this was quite limited, suggesting that participants had built distinct motor programs for each tool. A subsequent study from the same lab showed that separate motor programs could also be constructed for interactions based on different locations of a single object when these required different ways of interacting with the object (Heald et al., 2018).

Similar results have been found in neuroimaging studies investigating the formation of novel internal models of action in the cerebellum. As discussed in chapter 4, internal models of action are an important aspect of motor control (Wolpert et al., 1995), taking into account the biomechanical properties of the body and allowing estimation of motor commands needed to produce a desired goal state (inverse model) or prediction of the sensory consequences of a planned action (forward model). Imamizu et al. (2000) investigated the formation of a new internal model for the use of a novel tool (a computer mouse that featured a novel rotational transformation). They measured fMRI responses while participants learned to use this new tool. Responses across wide regions of the cerebellum showed responses during learning, which decreased as performance improved. Indeed, the magnitude of these responses was directly proportional to the magnitude of errors made by participants. These results thus appear to be related to the initial formation of a new internal model during motor learning. Another more focal response was found near the posterior superior fissure of the cerebellum, which remained even when learning had reached asymptote, which Imamizu and colleagues suggest reflects the acquired internal model. Notably, the location of this response was different from that observed when participants used a normal mouse, suggesting that the novel mouse was controlled by a new, distinct internal model and had not been incorporated into an existing model. In a subsequent study (Imamizu et al., 2003), these researchers trained participants to use two different novel tools: a mouse with a novel rotational mapping, the other with a novel velocity mapping. After training, they again found responses in the cerebellum when participants used the novel tools, but with little overlap, suggesting that two distinct internal models had been created.

One natural reason why it makes sense for the brain to use distinct motor programs for tools compared to the hand is the simple reason that the weight of the tool and the torques that it produces mean that different patterns of muscular responses need to be produced to act skillfully with a tool than with the hand alone. While more abstract goal-directed representation may generalize across the hand and tools, as Umiltà et al. (2008) found in the PMC, at some point, the differential forces produced by the tool on the arm need to be taken into account.

Another reason why tools and hands need to be distinguished is that we commonly use tools exactly because we want to avoid using the hands for a particular task. For example, if we need to stir a pot of soup, we use a spoon instead of the hand for both hygienic and safety reasons. Similarly, when using a poker to shift logs in a fireplace, the poker is useful both because it extends the effective length of the arm and because we wouldn't want to put our hand into the hot flames. We thus do not want to confuse our arm and the poker; it is essential to keep the distinction between them clear.[1] Interestingly, chimpanzees also judge appropriately when to reach with their hand and when it would be hazardous to do so and a tool ought to be used instead (Povinelli et al., 2010).

Integration of Tools with Mental Body Representations
Several studies have shown that after a period using a tool, there are systematic changes in the kinematics of subsequently performed reaching and grasping actions (Cardinali et al., 2009, 2016). Cardinali et al. (2009) used high-resolution three-dimensional motion tracking to record the kinematics of arm and hand movements while participants reached and grasped objects, both before and after a training period in which they used a long mechanical grabber to pick up objects. After using the grabber, the kinematics of subsequent reaching movements (performed without the grabber) changed systematically. There was a reduction in the peak velocity and acceleration of arm movement as well as corresponding increases in their peak latencies. These changes were specific to the transport phase of reaching and were not seen for kinematic features related to the grasp component, such as maximum grip aperture. The specificity of these changes to the transport phase makes sense, given that the long grabber essentially has the effect of increasing the effective length of the arm. In another study (Cardinali et al., 2016), these researchers investigated the effects of tools that changed the effective properties of the hand, such as a pair of pliers held between the thumb and index finger, and a pair of sticks taped to these two fingers, extending their effective length. In this case, kinematic changes were again seen for subsequent reaches performed without the tools. Critically, however, these changes were restricted to the grasp phase (e.g., maximal grip aperture) and did not influence the transport phase. Together, these studies provide an elegant double dissociation, showing that tools are incorporated into both the reaching and grasping networks depending on the functional properties of the tool.

In addition to measuring movement kinematics, Cardinali et al. (2009) also obtained implicit measures of the perceived length of the arm and hand. Participants were asked to point to the perceived location of touches applied to the elbow, wrist, or index fingertip of their tool-using arm. Following use of the long grabber, the perceived locations of the elbow and wrist became farther apart, indicating a perceived lengthening of the forearm. In contrast, however, there was no significant change in the perceived length of the hand, indexed by the distance between the judged locations of the wrist and fingertip. This pattern of alterations in the perceived length of regions of the arm is consistent with the effects on kinematics, where changes were seen only for the transport and not the grasp phase of reaching.

Similar results have been found in tasks in which participants indicate the location in external space of the midpoint of their extended forearm (Bruno et al., 2019;

Romano et al., 2019; Sposito et al., 2012). For example, Sposito and colleagues obtained judgments of perceived arm midpoint before and after fifteen minutes of training using a 60 cm tool. At posttest, there was a distal shift of the perceived arm midpoint by approximately 5 percent of arm length.

Another approach to investigating the effects of tool use on perceived body size is by its effects on tactile distance perception. As described in chapter 2, experimental modulations of perceived body-part size produce corresponding effects on the perceived distance between two touches on the skin (de Vignemont et al., 2005; Taylor-Clarke et al., 2004). Several studies have shown that tool use produces similar modulations. For example, Canzoneri et al. (2013) asked participants to judge whether tactile distances applied to the forearm were bigger or smaller than distances on the forehead, both before and after a twenty-minute period of using a 1 m stick. At posttest, perceived tactile distances on the forearm were expanded in the mediolateral arm axis and compressed in the proximodistal axis. A subsequent study by Miller et al. (2014) replicated this result and showed further that the correspondence between the morphology of the tool and arm determined the changes produced. After training using a long, thin grabber, changes in tactile distance perception occurred on the forearm but not on the hand. Conversely, after training using a mechanical, claw-like tool, which mimicked the structure of the hand, the pattern was reversed and changes were found on the hand but not on the forearm. This body-part specificity mirrors the specific effects that Cardinali et al. (2009, 2016) found for movement kinematics, showing that different types of tool use produce highly specific effects on both motor and somatosensory processing.

11 The Evolving Hand

On the evening of February 9, 1849, an audience gathered in London at the Royal Institution of Great Britain for a lecture by the eminent Victorian biologist Sir Richard Owen. The topic of Owen's lecture, later published as *On the Nature of Limbs* (Owen, 1849), was the similarity in the bone structure in the limbs of animals, already mentioned in chapter 1. The hands of humans, the wings of bats, the paws of bears, the hooves of horses, and the flippers of dolphins are radically different in both structure and function. Despite these differences, Owen systematically highlighted common elements of structure in the bones of each of these limbs. Indeed, the bones could be put into an exact one-to-one correspondence, which suggested that they were in some sense *the same* bones. Owen introduced the concept of "homology" to signify this idea of sameness and suggested that this "unity of type" reflected the existence of an *archetype* of the vertebrate skeleton (Desmond, 1982). Owen suggested that the archetype was similar overall to the *Lepidosiren*, a South American lungfish that had recently been described.

The post-Darwinian explanation of unity of type in terms of historical descent from a common ancestor can mask in our mind what an extraordinary scientific discovery the existence of such unity was. Indeed, Darwin himself was deeply influenced by Owen's work on the archetype, which became some of the strongest evidence for the theory of natural selection. In his handwritten notes in his personal copy of *On the Nature of Limbs*, Darwin describes his own view of the archetype not as an abstract Platonic form but rather as a reflection of common descent (Desmond, 1982): "I look at Owen's Archetypes as more than ideal, as a real representation as far as the most consummate skill and loftiest generalizations can represent the parent-form of the Vertebrata" (p. 50). A decade later, in *On the Origin of Species*, Darwin (1859) put the matter straightforwardly:

> If we suppose that the ancient progenitor, the archetype as it may be called, of all mammals, had its limbs constructed on the existing general pattern, for whatever purpose they served, we can at once perceive the plain signification of the homologous construction of the limbs throughout the whole class. (p. 435)

In this chapter, we will discuss the evolution of the human hand, comparing changes in both the physical hand and the invisible hand. We will focus on four key transitions in the evolution of limbs: the transition from the fins of ancient fish to the limbs of the earliest ground-living tetrapods, the emergence of mammals, the emergence of primates, and finally the evolution along the hominin line since our last common ancestor with chimpanzees, which has resulted in us.

Transition 1: Fins into Limbs

The story of the human hand starts with fish. Humans, like all ground-living animals, including amphibians, reptiles, birds, and mammals, evolved from a group of fish—the tetrapods—that emerged from the seas and started to live on dry land around four hundred million years ago. As discussed in chapter 1, while some transitional animals between fish and tetrapods had larger numbers of digits (Coates & Clack, 1990), the five-digit pentadactyl pattern has been the default in all tetrapods dating back as far as the earliest known fully terrestrial animal, the 340-million-year-old *Casineria* (Paton et al., 1999). The word "tetrapod" means "four footed," and these four limbs evolved from fish fins. This, then, is our first key transition.

Fins and Limbs

The embryonic development of vertebrate limbs relies on a characteristic pattern of gene expression in the developing embryo (Schneider & Shubin, 2013). Of particular importance are genes from the *Hox* family, known to be critical to the formation of animal body plans (Arthur, 1997). Complex spatial and temporal patterns of *Hox* gene expression across the proximodistal and anteroposterior limb axes produce the overall form of the limb. Recent studies of gene expression in developing fish fins show patterns strikingly similar to those seen in mammals such as mice (Nakamura et al., 2016). Further evidence for the similarity of these genetic regulatory mechanisms across vertebrates comes from studies using transgenic mice in which DNA from other species is injected into the nucleus of a fertilized mouse egg. Remarkably, *Hox* regulatory mechanisms from fish such as zebrafish and gar are able to drive normal development of mouse limbs (Gehrke et al., 2015), suggesting that developmental processes of appendage formation are highly conserved across vertebrate fins and limbs. Activating specific regulatory genes in mutant zebrafish can produce the formation of supernumerary long bones in the fin, structurally similar those of tetrapod limbs (Hawkins et al., 2021), suggesting that fish have latent genetic potential to develop limbs.

Beyond these structural and developmental similarities, there is also evidence that fish fins are used for sensory functions broadly similar to those of tetrapod limbs (Hale, 2021). Fins of a range of fishes have been found to be responsive to pressure (Hardy et al., 2016), muscle stretch (Fessard & Sand, 1937), and fin displacement (Aiello et al., 2016). Studies of mechanoreceptive afferents in fins show RFs similar in size to those seen on tetrapod limbs (Hardy & Hale, 2020). Collectively, these mechanisms in fish have been shown to support both touch (Hardy et al., 2016) and proprioception (Williams et al., 2013). These sensory signals appear to be tightly integrated into behaviors such as hovering (Williams & Hale, 2015) and swimming (Aiello et al., 2020).

Tetrapods evolved from a group called the lobe-finned vertebrates, represented among modern fishes by the coelacanth and the lungfishes. Studies of fossils have identified a remarkable sequence, showing a clear progression from the limb-like fins of Devonian lobe-finned fish such as *Tiktaalik* (Shubin et al., 2006) and *Panderichthys* (Boisvert et al., 2008) to the fin-like limbs of early tetrapods such as *Acanthostega* (Callier et al., 2009) and *Parmastega* (Beznosov et al., 2019). Figure 11.1 shows the skeletal structure of fore appendages in living lobe-finned fishes and ancient animals on both sides of the fish/tetrapod boundary. Numerous progressive changes in structure are apparent, including the emergence of the parallel bones in the tetrapod forelimb (i.e., the radius and ulna) and increasing differentiation of the terminal digits. In parallel to these changes to the overall shape of the limb, studies of the body processes to which muscles attach show corresponding changes in the musculature of the legs (Callier et al., 2009), which would be critical, given the differential support needed for legs than for fins.

Changes to the Tetrapod Nervous System

Studies of the skulls of these animals have shown a range of changes occurring in parallel to the remodeling of the fins into limbs. The snout elongates (Clack, 2009), the eye sockets increase in size (MacIver et al., 2017), and the neck appears, allowing the head to rotate independently of the rest of the body (Downs et al., 2008). The fish-like hyomandibula bone becomes the tetrapod-like stapes bone in the middle ear, showing dramatic changes in the auditory system (Clack et al., 2003). The timing of changes in braincase shape appears to closely parallel changes in fin/limb shape (Ahlberg et al., 1996).

A recent study by MacIver et al. (2017) argues that changes in the visual system were also key to this transition. By comparing fossils across the fish/tetrapod transition, they found that the move to land was preceded by large increases in eye size by a factor of three. In water, the advantages of increasing eye size are limited, as water clarity quickly becomes the limiting factor in visual sensitivity. Early tetrapods likely had a visual range

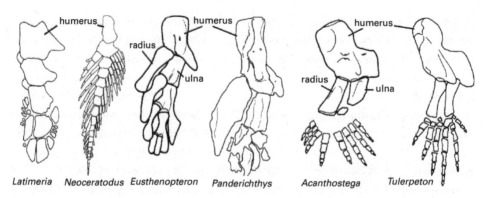

Figure 11.1
The transition from fins into limbs. The two leftmost appendages are from modern lobe-finned fishes, the coelacanth (*Latimeria*) and the lungfish (*Neoceratodus*). The two center appendages are fins from Devonian fishes (*Eusthenopteron* and *Panderichthys*), which begin to show features of the tetrapod forelimb, including separation of the bones of the forearm (radius and ulna) and differentiation of the terminal digits. The two rightmost appendages are limbs of early tetrapods (*Acanthostega* and *Tulerpeton*), which have a skeletal structure that is immediately recognizable as homologous to the human arm. From Clack (2009).

of about one body length when in the water. At this range, the benefits of vision are largely limited to rapid, reactive responses, and indeed many fishes have specialized receptors and neural circuits for rapid visually elicited escape responses that have disappeared in tetrapods (Lacoste et al., 2015). On land, however, increases in eye size would have produced major benefits in terms of the distance at which predators or prey could be identified. MacIver and colleagues estimate that the volume of space that could be visually surveyed was more than a million times greater in the air than in the water. Moreover, the benefits of vision would have been further enhanced by the evolution of the neck, which occurred around the same time, as shown by fossils such as *Tiktaalik* (Shubin et al., 2006), which would allow more efficient surveying of the visual surround without having to move the entire body. This opens up novel uses of vision focusing less on rapid reactions to immediate threats and based on precise and goal-directed visuomotor control, which have continued to evolve into the systems described throughout this book.

Transition 2: Mammals and the Cerebral Cortex

The first mammals emerged during the Triassic era, a little more than two hundred million years ago (Brusatte, 2022). The forelimbs of mammals are extraordinarily diverse, ranging from hands to paws to hooves to wings to flippers. As discussed above, however, the mammalian forelimb "archetype" that characterized the earliest mammals was, in several respects, very much like the modern human hand.

Mammals as a group are characterized by several features, including differentiated teeth for different functions, the ability to chew, warm-blooded metabolism, and, more relevant in the present context, large brains. The most salient feature of the brain in mammals compared to other vertebrates is the prominent cerebral cortex. A six-layered cortex appears to be a primitive characteristic of all mammals and is unique to the mammalian line, although structures such as the dorsal pallium in amphibians, the dorsal cortex in reptiles, and the Wulst in birds are believed to be homologous to the neocortex in mammals (Striedter, 2005).

The Mammalian Brain Archetype
Studies of fossil endocasts of the brains of Triassic mammaliforms such as *Morganucodon* and *Castorocauda* have linked increases in brain size (presumably including the emergence of the cortex) to the evolution of hair (Rowe et al., 2011), suggesting that tactile processing of signals from hair may have been an important contributor to the evolution of the cortex. Comparative studies of the three broad groups of mammals—the monotremes, marsupials, and placental mammals—have revealed a basic mammalian cortical plan, which likely characterized the brain of the earliest mammals (Kaas, 1987; Krubitzer, 1995). These animals likely had between fifteen and twenty distinct cortical areas (Kaas, 2013). These include both primary and secondary somatosensory, visual, and auditory cortices and a rostral somatosensory strip (i.e., area 3a) as well as more general-purpose association cortex between these regions.

Developmentally, this pan-mammalian cortical plan appears to be specified and maintained by molecular gradients of regional gene expression across the embryonic cortical sheet (Assimacopoulos et al., 2012; Grove & Fukuchi-Shimogori, 2003). This basic cortical plan can be—and has been—modified in a variety of ways, including alterations to the size of regions, the number of distinct regions, and their patterns of interconnectivity. However, the plan also places basic constraints on the types of alterations that can be made (Finlay & Darlington, 1995; Krubitzer, 2007). This can be seen in the preservation of the basic retinal-thalamo-cortical pathway in subterranean mammals, such as the naked mole rat (Cooper et al., 1993), despite these animals being almost entirely blind and with virtually the entire cortex, which is devoted to vision in other animals, instead being devoted to somatosensation (Catania & Remple, 2002). Indeed, despite dramatic increases in size and number of cortical areas, the human brain retains the basic characteristic mammalian organization.

Mammalian Sensorimotor Cortex

One intriguing feature of the early mammalian brain is that it appears not to have had a distinct motor cortex. Instead, as in living monotremes and marsupials, motor and somatosensory functions were controlled by a single sensorimotor amalgam, homologous to primary somatosensory area 3b in primates (Lende, 1963). A separate M1, along with the PMC, appears to have developed with the first placental mammals. While in marsupials the ventral lateral nucleus and the ventral posterolateral nucleus of the thalamus project to overlapping regions of the sensorimotor cortex, in placental mammals they project to M1 and S1, respectively (Striedter, 2005). Thus, the segregation of the primary motor and somatosensory cortices is mirrored by segregation of the inputs that each area receives from the thalamus.

The emergence of a distinct motor cortex in placental mammals was also paralleled by the emergence of more direct cortical control of spinal motoneurons within the corticospinal (pyramidal) tract by which motor commands are communicated to the periphery (Nudo & Masterton, 1990). In marsupials, projections from sensorimotor cortex terminate on spinal interneurons (propriospinal neurons) located in vertebrae C3 and C4 in the neck (Frost et al., 2000). This system of spinal interneurons then sends motor commands to motoneurons at lower levels of the spinal cord, which directly control muscles.

As discussed in chapter 3, in placental mammals, a direct CM pathway appears, which allows the motor cortex to send commands directly to spinal motoneurons, bypassing the spinal interneuron system (Bernhard et al., 1953; Porter & Lemon, 1993). This direct pathway has been linked to differences in the intensity of electrical stimulation of the cortex required to evoke movements, which is much higher in marsupials than in placental mammals (Beck et al., 1996; Frost et al., 2000), as well as to the dexterity of prey-catching forelimb actions in placental mammals such as rats versus marsupials such as opossums (Ivanco et al., 1996).

However, the direct pathway has continued to evolve within placental mammals as well. Heffner and Masterton (1975, 1983) compared sixty-nine species of mammal in measures of forelimb dexterity and several measures of the anatomy of the pyramidal tract—for example, the depth of termination of the pyramidal tract within the spinal cord. There were strong correlations between dexterity and the depth at which corticospinal neurons penetrated down the spinal cord. Iwaniuk et al. (1999) criticized Heffner and Masterton's analysis for not taking proper account of the phylogenetic relations among the species tested. However, even when this correction was made, the depth of penetration of corticospinal neurons continued to explain 40 percent of the variance in digital dexterity among primates. The CM pathway has developed to particular prominence in primates, as will be discussed below.

While all mammals appear to have a somatotopically organized somatosensory cortex, the exact features of these maps differ widely, depending on the physical organization of each animal's body and the ecological details of their mode of life. For example, Krubitzer et al. (1995) investigated the organization of the somatosensory cortex in the duck-billed platypus—a monotreme mammal whose last common ancestor with humans was likely in the Jurassic period, approximately 166 million years ago (Warren et al., 2008). As in other mammals, the platypus has clearly identifiable primary visual, auditory, and somatosensory cortices. Nevertheless, the features of these areas are very different from those seen in primates. A highly conspicuous feature of the platypus is its eponymous duck-bill, which is densely packed with sensory receptors and which it uses to detect and capture prey. Strikingly, three-quarters of the total cortical area devoted to sensory functions is devoted to the somatosensory system, and within the somatosensory system, 90 percent of cortical tissue is devoted to the bill.

We thus see cortical magnification at two levels in the platypus. First, there is magnification of the entire somatosensory system compared to other sensory systems. And second, within the somatosensory system, there is magnification of one specific sensory organ: the bill. Both forms of magnification are direct reflections of the importance of touch generally, and of the bill specifically, to the platypus's mode of life.

A similar pattern is seen for the star-nosed mole—a small North American animal with an extraordinary nose, featuring twenty-two dexterous appendages with densely packed and highly sensitive tactile receptors. The star-nosed mole uses its star to forage for food and has been found to have the fastest recorded time to identify and eat prey—as little as 120 ms (Catania & Remple, 2005). Neurophysiological studies have shown highly organized somatotopic representations in the mole's somatosensory cortex, with massive magnification of the star (Catania et al., 1993; Catania & Kaas, 1995, 1997a). Figure 11.2 shows the true proportions of the star-nosed mole (left panel) and the proportions seen in the somatosensory cortex (right panel). Although the star makes up only a modest proportion of the actual size of the animal's body, it takes up an outright majority of cortical territory.

In humans, of course, it is the representation of the hand that shows most conspicuous cortical magnification. But, at the same time, it is important to emphasize that this is not a uniquely human characteristic. Raccoons, for example, use their forepaws in highly dexterous ways to obtain and eat food (Iwaniuk & Whishaw, 1999), despite lacking the prehensile capabilities of primates. This striking behavioral difference to broadly similar members of the order *Carnivora*, such as cats and dogs, is mirrored by extremely high cortical magnification of the forelimbs in both S1 (Welker & Seidenstein, 1959) and S2

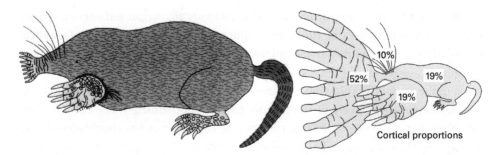

Figure 11.2
Cortical magnification in the somatosensory cortex of the star-nosed mole. *Left panel*: A drawing of the actual form of a star-nosed mole. From Catania and Kaas (1995). *Right panel*: A schematic showing the proportions of primary somatosensory cortex devoted to different body parts. Although the star is only a modest proportion of the actual surface area of the animal's body, it takes up a majority of the somatosensory cortex. From Catania and Kaas (1997a).

(Herron, 1978). The magnification of the raccoon forepaw is massive, substantially larger even than in humans, as a proportion both of S1 and of total cortical area.

The Emergence of New Cortical Regions

Mammalian brains differ not only in the proportionate sizes of cortical regions but also in the number of regions in the first place. Beyond the highly conserved primary sensory and motor regions, there are numerous distinct regions that appear in some mammalian lineages, but not in others. Brodmann (1909) noted more than a century ago, based on cytoarchitectonic data, that the number of distinct brain areas differs widely across animals. Indeed, the number of distinct brain areas appears to scale in a direct allometric relation with cortical volume (Changizi & Shimojo, 2005). In the human cerebral cortex, there are as many as 180 areas (in each hemisphere; Glasser et al., 2016). Just as the magnification of sensory regions in different species reflects the ecological importance of each region in each animal's mode of life, so too does the number of cortical areas devoted to each modality. For example, while the echidna, a monotreme from Australia, has only the two visual areas that appear to be primitive to all mammals, macaque monkeys have more than fifteen distinct visual areas (Krubitzer, 1995).

How do new cortical regions form during evolution? Two broad theories of the formation of new cortical areas have been proposed (Striedter, 2005). First, Leah Krubitzer has proposed a module aggregation hypothesis, in which a single brain area starts to receive a new type of input in addition to that it had previously received (Krubitzer, 1995; Krubitzer & Huffman, 2000). This causes the area to become separated into islands responsive to the new and the old inputs. A gradual process of aggregation may then sort these islands into two coherent regions, each responsive to one input type, at which point they function essentially as distinct cortical areas.

Evidence for the module aggregation hypothesis comes from studies showing that completely new brain areas can be created in individual animals by manipulating their developmental exposure to specific sensory signals (Kahn & Krubitzer, 2002; Rakic et al., 1991). For example, Rakic and colleagues surgically removed the retinas from macaque fetuses, which were subsequently born otherwise healthy. This led to a reduction in the volume of the LGN in these animals, reflecting the absence of visual input signals. More importantly, this alteration of thalamic inputs to the cortex led to the emergence of a completely novel cortical area (area X), which was distinct in both laminar and cytoarchitectonic structure from the (largely normal) areas 17 and 18.

A second theory of the formation of new cortical areas is the area duplication hypothesis, in which a brain area (Allman & Kaas, 1971), or potentially even an entire brain pathway (Chakraborty & Jarvis, 2015), is duplicated. Such duplication could result from the duplication of an entire gene, from altered expression of genes (e.g., analogous to the formation of new serially homologous body segments by expression of *Hox* genes), or from the presence of signaling molecules in the developing cortical sheet. An elegant example of the latter effect was provided by Fukuchi-Shimogori and Grove (2001). They investigated a protein called fibroblast growth factor 8 (FGF8)—a signaling protein that is involved in specifying the area identity in the anteroposterior axis of the cortex. In normal mice, FGF8 is concentrated at the anterior pole of the embryonic cortical sheet and is involved in the formation of the somatosensory "barrel" cortex representing the whiskers. Increasing or decreasing the concentration of FGF8 at the anterior pole produced posterior and anterior shifts of the eventual location of barrel cortex, respectively. More strikingly, introducing a second concentration of FGF8 at the posterior end of the cortex resulted in the formation of a second barrel cortex, spatially distinct from the original one but featuring the same highly organized topographic representation of individual whiskers. Subsequent work from the same lab showed that functional duplications of either S1 or V1 could be created by changing concentrations of FGF8 at different locations on the embryonic cortical sheet (Assimacopoulos et al., 2012). These results demonstrate that duplication of existing brain areas can be induced by changes in the location and concentration of signaling proteins within an early cortical "protomap" (Rakic, 1988).

An intriguing feature of the duplicated barrel cortex that Fukuchi-Shimogori and Grove created is that it is mirror inverted relative to the original barrel cortex. This presumably reflects the fact that the new concentration of FGF8 effectively "inverts" the anteroposterior axis of the cortical sheet, leading to the formation of an area as if the posterior end of the cortex were actually the anterior end. This reversal has an intriguing resemblance to the mirror inversions seen at the boundaries between both somatotopic maps in the primate somatosensory cortex (Blankenburg et al., 2003; Kaas et al., 1979) and retinotopic maps in the visual cortex (Sereno et al., 1995), discussed in chapter 2. This suggests that these somatosensory and visual regions may have emerged by successive duplication of existing regions. Once a region has been duplicated, one (or both) of the copies can then take on new functions. Recent research has provided evidence that such duplications may have been common in mammalian (and, indeed, vertebrate) brain development. For example, by comparing patterns of genetic expression and connection patterns in chickens, mice, and humans, Kebschull et al. (2020) argue that cerebellar nuclei have evolved by repeated duplication across hundreds of millions of years. Frangeul et al. (2016) have similarly argued, based on the highly conserved patterns of genetic control over the formation and plasticity of thalamocortical connection patterns seen in the mammalian visual, auditory, and somatosensory cortices, that these primary sensory cortices themselves evolved from a primitive, modality nonspecific sensory cortex.

Transitions 3 and 4: Primate and Hominin Evolution

The last common ancestor of living primates lived around sixty million to eighty million years ago. The earlier date comes from studies of fossils (Tavare et al., 2002), and the later date comes from comparisons of the DNA of living primates (Vanderpool et al., 2020). Figure 11.3 shows a clade diagram of living primates based on analysis of DNA similarity. Around sixty million years ago, one group of primates, the prosimians, split off. This group includes modern lemurs, lorises, and galagos. A few million years later, the ancestors of modern tarsiers split off from the simians, which gave rise to modern monkeys and apes. Just under forty million years ago, a group of African monkeys reached South America, presumably floating on driftwood, leading to the split between the (American) New World monkeys and the (African and Asian) Old World monkeys. Around thirty million years ago, the ancestors of modern apes split from the Old World monkeys. Finally, between five and ten million years ago, the hominin lineage leading to modern humans split from our last common ancestor with chimpanzees.

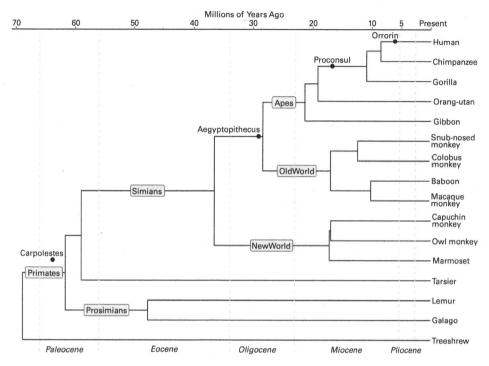

Figure 11.3
A clade diagram showing the phylogenetic relations between living primates, with tree shrews as an outgroup comparison. The location of each split indicates the estimated date of the last common ancestor. Estimated locations (temporally and phylogenetically) of fossil species discussed in this chapter (*Carpolestes*, *Aegyptopithecus*, *Proconsul*, and *Orrorin*) are indicated with black circles. Dotted lines indicate boundaries between geological epochs, which are named at the bottom of the figure. Phylogenetic relations and dates of divergence were estimated using genome discordance (Vanderpool et al., 2020).

Tree shrews, from the order *Scandentia*, are among the closest living relatives to primates (Janecka et al., 2007) and indeed, have in the past been classified as prosimian primates (Napier, 1970). Comparison of fossils with modern animals shows that tree shrews have changed remarkably little in the past thirty-four million years, leading to them being referred to as "living fossils" (Li & Ni, 2016). Accordingly, the tree shrew has been taken as a useful living model of the earliest primates in terms of both the physical structure of its hand (Napier, 1962b) and the organization of its brain (Kaas et al., 2018).

Traditionally, a constellation of features has been viewed as distinguishing primates from their closest relatives, such as tree shrews (Le Gros Clark, 1959; Wood Jones, 1916).

These include large brains for their body size, forward-facing eyes, and specializations of the limbs, such as large big toes, some level of thumb opposability and grasping ability, and the presence of nails rather than claws. This suite of features may have evolved to facilitate visually directed predation of insects (Cartmill, 1974), leaping through tree branches (Szalay & Dagosto, 1988), or grasping branches to eat fruits at their extremities (Bloch & Boyer, 2002).

The Hands of Primates

One notable feature of primate hands is that they generally preserve the basic primitive mammalian limb structure. The primate hand is, however, characterized by two key novelties: (1) the presence of fingernails instead of claws and (2) enhanced divergence and opposability of the thumb. The fifty-six-million-year-old fossil *Carpolestes simpsoni*, which appears closely related to the earliest primates, features a partially opposable big toe with a thumb-like saddle joint (Bloch & Boyer, 2002). Moreover, this big toe features a nail, although the other toes have claws. The emergence of the thumb nail alongside a key feature of bone structure supporting dexterous prehension suggests that the transition from claws to fingernails was an adaptation specific to the development of grasping abilities in the primate lineage.

In the tree shrew, in contrast, the hand remains essentially a paw (Napier, 1962b). By flexing the fingers, the hand can be shaped into a sort of bowl shape, which can be used to hold food—what Napier calls "convergence." Even in the tree shrew, however, some specializations of the anatomy and musculature of the thumb can be seen, which become more elaborated in primates.

Hamrick (2001) argued that primates are distinguished from tree shrews and other near relatives by systematic changes in hand proportions, including shorter metacarpals but longer phalanges. He linked these differences to comparative studies in marsupials, showing that they are features predictive of more highly dexterous use of the forepaws (Lemelin, 1999). Subsequent research has shown less clear separation between primates and related mammals (Boyer et al., 2013). Even in this latter study, however, all primates were distinguished from related species by adaptations to the hand joints allowing more effective grasps.

Throughout primate evolution, changes in the structure of the hands and forelimbs have been driven by changing patterns of locomotion, most of which have been primarily arboreal in nature. In his classic book *Arboreal Man*, Wood Jones (1916) argues that the arboreal environment was conducive to the emergence of functional segregation between the forelimbs and hindlimbs. When standing on a branch, the legs need to support the weight of the body, while the arms need to grasp food or branches to stabilize the body.

This functional differentiation results in "emancipation of the fore-limbs" (p. 7) from locomotor requirements. While the supporting role of the legs is broadly consistent across primates, different locomotor styles use the hands and arms in very different ways.

Napier (1970) divides primate locomotion into four categories. In the first, vertical clinging and leaping, the animal keeps its body oriented vertically when at rest, using its feet to hold onto tree trunks and branches, with the hands having a lesser supporting role (Gebo, 2011; Napier & Walker, 1967). Long, powerful legs are used to leap from one tree or branch to another. This is the dominant mode of locomotion in prosimians, such as lemurs and galagos. In other primates, we can see more advanced forms of prehensility. In tarsiers, for example, objects can be grasped by flexion of the fingers against the palm or by opposition of the thumb again the other fingers. This remains what Napier (1961) calls "pseudo-opposability," however, as the movement is restricted to the metacarpophalangeal joint.

In the second locomotor category, quadrupedalism, locomotion is accomplished with all four limbs, either on the tops of tree branches or on the ground. Quadrupedalism is the dominant locomotor style in monkeys. In some New World monkeys, occasional arm swinging is also observed—what Napier (1970) calls "semi-brachiation." The ability to use a separated thumb or big toe to support the body makes it easier to walk farther out on tree branches where fruit and leaves are more abundant but where the branch is narrower and springier. The ability to hold onto swinging branches was presumably related to the emergence of this characteristic feature of primates. In monkeys, there is a clear advance over prosimians in the opposability of the thumb and the dexterity of the hand (Napier, 1961). New World monkeys generally show pseudo-opposability, in which the thumb can be opposed against the sides of the fingers, forming a clamp that can hold objects but without direct contact between the fingertips of the thumb and index finger. True opposability appears for the first time with Old World monkeys. Their carpometacarpal joints show a well-developed saddle configuration, comparable to humans, which allows tip-to-tip contact of the tips of the thumb and index fingertips, facilitating precision grips.

Intriguingly, one type of New World monkey, the capuchin monkey, appears to have evolved an opposable thumb independently of the Old World monkeys. A range of anatomical features differentiate the hands of capuchin monkeys from closely related New World monkeys, including the depth of the carpal tunnel in the wrist and the degree of flexion allowed by the carpometacarpal joint of the thumb (Napier, 1961). Capuchin monkeys show highly dexterous use of the hands (Torigoe, 1985), including precision grips (Costello & Fragaszy, 1988) and tool use and manufacture (Mannu & Ottoni, 2009). It is for this reason that capuchins were the monkeys used traditionally by street musicians such as organ grinders to perform tricks and to collect money from spectators. Critically, these anatomical and behavioral specializations are paralleled by a range of differences in the brain compared to other New World monkeys, including direct corticomotoneuronal connections (Bortoff & Strick, 1993), expansion of the cerebral cortex generally (Rilling & Insel, 1999) especially of the PPC (Chaplin et al., 2013), and emergence of somatosensory area 2 (Padberg et al., 2007). There is thus a tight link between the anatomical features of the physical hand, the use of the hand in skilled manipulative behaviors, and the neural mechanisms in the brain controlling the hand.

The third locomotor category, true brachiation, involves the dominant use of the arm to suspend the body from branches, with propulsion of the body achieved by rapid alternating movements of the arms. Brachiation is a specialization of apes. Napier (1970) classifies the knuckle walking of the African great apes, chimpanzees, bonobos, and gorillas as a modified form of brachiation, although this interpretation has been challenged (Tuttle, 1969a). Regardless of this point, apes share a suite of features specialized for brachiation and knuckle walking, including very long fingers relative to thumb length, elongated metacarpals of digits 2–5, and reduced thumb flexor muscles. Napier (1970, p. 128) contrasts the "specialized grappling-hooks" of ape hands with the "very generalized, all-purpose, hand of monkeys."

Across these modes of locomotion, there is a progressive change from leg dominance to arm dominance. On its face, the final locomotor category, bipedalism, flouts this progression, with locomotion relying entirely on the legs, excluding the arms entirely. Of course, in doing so, bipedalism has allowed for the arm and hand to become dominant in functions separate from locomotion. While many other primates (indeed, many mammals) can walk bipedally for short durations, only humans use bipedalism habitually as a dominant mode of locomotion. By freeing the arms from locomotor functions, bipedalism has allowed the uniquely dexterous hands actions of humans (DeSilva, 2021). One prominent interpretation is that the relaxation of selective pressure on the hands for locomotion allowed them to become specialized for grasping instead (Alba et al., 2003; Wood Jones, 1916). It is also possible, however, that given that the hands and feet are closely linked, serially homologous structures, changes in hand shape could be secondary consequences of changes in foot structure related to bipedalism (Rolian et al., 2010).

The Hands of Hominins

The lineage leading to us split from that leading to our closest living relatives—the chimpanzees and bonobos—between six and eight million years ago. As with estimates of the last common ancestor of all primates, analyses based on fossils (Parins-Fukuchi et al., 2019) give somewhat older estimates than analyses based on DNA (Patterson et al.,

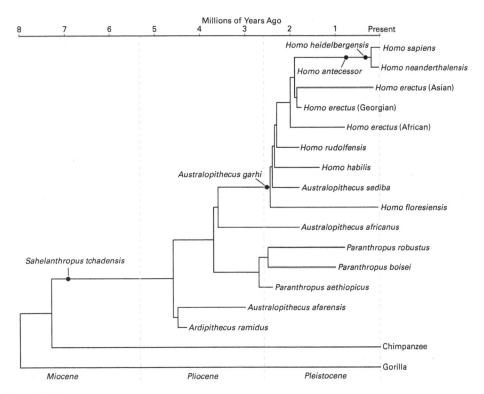

Figure 11.4
A tentative clade diagram of hominin evolution based on the model of Parins-Fukuchi et al. (2019).

2006). Interestingly, DNA estimates differ depending on where in the genome is analyzed, suggesting that even after the initial split between these populations, there were multiple occasions of interbreeding, potentially occurring over millions of years. The final transition we will consider will be the emergence of the hominin lineage since our last common ancestor with chimpanzees. Figure 11.4 shows a clade diagram of hominin evolution based largely on fossil evidence.

It is tempting to think of our last common ancestor with chimpanzees as being essentially similar to living chimps. But of course, evolution has been occurring just as surely on the lineage leading to chimpanzees as on our own. One important issue thus concerns whether the last common ancestor of humans and the African great apes had hands similar to those of extant apes, such as chimpanzees. Chimpanzee and gorilla hands are highly specialized for knuckle walking and are thus not intermediate between those of Old World monkeys and humans, but rather are specialized for a mode of life that humans do not share. Indeed, fossils of early apes, such as *Proconsul* from fifteen to twenty million years ago, show hands quite similar to those of Old World monkeys (Almécija et al., 2012; Napier & Davis, 1959) and nothing at all like modern African great apes.

During most of the twentieth century, it was believed that the lineage leading to humans had diverged early from the common lineage leading to all the other great apes (orangutans, gorillas, chimpanzees, bonobos). The specialized features of great ape hands for brachiation and knuckle walking were thus interpreted as homologous features that evolved after the lineage leading to humans had diverged from the great apes (Le Gros Clark, 1959; Napier, 1962b; Straus, 1949). On this view, the last common ancestor of humans and extant great apes would have hands totally unlike those of living chimpanzees. A typical statement of this view is given by John Napier (1962b) in *Scientific American*: "The extremely specialized form of the hand in the anthropoid apes can in no way be regarded as a stage in the sequence from tree shrew to man" (p. 59).

The emergence of molecular genetic tests over the past three decades led to reassessment of this view. It became apparent that the lineages leading to orangutans split off first, followed by the lineage leading to gorillas, before the split of the lineages leading to chimpanzees and humans. Thus, there was no common ancestor of gorillas and chimpanzees that was not also an ancestor of humans. Therefore, supposing that the common ancestor of chimpanzees and humans did not have the specialized features for brachiation and knuckle walking, they must have evolved independently in the lineages leading to gorillas and chimpanzees. Knuckle walking was seen by many as too bizarre a behavior to have risen independently in different genera. Accordingly, many researchers argued that from parsimony, it was more likely that the common ancestor's hand had these specialized features, which were lost specifically in the hominin lineage (Richmond et al., 2001; Richmond & Strait, 2000; Tocheri et al., 2008). On this view, human ancestors went through a "troglodytian" stage of knuckle walking, similar to modern chimpanzees, before the advent of bipedalism. The modern human hand would thus be highly derived, while the features seen in gorillas and chimpanzees would be primitive features of all apes.

Recent research, however, has provided evidence supporting the earlier view that the human hand is generally primitive and the hands of extant great apes derived. For example, studies of the anatomical basis of knuckle walking in gorillas and chimpanzees, and its developmental emergence, have provided strong evidence that this behavior has indeed evolved separately in these apes (Kivell & Schmitt, 2009). Another important piece of evidence comes from the discovery of fossils of *Ardipithecus ramidus* (nicknamed "Ardi") in the Awar region of Ethiopia (White et al., 2009). Ardi is an early hominin that predates the earliest known australopithecine (*Australopithecus anamensis*) by around two hundred thousand years. These fossils have been dated to 4.4 million years ago, relatively close to the last common ancestor of hominins and chimpanzees. Lovejoy et al. (2009) interpret the hand of *Ar. ramidus* as entirely lacking the specializations for brachiation and knuckle walking seen in extant great apes. In terms of the length of the metacarpals of D2–D5, the size and apparent musculature of the thumb, and other features, the hand of *Ar. ramidus* seemed strikingly unlike those of chimpanzees. Marzke (2013) suggests that the *Ar. ramidus* hand is most similar to that of Old World monkeys and, indeed, to the primitive ape *Proconsul*.

Similar conclusions have been drawn from the hand of *Orrorin tugenensis* (Senut et al., 2001) found in Kenya and dated to approximately six million years ago, very close to the split between the human and chimpanzee lineages. *Orrorin*'s hand shows several characteristics that appear strikingly humanlike and dissimilar to chimpanzees (Almécija et al., 2010). These include protrusions on the thumb bones, providing attachments for the FPL muscle, which increases thumb strength, and wide apical tufts at the tip of the phalanges, which increase fingertip surface area facilitating precision grips (Mittra et al., 2007). This latter feature is shown in figure 11.5, which shows distal thumb phalanges from extant African great apes, *Orrorin*, and modern humans. The apical tufts are very well developed in *Orrorin*, suggesting that it may have had a dexterous precision grip. Although the FPL muscle is absent in gorillas and chimpanzees, it has been found not only in *Orrorin* but also in virtually all fossil hominins, including *Ar. ramidus* (Lovejoy et al., 2009), *A. afarensis* (Ward et al., 2012), *A. africanus* (Ricklan, 1988), *Australopithecus sediba* (Kivell et al., 2011), *Paranthropus robustus* (Susman, 1988), *H. habilis* (Napier, 1962a), and *Homo naledi* (Kivell et al., 2015). Indeed, similar evidence for precision grip ability has been reported in the Miocene ape *Oreopithecus bambolii* (Moyà-Solà et al., 1999), which has been dated to eight million years ago and may even predate the last common ancestor of humans and chimpanzees. Thus, specialization of the thumb for grasping appears to have begun very early in the hominin lineage.

Two recent studies have used phylogenetically informed geometric morphometric analyses to understand the evolution of hand shape in apes (Almécija et al., 2015; Prang et al., 2021). These studies have, however, reached contradictory conclusions. Almécija et al. (2015) conducted a principal component analysis on hand proportions in 270 living simian primates, reflecting a wide range of monkeys, apes, and humans, as well as several fossil hands, including from *Proconsul* and *Ar. ramidus*. By considering the phylogenetic relations among species, they estimated the hand structure of the last common ancestor of humans and chimpanzees. This analysis found that the chimpanzee hand is highly derived compared to the last common ancestor, whereas humans (and gorillas) have hands that are more primitive. In fact, the estimated hand of the last common ancestor was quite similar to that of *Ar. ramidus*. On this interpretation, the human hand is

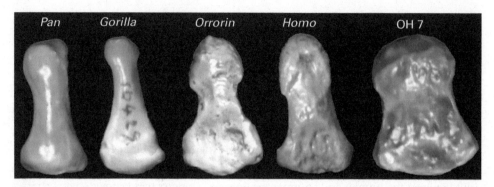

Figure 11.5
Distal phalanges from the right thumb of a chimpanzee (*Pan*), a gorilla, the Miocene hominin *O. tugenensis*, a modern human, and the type specimen of *H. habilis* from Olduvai Gorge (OH 7). All bones have been scaled to the same length. Adapted from Almécija et al. (2010).

only slightly modified from the last common ancestor with chimpanzees, with slightly shorter fingers and a slightly longer thumb—modifications that likely occurred with the emergence of habitual bipedalism and tool use.

In contrast, a more recent study using a similar logic (Prang et al., 2021) reached very different conclusions. Prang and colleagues argue that the hand of *Ar. ramidus* is more similar to that of chimpanzees than claimed by Lovejoy et al. (2009), showing features such as finger curvature consistent with brachiation. On this view, then, the last common ancestor may have had hands that were more chimpanzee-like. One intriguing result of Prang and colleagues' study is the finding of an evolutionary shift in hand and wrist structure between *Ar. ramidus* and all subsequent hominins. This suggests that major changes in hand structure may have emerged with the first australopithecines three to four million years ago, which is broadly consistent with the earliest reported evidence for hominin stone tool use (Harmand et al., 2015; McPherron et al., 2010), discussed in the previous chapter.

There are several features of the modern human hand that differ from all other living primates and that are commonly linked to the extraordinary dexterity of the human hand. These include the relative length of the thumb to the other fingers (Napier, 1961), wide apical tufts on the distal phalanges (Susman, 1988), a saddle structure in the joints that allows the thumb to rotate rather than merely press against the index finger (Napier, 1962a), numerous changes to the carpometacarpal and metacarpophalangeal joints (Lewis, 1977), wide and robust thumb metacarpal bones (Susman, 1994), radial orientation of the third metacarpal (Susman, 1979), and changes to the intrinsic muscles (Tuttle, 1969b) and tendons (Shrewsbury et al., 2003) of the thumb. Recent summaries of these features, linking them to evidence from fossil hominins, are given by Tocheri et al. (2008), Marzke (2013), and Kivell (2015).

Kivell (2015), reviewing this literature, argues that there are three key manipulative abilities that appear to be unique to the human hand: (1) the ability to rotate and manipulate objects with one hand using thumb and fingertips, (2) forceful precision grip with the finger pads forcefully stabilizing an object and withstanding large external forces, and (3) power squeeze gripping of cylindrical objects. How these abilities emerged during hominin evolution, along with the physical characteristics of the hand and nervous system, remains poorly understood. However, recent fossil discoveries have provided intriguing clues about this, even if they haven't provided a neat and tidy story.

To date, four relatively complete early hominin hands have been found, as well as more fragmentary remains from several other species (Kivell, 2015). All four of these mostly complete hands have been discovered in the past thirty years and are still being described and interpreted. I will describe these in broad chronological sequence. The earliest is *Ar. ramidus* (Lovejoy et al., 2009), which has already been described. The next mostly complete fossil hominin hand is from the "Little Foot" australopithecine skeleton found at Sterkfontein in South Africa (Clarke, 1999; Pickering et al., 2018), dated to 3.7 million years ago. Little Foot's hand shows a complex mix of features, some of which appear humanlike (e.g., a robust thumb), others more apelike (e.g., curved proximal phalanges), and others that appear unlike any extant species (e.g., the thumb metacarpal joints). This suggests that Little Foot had relatively limited manual dexterity, consistent with the absence of stone tools from the area it was found (Pickering et al., 2018).

While no complete hand has been recovered from *A. afarensis* (the species of the famous Lucy skeleton), numerous individual hand bones have been found, such as the three-million-year-old bones from Hadar, Ethiopia (Ward et al., 2012), although their interpretation has been controversial. Some authors have interpreted the *A. afarensis* hand as similar to that of gorillas in its proportions (Rolian & Gordon, 2013) and apelike in metacarpal structure (Susman, 1994), which would suggest that they were unable to produce humanlike precision grips or to use tools. Other researchers, however, have argued that *A. afarensis* likely had at least some level of ability to produce the precision grips needed to make Oldowan stone tools (Marzke, 1983, 1997) or even fully humanlike precision grip ability (Alba et al., 2003; Almécija & Alba, 2014). Indeed, it is very possible that the earliest evidence for stone tool use in Africa was from this species (Harmand et al., 2015; McPherron et al., 2010).

Similarly, no complete hand skeleton of *A. africanus* has been found. Studies of isolated hand bones suggested that *A. africanus* had some features characteristic of the modern human hand, including broad apical tufts on the distal thumb phalanx (Ricklan, 1988) and the proportions of the metacarpals (Green & Gordon, 2008), alongside other more primitive features such as the structure of its wrist bones (Kibii et al., 2011). This pattern has been interpreted as evidence for limited dexterous precision grip use (Marzke, 1997, 2013). A recent analysis of bone density in the metacarpals, however, has provided evidence that *A. africanus* may have used its hands in surprisingly humanlike ways (Skinner et al., 2015). Skinner and colleagues identified patterns of bone density resulting from the peak forces arising from habitual use of the hands for activities such as tool use, which produces characteristic patterns of a specific type of bone tissue called "trabecular bone." This pattern was clearly present in the metacarpal bones of *A. africanus* from Swartkrans, South Africa. Morphometric analyses of the shape of the metacarpals, however, have indicated that these animals would have been limited in the range of hand movement they could perform (Galletta et al., 2019; Marchi et al., 2017).

The third hominin species for which there is a nearly complete hand is *A. sediba* (Kivell et al., 2011) from Malapa, South Africa, just under two million years ago. This hand features a diverse array of primitive and derived features. Its wrist shows features unlike modern human hands in the moderate depth of the carpal tunnel and primitive joints, while the phalanges are somewhat curved. Similarly, the metacarpal bones are relatively thin, which would imply that *A. sediba* was unlikely to use tools according to the criteria proposed by Susman (1994). While, to date, no tools have been found associated with *A. sediba*, there is clear evidence for contemporaneous use of tools made of stone and bone at the Swartkrans (Backwell & d'Errico, 2001) and Sterkfontein (Kuman & Clarke, 2000) sites, both in the immediate vicinity of Malapa. Consistent with the interpretation that *A. sediba* used tools, its hands also show a range of features that are more humanlike, suggesting some level of manual dexterity. These include broad apical tufts on the distal phalanges, a long thumb, and well-developed thumb muscles.

Finally, a nearly complete hand was found of *H. naledi* from the Rising Star cave system in South Africa (Kivell et al., 2015), dated to between 236,000 and 414,000 years ago. As with *A. sediba*, the hand of *H. naledi* shows an intriguing mixture of humanlike and primitive features. Its thumb is long and robust and shows attachments for humanlike musculature such as the FPL. It thus likely was able to make the forceful precision grips that are critical to effective tool manufacture and use. *H. naledi*'s wrist is strikingly modern, within the range of variation seen in Neanderthals and modern humans. At the same time, the fingers are strongly curved, suggesting that *H. naledi* used its arms for climbing or suspending itself from branches. The seeming specialization of the hand for both object manipulation *and* arboreal locomotion is striking, and it calls into question the traditional view, going back to Darwin (1871), that manipulative skill could only emerge once the hand was freed from locomotor functions.

The humanlike wrist of *H. naledi* is especially notable, as it is part of a suite of anatomical features that are shared by the modern human hand and the hands of Neanderthals. Traditionally, the hands of Neanderthals had been viewed as less capable of precision grips than modern human hands (Musgrave, 1971). Such differences could be related to different tool kits, such as specialization in Neanderthals for stronger power grips for the control of hafted tools (Bardo et al., 2020; Niewoehner, 2001; Trinkaus & Villemeur, 1991). Recent results, however, have suggested both that Neanderthals were capable of dexterous precision grips (Niewoehner et al., 2003) and that such grips were commonly used in everyday activities (Karakostis et al., 2018). Indeed, a recent geometric morphometric analysis of hand shape found that Neanderthal hands fell within the range of modern humans (Galletta et al., 2019). This is, of course, consistent with the well-established fact that Neanderthals manufactured and used tools, described in the previous chapter, including highly advanced techniques such as the Levallois prepared core technique and multicomponent hafted tools using complex adhesives. Indeed, at Near East sites where Neanderthals and early modern humans coexisted between 140,000 and 50,000 years ago, it is nearly impossible to distinguish which people were at a given site from assemblages of tools and other artifacts (Churchill, 2001). Recent research has suggested that the hands of a recently discovered archaic hominin known as Denisovans may have been even more similar to modern humans than were Neanderthal hands (Bennett et al., 2019).

A number of isolated fossil hand bones have provided some insight into the emergence of this suite of modern hand features over the past two million years. One set of results have been found at the site of Sierra de Atapuerca in northern Spain. Fossils from the Sima de los Huesos (Pit of Bones) cave have been dated to 430,000 years ago and attributed to *H. heidelbergensis*—the species that was likely to have been the immediate ancestor of both Neanderthals and modern humans. These fossils show a hand very similar to that of modern humans, clearly capable of a range of precision and power grips (Arsuaga et al., 2015). This strongly suggests that the similar features of human and Neanderthal hands did not evolve separately but were inherited from a common ancestor. Older fossils from the Gran Dolina cave at Atapuerca, dated to nine hundred thousand years ago and attributed to *Homo antecessor*, also show broadly humanlike hand structure (Lorenzo et al., 1999). Finally, an even older proximal phalanx dated to 1.3 million years ago at the Sima del Elefante (Pit of the Elephant) cave site at Atapuerca again shows a broadly modern hand form (Lorenzo et al., 2015).

In Africa, evidence for modern hand structure can be seen at even earlier sites. Ward et al. (2014) describe a 1.42-million-year-old third metacarpal from Kaitio in Kenya with an essentially modern form. Similarly, Domínguez-Rodrigo et al. (2015) describe a 1.84 million-year-old proximal phalanx from Olduvai Gorge in Tanzania, which also appears modern. This latter result is particularly interesting for a couple of reasons. First, it is broadly contemporaneous with the earliest reported Acheulean hand axes, which, as discussed in the last chapter, have been found from several East African sites dated to around 1.7 million years ago (Beyene et al., 2013; Lepre et al., 2011), including Olduvai Gorge (Diez-Martín et al., 2015). Second, the presence of this member of *Homo* with a modern hand shape seems to have coexisted at Olduvai with at least two other hominin species: *H. habilis* (Leakey et al., 1964), whose hand appears much less modern (Napier, 1962a), and *Paranthropus boisei*, a robust australopithecine also known as "Nutcracker man" (Leakey, 1959).

Overall Brain Size and Organization

Primates collectively have larger brains than would be predicted for mammals of their overall body sizes, being on average about double what would be expected from the basic mammalian pattern (Schoenemann, 2006). This is true for tree shrews as well, suggesting that early mammals had brains that were small in absolute terms (as little as 2–3 g) but proportionately large as a function of body size compared to closely related mammals such as rodents and rabbits (Striedter, 2005). While there is substantial variability among the brains of living primates, brain size appears to have increased repeatedly during primate evolution.

The left panel of figure 11.6 shows brain volume as a function of total body size for a range of living primate species. While prosimians such as galagos have larger cortical sheets than tree shrews (Wong & Kaas, 2010), brain size is substantially increased in monkeys compared to prosimians. While brain size appears to scale with body size in a similar way in prosimians and monkeys (i.e., the slope of the allometric scaling function is similar), for any given body size, monkeys show a substantially larger brain than prosimians (i.e., the allometric scaling function is shifted upward). While Old World monkeys and apes tend to be larger than New World monkeys, their brain size to body size appears to scale with body size in a similar way.

In humans, this pattern is exaggerated even more, with humans having brains more than three times as large as would be expected for a primate of our body size (Passingham, 1975). The right panel of figure 11.6 shows brain volume as a function of body size for apes and hominins. In australopithecines, there is again an increase in brain size for a given body size compared to apes. In species within the genus *Homo*, however, a

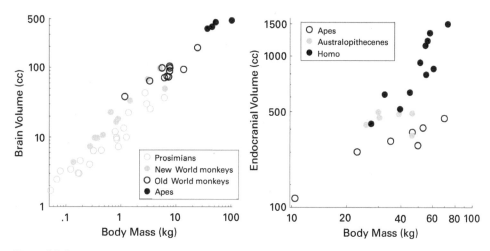

Figure 11.6
The relation between body size and brain size in primates.

more dramatic change occurs. The scaling function is not merely shifted upward but also increases dramatically in slope (Hofman, 1983; Pilbeam & Gould, 1974). This has led to an explosion in overall brain size in our lineage.

As brains increase in size, they do not just scale up homogenously. There are complex patterns of allometric scaling, such that different parts of the brain tend to increase or decrease in size as a proportion of total brain size (Jerison, 1973). There has been intense debate about whether different parts of the brain change in a concerted fashion as a result of developmental constraints on the embryological formation of the brain (Finlay et al., 2001; Finlay & Darlington, 1995) or in a more piecemeal mosaic fashion, with allometric scaling reflecting functional constraints in which brain regions need to scale differently to maintain functional equivalence (Barton & Harvey, 2000; DeCasien et al., 2022). Most likely, both types of process occur (Striedter, 2005).

An interesting approach to this issue was taken by de Winter and Oxnard (2001), who applied principal component analysis to measures of the proportionate size of different brain regions in 363 species of mammals. Just three variables were able to account for 85 percent of the variance in this data, providing striking evidence for the extent of linked regularities in the sizes of brain areas across scales. Within this three-dimensional space, major orders of mammals (e.g., primates, bats, insectivores) were clearly separated. Within primates, however, animals tended to cluster in terms of similarity of lifestyle rather than according to phylogenetic relationships. Specifically, there were clusters of primates reflecting a spectrum from hindlimb dominance (e.g., leaping and scurrying) to broadly equal limb usage to forelimb dominance (e.g., climbing and swinging). Increasing forelimb dominance was associated with proportionate increases in the size of the cerebral cortex, cerebellum, striatum, and diencephalon, and decreases in the size of the medulla and brain stem. As de Winter and Oxnard note, this is consistent with the forelimb-dominated lifestyle reflecting expansion of higher levels of the motor hierarchy, associated with an increase in voluntary over stereotyped motor control. The transition to forelimb-dominated lifestyles is also seen in patterns of somatotopy in the motor cortex and PMC. Whereas rodents show substantial overlap between responses in M1 and PMC for movements of the forelimbs and hind limbs, monkeys show a clear separation of arm and leg responses (Strick et al., 2021).

There is a long-running debate about whether distinctly human features of cognition result primarily from overall brain size (Herculano-Houzel, 2012; Sherwood et al., 2012) or from more specific features of brain organization (Smaers & Soligo, 2013). In parallel, there is a debate about whether changes in brain size (Falk, 1985) or organization (Holloway, 1981b) occurred earlier in hominin evolution. One recent study investigated the endocranial casts of fossil humans across the approximately five hundred thousand years since our lineage split from Neanderthals and Denisovans (Neubauer et al., 2018). Brain sizes reached modern sizes as much as three hundred thousand years ago, whereas brain shape appeared to undergo more gradual changes, reaching modern conditions only within the last hundred thousand years. These results indicate that changes in brain size occurred earlier in hominin evolution but that changes in the organization of specific brain regions are more closely linked to the emergence of modern human behavior and cognition associated with the Upper Paleolithic "revolution", as described in the previous chapter.

The Old and New Motor Cortex in Primates
As discussed above, across many mammals, manual dexterity is related to the presence of CM projections (Heffner & Masterton, 1975, 1983). The same pattern can be seen in detailed comparisons of species of monkey that differ in their manual abilities. For example, like Old World macaques, New World capuchin monkeys use a wide range of dexterous hand grips to interact with objects (Costello & Fragaszy, 1988) and regularly make and use tools in the wild (Mannu & Ottoni, 2009). In striking contrast, New World squirrel monkeys do not use precision grips, show much clumsier interactions with objects, and have not been observed using tools in the wild. Paralleling these behavioral differences, both macaque (Maier et al., 2002) and capuchin (Bortoff & Strick, 1993) monkeys have been shown to have robust CM projections to the hand and arm regions of the spinal cord (C6–T1), while these connections have been found to be much more modest for

squirrel monkeys (Bortoff & Strick, 1993; Nakajima et al., 2000). Bortoff and Strick (1993) used anterograde tracing to investigate the spinal projections of neurons in the hand and arm region of the primary motor cortex in capuchin and squirrel monkeys. In squirrel monkeys, only sparse CM connections were found, with most cortical connections terminating on spinal interneurons. In capuchin monkeys, in contrast, there were additional dense projections directly onto motoneurons in the ventral horn of the spinal cord. These CM connections were especially dense at the boundary between the cervical and thoracic vertebrae (C8–T1), in which they formed a ring encircling the group of motoneurons. Motoneurons at this level are involved in the control of the hand and arm, indicating that these CM projections are particularly involved in fine control of the hand.

Of course, directly specifying movements rather than relying on preset motor programs is complex and difficult. Accordingly, comparative studies have shown strong relations across species in the absolute sizes of the corticospinal tract and the cerebral cortex (Nudo & Masterton, 1990). By the same token, even in humans, the indirect pathway remains operative to some degree. While damage to direct CM fibers impairs fine manual motor control in macaque monkeys (Lawrence & Kuypers, 1968), subsequent studies have showed that monkeys can recover some level of fine manual control (Alstermark et al., 2011; Sasaki et al., 2004). There is evidence that propriospinal interneurons are necessary for this recovery, as no recovery is found when their function is selectively blocked (Tohyama et al., 2017), although recovery may also be subserved by a broader range of both cortical (Nishimura et al., 2007) and subcortical (Nishimura et al., 2009) structures. Similarly, use of a viral vector to selectively disrupt function of the propriospinal interneuron system in monkeys produced deficits in skilled reaching and grasping, as least in the immediate term (Kinoshita et al., 2012).

In a series of studies, Roger Lemon and his colleagues (Lemon et al., 2004, Maier et al., 1998, Nakajima et al., 2000) compared the CM and propriospinal pathways in different species of monkeys. Figure 11.7 shows a model of these systems in four species of mammal proposed by Nakajima and colleagues. In cats, signals to motoneurons controlling the forelimb at spinal segments C6–T1 are dominated by the reticulospinal pathway from the lateral reticular nucleus (LRN) in the brain stem, with more modest inputs from the corticospinal tract, which largely terminate on propriospinal interneurons. In squirrel monkeys, the corticospinal tract has taken on a larger role, but the influence of direct CM connections remains modest. In macaques, in contrast, the CM system starts to play a dominant role in controlling the hand—a trend that reaches an extreme in humans.

As discussed in chapter 3, the emergence of the CM system has been paralleled by changes in the organization of the motor cortex. Whereas old M1 uses the repertoire of spinal motor programs to construct actions, new M1 constructs actions more directly,

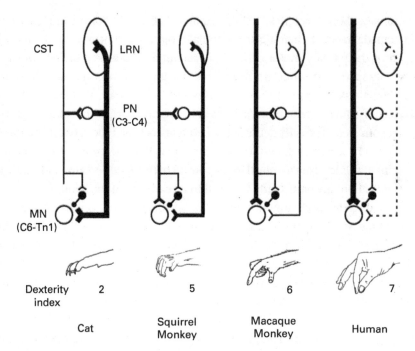

Figure 11.7
Differences in the organization of the corticospinal and reticulospinal pathways across species. CST, corticospinal tract; LRN, lateral reticular nucleus; PN, propriospinal interneuron; MN, motoneuron. From Nakajima et al. (2000).

allowing greater flexibility and control over precise manipulative actions. Old M1 is presumably common to all placental mammals, while new M1 may have developed in the first monkeys. It is an intriguing possibility that new M1 may have emerged as a duplication of old M1. Notably, separate rostral and caudal representations of the hand are also found in squirrel monkeys (Strick & Preston, 1982) and owl monkeys (Stepniewska et al., 1993), suggesting that the duplication of M1 hand areas occurred separately from the emergence of rich CM projections. This is consistent with findings that the propriospinal pathway may remain important for skilled action, even in species with a well-developed CM system (Kinoshita et al., 2012). The more extensive CM projections seen in capuchin monkeys are likely to be an instance of convergent evolution, as they are not seen in other New World monkeys such as squirrel monkeys, and there was no common ancestor of capuchin and macaque monkeys that was not also an ancestor of squirrel monkeys (see figure 11.3).

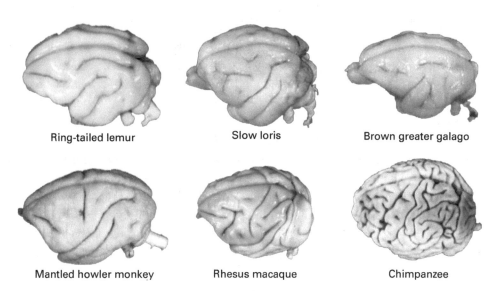

Figure 11.8
The emergence of the central sulcus in the simian cortex. Prosimians (top row), such as lemurs, lorises, and galagos, generally lack a central sulcus separating the frontal and parietal lobes. In the simian lineage (bottom row), in contrast, the central sulcus is a prominent feature of the brain, separating the primary motor and somatosensory cortices. This can be seen in New World monkeys (e.g., the howler monkey), Old World monkeys (e.g., the macaque), and in apes (e.g., the chimpanzee) as well as, of course, in humans. Brains are not shown to actual scale. From the University of Wisconsin, Madison Brain Collection, images from www.brainmuseum.org. Preparation of these images and specimens was funded by the National Science Foundation and the National Institutes of Health.

As discussed above, the cortical magnification of different body parts in sensorimotor cortices is systematically related across species to ethological aspects of each animal's lifestyle. Magnification of the sensorimotor cortex in simian primates can be seen by the formation of the central sulcus itself, separating M1 from S1. The sulcal fold allows more cortical territory to be devoted to these regions. While prosimians do not generally have a sulcus separating the primary motor and somatosensory cortices (and, hence, the frontal and parietal lobes), the central sulcus is one of the most salient landmarks in the brain of simians (Radinsky, 1975). The top row of figure 11.8 shows the brains of three prosimian species. In each case, there is no central sulcus separating the frontal and parietal lobes. Instead, there is a horizontal sulcus (the coronal sulcus), which separates the hand and face regions of the primary motor and somatosensory cortices (Radinsky, 1975). Notably, a central sulcus can be identified in the fossil endocasts of the twenty-five- to thirty-million-year-old monkey *Aegyptopithecus* (Radinsky, 1973), which predates the split between Old World monkeys and apes.

An analogous increase of magnification, but specific for the hand, can also be seen specifically in apes. Recall from chapter 3 that, in humans, the hand region of M1 is located on a "knob" of the precentral gyrus linked to the presence of a buried gyrus inside the central sulcus (the PPFM; Yousry et al., 1997). A similar landmark is present in chimpanzees and is also activated by hand actions (Hopkins et al., 2010). Hopkins et al. (2014) used structural MRI to investigate the surface area of the central sulcus and patterns of cortical folding in ten primate species, including humans, great apes, baboons, and Old World monkeys. Intriguingly, the PPFM was apparent only in great apes and did not occur in baboons or monkeys. The presence of the PPFM in great apes was linked to an increase in the surface area of the central sulcus as a proportion of overall cortical surface area. Notably, the PPFM has the effect of increasing cortical surface area—and, hence, cortical magnification—specifically in the hand region of the primary motor and somatosensory cortices. Thus, increased cortical magnification of the hand in the sensorimotor cortex appears to be a feature that distinguishes great apes as a group from other primates. However, it does not appear to be a feature that distinguishes humans from other great apes. The depth of the PPFM was actually less in humans than in bonobos, chimpanzees, and gorillas. Indeed, in humans, M1 is only 33 percent as big as would be expected for a primate of our brain size (Schoenemann, 2006). Note, though, that since the human brain as a whole is three times as big as would predicted based on our body size (Passingham, 1975), the size of M1 is very close to what would be predicted allometrically for a primate of our body size.

The Somatosensory Cortex in Primates

Just as we saw high levels of cortical magnification in the somatosensory cortex of the bill in the duck-billed platypus and of the star of the star-nosed mole, reflecting the ecological centrality of those body parts to the animal's mode of life, so too in primates we see magnification of the glabrous skin of the palm of the hand and the palmar surface of the fingers. These representations appear substantially expanded in monkeys compared to prosimians (Kaas et al., 2018). This is clearly apparent in the somatotopic map of the owl monkey shown in figure 2.3 in chapter 2 (Merzenich et al., 1978). Note that not only is the representation of the glabrous skin of the hand dramatically larger than that of hairy skin but also the entire somatotopic organization of the map is structured around the glabrous skin, the hairy skin being either interdigitated between glabrous representations (as in Area 1) or stuck in at odd places without preserving somatotopy (as in Area 3b). In monkeys with prehensile tails, such as the spider monkey, there is also a somatotopic representation of the tail with high magnification (Pubols & Pubols, 1971), but even here, the magnification of the tail remains less than that of the glabrous hand.

Differences in magnification are also seen at the level of individual digits. As discussed in chapter 2, in humans, fMRI studies have found that the magnification of the fingers is highest for the thumb and index finger and progressively smaller toward the little finger (Duncan & Boynton, 2007; Schellekens et al., 2021). This pattern is mirrored by the tactile spatial acuity of each digit (Duncan & Boynton, 2007; Vega-Bermudez & Johnson, 2001) as well as higher-level aspects of hand representation, including the size of fingers in proprioceptive hand maps (Longo & Haggard, 2010) and explicit judgments of finger length (Longo & Haggard, 2012). It is natural to interpret these differences in magnification across the fingers in terms of the importance of the thumb and index finger for precision grips. However, it must also be noted that a similar pattern is apparent in the representation of the fingers in area 3b of owl and squirrel monkeys (Merzenich et al., 1987)—species that do not use precision grips and are not noted for highly dexterous hand actions.

In area 3a, differences between species appear to have a closer connection to manipulative abilities. Macaque monkeys show clearly distinct representations of the thumb and index finger in area 3a, but neurons with RFs on the other three fingers are overlapping (Krubitzer et al., 2004). In contrast, for New World monkeys such as marmosets, none of the fingers have distinct representations in area 3a (Huffman & Krubitzer, 2001). These differences parallel the manual skill of these monkeys. While macaques show skilled precision grips (Torigoe, 1985) and tool use (Gumert & Malaivijitnond, 2013), marmosets tend to use only power grips (Torigoe, 1985).

Recall from chapter 2 that S1 in humans, apes, and Old World monkeys consists of four distinct areas: 3b, 3a, 1, and 2. Just as M1 and S1 separated from an earlier sensorimotor amalgam, so too has S1 itself separated into multiple distinct areas. Somatosensory areas 1 and 2 have emerged during primate evolution. Krubitzer and Calford (1992) described islands of neurons with in area 3b of flying foxes, which they interpreted as a primitive form of what would eventually become areas 1 and 2 in Old World monkeys and apes. Similarly, in prosimian galagos, there is a somatosensory region posterior to 3b, which is presumably comparable to area 1 but nothing resembling area 2 (Wong & Kaas, 2010). Clear differentiation of areas 3b and 1, with mirror reversal of somatotopic maps, has been found in New World monkeys, including owl monkeys (Merzenich et al., 1978), capuchin monkeys (Felleman et al., 1983), squirrel monkeys (Sur et al., 1982), marmosets

(Krubitzer & Kaas, 1990), and titi monkeys (Padberg et al., 2005), as well as in Old World monkeys such as macaques (Nelson et al., 1980) and, of course, in humans (Blankenburg et al., 2003).

The evolution of area 2 appears less clear. Area 2 is clearly present in Old World macaques (Nelson et al., 1980; Pons et al., 1985) and humans (Grefkes et al., 2001; Sanchez-Panchuelo et al., 2012). However, it appears to be either absent or poorly developed in many species of New World monkeys, such as marmosets (Krubitzer & Kaas, 1990), squirrel monkeys (Merzenich et al., 1978), and titi monkeys (Padberg et al., 2005). Strikingly, however, Padberg et al. (2007) showed a well-developed area 2 in New World capuchin monkeys that is very similar to that seen in macaques. This difference between capuchin monkeys and other New World monkeys is especially notable, given the dexterous hand actions and tool use seen in capuchin monkeys, which has already been mentioned. Padberg and colleagues suggest that area 2 has evolved separately in capuchin monkeys from the Old World monkeys, in both cases relating to the more dexterous hand movements and tool use of these animals. Indeed, in each case, the representation of the hand takes up a much larger proportion of area 2 than in areas 3b or 1 (Krubitzer & Disbrow, 2008).

The PPC in Primates

Comparison of prosimian galagos versus tree shrews shows that the PPC is greatly expanded in primates (Kaas et al., 2018). The PPC in galagos shows at least three subdivisions, of which the posterior-most and anterior-most divisions have been functionally characterized. The more posterior region of the galago PPC shows extensive connections with the visual system (Stepniewska et al., 2016). The more anterior PPC region, in contrast, is linked with manual actions, as shown by both microstimulation (Stepniewska et al., 2011) and inactivation with muscimol (Stepniewska et al., 2014). This suggests that the parieto-frontal reaching and grasping networks were likely present in the earliest primates (Kaas & Stepniewska, 2016).

Microstimulation studies show that similar parieto-frontal networks controlling reaching and grasping exist in New World owl and squirrel monkeys (Gharbawie et al., 2011). There is evidence that posterior parietal area 5 exists in a rudimentary form in titi monkeys (Padberg et al., 2005) and in a more developed form in capuchin monkeys (Padberg et al., 2007). Padberg et al. (2005) suggest that area 5 in monkeys is homologous to the rudimentary PPC seen in many mammals and that the refinement and development of this region in monkeys is linked to dexterous control of the hand.

The PPC has been well characterized in the Old World macaque, as described throughout this book. The macaque PPC appears to be further expanded over and above what is seen in New World monkeys. Chaplin et al. (2013) constructed three-dimensional surface-based models of the brains of marmosets, capuchin monkeys, and macaque monkeys obtained using MRI. They then quantified how the marmoset brain would need to be transformed to match the structure of the monkey brain, constrained by the identification of homologous areas in each brain. The results showed highly differential patterns of cortical expansion in both New World and Old World monkeys, showing that monkey brains are not simply scaled-up prosimian brains. Critically, the IPS and temporoparietal junction regions were among the areas of highest relative expansion (along with the prefrontal cortex), showing expansions of up to sixteen times in capuchin monkeys and thirty times in macaques.

Neuroimaging studies comparing macaques to humans have shown that the PPC has further expanded in our lineage (Hill et al., 2010; Kastner et al., 2017; Orban et al., 2006; Van Essen & Dierker, 2007). Hill and colleagues used an approach similar to that of Chaplin et al. (2013) described in the last paragraph. They quantified areas of relative expansion in transforming a macaque cortex into a human one. Again, the PPC was an area of particular expansion in the human brain. Although the cortex as a whole is about ten times as large in humans as in macaques, regions of the PPC are as much as thirty times larger.

Functional MRI studies have also provided evidence for a range of changes in the organization of the human PPC in comparison with macaques. For example, while in macaques just one region in the IPS appears sensitive to visual motion, four distinct regions have been identified in the human IPS (Orban et al., 2006). Kastner et al. (2017) suggest that the human homologues of the grasp-related AIP and reach-related PRR have shifted outside of the IPS, possibly making space for a set of new, human-specific areas involved in a variety of cognitive processes, including numerical processing (Piazza et al., 2007), tool use (Mruczek et al., 2013; Peeters et al., 2009), episodic memory (Cabeza et al., 2008), working memory (Todd & Marois, 2004), language (Lee et al., 2007; Mechelli et al., 2004), and social cognition (Saxe & Kanwisher, 2003).

It is possible that some of these areas may have resulted from duplications of areas that exist in macaques, as discussed above. For example, Orban and Caruana (2014) suggest that the tool-specific aSMG region may have evolved as a duplication of the grasp-related AIP. Konen and Kastner (2008) have argued that unlike in monkeys, the human dorsal visual pathway contains rich information about object categories, largely paralleling the hierarchical organization of the ventral visual pathway. This duplication of object processing may allow for manual actions to take more detailed account of semantic knowledge of objects, enabling actions to be richer and more flexible and allowing effective use of tools whose functions may not be transparent without detailed shape processing and integration with stored semantic knowledge.

In monkeys, the PPC appears to function largely as a sensorimotor hub for integrating visual and somatosensory information and using it to control actions. In humans, this function is retained, but this basic sensorimotor function has become integrated into a wider set of cognitive functions. Indeed, studies using network analyses of both structural (Hagmann et al., 2008) and functional (Buckner et al., 2009) MRI data have consistently identified the PPC as a "hub" region, centrally placed within the overall organization of the brain. This integration of sensorimotor processing for hand control with widespread semantic knowledge, executive control, and high-level decision-making allows our actions to be flexible, efficient, and situationally appropriate.

The PPC was a focus of the earliest descriptions of fossil hominin brain endocasts, including Raymond Dart's (1925) description of *A. africanus* (figure 11.9) and Franz Weidenreich's (1936) description of *H. erectus*. Multiple lines of evidence suggest that the parietal cortex has undergone major changes during human evolution. For example, more than a century ago, Grafton Elliott Smith (1904) noted morphological differences between humans and chimpanzees in the boundaries of the primary visual cortex. While chimpanzees and monkeys show a large, crescent-shaped sulcus (the lunate sulcus), which forms the anterior boundary of V1, in humans, this sulcus is shifted posteriorly or is absent entirely. Smith interpreted this change as reflecting a relative expansion of the parietal and temporal association cortices in humans at the expense of the primary visual cortex. This interpretation is consistent with modern neuroimaging research showing that there is no sulcus in humans homologous to the lunate sulcus (Allen et al., 2006). Indeed, while the absolute size of V1 is greater in humans than in other primates, it has not kept pace with the rate of increase in overall brain size during evolution, being substantially smaller than would be predicted for a primate of our brain size (de Sousa et al., 2010).

Raymond Dart's (1925) initial description of the fossilized brain endocast of the Taung child (*A. africanus*) focused on this difference. Dart interpreted the lunate sulcus as lying far back in the Taung endocast—a humanlike feature that would indicate an expansion of the PPC. In fact, it seems that Dart was mistaken about his identification of the lunate sulcus (Falk, 2011). There has been considerable controversy surrounding the interpretation of this feature of the Taung endocast, with some researchers interpreting the organization as humanlike (Holloway, 1981b) and others as apelike (Falk, 1980). Notably, a recent study of two other *A. africanus* endocasts found them to show major differences from modern humans, particularly in the parietal lobe (Beaudet et al., 2018). Thus, while it is clear that there has been a relative expansion of the PPC since the last common

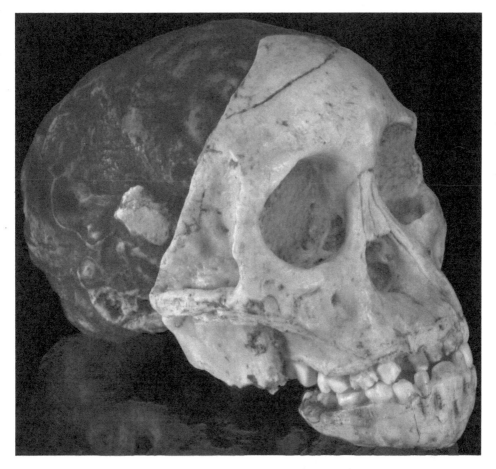

Figure 11.9
The fossilized skull and brain endocast of the Taung child, the type specimen of *A. africanus* (Dart, 1925). Photo by Didier Descouens.

ancestor with chimpanzees, it is not clear from fossil endocasts when this occurred during hominin evolution (Falk, 2014).

Another line of evidence for remodeling of the parietal cortex comes from studies using geometric morphometric tools to investigate shape differences between the human brain and those of nonhuman primates. The human skull and braincase have commonly been described as "globular," highlighting the more bulging, rounded shape of the back of the skull in humans compared to apes and early hominins (Lieberman et al., 2002). These studies have identified a bulging of the upper parietal surface as a key feature of this globular human brain morphology (Bruner et al., 2003; Bruner, 2004). Studies of the developmental progression of human brain development show that this parietal expansion occurs largely within the first two years of postnatal life, in contrast to the more protracted developmental expansion of the frontal lobes (Neubauer et al., 2009).

Interestingly, this parietal expansion is seen even when comparing modern humans with Neanderthals, both in terms of final adult form (Neubauer et al., 2018; Pereira-Pedro et al., 2020) and in terms of the developmental trajectory of brain development (Gunz et al., 2010; Ponce de León & Zollikofer, 2001). One recent study of modern humans showed that the presence of two specific genes associated with Neanderthal DNA are linked to reduced brain globularity (Gunz et al., 2019). *Australopithecus* endocasts appear apelike in this respect (Beaudet et al., 2018; Gunz et al., 2020). Indeed, Holloway (1981a) suggests that "the major changes in endocast surface morphology since the australopithecines have been in the superior parietal region" (p. 165). Comparisons of brain endocasts from *Homo sapiens* from the past three hundred thousand years show that while the earliest humans had brains of similar size to now, the process of globularization shows a more gradual progression, with modern form reached only within the past hundred thousand to thirty-five thousand years (Neubauer et al., 2018).

The PMC and Broca's area

The overall organization of the PMC appears to have shown less change during primate evolution than the PPC. All primates appear to have a broadly similar constellation of premotor regions, including dorsal and ventral premotor regions on the lateral surface of the cortex plus four regions on the medial wall of the frontal lobe: the SMA and three regions within the cingulate gyrus. This basic pattern has been found in prosimian galagos (Preuss & Goldman-Rakic, 1991; Wu et al., 2000), New World owl monkeys (Preuss et al., 1996), New World capuchin monkeys (Dum & Strick, 2005), Old World macaque monkeys (Dum & Strick, 2002), and humans (Fink et al., 1997; Picard & Strick, 2001). This similarity suggests that this organization has been broadly conserved among primates. In humans, the premotor cortices are only about 60 percent as large as would be expected for a primate of our body size on the basis of allometric scaling (Schoenemann, 2006).

One premotor region, the vPMC, however, has shown interesting patterns of change during primate evolution. This region appears to exist only among primates (Passingham & Wise, 2012; Preuss et al., 1996). While in galagos it is poorly connected with M1 (Wu et al., 2000), these regions are richly connected in capuchin monkeys (Dum & Strick, 2005) and macaques (Matelli et al., 1986). Moreover, there appear to be at least two distinct subdivisions of the vPMC in macaques (Matelli et al., 1985; Preuss & Goldman-Rakic, 1991) that are not present in galagos (Wu et al., 2000). Thus, this region seems to have shown progressive changes throughout primate evolution. In capuchin monkeys, Dum and Strick (2005) showed that the dPMC and vPMC are much more densely connected with M1 than are other premotor regions. They thus suggest that these three regions form a network responsible for the generation and control of hand actions.

Intriguingly, the vPMC appears limited to representations of the forelimb and orofacial movements, with no representation of the hind limb. This pattern appears consistently in galagos (Wu et al., 2000), owl monkeys (Preuss et al., 1996), and macaques (Rizzolatti et al., 1988). Rizzolatti and Arbib (1998) suggest that the more anterior ventral premotor region in macaques, area F5, may be homologous to Broca's area in humans—a region with deep links to language (Hagoort, 2005), as will be discussed in more detail in chapter 15.

Broca's area (as well as its homologue in the right hemisphere) is indeed a region of notable difference between the brains of humans and apes, including chimpanzees. While regions homologous to Broca's area in their cytoarchitectonic structure have been identified in all great ape species (Schenker et al., 2008) as well as in some Old World monkeys (Petrides et al., 2005), there has also been clear reorganization in the human lineage. In apes, this region is characterized by the fronto-orbital sulcus, which does not exist in humans (Connolly, 1950). In humans, expansion of this area has caused the gyri to cover the insula (i.e., to "operculate") and form the frontal operculum, consisting of the pars opercularis (Brodmann area 44) and the pars triangularis (Brodmann area 45). Brodmann areas 44 and 45 are more than six times as large in humans as in chimpanzees (Schenker et al., 2010), outpacing the growth of the cortex as a whole (~3×) or the frontal lobe (~4.6×).

The derived humanlike form has been consistently found in fossils attributed to the genus *Homo*, including *Homo rudolfensis* (Falk, 1983), *H. habilis* (Tobias, 1987), *H. naledi* (Holloway et al., 2018), and Neanderthals (Poza-Rey et al., 2017). In contrast, australopithecines have been found to retain the apelike pattern, with clear evidence for a fronto-orbital sulcus, including in *A. africanus* (Falk, 1980), *A. afarensis* (Gunz et al., 2020), *P. robustus* (Beaudet et al., 2019), *A. sediba* (Carlson et al., 2011), and the early australopithecine Little Foot (Beaudet et al., 2019). Interestingly, in the case of *A. sediba* (the most recent of these species by far), there is evidence that this region shows an intermediate stage of organization, combining a fronto-orbital sulcus with partial emergence of a frontal operculum (Carlson et al., 2011), despite the very small overall brain size. Holloway and colleagues thus suggest that a modern form of Broca's area was likely present in the last common ancestor of the genus *Homo*.

This interpretation, however, has been challenged by recent studies of the early *H. erectus* finds from Dmanisi in Georgia from around 1.8 million years ago, which show the primitive, apelike form (Ponce de León et al., 2021). By comparing these skulls to more recent *H. erectus* fossils from across Africa and Asia, this study showed that the derived, humanlike form of Broca's area appeared within *H. erectus* populations between 1.5 and 1.7 million years ago. Another interesting aspect of this study is that changes in the appearance of a humanlike Broca's area appear to be correlated with expansions of the PPC. This suggests that these changes evolved in tandem rather than at different times.

12 The Developing Hand

In 1932, a curious competition was held at the Columbia-Presbyterian Medical Center in New York City. The competitors were two one-month-old twin infant boys, Jimmy and Johnny Woods. The winner would be the infant to suspend themselves by holding onto a bar for as long as possible before falling. In the end, Jimmy emerged victorious with a dominant performance. While Johnny fell within five seconds, Jimmy continued to hold strong thirty-nine seconds in, when available film footage ends.

Jimmy's dominance would be short-lived. The Woods twins were participants in an intensive experiment conducted by the pioneering developmental psychologist Myrtle McGraw (1935), which aimed to determine whether motor skills emerged during infancy in a fixed series of stages as a result of neural development and whether this schedule could be altered by providing infants with practice. Setting up her lab in a nursery, McGraw provided Johnny with a rich set of activities to stimulate his motor abilities eight hours a day, five days a week, while Jimmy received routine care. Before two years of age, Johnny had acquired remarkable abilities in swimming, climbing, and even roller skating—activities that Jimmy was entirely unable to perform. Nevertheless, by the time the boys reached school age, Jimmy had caught up with his brother, and both were normal and healthy.

The effects of training on Johnny's abilities challenged views that argued that motor skills in infancy emerge purely as a result of brain maturation. At the same time, we are confronted with the very curious fact that at one month of age, both boys could support their entire body weight. Many adults would struggle to suspend themselves for thirty-nine seconds! The effects of training and experience that helped Johnny to perform such remarkable feats as an infant could only occur in the context of a nervous system that was innately equipped with an initial set of sensorimotor mechanisms. This chapter will discuss the development of the invisible hand. First, we will discuss the emergence of reaching and grasping before turning to the development of somatotopic maps of the body.

The Development of Grasping and Reaching

The Grasp Reflex

Jimmy and Johnny Woods were able to support themselves at one month of age due to a mechanism known as the "grasp reflex." Put your finger into a neonate's hand and move it gently across the skin of the palm, and the infant's hand will grasp your finger tightly, as shown in figure 12.1. This was first shown by Robinson (1891), who presented it as a critical test of Darwin's theory of evolution. Robinson noted that many primate mothers move by swinging between tree branches, and the infant needs to cling to the mother's fur. If humans have evolved from monkeys, Robinson reasoned, a strong vestigial tendency to grasp something put into the hand should be present in human infants. Sure enough, as Robinson reports:

> I have now records of upward of sixty cases in which the children were under a month old, and in at least half of these the experiment was tried within an hour of birth. . . . In every instance, with only two exceptions, the child was able to hang on to the finger or a small stick three-quarters of an inch in diameter by its hands, like an acrobat from a horizontal bar, and *sustain the whole weight of its body* for at least ten seconds. In twelve cases, in infants under an hour old, half a minute passed before the grasp relaxed, and in three or four nearly a minute. (p. 838)

McGraw (1940) used this fact to measure the developmental trajectory of the phenomenon by measuring longitudinally how long infants would remain suspended before falling. She found that suspension time peaked at around one month of age and then gradually declined over the next three months. A potential confound in this approach is changes in infant weight. Older infants may hang for less time not because the grasp reflex has gotten weaker but because their body has gotten heavier. Thelen et al. (1984) elegantly demonstrated that the analogous disappearance of the neonatal stepping reflex is a result of changes in leg weight by showing that the reflex reappears when the infant is submerged in water, making the legs effectively lighter. That something similar is happening with the grasp reflex is supported by McGraw's own observation that even while suspension duration declines, infants continue to grasp strongly onto a finger placed on their palm. McGraw interprets this dissociation as indicating that the two behaviors reflect distinct underlying mechanisms. However, it could equally be that only the suspension behavior is affected by the infant's changing body weight.

Halverson (1937) argued that the grasp reflex consists of two distinct components: finger closure in response to light tactile stimulation and tight gripping in response to proprioceptive displacement of the finger tendons. Subsequently, on the basis of a study of five hundred infants, Twitchell (1965) argued that there are three distinct elements of the grasp response, which come online at different ages. The first element, and

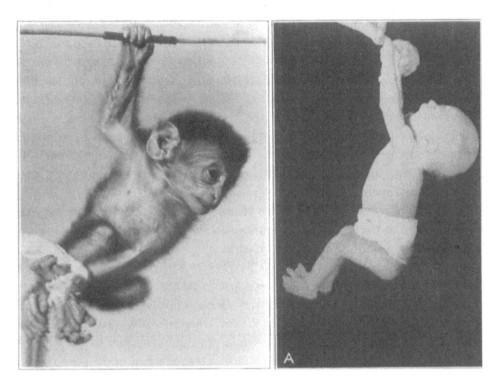

Figure 12.1
The neonatal grasp reflex. A newborn macaque monkey (left panel, from Richter, 1931) and a newborn human infant (right panel, from McGraw, 1940) suspending themselves with the grasp reflex.

the only one present at birth in full-term infants, is what Twitchell calls the "traction response"—a synergy involving combined flexion throughout the arm, including the shoulder, elbow, and wrist, as well as the fingers. Unlike what Twitchell considers the grasp reflex proper, the traction response is not induced by tactile stimuli but rather is triggered by proprioceptive cues involving stretching of the flexor muscles in the arm.

Around two weeks of age, the fully formed grasp reflex appears, in which the finger flexion seen in the traction response can be induced by purely tactile stimulation of the infant's palm. The optimal stimulus is a distally moving stroke on the radial (thumb side) of the hand. Over the next few weeks, the strength of response evoked by purely tactile stimulation increases, while that induced by purely proprioceptive stimulation decreases. Finally, between three and five months of age, tactile stimulation of the infant's palm, especially near the thumb, produces rotations of the hand and groping movements. Twitchell refers to this as the "instinctive grasp reaction."

Collectively, these three reactions reveal the existence of a highly automated, and presumably evolutionarily ancient, set of synergies for controlling hand and arm movements to grasp objects. Twitchell argues that these reflexive mechanisms form the mechanistic substrate for the later development of intentional grasping behaviors and identifies several parallels between the developmental emergence of reflexive and deliberate grasping, as will be discussed below. One intriguing feature of the grasp reflex linking it to later fine manual control is the fact that once it is fully developed, around six months, the reflexive movements evoked become increasingly fractionated. Tactile stimulation at the base of one finger increasingly produces flexion only of that single finger rather than of all the fingers (Twitchell, 1965).

Although the reflexive features of this behavior recede as the infant learns to grasp intentionally over the first year, the continued presence of the underlying mechanisms is shown by the reemergence of all three components of the grasp reflex in neurological patients with brain damage (Adie & Critchley, 1927; Seyffarth & Denny-Brown, 1948). Release of grasping and groping behaviors is specifically linked to damage to the frontal lobes and is likely related to visually driven behaviors involving automatic and compulsive use of objects, such as the anarchic hand syndrome, utilization behaviors, and William James's after-dinner fruit consumption, described in chapter 4.

As suggested by Robinson (1891), a prominent interpretation of the grasp reflex is that it relates to carrying behaviors in which an infant primate clings to its mother as she locomotes (Bishop, 1962). Across primates, there are two broad ways in which mothers carry young infants (Ross, 2001). In some species, infants are kept in a nest or tree hole for several weeks and are occasionally transported by being held in the mother's mouth (oral transport). In other species, infants cling effectively to the mother's fur and are carried on her back (fur grasping)—an ability presumably related to the grasp reflex. Given the prevalence of oral transport in unrelated mammals (e.g., rodents, carnivores), it likely reflects the primitive mammalian condition. Nevertheless, fur grasping appears to have evolved several times independently in different groups of primates (Nakamichi & Yamada, 2009). A recent study investigated the covariation across species in infant carrying behavior and later manual grasping in twenty-one species of prosimians (Peckre et al., 2016). Adults of each species were presented with food items, and their manner of grasping it was coded. Among fur-grasping species, 52 percent of items were grasped with one hand compared to only 11 percent in oral-transporting species. Thus, there is a clear evolutionary relation between early reflexive grasping behaviors in newborns and later adult manual grasping behaviors.

Early Hand–Mouth Behaviors

Human manual behavior does not start at birth. In a pioneering series of studies, Heinz Prechtl and his colleagues (de Vries et al., 1982, 1985; Prechtl, 1985) used ultrasound recording to show that fetuses with gestational age as young as ten weeks make a range of movements, including flexions of the arms and fingers. By fourteen weeks, they make hand movements contacting several parts of their own body and the uterine wall (Sparling et al., 1999), and by fifteen weeks, they bring their hand to their mouth and suck their thumb (Hepper et al., 1991). Indeed, it is not uncommon for neonates to have self-inflicted bruises on their fingers and hand from intense fetal sucking (Murphy & Langley, 1963). Importantly, by nineteen weeks of gestational age, there is evidence that fetuses open their mouth in anticipation of the thumb's arrival (Myowa-Yamakoshi & Takeshita, 2006)—a behavior that increases progressively throughout fetal development (Reissland et al., 2013). This coordination between hand and mouth demonstrates that these hand movements are not merely accidental but appear to be goal directed and deliberate. A similar conclusion was reached by a study that analyzed the kinematic pattern of fetal reaches (Zoia et al., 2007). Through eighteen weeks, hand movements appeared to be random, with jerky and zigzag movements. From twenty-two weeks, however, hand movements became smoother and straighter, showing patterns of acceleration and deceleration broadly similar to those seen in later reaching movements. These results suggest that movement not only emerges before birth but also features increasing complexity and anticipatory control of movement.

Shortly after birth, infants move their hands to their mouths—a behavior that frequently results in them sucking on their fingers (Brazelton, 1956). Kravitz and Boehm (1971) found that all 140 healthy newborn infants they studied managed to suck on their hand within six hours of birth. As with hand-to-mouth movements in fetuses, there is also evidence that newborn infants start to open their mouth in anticipation of hand contact (Butterworth & Hopkins, 1988), providing evidence that these are coordinated actions and not random or haphazard movements.

Such movements may be potential precursors of later hand movements to bring food to the mouth for feeding, which are among the earliest functional behaviors seen in human infants (Sacrey et al., 2012), and which are richly represented in the primate motor cortex, as shown by Graziano's (2016) stimulation studies described in chapter 3. Feldman and Brody (1978) compared infants before and after they were fed, finding that movements of the hand to the mouth were more common when the infants were hungry. Other studies have shown that oral administration of sucrose produces stronger traction grasp responses (Buka & Lipsitt, 1991) and more frequent movements of the hand to the mouth (Blass et al., 1989; Rochat et al., 1988), providing further evidence of a link between these movements and feeding behaviors.

Prereaching

There is also evidence that hand and arm movements in humans are already integrated with vision at birth, although successful reaches for objects will not emerge for several months (McDonnell, 1979; Rader & Stern, 1982; von Hofsten, 1982). Claes von Hofsten (1982) showed that the arm movements of neonates were more likely to be forward directed when a graspable object was placed in front of them than when the object was absent—a behavior he described as "prereaching." Similarly, the infant's arm movements ended closer to the object when the infant was fixating on the object than when looking elsewhere. In somewhat older infants, forward "swiping" movements of the arm are more common when an object is present of the right size for the infant to grasp than when the object is too big to grasp (Bruner & Koslowski, 1972). Longitudinal studies have shown that these prereaching movements change dramatically at around two months of age (von Hofsten, 1984) and become progressively smoother, straighter, and less jerky as the infant approaches the emergence of successfully object-directed reaching (Bhat & Galloway, 2006; Lee et al., 2011).

While the earliest prereaching movements do not succeed in allowing the infant to obtain objects, there is evidence that actions can be used purposefully to achieve desired outcomes in the world. When a newborn infant's head is turned to one side, the arm the infant is facing tends to extend, while the other arm flexes—a behavior known as the "asymmetric tonic neck reflex" (Prechtl, 1977), sometimes called the "fencer's reflex" because of the similarity to the posture adopted by fencers at the start of a match. In an elegant study, van der Meer et al. (1995, 1996) showed that this response is not a purely mechanical response to neck rotation but rather is controlled purposefully to keep the hand in the infant's field of view. As shown in figure 12.2, these authors attached wrist bands to the infant's arms and used a pulley system to apply perturbing forces to each hand. Different infants were given different visual information about their arms. One group had direct vision of the hand they were facing (figure 12.2, left panel). In a second group, vision was blocked by two occluders on either side of the infant's hand (figure 12.2, center panel). Finally, in a third group vision of the opposite arm was provided indirectly through a video feed (figure 12.2, right panel). Babies applied opposing forces to keep the hand in place when it was pulled, but only in the conditions in which this kept the hand in their field of view. For example, in the direct view condition, infants opposed forces applied to the hand they were facing but not the contralateral hand, while the converse pattern held in the video condition where they had visual information about the

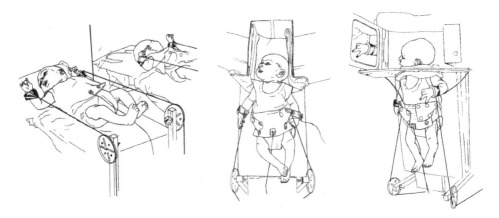

Figure 12.2
Integration of reaching movements with vision in newborn human infants. Adapted from van der Meer et al. (1996).

contralateral hand. This provides clear evidence that early reaching movements can be goal directed and integrated with visual feedback about the consequences of the action.

The Emergence of Reaching

Infants are first able to successfully reach for seen objects from around four months of age (Berthier & Keen, 2006; Thelen et al., 1993; White et al., 1964), although there is considerable variability across individuals. The earliest reaches, however, are uncoordinated, jerky, and curved (Berthier & Keen, 2006; Halverson, 1931). They are composed of sets of discrete sub-movements, with the arm moving in different directions and at different speeds (von Hofsten, 1991). When infants first reach, they tend to rely on movements of the shoulder, with little change in elbow angle (Berthier et al., 1999; Spencer & Thelen, 2000). Reaches become progressively smoother, straighter, and more coordinated across the first three years of life (Berthier & Keen, 2006; Konczak & Dichgans, 1997). Even after three years, however, the smoothness and fluency of reaching continues to develop, not reaching fully adult levels until eight to ten years of age (Schneiberg et al., 2002).

Remarkably, however, as soon as infants are able to reach successfully for stationary objects, they are also able to reach for moving objects. One study followed infants longitudinally from twelve to thirty-six weeks, finding that infants as young as eighteen weeks were able to reach for balls moving in circular trajectories at 30 cm/s (von Hofsten & Lindhagen, 1979). This ability is impressive, as it requires the infant to reach not toward where she sees the ball but rather at an anticipated location requiring prediction of the spatiotemporal trajectory of both the ball and arm (von Hofsten, 1980). Another study of somewhat older infants, between eight and nine months, showed that these reaches are highly precise, the standard deviation of timing errors in intercept being about 40–50 ms (von Hofsten, 1983). The extrapolation of object motion appears to be based on a principle of inertia, such that objects moving on straight paths are inferred to continue traveling straight, even after the infant has been presented with several example of movement paths violating this principle (von Hofsten et al., 1998). Interestingly, when infants are tested on "violation of expectancy" tasks involving looking times, rather than reaching, they do not show understanding of the principle of inertia until months later (Spelke et al., 1992). This dissociation suggests that physical knowledge implicit in sensorimotor processes may be distinct from, and inaccessible by, more conceptual understanding of physical principles.

In addition, as soon as infants reach successfully with full vision, they are also able to reach in the dark. Clifton et al. (1993) tested infants longitudinally from six to twenty-five weeks of age in conditions in which attractive objects covered with luminous paint were presented in full light or in the dark. The first successful contacts with these toys were found on average at similar ages in the light and dark (12.3 weeks vs. 11.9 weeks). While the infants in the dark condition are able to see the glowing objects, they are not able to see their hand or arm. That reaching in this condition develops in parallel to reaching in the light thus suggests that visual guidance of the hand is not necessary for successful reaching. Indeed, infants are also able to reach successfully to nonluminous objects in darkness based on sounds they make (Perris & Clifton, 1988). However, reaches to heard objects occur later in development (Stack et al., 1989) and are less accurate (Clifton et al., 1994) than reaches to glowing objects. This suggests that whereas vision of the *hand* is not important for early reaching, vision of the *target* is.

While real-time vision of the hand does not appear critical for the earliest reaches, it may nevertheless be that a developmental history of seeing the hand is important. Evidence for this in monkeys comes from a study by Held and Bauer (1967), who raised newborn macaques in an apparatus such that a cloth bib prevented them from seeing their arms and body, although the arms were not restrained. Macaques reared under more normal (laboratory) conditions make successful reaches for seen objects before one month of age (Mason et al., 1959). The monkeys reared without vision of their arms, however, showed poor reaching coordination when first allowed to see their arms at thirty-five days of age. After twenty days of daily one-hour sessions with vision of the arm, however, they became quite proficient at reaching. In a subsequent study (Held & Bauer, 1974), the same authors compared monkeys raised with bibs that were either opaque or transparent. Testing of reaching was done in a situation in which vision of the hand and arm was blocked for all monkeys, forcing them to use haptic feedback to guide the reaches. Performance was again dramatically worse for the monkeys raised without vision of their arm.

The Emergence of Grasping

Being able to move one's arm to an object doesn't imply being able to successfully grasp it. The development of grasping appears to lag that of reaching, such that the reaches of a three-month-old consist essentially of swatting at objects rather than skillfully grasping them. Before around six months of age, as Halverson (1932) puts it, "the hand of the infant moves about, paw-like, in so fixed a manner that this member appears to be only a distal hinge of the forearm" (p. 35). One study suggested that there may be an abrupt transition at around five months from reaching without effective use of the fingers to being able to grasp as well (Wimmers et al., 1998). The ability to grasp appears to be a major bottleneck in the infant's developing ability to engage with the world. In an influential series of studies, Amy Needham and her colleagues have provided prereaching infants with the ability to pick up objects using Velcro-covered "sticky mittens." This training produced systematic changes in the infant's visual attention and exploration of objects (Libertus & Needham, 2010; Needham et al., 2002). Remarkably, experience reaching with sticky mittens at three months of age had effects on object exploration when infants were tested a year later (Libertus et al., 2016).

In his pioneering studies using cinematographic recording, Halverson (1931, 1932) tested infants longitudinally from twelve to sixty weeks of age, investigating the grips they used to try to pick up several types of objects (figure 12.3). Early grasps within the first six months involved holding objects against the palm mainly by the middle, ring, and little fingers. By nine months, precision grips began to predominate. Thus, the overall pattern described by Halverson is of infants initially relying on power grips and eventually transitioning to precision grips[1] over the course of the first year. This pattern has been replicated by several subsequent studies (Butterworth et al., 1997; Hohlstein, 1982; Touwen, 1976).

It is notable that the emergence of precision grasps starting around six months of age parallels changes in the grasp reflex, which starts to show fractionated responses for individual fingers around the same age (Twitchell, 1965). Intentional precision grips likely result from the increasing fine-grained cortical control over the innate motor synergies of the spinal cord. As discussed in chapters 3 and 11, this control is provided by neurons with direct CM projections in new M1. The final section of this chapter will discuss the development of the CM system. For now, I will simply note that in monkeys, the emergence of functional CM connections clearly mirrors the onset of fractionated finger

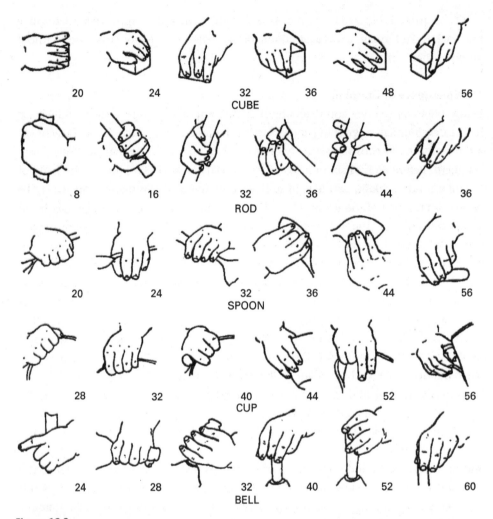

Figure 12.3
The development of grasping across the first year of life. The numbers indicate the infant's age in weeks. For each object, there is a progression from classic power grips in the youngest infants to increasing use of precision grips in the older infants. From Halverson (1932).

movements and precision grips (Lawrence & Hopkins, 1976). The reflexive mechanisms underlying early grasping behaviors thus serve as what Twitchell (1965, p. 254) called the "physiological substrata" on which later-developing processes, allowing flexible and intentional control of dexterous hand movements, can emerge.

An interesting limitation of Halverson's studies was pointed out by Newell et al. (1989). The same physical objects were used for infants across ages, although the physical size of the infant's hand naturally increases across the first year of life. The affordances of how an object—for example, the one-inch cube in the top row of figure 12.3—can be grasped will thus differ for a four-month-old and a twelve-month-old simply due to changes in the size of the infant's hand. Newell and colleagues presented a range of objects, including some smaller than those used by Halverson, and found evidence that even infants as young as four months produced grips that were appropriately scaled to object size. Indeed, in a subsequent study, Newell et al. (1993) argued that a common dimensionless ratio of hand size to object size determines grip configuration in both infants and adults.

Even in the studies of Newell and colleagues, however, there was clear evidence for major developmental changes in grasping over the first year. While eight-month-old infants adopted appropriate grip configurations prospectively based on visual information, four-month-olds did so only after contacting the object (Newell et al., 1989). It is not until around eleven months of age that there is a close correspondence between finger aperture and object size (Fagard, 2000). Similarly, Butterworth et al. (1997) found that six- to eight-month-olds nearly always physically displaced the object before successfully grasping it. Another study found that while five-month-olds anticipate the timing of object contact with the hand based on visual information, it is not until nine months of age that they pre-shape grip aperture based on object size (von Hofsten & Rönnqvist, 1988).

Similar development changes are seen in anticipatory control of hand shape based on object orientation (Jung et al., 2015; Lockman et al., 1984; Ossmy et al., 2022; von Hofsten & Fazel-Zandy, 1984; Witherington, 2005) as well as in other tasks such as fitting blocks through appropriately shaped holes (Ossmy et al., 2020). For example, Lockman et al. (1984) presented five- and nine-month-old infants with horizontally and vertically oriented dowels to grasp. The older infants tended to rotate their hands during the approach to the dowel to match its orientation. In the five-month-olds, however, this adjustment tended to happen after contact with the object had already been made.

As with reaching, however, grasping ability improves continuously into adulthood. Progressive improvements throughout childhood have been described in a wide range of features of precision grasping, including the coordination of grasping and lifting objects (Forssberg et al., 1991), the smoothness of movements (Kuhtz-Buschbeck et al., 1998), and rapidly adjusting to unexpected perturbations (Eliasson et al., 1995).

One particularly notable change concerns the ability to perform different actions with the two hands. Fagard and Jacquet (1989) provided infants between seven and twenty-four months of age with a range of objects to manipulate, some of which could be performed using symmetrical movements of the two hands while others required that the two hands do different things. Manipulations involving symmetrical hand movements were performed successfully earlier than those that required differentiated movements. This may relate to difficulties in suppressing movements of the contralateral hand when the infant tries to move a single hand. Indeed, when children move one hand, involuntary movements of the contralateral hand are often seen—a behavior that persists through late childhood (Cohen et al., 1967; Kakebeeke et al., 2018). This motor overflow, also known as "associated movements" or "mirror movements," reappears in a variety of neurological conditions, and aspects of it can be seen even in healthy adults, as will be discussed in the next chapter. It appears that the default state of the sensorimotor system is for motor commands to be sent symmetrically to both sides of the body—a tendency that may need to be actively inhibited, even in adults (Duque et al., 2005).

As already mentioned in chapter 3, one recent study used EEG to investigate patterns of brain connectivity underlying skill in a precision grip task in children from the age of eight through adulthood (Beck et al., 2021). Participants completed a force-matching task in which they had to press a device between the tips of their thumb and index finger to track a horizontal target line. Performance on this task increased progressively with age, showing continued improvement of grasping ability even beyond adolescence. Analyses of neural connectivity during the task showed increases with age in coupling from the SMA to the inferior parietal cortex. This suggests that adult-level performance involves top-down influences of brain regions involved in internally generated volitional control onto regions involved in more stimulus-driven functions. In addition, across ages, individual differences in task performance were associated with altered connectivity between the inferior parietal cortex and the vPMC. Interestingly, higher levels of precision grip skill were associated with increased connectivity from the PMC to the parietal cortex but *reduced* connectivity in the opposite direction. This again suggests that skilled performance is linked to feedback from the frontal to the parietal cortex.

The Development of Somatotopic Body Maps

So far, we have focused almost exclusively on the development of grasping and reaching behavior. Less is known about the development of brain mechanisms of hand control, especially in humans.

Development of Somatotopic Maps in Animals

Research in rodents has shown substantial changes in the organization of somatotopic maps in S1 across the first three weeks of postnatal life. Seelke et al. (2012) measured RF structure and somatotopic organization in S1 of rats at five, ten, fifteen, and twenty days following birth. By day 5, most neurons in S1 responded to stimuli on the face, even neurons in regions that would later represent other body parts. By day 10, a larger number of neurons was found responding to stimuli on the limbs, although these frequently showed responses to stimuli on multiple body parts. Moreover, the topographic organization of these neurons was imprecise, showing little relation to the somatotopic pattern that would emerge later. By day 15, many more neurons showed responses to individual body parts, and a somatotopic pattern had emerged broadly similar to that seen in adults. Finally, by day 20, responses were highly similar to those seen in adult rats.

One early study found that somatotopic maps of the body were strikingly similar in newborn kittens and adult cats (Rubel, 1971). Rubel mapped the responsiveness of neurons in the sensorimotor cortex to tactile stimuli in kittens between six and twenty-four hours after birth. The somatotopic arrangement of neurons was highly similar in kittens and adult cats. Moreover, kittens also showed the same gradient across the limbs, with RFs on more proximal regions being much larger than those near the toes. At the same time, there were also some differences. The responses of kittens showed substantially longer latencies, possibly related to developmental changes in the myelinization of the peripheral nerves. Differences were also apparent in the organization of RFs. Adult cats showed substantial variability in RF structure. For example, in adult cats, RFs of individual neurons might be very large, covering the entire paw or foreleg. Similarly, a neuron's RF might include noncontiguous regions, such as two adjacent toes. In kittens, however, RFs tended to be fairly homogenous in structure, with little overlap in RFs of adjacent neurons, and never included noncontiguous regions. These results suggest that the basic somatotopic structure of maps is in place at the time of birth, but that this structure is later elaborated, presumably on the basis of the statistical regularities of postnatal sensory inputs and behaviors.

In another study, Krubitzer and Kaas (1988) recorded from areas 3b and 1 of S1 in newborn New World monkeys (marmosets and squirrel monkeys) and Old World monkeys (macaques). In marmosets and squirrel monkeys, adultlike somatotopic maps were clearly present in both areas 3b and 1, as shown in figure 12.4 for a one-day-old squirrel monkey. In macaques, responses to tactile stimuli were not present in either area 3b or 1 in two newborn monkeys tested. By one month of age, however, responses were present and again showed an adultlike organization. The lack of responses to touch in

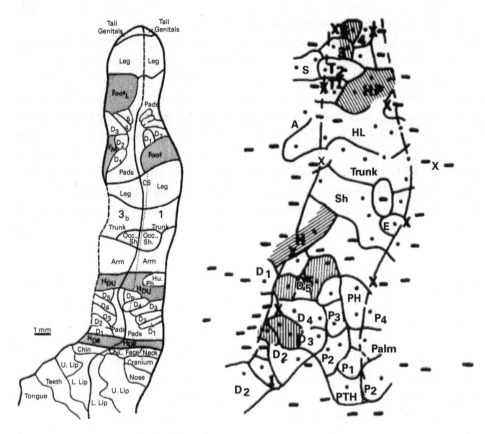

Figure 12.4
Left panel: The somatotopic structure in areas 3b and 1 of adult squirrel monkeys (Sur et al., 1982). *Right panel*: The somatotopic structure of area 3b in a one-day-old squirrel monkey (Krubitzer & Kaas, 1988). The organization is extremely similar.

newborn macaques is surprising and is likely a result of the fact that the monkeys in this study were tested under anesthesia. In any case, these results show that in both New World and Old World monkeys, as soon as responses are found in S1, they show an essentially adultlike pattern of somatotopy.

A recent study used fMRI to investigate somatotopic maps in response to tactile stimuli in newborn macaques (Arcaro et al., 2019). Tactile stimuli were applied to the feet, hands, and face of awake monkeys, ranging in age from newborns (as young as eleven days old) to nearly three years. An adultlike pattern of somatotopy in the sensorimotor cortex was apparent in all animals. Moreover, similar somatotopic maps were also found in several other brain regions, including the PMC, the PPC, the insula, the basal ganglia, and the thalamus. The magnification of body parts also appeared broadly adultlike, with the hand representation being more than three times as large as the foot representation. These results show a high level of continuity across development, with the basic structure of maps in place at least by one week of age. Notably, however, when these authors investigated somatotopy at the more fine-grained level of individual finger representations, clearer developmental changes were apparent. In the youngest monkeys, stimulation of individual fingers produced weak responses, which showed little spatial differentiation between fingers. Increasingly robust and spatially differentiated responses were found across the first year and a half of life.

Development of Somatotopic Maps in Humans
Somatotopic maps in human infants have also been identified using EEG (Meltzoff et al., 2019; Saby et al., 2015) and fMRI (Allievi et al., 2016; Dall'Orso et al., 2018, 2022). For example, Dall'Orso et al. (2018) tested preterm infants, such that they were younger than full term in gestational age at the time of testing. Tactile stimuli were applied to the ankles, wrists, and mouth using a specially designed robotic device. An adultlike somatotopic pattern was apparent, with the legs, hands, and face represented in sequence from top to bottom in the primary somatosensory cortex. Moreover, the cluster of activation produced by wrist stimulation was significantly larger than that for ankle stimulation, indicating that increased cortical magnification of the hand is already present. These results provide dramatic evidence that in human infants, the basic organization of the somatosensory cortex is in place from before the expected time of birth (i.e., full term).

A recent study from the same lab has further investigated early development of the somatosensory cortex by measuring resting-state functional connectivity (Dall'Orso et al., 2022). The idea here is to record fMRI responses without any stimulus at all and to measure correlations between brain regions in the time series of neural responses. In adults, the resting-state sensorimotor network shows somatotopic organization (Long et al., 2014), with subnetworks corresponding to distinct body parts, as well as strong links between homologous representations in the left and right hemispheres (van den Heuvel et al., 2015). Dall'Orso et al. (2022) found a crude, but spatially organized somatotopic pattern in the functional connectivity of preterm infants with a gestational age as young as thirty-two weeks. This pattern matures rapidly during the perinatal period, becoming approximately adultlike when infants are full term. Notably, preterm infants tested at term-equivalent age showed a very similar pattern to newborn full-term infants, despite having substantially more postnatal experience. This pattern suggests that the connectivity patterns are largely under genetic control and mature at similar rates before and after birth.

Another approach to investigating the early development of somatosensory cortical processing in humans is the use of EEG to measure SEPs. The early P1 SEP component, which in adults has been linked to area 3b (Allison et al., 1989), has been identified in preterm human infants with a gestational age from twenty-nine weeks (Hrbek et al., 1973; Pike et al., 1997). This shows that the basic spinothalamocortical pathway is in place and functioning well before birth, consistent with the fMRI studies described above. In contrast, subsequent stages of hierarchical cortical processing appear to develop later. A recent study by Whitehead et al. (2019) applied mechanical taps to the hands and feet of newborn infants that had been both preterm, early term, full term, and late term. There is a successive emergence during this period of the P1 component (arising from area 3b), the N2 component (area 2), the P2 component (PPC), and the N3 (top-down reactivation of S1). This developmental progression of the emergence of successive stages in hierarchical organization appears structured to result in all stages being in place by the expected time of birth at full term.

Theories of Somatotopic Map Formation
Two broad classes of explanation have been advanced for the development of orderly topographic maps, whether somatotopic maps of the body in the sensorimotor cortex or retinotopic maps of the visual field in the visual cortex (Kaas & Catania, 2002). These explanations are not mutually exclusive, and both types of mechanism are likely involved in the formation of somatotopic maps of the body. One class of explanation arises from Roger Sperry's (1963) chemoaffinity hypothesis and suggests that different brain areas have complementary molecular gradients that guide and maintain orderly patterns of connections between the two areas. Such gradients would reflect position-dependent patterns of gene expression and are related to the research discussed in the previous chapter in which concentrations of signaling molecules produce distinct brain areas in the developing cortical sheet (Grove & Fukuchi-Shimogori, 2003). Altered concentrations of the FGF8 protein can produce duplications of area S1, which have different topographic properties depending on the exact density (Assimacopoulos et al., 2012). This suggests that signaling molecules are involved not only in producing the large-scale pattern of cortical areas but also the more fine-grained topography of somatotopic maps.

The second class of explanation focuses on patterns of activation within a brain region and patterns of inputs that it receives from other brain areas or the periphery. In vision, this approach has been strongly influenced by studies showing that early visual deprivation results in major alterations in the structure of retinotopic maps in the visual cortex (e.g., Wiesel & Hubel, 1963). Research in rodents has provided evidence that the patterns of sensory inputs to the brain in early life are critical for map formation. In one study, individual whiskers were removed from newborn mice (Van der Loos & Woolsey, 1973). When somatotopic maps in the barrel cortex of the somatosensory cortex were measured, two to six weeks later, they were well organized, with one barrel representation for each whisker. Critically, there was no barrel corresponding to the whisker that had been removed. In a complementary study, mice were selectively bred to have additional, supernumerary whiskers (Welker & Van der Loos, 1986). The barrel representation in S1 again showed clear organization, with the additional whiskers mirrored by additional barrel representations, which appeared fully functional. Together, these studies show that the number of barrels that develops within these maps is driven by the inputs coming from the periphery.

Such effects are not specific to whiskers or to the barrel cortex. Recall from the last chapter that the star-nosed mole has an extraordinary nose, with eleven distinct appendages on each side. Catania and Kaas (1997b) found a single mole with twelve appendages on each side of its nose. Somatotopic maps of the star in S1 perfectly reflected the supernumerary appendage. One recent study has extended this logic to humans. As mentioned in chapter 1, Mehring et al. (2019) studied several members of a Brazilian family, many of whom have a sixth finger on each hand. Not only do these supernumerary fingers have normal sensorimotor function, but fMRI showed that finger maps in the sensorimotor cortex had distinct representations of all six fingers.

These results suggest that the detailed features of somatotopic maps are determined by the statistical structure of the inputs that an area receives. Several computational models using methods such as self-organizing "Kohonen" maps have shown that such maps naturally arise given appropriately structured inputs (e.g., Aflalo & Graziano, 2006; Durbin & Mitchison, 1990; Kohonen, 2006). It is, of course, possible that the genetic mutations that result in a supernumerary body part are paralleled by corresponding mutations in genes controlling signaling gradients so that the resulting somatotopic maps mirror the physical body. This, however, would likely take a long time to evolve and would often fail. This point is put well by Kaas and Catania (2002):

> It seems extraordinarily unlikely that a number of genes have mutated in consort so that independent intrinsic guidance factors in the face and at each of several levels of the nervous system have all been changed to precisely match modular components in a single or a few generations. Instead, it seems more reasonable to propose that information from the sensory sheet instructs the formation of central sensory representations in a cascading fashion. (p. 341)

Direct evidence that the formation of maps of supernumerary body parts is not coded genetically comes from a study in which supernumerary whiskers were induced experimentally by injection of the Sonic hedgehog protein into mouse embryos (Ohsaki et al., 2002). In this case, it is known that the supernumerary whisker is not genetically specified. Nevertheless, these whiskers were again represented by well-organized barrels in S1.

The Development of Direct CM Projections

Recall from the previous chapter that one important mechanism underlying precise control of the hand and fingers of precision grips is the monosynaptic projections from the cortex to motoneurons in the ventral horn of the spinal cord, or CM projections. Several studies have used tracing methods to investigate the postnatal development of CM connections in infant monkeys. Kuypers (1962) used an anterograde silver staining method that allowed him to identify sites of projection of neurons that were destroyed. He ablated the sensorimotor cortex in macaques ranging from four days to eight months of age, which had the effect of staining motor neurons that received direct projections from this region. Direct CM connections were absent in the youngest infants and increased to near-adult levels by eight months of age. Similar results were found by Galea and Darian-Smith (1995), who injected retrograde tracers into several sites of the cervical spinal cord to identify which cortical regions projected to them. At birth, several cortical areas, including M1, SMA, and the PPC, had already established projections to all cervical spinal segments tested. Critically, however, these projections did not terminate in the gray matter of the ventral horn, indicating that they did not reflect monosynaptic CM connections. Armand et al. (1997) injected anterograde tracers into M1, again finding few, if any, direct CM projections at birth. In this study, there was evidence for a rapid increase in CM connections over the first five months, followed by a more gradual increase that extended at least into the second year of life. Collectively, these studies show that CM projections are absent or only weakly represented at birth but emerge during the first year of life.

Developmentally, the emergence of fine finger control parallels the emergence of CM projections (Galea & Darian-Smith, 1995; Kennard, 1938, 1940; Lawrence & Hopkins, 1976). The monkey's earliest successful reaches for food occur between three and four weeks of age and involve wrapping all the fingers around the food item, pressing it against the palm. This is highly similar to Halverson's descriptions of the earliest grasps in human infants. These early reaches are clumsy and unsteady. By three months, however, the monkey's reaching is smooth and accurate but still does not show clearly differentiated use of the individual fingers. Clear precision grasps using the thumb and index finger are seen between four and six months, and by seven to eight months, performance approximates to adult performance. Similarly, grooming behaviors of other monkeys, which involve precise and individuated finger movements, are rarely seen before around six months of age (Hinde et al., 1964).

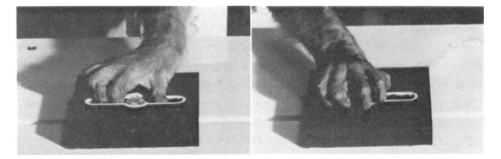

Figure 12.5
Attempts at precision grips in monkeys with brain lesions to M1 made during early infancy. A food item was placed into a small hole, which could be accessed by placing two fingers into grooves on either side of the hole using a precision grip. *Left panel*: This monkey had a lesion that covered most of M1 but spared a small region, including the hand area. Once mature, the monkey remains able to use a precision grip with the thumb and index finger to obtain the food. *Right panel*: This monkey, in contrast, had a complete lesion of M1. Unlike the first monkey, this animal was unable to use a precision grip at all, even years after the lesion. The monkey uses all four fingers flexed together—an approach that is inadequate for this task. From Passingham et al. (1978).

Lesion studies in infant monkeys have also linked precise finger control and the CM system. Lawrence and Hopkins (1976) lesioned the corticospinal tract bilaterally in infant macaques between five days and four weeks of age, which were then observed until three years of age. The overall motor behavior of these animals was surprisingly normal. They continued to reach for food and moved the arm in a smooth and accurate manner. However, when trying to grasp objects, they did not use precision grips with individuated movements of the fingers. Rather, "the semiflexed fingers moved in phase and closed all together" (p. 244).

Similar results have been found from studies of cortical lesions. In her classic studies of the effects of brain lesions in infant monkeys, Margaret Kennard (1936, 1938, 1940) showed that complete removal of M1 and the PMC in one hemisphere of infant macaques produced little long-term impairments in motor performance.[2] Notably, however, these animals remained impaired in fine control of the fingers, showing an "inability to perform fine purposeful prehensile movements of the distal portions of the extremities" (Kennard, 1940, p. 381). Similar results were reported by Passingham et al. (1978), who measured precision grip performance by placing food into a small hole, which required the monkey to place two fingers into narrow grooves, as shown in figure 12.5. Monkeys with unilateral removal of M1 (or M1 and S1) in the left hemisphere between one week and three months of age were unable to retrieve the food, even when tested more than two years after their lesion.

These lesion studies of the corticospinal tract and M1 do not specifically implicate direct CM projections, since there are many non-CM fibers in the corticospinal tract and many non-CM projections from M1. A more direct approach is to stimulate the motor cortex and measure the latency of evoked responses in the peripheral muscles using EMG. In adult macaques, the earliest evoked response to TMS recorded in hand muscles has been directly linked to CM projections, as its latency is too short to be due to multi-synaptic projections (Baker et al., 1994; Edgley et al., 1990). In infant macaques, in contrast, such short-latency responses to M1 TMS are not found before three to five months of age (Flament et al., 1992; Olivier et al., 1997). Moreover, the threshold at which TMS evokes responses decreases until five to eight months of age (Flament et al., 1992).

Broadly similar results have been found in studies of human infants, with both latency and threshold values for TMS decreasing across infancy (Eyre et al., 1991; Koh & Eyre, 1988). Indeed, when correction is made for changes in body size, conduction velocity within the corticospinal pathway continues to increase until around ten years of age (Koh & Eyre, 1988; Müller et al., 1994). Neuroimaging studies using DTI to track the development of the corticospinal tract show a similar time course, with changes in fiber volume well into the second decade of life (Lebel & Beaulieu, 2011; Yeo et al., 2014). This prolonged developmental time course is consistent with the similarly extended development of precision grip skill described above. However, changes in conduction velocity are likely to result from several factors beyond direct CM projections, including patterns of myelinization.

One striking finding came from a study by Eyre et al. (2000), who conducted a postmortem morphological study of the corticospinal tract in preterm and term humans. They identified cortical projections in the lower cervical spinal cord by a gestational age of twenty-four weeks, which began to innervate the gray matter of the ventral horn a few weeks later. Thus, full-term newborn humans appear to have at least some CM projections in place, which appears to be an important difference between humans and monkeys, as discussed above, although these results certainly do not mean that all CM projections, let alone the complete corticospinal tract, are fully mature at birth. Eyre and colleagues suggest that this early emergence of CM projections is "so that the cortex can be intimately involved in spinal motor center development from an early stage, reflecting the uniquely dominant role of the corticomotoneuronal system in human movement control" (p. 60).

Some evidence consistent with this interpretation comes from a study by Wallace and Whishaw (2003). They described several types of spontaneous hand movements that human infants make during the first five months and which do not appear to be object directed. These include fists, "pre-precision" grips in which the thumb presses against the side of the index finger, precision grips of the fingers against each other, and self-directed grasps. These movements are similar to those described by Piaget (1936) as primary circular reactions of hand movements, which he describes evocatively as "an empty use of the grasping mechanism" (p. 90). Wallace and Whishaw suggest that these goalless behaviors seen in the first few months of life reflect a form of "hand babbling," which, like vocal babbling, sets the stage for later more sophisticated behavior. The early development of direct CM connections in humans, on this interpretation, could function to provide an early period by which the basic motor components of later skilled action can be strengthened during early infancy.

13   The Bilateral Hand

During the course of his experimental work in the 1850s, Gustav Fechner, one of the founders of experimental psychology and perceptual psychophysics, made a curious observation about his own behavior. He was running studies that required him to manipulate equipment with one hand while recording observations with the other hand (Fechner, 1858). He preferred to use his dominant right hand for adjusting the delicate equipment, leaving his left hand for writing. As he was making fifty to a hundred observations daily, and as the observation values were almost all between nine and ten, he wrote many '9's, with different numbers coming after the decimal. One day, having completed the hundredth measurement and having only to write down the final observation, he moved the pen to his now-free right hand. Looking at what he had written, Fechner was surprised to find the 9 written backward, as if reflected in a mirror. Somehow, his experience of writing 9's with his *left* hand had disrupted his ability to write 9's with his *right* hand. Nor was this an isolated lapse. After several months, Fechner found that he naturally wrote 9's backward with his dominant right hand, although only briefly following left-handed data recording. The effect, moreover, was specific to 9's, which were by far the most common number written with the left hand.

Fechner's observations suggested to him—and suggest to us today—that there are deep connections between the representations of our right and left hands. Our bodies, like those of all vertebrates, are bilaterally symmetric, making our hands mirror images of each other. Our hands have the same bone structure, are controlled by the same systems of muscles and tendons, and have the same sensory apparatus. Moreover, the sensorimotor networks of the right and left cerebral hemispheres are also mirror images. This makes it natural to ask how the neural representations of the two hands are linked. This chapter will explore this question, focusing first on the motor system and then on touch.

Bilateral Motor Representations

That movements of each side of the body are controlled by the contralateral cerebral hemisphere—the Valsalva doctrine—is "textbook" knowledge, which all undergraduates in psychology and neuroscience learn as fact. It is the sort of claim for which no citation is needed, and indeed it isn't clear what citation one would give if asked. These sorts of textbook facts, enshrined by repetition and unanimous acceptance, can be dangerous things. They can prevent fields from questioning foundational assumptions and identifying unsuspected discoveries. Thus, before considering ways in which the representations of the two hands interact—ways that may suggest that the textbook view is wrong, or at least incomplete—it is worth considering where the textbook view comes from and what the evidence for it is. I will thus start with a brief overview of this evidence. My presentation is part historical survey and part "prosecutor's case" for this view.

Evidence for Dominant Contralateral Representation in the Motor Cortex

The idea that each side of the body is controlled by the brain on the opposite side dates to antiquity. The ancient Greeks had understood this principle from their observations of, and efforts to help, people with head wounds. In Part 19 of Hippocrates's (1849a) *On Injuries of the Head*, we are told: "For the most part, convulsions seize the other side of the body; for, if the wound be situated on the left side, the convulsions will seize the right side of the body; or if the wound be on the right side of the head, the convulsion attacks the left side of the body" (p. 464). In Part 13 of the same work, doctors are advised to avoid making incisions in the region of the temple and above, as this may induce convulsions on the opposite side of the body. This shows a practical understanding not only of the fact that each hemisphere predominantly controls the contralateral side of the body but also of the approximate location of what we would now call the motor cortex.

The Hippocratic writers discussed contralateral control in the context of traumatic injury. A more general association of brain damage of any type with weakness of the contralateral body emerged from the work of Antonio Valsalva and Giovanni Morgagni in the early eighteenth century (Louis, 1994). Valsalva and Morgagni observed the movement problems of patients with strokes or brain tumors while alive and then dissected their brains after they had died. They observed a strong link between motor weakness or paralysis on one side of the body and damage of the contralateral hemisphere, developing this into a general principle of contralateral control of movement, which Morgagni labeled "Valsalva's doctrine." This link was widely confirmed by subsequent studies. This principle was implicit in discussions in the mid-nineteenth century about the association between aphasia and hemiplegia of the right hand, which was taken to support the link that Dax and Broca drew between language and the left hemisphere (Jackson, 1864). By the end of the nineteenth century, Charcot and Pitres (1895) could write:

> In the vast majority of cases, cerebral paralysis is crossed in relation to the lesions that determine it, that is, lesions of the right hemisphere of the brain give rise to paralysis on the left side of the body, and lesions of the left hemisphere to paralysis of the right side. This is the law. It has been established in innumerable observations and finds its natural explanation in the fact of the decussation of the pyramidal bundles. (p. 184)

Contralateral control of movements was dramatically confirmed by the studies of electrical brain stimulation of the motor cortex, first in dogs by Fritsch and Hitzig (1870, 2009) and shortly afterward in a variety of animals by Ferrier (1873, 1876). This approach was soon extended to humans, first in the study by Roberts Bartholow (1874) of his patient Mary Rafferty, and eventually in the systematic investigations of Foerster (1926, 1936) and Penfield (Penfield & Boldrey, 1937; Penfield & Jasper, 1954; Penfield & Rasmussen, 1950). When stimulating in the precentral gyrus, Penfield found exclusively contralateral motor responses. Penfield and Jasper (1954, p. 66) report one single occasion in which stimulation of the precentral gyrus evoked bilateral leg movements. They suggested that this might have been due to an effect on the adjacent SMA, which has more bilateral representation, as I will discuss below. Penfield and Jasper (1954) concluded that "bilateral representation of movement of either extremity does not exist in man as far as our experience of Rolandic stimulation goes" (p. 66). The subsequent development of noninvasive methods to stimulate the motor cortex in healthy humans using electrical currents (Merton & Morton, 1980) or TMS (Bestmann & Krakauer, 2015) has made such demonstrations routine. Application of such stimulation above the motor cortex on one side of the body evokes twitches or activity in muscles of the opposite side of the body. I have a vivid memory from my own undergraduate class on cognitive neuroscience of my professor, Richard Ivry of the University of California, Berkeley, bringing a TMS machine to the lecture hall and holding it to his own head to evoke movements of the fingers of his contralateral hand.

The stimulation studies of Fritsch, Hitzig, and Ferrier were quickly followed by lesion studies, showing that experimental ablations of the motor cortex produce weakness, paralysis, and a lack of coordination in specific parts of the contralateral side of body in many animal species (Ferrier, 1876; Kennard, 1936; Lashley, 1924; Leyton & Sherrington, 1917). In humans, much the same conclusion can be drawn from the so-called Wada test in which the sedative drug sodium amytal is injected into the carotid artery on one side, effectively putting one hemisphere of the brain to sleep. This procedure rapidly induces paralysis of the contralateral side of the body in both monkeys and humans (Wada & Rasmussen, 1960). Indeed, contralateral hemiparesis is so invariant a consequence of the Wada test that it is used as a manipulation check that the procedure has worked.

Evidence for Bilateral Motor Representations

The pieces of evidence described in the previous section are individually convincing, and collectively provide overwhelming evidence for predominantly contralateral motor function. "Predominantly," however, does not mean "exclusively," and there are hints that there may be some level of bilateral function in the sensorimotor system. For example, individuals who have had one entire cerebral hemisphere surgically removed, typically as a treatment for epilepsy, often retain some level of somatosensory and motor function on the contralateral side of the body (Holloway et al., 2000). In some cases, this may reflect compensatory reorganization, particularly in children. In other cases, however, residual function is also seen following removal of one entire cerebral hemisphere in adults. Patient A.P. had her right hemisphere removed at the age of eighteen to treat intractable epilepsy (Müller et al., 1991). She showed left-sided paralysis for one week following surgery, but she eventually regained the ability to move and led a relatively normal life. Formally tested nearly thirty years later, she showed some impairment with individuated movements of the fingers of her left hand, but had good control over more proximal arm movements.

Other evidence for bilateral representation of movements comes from studies of so-called split-brain patients, in which the corpus callosum has been cut, generally as a treatment for epilepsy. In the immediate period following surgery, these patients are generally unable to perform simple actions with their left hand to verbal instruction (e.g., "raise your left arm"; Sperry et al., 1969). Because in most people the left hemisphere is much stronger than the right hemisphere in speech comprehension, this pattern is consistent with contralateral control of motor function. The right hemisphere is able to control the left hand, but lacks linguistic proficiency; the left hemisphere has the linguistic ability, but is unable to control the left hand.

In most patients, ipsilateral control of motor function improves gradually. By presenting visual instructions in only one visual field, the motor abilities of each brain hemisphere can be tested. In one study of nine split-brain patients (Gazzaniga et al., 1967), each hemisphere appeared able to perform simple movements of both sides of the body. This included actions such as pointing to objects, tracing shapes, and even writing.

While such results suggest a degree of ipsilateral motor function, this appears different from the more robust contralateral control. Ipsilateral control is less proficient, more delicate, and more easily disrupted than contralateral control. While ipsilateral control is fairly robust for midline structures, such as the face, neck, and torso, it is substantially weaker for the hand and fingers—a pattern found in patients with surgical section of the corpus callosum (Gazzaniga et al., 1967) and with congenital callosal agenesis (Jakobson et al., 1994), as well as after experimental removal of the corpus callosum in monkeys (Brinkman & Kuypers, 1972).

One potential hypothesis regarding such effects is that they reflect mechanical coupling between the two arms. For example, when walking, there is evidence that arm swing is produced passively by movements of the pelvis and torso resulting from leg movements (Pontzer et al., 2009). Since both arms are connected to one and the same torso, movements of one arm may lead to changes in posture or muscle tone in the torso, which may have physical effects on the contralateral arm. This possibility has been eliminated, however, in several intriguing cases. As discussed in chapter 6, bimanual interference has been found on an intact hand when coupled with "movements" of a phantom hand in amputees (Franz & Ramachandran, 1998) and with movements of a paralyzed hand in patients with anosognosia for hemiplegia (Garbarini et al., 2012). Similarly, bimanual interference can be found when a healthy participant moves one hand while merely imagining that the other hand is also moving (Heuer et al., 1998). In each of these cases, interference effects are found between the two hands, even though only one hand physically moves. The coupling between the hands cannot be due to mechanical links between the physical hands. Instead, it is the two *invisible hands* that are coupled.

Neural Mechanisms of Bimanual Actions

Although the majority of neurons in M1 respond in relation to movements of the fingers of the contralateral hand, single-unit recordings in monkeys have identified a minority of neurons that respond to ipsilateral movements (Evarts, 1966; Tanji et al., 1987, 1988). For example, Tanji et al. (1988) recorded from 378 neurons in macaque M1 that responded to finger movements. Of these, 323 (85 percent) responded in advance of contralateral hand movements, thirty-one (8 percent) to ipsilateral hand movements, and a further seventeen (4 percent) to movements of either hand (see figure 13.1). Intriguingly, one region located at the boundary between the representations of the hand and face appears to have an unusually large number of neurons with bilateral responses (Aizawa et al., 1990). Similar ipsilateral responses have also been observed in intracranial recordings from human patients undergoing brain surgery (Ganguly et al., 2009; Goldring & Ratcheson, 1972).

In contrast to the studies just described, which investigated precise movements of the monkey's fingers, studies that have investigated more proximal reaching movements of the entire arm have found stronger evidence for bilateral responses in M1

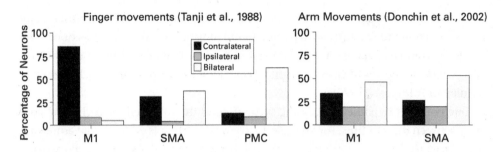

Figure 13.1
Contralateral, ipsilateral, and bilateral responses of neurons in the motor cortices of macaque monkeys. *Left panel*: Data from Tanji et al. (1988) using a button-pressing task involving precise movements of individual fingers. Note that the bars do not sum to 100 percent because of the presence of an additional subset of neurons with more complex response properties. *Right panel*: Data from Donchin et al. (2002) using whole-arm reaching movements.

(Donchin et al., 1998, 2002). The right panel of figure 13.1 shows data from a "center-out" task in which monkeys held a manipulandum and made whole-arm movements in different directions (Donchin et al., 2002). In M1, there is substantially more robust bilateral representation than found for finger movements by Tanji and colleagues. This pattern fits well with the studies of human split-brain patients, described above, showing stronger ipsilateral control over reaching than over grasping.

In higher-level regions of the frontal motor system, there is stronger evidence for bilateral representations, including in the SMA (Donchin et al., 2002; Tanji et al., 1988) and the PMC (Cisek et al., 2003; Tanji et al., 1988). Tanji and colleagues found a clear progression of increasing abstraction of motor representations from M1 to SMA to PMC, with the proportion of neurons showing contralateral responses decreasing and the proportion of bimanual responses increasing. This parallels the more general hierarchical organization of the motor system, as discussed in chapter 3.

Neuroimaging studies in humans using fMRI have also found changes in brain activation in M1 ipsilateral to hand movements. Consistent with results described above, these activations are larger for movements of the proximal than the distal arm muscles (Nirkko et al., 2001). In many cases, this reflects a *reduction* in activation compared to baseline (Hamzei et al., 2002), suggesting that it is actively inhibited by the active contralateral M1.

Other studies have shown that activations of ipsilateral M1 occur only with complex actions (Buetefisch et al., 2014; Schluter et al., 2001; Verstynen et al., 2005) or when high levels of force are applied (Andrushko et al., 2021). For example, Verstynen et al. (2005) asked participants to produce three types of finger movements: simple repetitive taps with one finger, multifinger sequences of taps, and "chords" in which multiple fingers tapped simultaneously. Ipsilateral M1 was activated in the latter two conditions but not when participants made simple movements of one finger. Thus, ipsilateral representations appear to be selectively engaged when processing demands are high, suggesting that they may play a functional role in augmenting the resources available to the motor system.

Another interesting feature of the study of Verstynen and colleagues is that ipsilateral responses were much stronger in left M1 following left-hand movements than the converse—a pattern seen consistently in many other studies (e.g., Kim et al., 1993; Nirkko et al., 2001). This pattern is stronger in right-handed individuals, which will be discussed further in the next chapter. It is also presumably related to the dominance of the left hemisphere for complex actions, as seen in patients with apraxia, discussed in chapter 9.

Studies using both electrocorticography and fMRI have investigated the distributed patterns of brain activation produced by contralateral and ipsilateral hand and arm movements. In general, these patterns show a high degree of similarity, and classifiers trained on movements of one arm can decode movements by the contralateral arm. Such similarity has been found for the direction of arm movements (Haar et al., 2017), kinematic features of movement such as velocity (Bundy et al., 2018; Merrick et al., 2022; Willett et al., 2020), and which body part (Fujiwara et al., 2017) or finger is moving (Diedrichsen et al., 2013; Scherer et al., 2009).

At the same time, other research indicates that representations of the contralateral and ipsilateral arms are at least partly decoupled. For example, while premotor neurons in monkeys show a common preferred direction of arm motion for both arms, M1 neurons show different directional preferences for each arm (Cisek et al., 2003). Other studies have used principal component analysis to characterize the population-level representational spaces for ipsilateral and contralateral arm movements, finding that these inhabit largely orthogonal subspaces (Ames & Churchland, 2019; Heming et al., 2019). Thus, while each individual neuron may respond to both contralateral and ipsilateral arm movements, the population-level representation of the two arms may nevertheless be independent. One recent study using intracranial recordings in humans found that while the representation of whole-arm reaching movements showed independent representations of ipsilateral and contralateral actions, representation of grasping showed correlated representations for the two hands (Downey et al., 2020).

Bilateral integration has also been described in the PPC. In their initial report of PPC "hand manipulation" neurons described in chapter 3, Mountcastle et al. (1975) noted that many of these neurons responded to movements of either hand, particularly those in area 7. Rather than asking whether a single neuron responded to movements of both arms, a recent study by Mooshagian et al. (2021) recorded simultaneously from the PRR of both hemispheres while monkeys made unilateral arm movements. While the population average response was dominated by contralateral arm movements, clear bilateral integration between the two hemispheres was identified in beta-band oscillations. Finally, as discussed in the context of motor equivalence in chapter 3, neuroimaging studies in humans have identified areas of the PPC responding during movements of either arm (Gallivan et al., 2013).

Interhemispheric Inhibition and Excitation

Evidence for excitatory relations between homologous muscles of the two hands comes from studies showing that contracting the muscles of one hand produces an increase of the MEPs evoked in muscles of the opposite hand following TMS. Hess et al. (1986) recorded MEPs from the right abductor digiti minimi (ADM) muscle following TMS pulses to left M1. The amplitude of these responses increased when participants contracted their *left* ADM. This indicates that movement of the left hand increased cortical excitability in the ipsilateral left hemisphere. In contrast, contraction of a different muscle in the left hand did not produce such facilitation. Thus, the faciliatory effect of muscle contraction was specific to the homologous muscle on the opposite hand. Other studies have parametrically varied the force of muscle contraction, finding that the more force produced in the muscles of one hand, the greater the facilitation of MEPs found in homologous muscles of the opposite hand (Perez & Cohen, 2008).

There is also evidence of inhibition between the two motor cortices. One common approach is the use of paired-pulse TMS, in which a "conditioning" TMS pulse is applied to M1 on one side shortly before a "test" pulse is applied to the opposite M1. Ferbert et al. (1992) found that a conditioning pulse applied more than 5–6 ms before a test pulse reduced the MEPs evoked. However, at shorter latencies between conditioning and test stimuli, this interhemispheric inhibition switches to facilitation (Hanajima et al., 2001). Thus, a single TMS pulse applied to M1 produces successive waves of excitation and inhibition on the contralateral hemisphere. There is evidence that the balance between inhibition and excitation between the two M1s varies in complex ways, depending on task complexity (Tinazzi & Zanette, 1998), and at different phases in the preparation and execution of actions (Tazoe & Perez, 2013).

The excitatory and inhibitory effects of M1 TMS have different patterns of spatial spread, providing important insight into their underlying mechanisms. Hanajima et al. (2001) found that the early facilitatory effect required that the conditioning pulse be applied directly over the M1 hand region, whereas the later inhibitory effect also occurred following stimulation to nearby locations. Similarly, the facilitatory effect of finger movements of one hand is muscle specific, as it only affects MEPs recorded from the homologous muscles in the other hand. In contrast, the inhibitory effect is not muscle specific, as it also occurs for other hand muscles (Sohn et al., 2003). Such effects suggest that activation of a motor representation in M1 produces a highly specific excitatory projection to the homologous representation in the other hemisphere, as well as a more diffuse pattern of inhibition. This "surround inhibition" (Beck & Hallett, 2011) is similar to the "center surround" organization of RFs in sensory systems, which serves to sharpen spatial resolution. Carson (2020) argues that interhemispheric inhibition, similarly, serves to "sculpt" highly precise motor outputs.

Mirror Movements

Other researchers have proposed that interhemispheric inhibition might serve to prevent unwanted "mirror" movements of the contralateral limb (Duque et al., 2005). Indeed, such mirror movements are a salient feature of many clinical disorders, including congenital diseases such as Klippel–Feil syndrome (Matthews et al., 1990) and cerebral palsy (Carr et al., 1993), neurodegenerative diseases such as Parkinson's disease (Cincotta et al., 2006) and focal hand dystonia (Beck et al., 2009), and following stroke (Ejaz et al., 2018). In other cases, mirror movements occur congenitally in individuals with no other apparent neurological abnormalities (Schott & Wyke, 1981). Such congenital mirror movements frequently run in families and have been linked to mutations in a specific gene, *DCC* (Srour et al., 2010), known to be involved in development of the corpus callosum (Marsh et al., 2017).

Mirror movements are commonly seen in healthy children. One recent large-scale study of children (Kakebeeke et al., 2018) found a progressive reduction in mirror movements in healthy children from three to eighteen years of age. Even when no overt mirror movements are made, EMG responses can be recorded in contralateral muscles, both in children and adults (Mayston et al., 1999). Indeed, mirror movements can occur in healthy adults during particularly effortful actions (Armatas et al., 1994).

An intriguing example was reported by Haerer and Currier (1966). Their patient had made mirror movements all his life, like several other members of his family. At the age of sixty-five, he had a left-hemisphere stroke, which left him with hemiplegia of his right hand. In contrast to his inability to make voluntary movements of his right hand, when he moved his left hand, his right hand continued to make mirror movements like before his stroke. These remarkable mirror movements of an otherwise paralyzed hand suggest that the mirroring is not due to transcallosal signals, since this patient's left M1 is unable to generate hand movements on its own. Rather, the mirroring likely results from direct ipsilateral projections from the right hemisphere. A recent longitudinal study of mirror movements in hemiplegic stroke patients has similarly concluded that this phenomenon arises through subcortical pathways (Ejaz et al., 2018). However, the occurrence of mirror movements in individuals with congenital malformations of the corpus callosum (Marsh et al., 2017) suggests that transcallosal signals are involved in some cases.

Mirror Writing

Another likely related phenomenon is mirror writing—the ability of people to write text in a mirror-imaged way, invariably with their nondominant hand (Critchley, 1928; Schott, 1999). This phenomenon has traditionally been linked to damage to the left parietal lobe and to hemiparesis of the right hand. Balfour et al. (2007) tested eighty-six consecutive patients with strokes affecting just one hemisphere, finding mirror writing in fifteen (17 percent). The overwhelming majority of these were patients with left-hemisphere lesions who started writing with the left hand due to hemiparesis in their dominant right hand.

Mirror writing in these patients may not actually result from brain damage, as they may never have previously tried writing with their left hand. As Allen (1896) noted: "Mirror writing is often a symptom of nerve disease; but the disease need not be the cause of the existence of the faculty, but only the cause of its discovery" (p. 385). Consistent with this interpretation, Balfour et al. (2007) found mirror movements in six of eight-six healthy elderly participants. Similarly, mirror writing can occur following peripheral injuries to the right arm. Like stroke patients with right hemiplegia, these individuals switch to left-handed writing but do not have any neurological issues. Schott (1980) tested ten patients who had switched from right- to left-handed writing following peripheral injuries to the right arm. Remarkably, seven of the ten reported some form of mirror writing. For example, J.H. switched to left-handed writing after an accident at the age of seven, and went through a ten-year phase in which she would inadvertently produce mirror writing. Interestingly, several of these patients also reported mirror reversing other object-directed actions, including turning screwdrivers and door handles, combing their hair, and turning the pages of magazines.

In other cases, apparently healthy people have realized spontaneously that they can mirror write with their nondominant hand (Allen, 1896; Hughes, 1908; Smetacek, 1992). Allen (1896), for example, discusses his own mirror-writing ability, which he discovered at the age of thirteen. Remarkably, he could do so on his first attempt without error or hesitation, an ability he retained throughout his life. Figure 13.2 shows some examples of Allen's mirror writing. The style of writing is virtually identical for normal writing with his dominant right hand and mirror writing with his left hand. This is a twist on the concept of motor equivalence but, as with the examples discussed in chapter 3, provides clear evidence that the same motor programs are being utilized in each case.

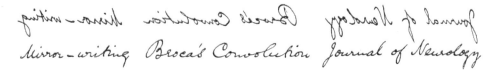

Figure 13.2
Top row: Examples of Allen's (1896) mirror writing with his left hand. *Bottom row*: The same phrases written normally with his dominant right hand.

Contralateral Transfer of Motor Learning: Cross-Education

Finally, practice or training with one hand can transfer to the other hand. This effect is generally known as "cross-education." In a classic study, E.W. Scripture et al. (1894) investigated how practicing an action with one hand affects performance with the other hand. In one experiment, the participant (and co-author) Miss Smith completed a task in which she had to insert a needle into a small hole (3.26 mm in diameter) without touching the sides of the hole. On day 1, she succeeded on 50 percent of trials using her left hand (figure 13.3, left panel). For ten days, she completed two hundred trials per day using her *right* hand, over which her performance gradually improved. The critical question was how she would now perform with her left hand, which she had not used since the first day of testing. Remarkably, she succeeded in 76 percent of trials, an improvement of more than 50 percent since her last attempt.

Scripture et al. (1894) applied the same logic to the training of hand strength. They constructed a device in which a rubber bulb was attached to a mercury dynamometer (essentially a barometer), allowing them to measure the force produced when the bulb was squeezed. Another participant/co-author, Miss Brown, first squeezed the bulb as hard as she could with her left hand, producing 29.6 inches of mercury of pressure (figure 13.3, right panel). Across nine training sessions with her right hand, her strength gradually increased. This improvement again transferred to the untrained left hand, which produced 42.3 inches of mercury of force—an increase of 43 percent.

There are several weaknesses to the study of Scripture et al. (1894), not least of which is the use of only a single participant in each experiment. Nevertheless, their basic effects have been widely replicated over the last 130 years. The effects of strength training have been more heavily studied, and recent meta-analyses have consistently found clear evidence for contralateral transfer (Green & Gabriel, 2018; Manca et al., 2017). Green and Gabriel identified sixty-seven studies involving 1,571 participants. Overall, there was a 30 percent increase in strength from baseline for the trained arm and a 17 percent increase in the untrained arm—transfer of more than 50 percent.

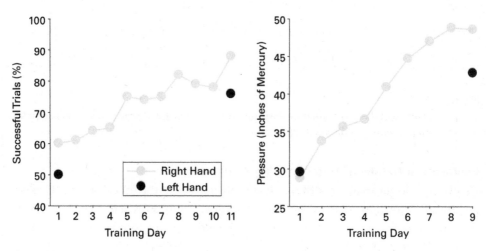

Figure 13.3
Transfer of motor training between hands (cross-education) in the study of Scripture et al. (1894). *Left panel*: Miss Smith's performance placing a needle through a small hole. Progressive improvement was seen across eleven days of training with the right hand (gray dots and line). This improvement transferred to the untrained left hand (black dots). *Right panel*: Miss Brown's performance squeezing a rubber bulb as hard as possible. Strength progressively increased across nine training sessions with the right hand (gray dots and line). This improvement again transferred to the untrained left hand (black dots).

These effects also appear to be specific to the contralateral homologues of the specific muscles exercised. Mason et al. (2017) measured TMS-induced MEPs from two muscles of the left arm, the biceps brachii and the flexor carpi radialis, before and after three weeks of training of the right biceps. MEPs in the left arm increased only for the biceps, the homolog of the muscle that had been trained in the right arm.

Cross-education of motor skill has also been widely replicated. Training of one hand has been found to produce enhanced performance for the untrained contralateral hand in a wide range of tasks, including rapid finger tapping (Davis, 1898; Laszlo et al., 1970), sequences of finger movements (Grafton et al., 2002; Ossmy & Mukamel, 2016), catching balls (Morton et al., 2001), and the board game *Operation* (Pearcey et al., 2022).

Adaptation to altered sensorimotor situations also transfers between the two hands. Such transfer occurs for a variety of perturbations, including visuomotor adaptation to prism displacement (Taub & Goldberg, 1973), to Coriolis forces in a rotating room (DiZio & Lackner, 1995), to visuomotor rotations (Wang & Sainburg, 2004), and to novel force-field opposing hand movements (Criscimagna-Hemminger et al., 2003; Malfait & Ostry, 2004; Taylor et al., 2011).

An important feature of this adaptation is that it occurs in an extrinsic rather than intrinsic reference frame (Criscimagna-Hemminger et al., 2003; DiZio & Lackner, 1995). That is, the adaptation affects the hands in the same absolute direction in space rather than in the mirror-imaged pattern that might be expected from phenomena such as mirror movements and mirror writing. This pattern makes sense in terms of opposing perturbations that exist independent of the observer. For example, for a participant in a room rotating clockwise (DiZio & Lackner, 1995), both of the participant's hands are also moving clockwise. If adaptation to the resulting Coriolis forces transferred to the untrained hand in a mirror symmetric way, adaptation would *double* the perturbation rather than correcting for it. Thus, transfer in an extrinsic reference frame not only makes functional sense but also suggests that transfer of short-term adaptation may rely on different mechanisms than other forms of cross-education.

Two broad classes of theory have been advanced for the mechanistic basis of cross-education (Calvert & Carson, 2022). These theories are not mutually exclusive, and both may operate simultaneously or under different circumstances. The bilateral access theory (Taylor & Heilman, 1980) proposes that training results in the formation of motor programs in the trained hemisphere, which can subsequently be accessed for control of either hand. On this interpretation, there may be no transfer of information between hemispheres at all during the process of learning itself. Instead, it is the *results* of learning that can be transferred between the two hemispheres when the participant attempts the trained action using the untrained hand.

In contrast, the cross-activation theory suggests that there is bilateral activation of both hemispheres during the process of learning itself (Parlow & Kinsbourne, 1989). As the contralateral hemisphere controls the hand during training, the ipsilateral hemisphere also becomes activated and can learn from the training. On this account, no transfer of information between hemispheres need occur when the participant first tries the task with the untrained hand, as changes in the ipsilateral hemisphere had already occurred during the learning process.

Evidence supporting cross-activation theory was provided by Lee et al. (2010). They trained participants to perform index finger abductions with their right hand as quickly as possible. Before and after training, they measured MEPs from both the trained right hand and the untrained left hand after TMS to contralateral M1. After training, MEPs increased by 63 percent in the trained right hand and by 35 percent in the untrained left hand. This result shows that the training increased corticospinal excitability in the untrained right hemisphere. This result is consistent with research using TMS finding reduced interhemispheric inhibition between the motor systems in the two hemispheres immediately following unilateral training (Hortobágyi et al., 2011; Perez, Wise, et al., 2007).

Lee et al. (2010) provided further evidence for cross-activation theory by using rTMS to induce virtual lesions of the motor cortex after training was complete. According to the bilateral access theory, the results of training the right hand should be reflected in the left hemisphere, meaning that virtual lesions of left M1 should reduce training-induced gains for the ipsilateral left hand. In contrast to this prediction, rTMS to each hemisphere only affected performance for the hand contralateral to the virtual lesion. This result indicates that the results of motor learning had transferred to the untrained hemisphere during the training.

An increasing body of research has suggested that cross-education is driven less by transcallosal communication within M1 than by more anterior regions of the motor system, including the dPMC and—especially—SMA. Perez, Tanaka, et al. (2007) used fMRI to measure brain activity before and after participants trained a sequential finger movement task with their right hand. SMA activation after training was correlated with the degree to which training transferred to the left hand. In a second experiment, Perez and colleagues used rTMS to disrupt processing in the SMA, finding that this prevented transfer of learning to the untrained hand but critically did not affect learning in the trained hand.

Cross-education has a variety of potential clinical applications. For example, limb immobilization has been found to produce a range of negative effects on hand function. Training of the contralateral, mobile hand can attenuate these effects. Farthing et al. (2009) asked participants to wear a cast on their left arm for three weeks. In one group of participants who didn't do any special exercise, strength of the immobilized left arm decreased by nearly 15 percent. Another group underwent strength training with their right arm for five days per week. This latter group showed no reduction in the strength of their left arm. In a subsequent study, Farthing et al. (2011) measured neural responses to unimanual contractions using fMRI before and after immobilization of the left arm, showing that right-arm exercise was linked to increased activation in right M1 at posttest.

Such effects provide important insights for developing effective clinical methods for alleviating functional loss in patients with injury or neurological damage (Farthing & Zehr, 2014). For example, Magnus et al. (2013) conducted a randomized controlled trial to assess the effects of cross-education in women with unilateral distal radius fracture. One group of patients underwent handgrip strength training of their nonfractured hand for three days a week, while a control group did not undergo any special training. While both groups showed large reductions in both grip strength and range of wrist motion at nine weeks following the fracture, the training group showed large improvements on both measures by twelve weeks, which were not seen in the control group until substantially later.

For different reasons, patients with hemiparesis following stroke may also be unable to move one of their hands. Recent evidence suggests that cross-education may also be a useful clinical intervention for these patients. Ausenda et al. (2014) asked stroke patients with unilateral hemiparesis at the end of their period of rehabilitation to practice a pegboard task in which small pegs need to be rapidly placed into a grid of small holes with their non-paretic hand ten times per day for three days. After training, performance at this task with the untrained paretic hand was 22.6 percent faster than it had been at baseline. In a recent randomized controlled trial, Dehno et al. (2021) provided stroke patients with a four-week regime of training wrist strength in the non-paretic hand. Compared to a control group, the training group showed greater grip strength in their paretic hand as well as larger MEPs in the paretic hand following TMS applied to contralateral M1.

Bilateral Somatosensory Representations

Anatomical studies in monkeys show a progressive increase in the number of callosal projections between the left and right hemispheres from area 3b to area 1 to area 2 (Manzoni et al., 1980). There are few such connections in area 3b and a modest number in area 1, while area 2 is richly connected. Within each region, moreover, there are richer connections between areas representing proximal body parts than for the distal extremities, such as the hands. Consistent with these anatomical connections, several early studies reported the existence of neurons in S1 with RFs on bilaterally symmetric parts of the monkey's body (Conti et al., 1986). However, such responses were believed to be limited to skin regions near the body midline, such as the face, neck, or torso. These bilateral RFs would have the effect of stitching together the two sides of the body, which were predominantly represented in different hemispheres of the brain, producing "midline fusion" to create a continuous map of the whole body (Manzoni et al., 1989).

Bilateral Tactile RFs
Subsequently, however, Iwamura et al. (1994) provided dramatic evidence that there are also bilateral RFs in S1 for the hands. They found that many neurons in areas 2 and 5 of the monkey's brain had bilateral RFs (figure 13.4). These neurons were of a variety of types, representing both the glabrous and hairy skin, as well as cutaneous, deep, and joint movement responses.

By damaging S1 in one hemisphere, Iwamura et al. (1994) further showed that these bilateral responses are driven by transcallosal projections. When the finger region of S1 in one hemisphere was destroyed, bilateral RFs vanished in the other, intact hemisphere.

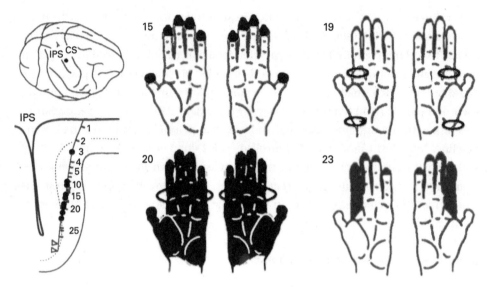

Figure 13.4
Bilateral RFs of S1 neurons in the study of Iwamura et al. (1994). The black spot on the macaque brain (top left) shows the location of the electrode penetration (bottom left) on the postcentral gyrus between the central sulcus (CS) and the IPS. Of twenty-six neurons recorded, nine showed bilateral RFs (indicated by black circles). Tactile RFs are shaded in black, and responsiveness to joint movement is indicated by black circles. Neurons 15 and 23 show bilaterally symmetric RFs on the fingers of both hands. Neuron 19 shows symmetric responsiveness to joint movements of the index finger and wrist. Neuron 20 has symmetric responses to both touch and joint movements. Adapted from Iwamura et al. (2001).

This shows that the bilateral responses are not driven by inputs received directly by the ipsilateral hemisphere without first going through contralateral S1.

The RFs on each hand were generally bilaterally symmetric. However, the intensity of response often differed. Among 105 bilateral neurons recorded in one monkey, sixty-one (58.1 percent) showed stronger responses to contralateral touch and nine (8.6 percent) to ipsilateral touch. Recall from chapter 2 that there is a gradient in the spatial extent and complexity of RFs from areas 3b to area 1 to area 2 to area 5. These neurons with bilateral RFs sit at the top of this hierarchy and reflect a natural extension of the spatial generalization seen at successive steps in this processing continuum (Iwamura, 1998).

Similar bilateral representations appear to exist in humans. In an elegant study, Tamè et al. (2012) used fMRI adaptation to investigate whether there are similar bilateral RFs in humans. They measured responses to tactile "probe" stimuli that were preceded by "adaptor" stimuli, which occurred on either a homologous or a nonhomologous finger.

In some cases, the adaptor and probe stimuli were on the same hand. For example, on a homologous trial, both the adaptor and probe stimuli might be applied to the right index finger (same finger). On a nonhomologous trial, the adaptor might be applied to the right middle finger and the probe to the right index finger. They found a classic repetition suppression effect, with reduced activations in both contralateral S1 and S2 in response to the probe stimulus when stimuli occurred on the same compared to the different finger.

The critical question is what would happen when the adaptor was presented on the opposite hand as the probe. For example, on a homologous trial, the adaptor might be presented on the *left* index finger and the probe on the *right* index finger. On a nonhomologous trial, the adaptor might be on the left middle finger and the probe on the right index finger. There was again clear adaptation in these bilateral trials, in both S1 and S2, although more prominent for the homologous compared to the nonhomologous fingers' stimulation in S1. Such results provide evidence for bilateral integration of tactile stimuli in S1. In a subsequent study, Tamè et al. (2015) used MEG to assess whether the integration of contralateral and ipsilateral tactile information occurs in S1 at early or late stages of tactile processing. In this study, which used a similar design as the fMRI work, contralateral adaptors on homologous fingers produced adaptation in S1 when they preceded probe stimuli by 25 ms and occurred on homologous fingers of the two hands, but not when they preceded probe stimuli by 125 ms and occurred on nonhomologous fingers. Given estimates of transcallosal conduction time (Aboitiz et al., 1992), with a 25 ms delay between stimuli, the two touches should reach S1 contralateral to the probe stimulus at approximately the same time, assuming transfer happened transcallosally between S1 of the two hemispheres.

Direct Ipsilateral Tactile Projections

While such results suggest that tactile signals from the hand reach ipsilateral S1 via transcallosal connections, other evidence suggests that there may also be more direct ipsilateral pathways not involving the corpus callosum. Noachtar et al. (1997) recorded from subdural electrodes placed directly on the surface of the brain in patients undergoing evaluation for surgery to treat epilepsy. They identified clear ipsilateral responses over S1 to electrical stimulation of the median nerve at the wrist. These responses were seen earlier than 30 ms following stimulus onset—much too quick for them to reflect top-down processing, such as from S2 or even from contralateral S1. While ipsilateral responses were delayed by 1–18 ms compared to contralateral responses, this is too short to reflect transcallosal signals.

In another study, Forss et al. (1999) used MEG to measure responses in the ipsilateral left hemisphere to median nerve stimulation of the left hand. In patients with right-hemisphere damage following stroke, contralateral responses in right S1 and S2 were closely related to sensory impairments. When right S1 responses were absent, so too were S2 responses, consistent with a serial organization, as discussed in chapter 2. However, responses of ipsilateral left S2 appeared in all patients, irrespective of the level of sensory loss on the left hand or the level of responsiveness of the right somatosensory cortex. This pattern again indicates that the ipsilateral somatosensory cortex receives direct signals from the periphery that are not transcallosally mediated.

Other evidence for a direct, ipsilateral pathway comes from split-brain patients. Because control of spoken language is lateralized to the left hemisphere in most individuals, the ability of these patients to verbally describe tactile stimuli applied to the left hand provides a test of whether tactile information reaches the ipsilateral left hemisphere by a non-callosal route. Indeed, Sperry et al. (1969) reported that these patients could verbally report the onset of touch in the left hand as well as basic localization, such as whether the thumb or little finger had been touched. A recent, detailed study of patient D.D.C., who underwent complete section of the corpus callosum around the age of twenty, confirmed this pattern (de Haan et al., 2020). In both tactile detection and localization tasks, D.D.C. showed similar levels of performance whether he responded with the stimulated hand or with the contralateral hand, indicating that each hemisphere receives signals from both hands without involvement of the corpus callosum.

At the same time, these ipsilateral projections appear quite impoverished, in some cases appearing to be absent entirely (Gazzaniga et al., 1963). For example, while patients can verbally detect objects placed in the left hand, they show greater difficulties in naming them (Fabri et al., 2005; Sperry et al., 1969; Zaidel, 1998). In one study, patients correctly named 93 percent of objects held in the right hand but only 30 percent in the left hand (Fabri et al., 2005). These ipsilateral projections appear weakest for the hands and feet, with other regions of the body showing more bilaterally symmetric responses (Sperry et al., 1969), consistent with the anatomical studies of the corpus callosum mentioned above.

Tactile Interference Between the Two Hands

Another piece of evidence for tactile interactions between the two hands comes from the phenomenon of extinction, which was mentioned in chapter 7. In tactile extinction, a patient shows awareness of touch when the hands are tested individually but fails to perceive stimuli on one hand when both hands are touched simultaneously. For example, patient H.W. was a thirty-nine-year-old woman who suffered a right-hemisphere stroke (Bender, 1952). Despite paralysis of her left hand, there were no differences in tactile sensitivity between her hands when tested individually. Nevertheless, when touched simultaneously on both hands, she failed to perceive the stimulus touching the left hand. If the right-hand stimulus was then removed, she felt the touch on the left hand after a delay of one to twenty seconds.

Extinction appears to be a common symptom of stroke, at least in the acute phase. Of fifty stroke patients with hemiplegia tested by Bender et al. (1949), forty-four (88 percent) showed some level of tactile extinction. Among 159 right-hemisphere stroke patients tested by Vallar et al. (1994), thirty-one (19 percent) showed tactile extinction. Although extinction is more frequent for the left hand following a right-hemisphere stroke, the converse pattern also occurs (Gainotti et al., 1989). Similar extinction phenomena are also seen for visual, auditory, and multisensory stimuli (Bender, 1952). Extinction for different sensory modalities frequently co-occur but can also be doubly dissociated. For example, Vallar and colleagues identified twenty-one patients with unimodal tactile extinction, fifteen with unimodal visual extinction, and ten with extinction in both modalities. As discussed in chapter 7, extinction can also occur across modalities, such that a stimulus in one modality can be extinguished by a contralateral stimulus in another modality—a phenomenon that has been widely used to investigate representations of peripersonal space (Farnè et al., 2000).

Extinction, like hemispatial neglect, has traditionally been associated with damage to the contralateral PPC. Lesion overlap analyses have indeed implicated the inferior parietal lobe when extinction co-occurs with neglect (Vallar et al., 1994). However, the same analysis also found that extinction without neglect tended to be associated with subcortical lesions. One PET study of patients with left-hand tactile extinction showed that left-hand stimulation produced normal activations in right S2 but no activity in right S1 (Remy et al., 1999). Activation of S2 may be sufficient to support stimulus detection when a unilateral stimulus is presented but may be too weak in the context of a competing contralateral stimulus.

Other studies have shown that extinction can be induced experimentally. In one study, the Wada test was used to put one cerebral hemisphere to sleep (Meador et al., 1988), which induced tactile extinction in many cases. The incidence of extinction was greater following right-hemisphere inactivation. Similarly, extinction-like deficits of tactile detection can be induced by TMS to the right PPC but not to the left PPC (Oliveri et al., 1999). In both cases, disruption of the right hemisphere produced extinction on the left hand, whereas there was no comparable effect of left-hemisphere disruption. This is likely related to the increased prevalence of extinction following right-hemisphere damage—an asymmetry that is also clear in visual hemineglect (Kinsbourne, 1987).

Extinction has also been interpreted in the context of hemispheric rivalry, in which the two hemispheres are in competitive and mutually inhibitory relations (Kinsbourne,

1987). On this account, conditions such as extinction and neglect can be interpreted as a dominance of the contralateral orienting bias of one hemisphere over the other. Consistent with this view, disrupting activity of either the left frontal or parietal lobes can reduce extinction of tactile stimuli on the right hand (Oliveri et al., 2000). This effect in patients is consistent with results in healthy participants showing that TMS to the right PPC can enhance detection of tactile stimuli on the ipsilateral right hand (Seyal et al., 1995). Seyal and colleagues interpret this result in terms of disruption of the right PPC releasing the left PPC from inhibition, thus allowing it to process stimuli on the right hand more efficiently. This interpretation was challenged, however, by a subsequent study that found that disruption of right PPC led to enhanced detection on the ipsilateral right hand (replicating the main result of Seyal and colleagues) but *impaired* performance on a discrimination task (Eshel et al., 2010). This more complex pattern suggests that the relations between the two parietal lobes are more complex than the "seesaw" pattern predicted by simple models of hemispheric rivalry.

There are also interference effects in healthy humans that show functional connections between tactile stimuli on the two hands. Hirsh and Sherrick (1961) presented vibrotactile stimuli to each hand and asked participants to judge which hand had been touched first. They observed incidentally that on a surprising number of trials, participants reported feeling only one of the touches—a phenomenon potentially analogous to extinction. Following up on this observation, Sherrick (1964) showed that the threshold to detect a vibrotactile stimulus on a finger of one hand is substantially increased when another vibrotactile stimulus is presented to the other hand, especially when the two homologous fingers are stimulated. Similar interference, or masking, effects from contralateral stimuli have been described in numerous other studies (e.g., D'Amour & Harris, 2014; Laskin & Spencer, 1979b; Tamè et al., 2011).

Sensory Mislocalizations between the Two Hands
In extinction, tactile stimuli on one hand suppress awareness of touch on the other hand, suggesting competitive, inhibitory links between the two hands. In other cases, the reverse occurs, and touch applied to one hand is actually perceived as being located on the other hand. In allochiria, a touch applied to one side of the body is perceived on the corresponding location on the opposite side of the body. In the much rarer condition of synchiria, a touch to one side is perceived on both sides of the body.

Allochiria was first systematically described by Obersteiner (1882) who emphasized that in many cases, the patient's ability to detect touch was intact or only mildly impaired. This shows that mislocalizations cannot be interpreted as reflecting unawareness of the stimulus or uncertainty about its location. For example, Ferrier (1882) described the case of James S., a twenty-nine-year-old carpenter who suffered a head injury when a scaffold pole fell on his head. On testing, he localized touch quickly and with "unhesitating precision," yet systematically localized touch on the wrong side of the body.

In other cases, however, allochiria occurs on a hand in which ordinary tactile sensation is disrupted or absent. For example, in some amputees, touch applied to the intact hand is perceived on the corresponding location of the contralateral phantom hand (Flor et al., 2000; Ramachandran et al., 1995). Similarly, mislocalization may occur in patients with sensory loss following stroke (Young & Benson, 1992) or nerve injury (Pourrier et al., 2010).

Allochiria has received little attention in recent decades. However, it may not be especially rare. Of 123 stroke patients tested by Kawamura et al. (1987), twenty (16 percent) showed allochiria. Similarly, of nineteen right-hemisphere stroke patients tested in another study, four (21 percent) showed allochiria (Antoniello & Gottesman, 2013). Sensory testing of patients may fail to reveal such mislocalizations if not specifically tested for. If a patient who is asked to indicate where she was touched responds, "On my index finger," this may be taken as a precise and accurate response, without follow-up about *which* index finger she felt was touched. Even if the patient were to indicate the opposite hand, this might understandably be attributed to left/right confusion, which anyone might make. Indeed, this possibility was noted in Obersteiner's (1882) original report: "I believe that the phenomenon is less rare than overlooked, as errors on the part of the patient are apt to be regarded as merely a *lapsus verbi*" (p. 153).

In other patients, rather than being referred to the contralateral side of the body, touch is felt *both* at the stimulated and contralateral locations—a phenomenon known as "synchiria" (Bender, 1952; Jones, 1909). For example, Sathian (2000) reported six stroke patients with sensory loss to one hand who experienced pressure sensations on their impaired hand following application of pressure to their intact hand. These referred sensations started three to five seconds after touch had been applied and were localized on the finger homologous to the touch on the intact hand.

In Sathian's patients, synchiria appeared specific to pressure and did not occur for pinprick or thermal sensations. However, such stimuli did produce feelings of pressure or tingling in the contralateral limb, suggesting some degree of referral across a wide variety of stimuli. Synchiria can also be experienced for perception of limb movement, such that passive movements of one arm produce feelings of both arms moving simultaneously (Jones, 1909; Schilder, 1935). In other cases, passive movements of one hand may produce vaguer sensations of "pulling" or "stretching" in the contralateral hand (Sathian, 2000).

Patient D.L.E. was a seventy-one-year-old man who suffered a stroke affecting large regions of the left frontal and parietal lobes (Medina & Rapp, 2008). His ability to detect stimuli on the right hand was preserved, but he made large localization errors. When touched on his intact left hand, he always felt the touch on the left hand but felt additional synchiric sensations about half the time. In striking contrast to the poor localization of unilateral right-hand touches, synchiric sensations were precisely localized to the location on the right hand homologous to the touched location on the left hand.

Synchiria can be reflected not only in illusory contralateral percepts to unilateral stimulation but also in terms of capture of perceived location of touch on one hand by the location of touch on the other hand. For example, in one experiment, D.L.E. received touches on both hands on every trial: half delivered to homologous locations (e.g., the index finger of each hand) and half to nonhomologous locations (e.g., the right index finger and the left little finger). On nonhomologous trials, D.L.E. reported touch on both hands at the location that had been touched on the left hand on 82 percent of trials (Medina & Rapp, 2008). Similar results were reported by Ricci et al. (2019), who tested patients with unilateral stroke. They developed a Tactile Quadrant Stimulation test, in which touch was applied to one of four quadrants on the back of each hand. Nearly half of patients showed systematic mislocalizations of touch on their contralesional hand, the perceived location being shifted to the location homologous to the stimulated location on the ipsilesional hand. While full-blown synchiria has only been reported infrequently in the literature, such results suggest that more subtle forms of this phenomenon may be much more common than previously thought.

An intriguing aspect of both allochiric (Kawamura et al., 1987; Nathan, 1956) and synchiric (Sathian, 2000; Schilder, 1935) sensations is that they are frequently experienced only after a delay following stimulus onset. For example, Kawamura and colleagues reported that allochiria occurred about half a second following stimulus application. Similarly, synchiric sensations in Schilder's patient Barbara M. emerged four to ten seconds after stimulus delivery, while Sathian reported a delay of three to five seconds. Similarly, referred sensations are abnormally persistent and may continue long after the actual stimulus has been removed. Jones (1909) describes persistence of six to sixty seconds, while Nathan found them to sometimes last for minutes.

Conditions seemingly analogous to both allochiria and synchiria can be induced experimentally in monkeys. Mott (1892) showed that allochiria can be induced by unilateral damage to the spinal cord. As monkeys obviously cannot verbally report where they experience touch, Mott use what he called the "clip method," in which the monkey is placed under general anesthesia, and an uncomfortable clip is placed on one of its feet. When the monkey awakes, it is prevented from seeing its feet, but naturally reaches to try to remove the clip. Mott found that when the clip had been placed on the paralyzed foot, monkeys frequently reached for their *non-paralyzed* foot and scratched and examined the region of the foot homologous to where the clip was applied on the other foot. It seems that the sensations produced by the clip on the paralyzed foot were experienced by the monkey as occurring on the other side of the body, similar to the results of Nathan (1956) in human patients with spinal cord damage.

Similarly, Eidelberg and Schwartz (1971) showed that synchiria can be induced in monkeys by damage to several parts of the nervous system, including the spinal cord, midbrain, and inferior parietal cortex. They trained monkeys to perform a forced-choice task in which they pressed one of three buttons, depending on whether the left hand, the right hand, or both hands were touched. This allowed them to infer that the monkeys had felt bilateral sensations following unilateral stimulation.

There is also evidence that synchiria can be induced in humans from peripheral nerve block. Case et al. (2010) tested six patients undergoing unilateral block of the brachial plexus in preparation for orthopedic surgery, which resulted in the entire arm being anesthetized. Four of the six patients reported feeling touch in their anesthetized arm during a period in which their contralateral arm was being stroked.

There are also several pieces of evidence suggesting that experiences of analogous referred sensations can be induced in healthy adult humans. For example, Petkova and Ehrsson (2009) used a variant of the rubber hand illusion, shown in figure 13.5. They applied synchronous touch to the participant's left hand and to a *right* rubber hand that rested next to the participant's unseen right hand while participants looked toward the rubber hand. Unlike the traditional version of the rubber hand illusion, discussed in chapter 6, the touch felt (on the left hand) and the touch seen (on the right rubber hand) clearly do not correspond to a single event. Nevertheless, the majority of participants experienced ownership over the rubber hand, as measured by the usual array of questionnaire, proprioceptive, and autonomic measures discussed in chapter 6. More strikingly, participants reported actually feeling touches on their right hand, which was never actually touched. Neither ownership of the rubber hand nor felt touches on the right hand occurred when the two touches were asynchronous. This shows that neither feeling touch on the left hand nor seeing touch on the right rubber hand is sufficient on its own to induce the illusion of touch on the right hand. The feeling of touch on one hand may automatically elicit sub-threshold synchiria on the opposite hand. In this illusion, the visual perception of the rubber hand being touched may have raised this normally implicit processing above the threshold for conscious experience.

In another set of studies, Marcel and colleagues (Marcel, Postma et al., 2004; Marcel et al., 2006) have described what they call "migrations" of tactile sensations between the two hands in some healthy participants. They used a task in which tappers presented one of two tactile patterns to each hand. "Tap" stimuli consisted of a continuous

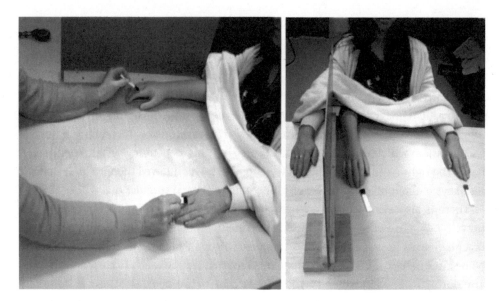

Figure 13.5
Petkova and Ehrsson's (2009) procedure for experimental induction of synchiria using the rubber hand illusion.

250 ms touch, while more salient "drum" stimuli consisted of a series of three 50 ms touches separated by 50 ms. On each block of trials, participants were asked to report which stimulus was delivered to one hand while ignoring the stimulus on the other hand. About a quarter of participants showed very high rates of errors when salient drum stimuli were presented on the unattended hand while tap stimuli were presented on the attended hand. The more salient stimulus on the unattended hand appeared to fuse with the attended stimulus. This has a striking resemblance to the synchiric capture effects in neurological patients, described above (Medina & Rapp, 2008; Ricci et al., 2019). Marcel and colleagues suggest that some individuals have a premorbid disposition to allochiric-type experiences.

Contralateral Transfer of Tactile Learning

Recall the phenomenon of cross-education discussed above in which training of one hand improves both strength and skill in the untrained contralateral hand. Similar findings have also been found for tactile learning. In a seminal early study, Volkmann (1858) trained participants on a two-point discrimination task, showing that improvements in spatial sensitivity were found not only on the trained finger but on other fingers of the same hand and contralateral hand. Many subsequent studies have replicated this pattern of contralateral transfer for a variety of tactile judgments, including left/right location judgments (Sathian & Zangaladze, 1998), temporal duration judgments (Nagarajan et al., 1998), texture judgments (Harris et al., 2001), and orientation judgments (Harrar et al., 2014).

There is also evidence for spread of tactile learning between the hand and other body parts, such as the lips (Muret et al., 2014) and the feet (Frank et al., 2022). This raises the question whether the effects described in the previous paragraph reflect a specific link between the two hands or a broader relation between many body parts. This sort of link could be thought of as a sensory analogue of the concept of motor equivalence, which I have highlighted throughout this book—what Luigi Tamè and I (2023) have recently called "tactile equivalence." There is indeed evidence that this transfer of tactile learning is particularly strong between homologous fingers, with almost complete transfer between the trained finger and its contralateral homologue (Harrar et al., 2014; Harris et al., 2001). Strong transfer is also found for the fingers adjacent to the trained finger, suggesting that transfer may relate to the degree of overlap between body parts in early somatotopic maps, in which multidigit and bilateral tactile RFs emerge (Iwamura, 1998).

Studies in cats have provided evidence that the corpus callosum is critical for contralateral transfer of tactile learning. Stamm and Sperry (1957) showed clear transfer of tactile learning on one forepaw to the contralateral forepaw. This transfer was completely absent in split-brain cats following section of the corpus callosum, although these animals were able to learn on the trained paw. Broadly similar results have also been found in primates, including monkeys (Ebner & Myers, 1962), chimpanzees (Myers & Henson, 1960), and humans (Russell & Reitan, 1955; Smith, 1951). In a later study, Sechzer (1970) showed that split-brain cats also showed impairments on the trained forepaw itself, showing much slower tactile learning than intact animals. This result raises the possibility that the transfer of learning is an artifact of the recruitment of a bilaterally symmetric sensorimotor network for unilateral tactile learning. If you try to learn a new tactile discrimination with your right hand, learning can be speeded up if somatosensory regions in both hemispheres are recruited. This may result in transfer to the contralateral hand as a purely passive byproduct of bilateral recruitment.

14 The Dominant and Nondominant Hands

The Cueva de las Manos (Cave of the Hands) is a UNESCO World Heritage site in the province of Santa Cruz in southern Argentina. It is named for the spectacular collages of hand paintings on its walls, which were studied over thirty years by Carlos Gradin of the University of Buenos Aires. The hands are stencils, made by holding one hand against the rock and using a bone tube held in the other hand to blow paint, as shown in figure 14.1. By dating these tubes and the pigments used as paints, Gradin (1994) showed that the paintings were made in multiple waves from around 7,300 BCE to 700 CE—an extraordinary span of eight thousand years. Little is known of the purpose of these paintings or the culture of the people who made them. But their nature and the continuity of style across thousands of years suggest they had deep symbolic and religious significance. What is certain is that they are among the best records we have of the early peoples of South America and a glorious expression of human artistic creation.

There is, however, an oddity about these paintings. Nearly all of them depict left hands. Of 360 hand stencils for which laterality can be clearly assigned, fully 329 show left hands—more than 91 percent (Gradin, 1994). To understand why this might be, we must consider the process by which they are made. One hand is held against the rock, while the other hand holds the tube used to blow paint. It is most natural to hold the tube with the dominant hand, leaving the nondominant hand to serve as a stencil. The laterality of the hand paintings tells us which hand the artist preferred to use to hold a tool. The overwhelming prevalence of left-hand stencils in the Cueva de las Manos thus has a simple explanation. The people who made them were right-handed.

Handedness in Humans

In the previous chapter, we discussed the bilateral symmetry of the body and ways in which the representations of the left and right hands are similar. It is an obvious fact, however, that most people have a strong preference to use one hand over the other, and

Figure 14.1
Hand stencils from the Cueva de las Manos (Cave of the Hands) in Santa Cruz, Argentina. Even from this single image, the prevalence of left-hand stencils is clear. Photo by Pablo Gimenez.

that it is usually the right hand that is preferred. This difference is known as "handedness," defined by McManus et al. (2016) as "that difference between the hands of which every right-hander and left-hander is entirely aware from their own behaviour, but for which we have almost no adequate scientific explanation" (p. 24).

The Consistency of Right-Handedness in Human Populations

The consistency of populational-level right-handedness in humans is striking. There is no known human population, now or at any time in the past, in which left-handers have exceeded (or even gotten close to) 50 percent of the population. One modern study (Perelle & Ehrman, 1994) surveyed twelve thousand participants from seventeen countries and found no society in which the percentage of left-handers exceeded 13 percent. Similar results have been found in studies of traditional societies across several continents (Faurie et al., 2005; Marchant et al., 1995). If the preferential use of one hand were a cultural convention, we should expect to find societies with more left-handers than right-handers. That this is not the case suggests that handedness reflects something fundamental about the neural mechanisms of hand control.

Indeed, the prevalence of right-handedness is not a new phenomenon. Historical sources reveal population-level right-handedness as far back as we have written records.

The Old Testament book of Judges (3:12–26) tells how Ehud managed to assassinate Eglon the King of Moab. As a left-hander, Ehud wore his sword on his right hip, allowing him to smuggle it into Eglon's presence, since the guards who searched him expected it on the opposite hip. Elsewhere in Judges (20:15–16), the Benjamite army includes twenty-six thousand right-handed and seven hundred left-handed soldiers, or 2.6 percent left-handers. Whatever the historical accuracy of these two stories, they show that the authors of Judges, writing nearly three thousand years ago, expected a preponderance of right-handers. Similarly, both Plato (1875) in the *Laws* and Aristotle (1908a) in the *Metaphysica*, writing in the fourth century BCE, take the dominant role of the right hand for granted, although Plato views it as a bad habit that we should work to overcome.

Similar conclusions come from artistic depictions of hand use. Dennis (1958) investigated handheld tools in wall paintings from ancient Egyptian tombs. At Beni Hasan, 111 out of 120 examples (93 percent) show use of the right hand, as do 100 of 105 (95 percent) at Thebes. Coren and Porac (1977) extended this analysis to art from around the world across more than five thousand years. There was a remarkably consistent bias for right-handed tool use, with little change across time and space. Of thirty-nine instances of tool use from before 3,000 BCE, 90 percent showed the right hand being used—nearly identical to the 89 percent in the twentieth century.

Several other lines of evidence have projected right-handedness even farther into prehistory. One line comes from analyses of wear on stone tools. Posnansky (1959) analyzed patterns of asymmetry in Paleolithic hand axes from the Trent Valley and Furze Platt in England. More than twice as many hand axes showed asymmetries conducive to use with the right hand than the converse. Similar evidence comes from analyses of head wounds. For example, Bennike (1985) assessed skulls that suffered from head wounds from Neolithic burials in Denmark dated to between five thousand and six thousand years ago. An overwhelming preponderance of these wounds were on the left side, consistent with a forehand blow inflicted by a right-handed foe.

Finally, evidence comes from studies of cave paintings, as in the Cueva de las Manos. Images of hands are among the most common themes in ancient cave art and are commonly made using two techniques. "Negative" hand stencils are made by holding a paint tube with one hand and using the other as a stencil. In contrast, "positive" handprints are made by applying paint to the palm of the hand and pressing the hand against the wall. It is most natural to make prints of the dominant hand but stencils of the nondominant hand. Thus, a preference to use the right hand should be expressed in cave paintings by a predominance of left-hand stencils but of right handprints.

This pattern has consistently been observed for cave paintings from around the world, as recently reviewed by Uomini and Ruck (2018). For example, Faurie and Raymond

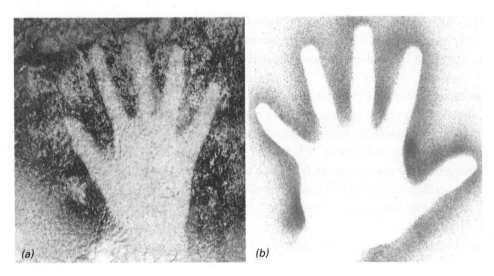

Figure 14.2
Painted hand stencils made by (a) a Paleolithic human more than twenty thousand years ago in the cave of Le Pech-Merle, France, and (b) a modern French university student. From Faurie and Raymond (2004).

(2004) identified 343 painted hand stencils from the Upper Paleolithic (ten thousand to thirty thousand years ago) in France and Spain. The left hand was depicted in 77 percent, consistent with the artists having been mostly right-handed. A nearly identical percentage of left-hand images (77.1 percent) was obtained when modern French participants made similar stencils (figure 14.2).

Despite continuity in the general pattern of handedness across time and cultures, there is also geographical and cultural variance in the exact prevalence of left handedness. In the large study of Perelle and Ehrman (1994), mentioned above, the prevalence of left-hand writing ranged from 2.5 percent to 12.8 percent across countries. Similarly, a review of eighty-one studies of hand preferences for throwing and hammering (Raymond & Pontier, 2004) found rates of non-right-handedness (combining left-handed and ambidextrous patterns) from 5 to 26 percent. Thus, while left-handers are a clear minority in all societies, their prevalence nevertheless ranges across a factor of five.

Connolly and Bishop (1992) tested participants in England and in an isolated population in the highlands of Papua New Guinea on several measures of both hand preference and proficiency. Overall, similar rates of right-handedness were observed in the two cultures. They then conducted a principal component analysis, finding support for a two-component model. The first component had strong loadings for all measures and accounted for more than 57 percent of the variance in the data. This component showed no difference between groups. The second component, accounting for about 11 percent of variance, showed opposite loadings for preference and proficiency measures and differed systematically across cultures. This led to the Papuan group being slightly less right-handed than the English group on proficiency tests but slightly more right-handed for preference tests. Overall, the results suggest that the main component of interpersonal variance in handedness is impervious to cultural influences, while culture may influence a secondary component.

Some cultural differences likely reflect social pressure for left-handers to conform with right-handed social norms. For example, in East Asian cultures, there has been social pressure for people to write and use chopsticks with the right hand. Studies in Taiwan (Teng et al., 1976) and Hong Kong (Hoosain, 1990) have identified substantial populations who use their right hand for those two activities but their left hand for most other activities. The same has been true in the West. For example, Ireland (1880) describes the procedures at one Scottish primary school in the late nineteenth century: "They were not allowed to use their left hands in writing or ciphering. Great trouble had, in fact, been taken to make them desist from using their left hands. Some of them had to be kept near the teacher at their writing, but in the playground they threw stones and played at bowls with their left hands" (p. 208). One study of hand waving in nineteenth-century films from England showed that the incidence of left-handedness declined precipitously during the Victorian era, presumably due to social stigmatization, before gradually returning to earlier levels during the twentieth century (McManus & Hartigan, 2007). Questionnaire studies have shown a generational increase in the prevalence of left-hand writing across the twentieth century in several countries, including France (Dellatolos et al., 1988), Italy (Salmaso & Longoni, 1985), Brazil (Martin & Freitas, 2002), and the United States (Gilbert & Wysocki, 1992). Such changes likely reflect increasing social acceptance of left-handedness and a reduction of efforts to force children to use their right hand.

Asymmetries of Hand Preference and Hand Skill
The evidence presented thus far has mostly concerned which hand people prefer to use for specific tasks. This does not, of course, mean that people would be unable to do these things with their non-preferred hand. Should they have wished to, the artists of the Cueva de las Manos could, no doubt, have held the paint tube in their left hands to make right-hand stencils. As discussed in chapter 3 in relation to the principle of motor equivalence, most of us can pick up a pencil and write with our non-preferred hand, producing text that is recognizably our own. While this shows that the motor programs involved are to some degree abstracted from the specific effector used, it is also obvious

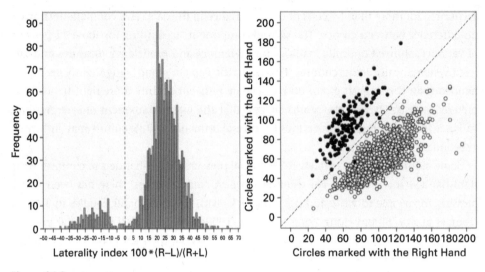

Figure 14.3

Lateral asymmetry of hand skill in Tapley and Bryden's (1985) circle-marking task. *Left panel*: Histogram showing laterality index scores, with positive numbers indicating better performance with the right hand and negative numbers with the left hand. There is a clear bimodal distribution. *Right panel*: Scatterplot showing the association between performance with the right and left hands. The color of points indicates self-reported handedness, with gray circles indicating right-handedness and black circles left-handedness. Figure from McManus et al. (2016).

that, for most of us, writing with the non-preferred hand will be less neat and legible than with our preferred hand.

Several measures of differences in hand skill have been developed over the years. Annett (1970) measured the time needed to place ten pegs into a row of small holes using each hand. This "pegboard" task has been widely used and shows that people tend to be about 10 percent faster when using their preferred hand. Similarly, Tapley and Bryden (1985) gave participants arrays of small circles and asked them to make a dot in as many as they could in twenty seconds using each hand. Figure 14.3 shows results from 1,556 Canadian undergraduates tested in class. The histogram in the left panel shows a bimodal pattern in skill. Most people perform better with their right hand, but a smaller group perform better with their left hand. The scatterplot in the right panel shows that these performance asymmetries correspond well with self-reported hand preferences.

Nevertheless, hand preference and skill do not always align. For example, Angstmann et al. (2016) tested a group of strongly right-handed children and adolescents on a task in which they had to draw repeated concentric circles on a digitizing tablet. Manual skill was assessed by quantifying the consistency in shape of the circles produced. While performance was higher, on average, for the right hand, many individual participants performed better with their non-preferred left hand. Similarly, handedness can be discordant across tasks. Gilbert and Wysocki (1992) obtained measures of hand preference for writing and throwing from more than one million Americans aged from ten to eighty-six. Overall, nearly 5 percent of people showed mixed hand preferences for the two tasks. A recent meta-analysis estimated mixed-handedness at 9.3 percent (Papadatou-Pastou et al., 2020).

Neural Signatures of Handedness

Handedness by its very nature reflects a laterality in the abilities of the two hands. This behavioral laterality must correspond to lateralized differences in the brain. Identifying these differences in the brain, however, has been surprisingly difficult, given how obvious the behavioral differences are. One natural hypothesis is that motor regions, such as M1 or the reaching and grasping networks, are simply bigger in the hemisphere contralateral to the dominant hand. While some studies using structural MRI have reported that the depth of the central sulcus is greater in the hemisphere contralateral to the preferred hand (e.g., Amunts et al., 1996), these results have been inconsistent and weak. Sometimes they occur for right-handers but not left-handers, sometimes for men but not women. One recent study tested a large sample of 1,960 right-handers and 106 left-handers using automated parcellation methods to quantify cortical surface area throughout the brain (Guadalupe et al., 2014). There were no brain regions that showed statistically reliable associations with self-reported handedness.

Studies using functional MRI have identified clearer hemispheric differences. As noted in the previous chapter, many studies have found that left M1 becomes active during movements of both hands, whereas right M1 responds only during left-hand movements (Kim et al., 1993; Verstynen et al., 2005). Tzourio-Mazoyer et al. (2015) tested 284 individuals, half right-handers and half left-handers, on a finger-tapping task. In right-handers, clear deactivation was found in ipsilateral right M1 during right-hand tapping, while much more modest ipsilateral deactivation was found during left-hand tapping. In left-handers, in contrast, symmetrical deactivation of ipsilateral M1 was found for movements of both hands.

Such deactivations are clear evidence for interhemispheric inhibition in the motor system. As discussed in the last chapter, there are different conceptions of the function of such inhibition. On one view, inhibition of the ipsilateral hemisphere has the function of preventing unwanted mirror movements (Duque et al., 2005), while on other views, inhibition is less about suppression than about sharpening neural representations to "sculpt" movements (Carson, 2020). To investigate how ipsilateral deactivation relates to manual skill, Tzourio-Mazoyer et al. (2015) additionally measured manual skill by having participants make as many taps as possible in ten seconds with the index finger of each hand. Across participants, there was a clear relation between lateral asymmetries in manual skill and ipsilateral M1 deactivations. These results indicate that inhibition between the hemispheres not only suppresses unwanted overflow but also has a deeper functional role in skilled actions.

Another study used dynamic causal modeling of functional connectivity between brain regions to investigate broader neural networks involved in hand control (Pool et al., 2014). For right-handers, movements of the dominant right hand led to increased functional connectivity between the left SMA, right SMA, and PMC compared to left-hand movements. The strength of this effect was correlated across individuals with right-handedness assessed via questionnaire. Thus, the dexterity of the dominant right hand appears linked to the recruitment of a wider neural network that is not recruited for movements of the nondominant left hand.

The predominance of right handedness is likely also related to the key role of the left hemisphere for skilled manual action, discussed in chapter 9 in relation to apraxia. Following Liepmann's (1905) classic report that patients with left-hemisphere—but not right-hemisphere—lesions showed apraxia in their ipsilesional hand, many studies have confirmed that left-hemisphere damage disrupts skilled control of both hands (Geschwind, 1965; Hanna-Pladdy et al., 2002; Stoll et al., 2022). Similarly, cutting the corpus callosum in split-brain patients can produce apraxia specific to the left hand (Gazzaniga et al., 1962, 1967). These results suggest a left-hemisphere specialization for representations of high-level, skilled actions that is used by both hands. While this information can be transferred to the right hemisphere to control the left hand, this may simply be less rapid and efficient than transferring it within the left hemisphere to control the right hand.

Is There Handedness for Somatosensory Function?
Given how obvious the differences in motor function between the dominant and nondominant hands are and how easy they are to demonstrate, it is natural to ask whether there are equivalent differences in somatosensory function on the two hands. Is the sense of touch more effective on the dominant hand?

In general, the answer seems to be that it isn't. For example, Vega-Bermudez and Johnson (2001) tested right-handed participants on two tasks of tactile spatial ability, judging the orientation of a square-wave grating applied to the fingertip or recognizing letters embossed in plastic. While there were clear differences in spatial acuity between fingers, with performance decreasing across the hand from the index finger to ring finger, there were no differences at all between the two hands. Consistent with this result, a recent study that mapped somatotopic finger maps in S1 using fMRI found virtually identical maps of the dominant and nondominant hands (Schweisfurth et al., 2018). The same is true for higher-level haptic tasks, in which similar performance in object recognition has been found for the left and right hands (Craddock & Lawson, 2009; Fagot et al., 1993).

When differences between the hands in tactile performance have been observed, they generally relate to the nature of the stimuli rather than to the ability of the hands as such. Summers and Lederman (1990) reviewed a large number of studies comparing the two hands on a variety of tactile tasks. Overall, there appears to be a modest left-hand advantage for tasks involving high-level spatial information and a right-hand advantage for tasks involving linguistic information. Notably, however, the same patterns have been found for visual and auditory stimuli, and the advantage seems to reflect the more general hemispheric specializations of the right hemisphere for spatial information and the left hemisphere for linguistic information, rather than anything specific to the somatosensory system.

The Evolution and Development of Handedness

Why the right hand? There is no obvious feature of the environment that makes population-level right-handedness advantageous compared to left-handedness. Indeed, as the Victorian critic Thomas Carlyle noted, "Why that particular hand was chosen is a question not to be settled, not worth asking except as a kind of riddle" (Froude, 1884, p. 408). It is more likely that the advantage of handedness was simply that there be a population-level preference for *some* hand. That the right hand ended up winning this competition would then be just one more in the innumerable number of chance accidents that have shaped our evolution.

Handedness in Primates

Lateral asymmetries in behavior are not specific to humans but rather are widespread throughout the animal kingdom. One systematic review of limb preferences in vertebrates (Ströckens et al., 2013) identified data from 119 species and found that a majority (sixty-one) showed a population-level asymmetry. While such results indicate that lateralization is widespread among animals, it is also true that not each of these asymmetries reflects a right-limb preference, and virtually none is of comparable strength to human handedness. A possible exception is the strong rightward biases for behavior shown by marine mammals, such as cetaceans (e.g., whales and dolphins) and pinnipeds (e.g., seals and walruses; MacNeilage, 2014).

In an influential paper, MacNeilage et al. (1987) proposed what has become known as the "postural origins theory" of handedness. Recall from chapter 11 that the earliest primates were arboreal and supported themselves on tree branches while using their hands to obtain food, such as insects (Cartmill, 1974) or fruit (Bloch & Boyer, 2002). MacNeilage and colleagues argue that early primates used their left hand to reach for food, while their right hand was used for stabilizing the body on branches. This pattern continues to be seen in modern prosimians, such as lemurs, which show strong biases to reach for food with their left hand (Ward et al., 1990). With the move to a terrestrial lifestyle, the right hand was freed from its role of supporting the body and gradually became specialized for tasks requiring precision manipulation, and eventually for tool use.

A substantial literature has investigated hand-use preferences in primates using simple unimanual tasks such as reaching for objects, but no clear evidence has been produced for population-level lateral biases. In the first systematic study of handedness in nonhuman primates, Finch (1941) tested thirty chimpanzees on four simple manual tasks, measuring which hand they preferred to use. Finch classified fourteen of the chimpanzees as left-handed, eleven as right-handed, and five as ambidextrous. Thus, there was no evidence at all for population-level right-handedness in these chimpanzees and, indeed, on the contrary, a modest trend in the opposite direction. Although hundreds of subsequent studies have investigated this issue in numerous primate species, the overall pattern of results is consistent with Finch's findings that chimpanzees, and other nonhuman primates, do not show consistent population-level hand preferences on simple unimanual tasks such as reaching (Hopkins, 2006; McGrew & Marchant, 1997).

Fagot and Vauclair (1991) argued that it is critical to consider the complexity of the tasks being performed when assessing handedness. They distinguished between "low-level" tasks, which involve simple, familiar, or highly practiced behaviors, and "high-level" tasks, which involve novel, cognitively complex behaviors. In their task complexity hypothesis, Fagot and Vauclair argued that laterality effects should be more apparent on high-level tasks, while performance should be more symmetrical for low-level tasks. Critically, simple reaching and similar unimanual behaviors, such as those used by Finch (1941), are low-level tasks for which laterality preferences might not be expected. For example, if you are asked to perform an easy task such as picking up an apple, there is hardly any reason to prefer your dominant hand. In such a trivially easy task, the selection of which hand to use is likely to be determined by which hand is closer to the apple or is in a more convenient posture to grasp it. Similarly, studies measuring hand preferences in nonhuman primates using simple reaching tasks may have missed preferences by not giving animals tasks for which they were strongly motivated to use their preferred hand.

Figure 14.4
The tube task being performed by nonhuman primates, including an orangutan (left; Hopkins et al., 2011), a baboon (center; Vauclair et al., 2005), and a snub-nosed monkey (right; Zhao et al., 2012). The orangutan and snub-nosed monkey are using their left hand to retrieve the food, while the baboon is using its right hand, consistent with the population-level preferences found in each species.

Consistent with the task complexity hypothesis, studies using complex bimanually differentiated tasks have provided clearer evidence of manual asymmetries in primates. An important advance came from Hopkins's (1995) development of the "tube task," which requires coordinated action of both hands. A bit of food is placed inside a long thin tube such that it can be extracted by inserting one finger inside. This requires that one hand be used to hold the tube in place while the other hand retrieves the object (figure 14.4). Using this task, clear individual and population-level hand preferences have been described in a wide range of primate species, and it is widely agreed to be a more reliable measure of handedness than simple reaching (Meguerditchian et al., 2013; Nelson, 2022). Indeed, hand preferences on the tube task are associated with asymmetries in brain structure measured using MRI in chimpanzees (Hopkins & Cantalupo, 2004), baboons (Margiotoudi et al., 2019), and capuchin monkeys (Phillips & Sherwood, 2005).

Intriguingly, population-level hand preferences among primates measured with the tube task differ systematically between arboreal and ground-living species. Ground-living species tend to show preferences to use their right hand to obtain the food, including chimpanzees (Hopkins, 1995; Hopkins et al., 2004), gorillas (Hopkins et al., 2011), baboons (Vauclair et al., 2005), macaque monkeys (Regaiolli et al., 2018), and, of course, humans. In striking contrast, primates with primarily arboreal lifestyles tend to show a left-hand preference, including orangutans (Hopkins et al., 2011) and snub-nosed monkeys (Zhao et al., 2012).

Other studies have suggested that the African great apes show stronger right-hand preferences when grasping objects using precision grips involving the index finger and thumb than when using less precise grips (Llorente et al., 2009; Meguerditchian et al., 2015). Such right-hand biases for precision grips have been reported in chimpanzees, gorillas, and bonobos. In contrast, no such bias appears in species from outside of Africa, such as orangutans (Meguerditchian et al., 2015) and capuchin monkeys (Spinozzi et al., 2004), despite the ability of both species to use precision grips effectively. This pattern is consistent with results from the tube task and suggests that a right-hand bias for precision grasping emerged in the common ancestor of the African great apes, from which humans evolved.

Other studies have investigated laterality for species-specific communicative gestures. Across species, there is strong evidence for right-hand biases for communicative gestures, including in chimpanzees (Meguerditchian et al., 2012), bonobos (Hopkins & Vauclair, 2011), gorillas (Shafer, 1993), and baboons (Meguerditchian et al., 2011). This pattern suggests that there may be important links between handedness and language—a point we will return to in the next chapter.

Handedness in Hominins

As discussed above, a predominance of right-handedness is apparent in anatomically modern humans as far back as we have written records, and even farther back in analyses of art. Another intriguing approach to understanding the evolution of handedness is the reconstruction of handedness in extinct hominins from archaeological evidence (Uomini & Ruck, 2018). As in sites inhabited by anatomically modern humans, Neanderthal cave paintings show a predominance of left-hand stencils. Among the sixty-four-thousand-year-old Neanderthal hand stencils at Maltravieso cave in Spain, for example, there are twenty-six left hands and eleven right hands (Groenen, 2011).

Another strategy is to investigate the patterns of wear on stone tools or the debitage left over from the manufacture of such tools (Bargalló & Mosquera, 2014; Dominguez-Ballesteros & Arrizabalaga, 2015). Keeley (1977) examined the patterns of wear on stone tools used for boring holes into wood found at Clacton-on-Sea in eastern England, dating to around four hundred thousand years ago and thought to have been made by *H. heidelbergensis*. The pattern shows that the tools had been turned in a clockwise rotary motion while downward pressure was applied. This is the direction of motion that would be used by the right hand. Toth (1985), similarly, found that the flakes produced by right- and left-handed knappers differ, on average, due to the direction in which the hammerstone tends to hit the core when held in the right versus left hand. By comparing the patterns of flake debitage from Oldowan and Acheulean sites at Koobi Fora in

Kenya to that produced by modern right-handed archaeologists, Toth argued that the Koobi Fora knappers were predominantly right-handed.

A third line of evidence comes from scratch marks on the frontal surfaces of teeth, which, while uncommon in modern humans, are ubiquitous in Neanderthals and other hominins. These scratches are believed to come from the use of stone flake tools to cut pieces of meat held between the teeth and one hand (figure 14.5, top left). This solves a basic problem in cutting meat using a flake that two hands are needed to hold each side of the meat taut and a third to wield the flake. Bermúdez de Castro et al. (1988) showed experimentally that virtually identical scratch marks were produced on teeth embedded in a mouth guard when a stone flake was used for cutting. Bruner and Lozano (2014) have gone so far as to claim that the mouth constituted a "third hand" for Neanderthals. The relevant point here is that these scratch marks occur in different orientations, depending on which hand is used to hold the cutting flake. With the right hand, the scratches run from top right to bottom left; with the left hand, they run in the opposite direction (figure 14.5, bottom left). By analyzing the direction of scratches, researchers have thus been able to determine hand preferences, even when only isolated fossil teeth are found.

These analyses have consistently found a high prevalence of right-handedness. The oldest individual whose handedness has been identified is the *H. habilis* specimen OH-65 from Olduvai Gorge (Frayer et al., 2016), shown in figure 14.5 (top right). Of 559 scratches on the frontal surface of these teeth, 260 (46.5 percent) run from top right to bottom left, whereas only sixty-three (11.3 percent) run in the opposite direction. This provides strong evidence that this individual was right-handed. While population-level data are unfortunately not available for *H. habilis*, much more data is available for Neanderthals and pre-Neanderthal humans at European sites. For example, at the 430,000-year-old Sima de los Huesos (Pit of Bones) cave in Spain, all fifteen Neanderthals whose teeth had enough scratches for analysis were right-handed, as did nine of eleven Neanderthals at Krapina in Croatia (Lozano et al., 2017). Overall, of thirty-six European Neanderthals whose handedness has been assessed from dental scratches, thirty-two (88.9 percent) appear to have been right-handed—a prevalence in line with modern human populations.

Compiling evidence on handedness from all of these sources, Uomini and Ruck (2018) show that population-level right-handedness is a long-standing feature of human populations but may also have undergone evolutionary change. Among anatomically modern humans whose handedness was identified through the same types of archaeological evidence, 87 percent were right-handed compared to 77 percent among Neanderthals and 69 percent among pre-Neanderthals (*H. heidelbergensis*, *H. antecessor*). Importantly, however, while right-handedness appears to have increased during hominin evolution,

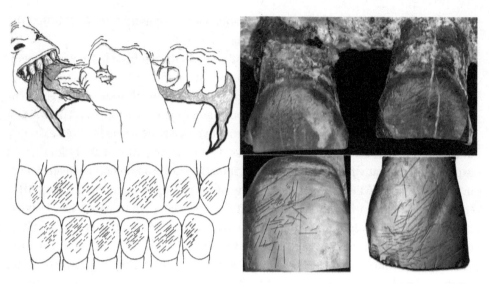

Figure 14.5
Assessing handedness in fossil hominins from scratches on teeth. *Top left*: Depiction of a hominin holding a piece of meat between her teeth and left hand while using her right hand to cut the meat using a flaked tool (from Frayer et al., 2016). *Bottom left*: The pattern of scratches produced on the front surface of the teeth by a tool held in the right hand, oriented from upper right to lower left (from Bermúdez de Castro et al., 1988). *Top right*: The front teeth of OH-65 (*H. habilis*) from Olduvai Gorge. The scratch marks suggest this individual was right-handed (from Frayer et al., 2016). *Bottom right*: Incisors from Neanderthals from Krapina (left) and Vindija (right), both in Croatia. Both of these individuals were right-handed. From Lozano et al. (2017).

left-handedness has not decreased. Instead, the prevalence of mixed handedness has declined alongside the increase in handedness.

Development of Handedness in Humans

Handedness has long been known to run in families. Charles Darwin (1868), for example, comments that left-handedness is "well known to be inherited" (p. 12). Modern studies have found that a child with a left-handed parent is at least twice as likely to end up left-handed as a child with two right-handed parents (McManus, 2002). Until the early twenty-first century, there was hope that handedness could be explained using simple, single-gene models, such as the right shift theory of Annett (2002) and the dextral/chance theory of McManus (1985). Large-scale genome-wide association studies, however, have proved inconsistent with such simple models (Armour et al., 2014). While there is clear evidence for a genetic contribution to handedness, this contribution is fairly modest. One meta-analysis of twin studies comparing concordance in pairs of identical and fraternal twins estimates that about 25 percent of the variance in hand preference is explained by genetic variance (Medland et al., 2006).

Rather than being a result of the genes themselves, some evidence suggests that handedness may be linked to early asymmetries during fetal life in epigenetic modulation of gene expression in the cortex (Sun et al., 2005) or spinal cord (Ocklenburg et al., 2017). Indeed, manual preferences have been observed even before birth. As described in chapter 12, arm movements are seen in fetuses from nine to ten weeks' gestation (de Vries et al., 1985). Even these earliest movements show strong lateral preferences. Hepper et al. (1990) used ultrasound to record instances of thumb-sucking in 224 fetuses, of which nearly 95 percent (212) were of the right thumb. Other studies have reported similar right-arm preferences even for simple movements (Hepper et al., 1998; Kurjak et al., 2002). These preferences appear to be related to later handedness: there is strong correspondence between the hand used for thumb-sucking as a fetus and handedness measured at ten to twelve years of age (Hepper et al., 2005).

Critically, such manual asymmetries are present during the first trimester, before the corticospinal tract starts to connect to the spinal cord. In one study of seventy-two fetuses of ten weeks' gestation, fifty-four showed more movements of their right arm than of their left, with only nine showing a left preference (Hepper et al., 1998). This early emergence of manual asymmetries suggests that they are not caused by asymmetries within the central nervous system. Indeed, the causal link may go in the opposite direction. Several authors have linked lateral asymmetries, including handedness, to the orientation of the fetus in the uterus (Goodwin & Michel, 1981; Previc, 1991). In most pregnancies, the fetus is positioned head down, facing the mother's right side, resulting in their right side facing outward. This leaves the right hand freer to move within the confined space of the uterus and the right ear facing outward, resulting in asymmetric inputs to the auditory and vestibular systems (Previc, 1991).

Over several decades, Michel (2021) has developed a theory of human handedness emerging from a developmental cascade of events, snowballing from such early biases. The uterine position of the fetus produces manual and vestibular biases that lead the neonate to prefer orienting its head to the right. This, in turn, leads to the right hand being viewed more than the left hand and consequently used preferentially when unilateral reaching and grasping emerges over the first year of life. The preferred use of the right hand for simple grasping leads it to take the dominant position in role-differentiated bimanual movements, such as the tube task, which emerge at the end of the first year, and eventually in handedness more generally.

15 The Communicating Hand

A full moon rises over the forest clearing, where four people sit, their dinner sizzling over the campfire. One of the people, a stranger to these parts, differs strikingly from the others in appearance and attire. She clearly struggles to understand what the others are saying but listens politely and with interest, frequently pausing to write in a notebook. As the conversation wanes, a small woodland mammal hops tentatively out of the trees at the edge of the clearing, eyeing the people warily. One of the locals extends his arm, pointing in the animal's direction. He utters just a single word—*gavagai*—emphasizing each syllable, his voice echoing through the forest. The two other locals nod in agreement. The stranger repeats the novel word, immediately understanding that *gavagai* is the word for "rabbit" in the local dialect. She lifts her pencil to make a note of the word.

But then she hesitates. Doubt creeps over her. How can she know that *gavagai* means "rabbit" as she understands that English word? Perhaps *gavagai* refers to anything that would be good to eat, or to a collection of undetached rabbit parts, or to a three-dimensional "slice" of a rabbit in four-dimensional Einsteinian space-time. As Quine (1960) argued in his famous thought experiment, from a logical standpoint, each of these possibilities is equally consistent with the local's utterance—a fact Quine refers to as the "indeterminacy of translation."

The stranger recalls her undergraduate philosophy lectures and ponders. She glances back at the local's outstretched hand and is immediately struck by the clarity of his communicative meaning. The abstract logical argument loses its force in the face of his pointing hand. She picks up her pen and writes the word down in her notebook with satisfaction. This story is, of course, fictional. But it is clear that in addition to its role in perceiving the world, in everyday actions, and in tool use, the hand is also a powerful means of communicating with other people.

Using the Hand to Communicate

Our hands can communicate in a variety of ways. One powerful communicative role of the hand is pointing. As I write this, I am currently in Japan, where I have been pointing a lot. It has been invaluable at helping me manage my daily activities. But pointing can do more than help you obtain delicious takoyaki. There is also evidence that the understanding of pointing is related to wider aspects of cognition, such as language (Colonnesi et al., 2010). For example, the size of later vocabulary is predicted both by infants' production of points (Brooks & Meltzoff, 2008; Rowe et al., 2008) and their perception of pointing by adults (Iverson et al., 1999; Pan et al., 2005). Understanding of pointing is also disrupted in developmental disorders involving altered social cognition, such as autism (Baron-Cohen, 1989).

In the story at the start of this chapter, the local's point was highly effective at drawing a link between the spoken word *gavagai* and the mammal hopping across the clearing. The same is true for children learning their first words, heard while their parents point out the objects referred to. For this reason, Butterworth (2003) has called pointing the "royal road to language" (p. 9).

Development of Understanding of Points

Several studies have suggested that infants' understanding of the referential nature of points emerges around the end of the first year of life (Carpenter et al., 1998; Leung & Rheingold, 1981; Woodward & Guajardo, 2002). Woodward and Guajardo used a habituation paradigm to investigate understanding of the referential nature of pointing in nine- and twelve-month-old infants. During habituation, infants saw a display with two objects (a ball and a teddy bear). An actor's hand repeatedly entered and pointed to one of the objects (e.g., the ball). Across multiple presentations, infants looked for progressively less time at this event. Once the infant had habituated, test trials were presented in which the location of the two objects had switched. The key question was whether infants would dishabituate (i.e., show surprise) when they saw the hand point to the new location where the old object was now located (new side trials) or to the old location where the other object was now located (new referent trials). Twelve-month-old infants showed longer looking to new referent trials than to new side trials, indicating that they had interpreted the points as being directed toward a particular object rather than a particular location in space. In contrast, nine-month-olds showed no difference in looking times to the two types of test trials. In a second study, Woodward and Guajardo studied infants between nine and ten months of age and assessed whether they produced points in their own behavior, based on observation and parent interviews. While the pointing

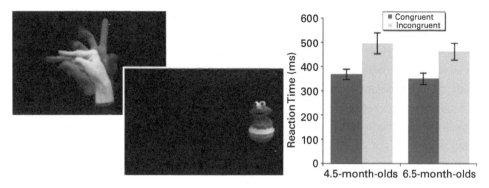

Figure 15.1
Attentional cueing from seeing pointing gestures in young infants. Infants saw a video of a hand pointing either to the left or right before an object appeared (left panel). Response time for infants to fixate the object was lower when the object appeared in the location that the hand had pointed to (right panel). Adapted from Rohlfing et al. (2012).

infants showed dishabituation to the new referent trials, the non-pointers showed no difference between test trials. This suggests that sensitivity to pointing emerges around the end of the first year of life simultaneously in the infant's production of points and understanding of other people's points.

Other research suggests that sensitivity to seeing points may emerge, at least to some degree, at even earlier ages. One study used EEG to record the brain responses of eight-month-old infants to the onset of objects that appeared at the location to which a hand was pointing or at a different location (Gredebäck et al., 2010). Differential responses were seen in the P400 component, indicating that infants had processed the stimuli differently depending on the location of the point. Other studies have reported that infants as young as four months of age show shifts in covert attention to spatial locations cued by pointing gestures (Bertenthal et al., 2014; Daum et al., 2013; Rohlfing et al., 2012). For example, Rohlfing and colleagues adapted the logic of the Posner cueing task and measured response times for infants to fixate objects that appeared either at a location congruent with a pointing hand or in an incongruent location (see figure 15.1). Infants at both four and a half months and six and a half months showed quicker reactions to objects appearing at locations congruent with the point. This suggests that infants had implicitly shifted their attention toward the distal reference of the pointing gesture. This potentially implicit covert orienting may contribute to later conceptual understanding of the social and communicative functions of pointing.

There are even hints of potential precursors of pointing movements in infants as young as two months of age (Fogel & Thelen, 1987). Two- to three-month-old infants have been observed to make extensions of the index finger, structurally similar to points, when in face-to-face interactions with their mother (Fogel & Hannan, 1985; Trevarthen, 1977). Notably, these movements are more frequent when the mother adopts active social responses toward the infant (Legerstee et al., 1990; Masataka, 1995). A longitudinal study of fifteen infants found that such pointing movements are found continuously across the first year of life (Hannan, 1987). While such movements are clearly not full-blown referential and communicative gestures, these early behaviors suggest the embryonic components of pointing are present from early in life (Butterworth, 2003).

Even once clearly intentional pointing emerges near the end of the first year of life, some authors have taken deflationary interpretations to the significance of these actions. Moore and colleagues, for example, have argued that infants point in order to produce positive emotional reactions from an adult rather than to produce joint awareness of objects (Moore & Corkum, 1994; Moore & D'Entremont, 2001). An elegant series of studies by Michael Tomasello et al. (2007), however, has provided evidence for understanding of the referential and communicative functions of pointing from around one year of age. For example, Behne et al. (2005) engaged children in a hiding task in which they needed to find a toy underneath one of two buckets on a table. Children as young as fourteen months were able to use points to one of these locations by an adult to find the toy. In contrast, in a condition in which the adult held her hand in a similar posture but with a facial expression suggesting that she was preoccupied with her wristwatch, children did not search under the bucket the finger was pointing toward. This suggests that the children were not only able to follow points but also further understood that the adult intended to provide information relevant to the hide-and-seek game.

Similar awareness is also evident in the child's own points. For example, one study manipulated adults' responses to the points of twelve-month-olds (Liszkowski et al., 2004). When the adult responded to the child's point by attending directly to the infant and ignoring the referent of the point, children showed dissatisfaction, repeatedly pointing to the object. This indicates that twelve-month-olds use points to create joint attention with the adult about an object and not merely to elicit positive affect. Similarly, when an adult responded by attending to a different object than the one the child was pointing at, the child again repeated pointing to redirect the adult's attention (Liszkowski et al., 2007b). This indicates that the child is not only trying to engage in joint attention with the adult but also to focus this joint attention on a specific object of interest. Intriguingly, twelve-month-old infants also show points toward the location where now-absent objects previously were (Liszkowski et al., 2007a), showing a sophisticated understanding of what adults know and how to use this communicatively.

Pointing in Nonhuman Primates and Other Animals

Even on the most skeptical interpretation, by two to three years of age, human children are able to use pointing gestures flexibly to direct and manipulate the attention of other people to objects in their environment. This has resulted in intense interest in whether similar use of pointing occurs in nonhuman primates. Despite extensive observations, however, remarkably few examples of pointing have been described among apes in the wild. Only a handful of cases of pointing by wild chimpanzees and bonobos have been reported (Hobaiter et al., 2014; Veà & Sabater-Pi, 1998). For example, Veà and Sabater-Pi report a single instance of a wild bonobo pointing toward the location of two camouflaged human observers. There are also some reported examples of referential gestures in the wild that bear at least some resemblance to pointing, such as "directed scratches" to request that other chimpanzees groom specific body parts (Pika & Mitani, 2006) and "beckoning" gestures in bonobos to summon potential sexual partners (Genty & Zuberbühler, 2014).

In striking contrast to the scarcity of pointing in wild apes, pointing has been widely reported in apes living in captivity. Indeed, sensitivity to pointing has been described in every species of great ape, including chimpanzees (Krause & Fouts, 1997; Leavens et al., 2004), bonobos (Savage-Rumbaugh et al., 1986), orangutans (Byrnit, 2004; Call & Tomasello, 1994), and gorillas (Byrnit, 2009; Peignot & Anderson, 1999), as well as in gibbons (Inoue et al., 2004). In one study of 115 chimpanzees, bananas were left just out of reach of the chimpanzee's cage (Leavens & Hopkins, 1998). Nearly half of the chimpanzees (53/115) spontaneously pointed to the banana to try to get the human experimenter to give it to them. In most cases, apes tend to point using their whole hand, with all fingers extended (Leavens, Hopkins, et al., 2005),[1] although index finger points have been reported as well (Leavens & Hopkins, 1999). Nevertheless, the communicative nature of the whole-hand pointing gesture is demonstrated by the fact that chimpanzees will point toward food that is out of their reach only when a human is there to give it to them (Leavens et al., 2004). This indicates that the act of pointing is an intentional and flexible behavior, appropriately taking the presence and attentional state of the other into account. Similarly, when the human observer responds by giving a chimpanzee or orangutan an object other than the one being pointed at, the animals persist in pointing (Cartmill & Byrne, 2007; Leavens et al., 2005). It is notable that this measure of persistence is the same that Liszkowski et al. (2004) used to infer that human infants' points were sophisticated and mentalistic attempts to create joint attention.

That sensitivity to points is more common in apes living in captivity suggests that it may emerge as a result of interactions with humans. For example, Call and Tomasello (1994) compared two orangutans, Chantek and Puti, which had been raised in different environments. Chantek was raised in an enculturated environment, with rich interactions with people as part of an ape sign language training program. Puti, in contrast, although raised in captivity and familiar with humans, had less social interaction and no experience using language. Chantek outperformed Puti across a range of measures of point production as well as comprehension of points by humans and also demonstrated greater sensitivity to the attentional state of the human observer, such as whether their eyes were open or closed. It is not only the sensitivity to pointing that changes depending on cultural experience but also the specific form that pointing takes. Leavens and Hopkins (1999) found that captive chimpanzees raised without exposure to language tended to use their whole hand to point, whereas language-trained chimpanzees were more likely to use their extended index finger.

The ability to produce and comprehend points is not limited to primates (Krause et al., 2018). Sensitivity to points or similar referential gestures has been shown in a striking range of animal species, including horses (Maros et al., 2008), dogs (Miklósi et al., 1998), goats (Kaminski et al., 2005), dolphins (Xitco et al., 2001), fish (Vail et al., 2013), birds (Kaplan, 2011), and bats (Hall et al., 2011). The sensitivity of dogs to human social cues has received particular interest, given the intriguing possibility that their sensitivity to human social signals is a result of selective breeding during the domestication of wild wolves into dogs. Indeed, the sensitivity of dogs to human points is impressive. In a review by Krause et al. (2018), sensitivity to points was found in all fifty-three published studies of domestic dogs. In contrast, in chimpanzees, sensitivity to pointing was identified in ten of fourteen studies. In head-to-head comparisons using matched methods, dogs clearly outperform chimpanzees (Hare et al., 2002). Intriguing evidence regarding the origins of this sensitivity comes from studies comparing dogs to wolves, the ancestral population from which dogs were domesticated (Hare et al., 2002; Virányi et al., 2008). While wolves can be trained to recognize the significance of human gestures such as points, this appears strikingly different from the sensitivity of domestic dogs that show this sensitivity from as young as six weeks of age without special training (Riedel et al., 2008). These results suggest that the sensitivity of dogs to human communicative cues may have evolved as a direct result of the ancient process of domestication in which more docile and friendly animals were selectively bred (Hare & Tomasello, 2005; Miklósi et al., 2003).

Dramatic support for this hypothesis comes from one of the world's longest-running genetic experiments. In 1959, Dmitry Balyaev initiated a study of the domestication of silver foxes at the Institute of Cytology and Genetics in Novosibirsk, Siberia, which continues to the present day (Balyaev, 1979; Trut, 1999). An experimental population of foxes has been maintained continuously for decades, selectively bred based on whether they fearlessly approached a human without showing aggression. Hare et al. (2005) found that fox kits bred for many generations to approach humans showed sensitivity to human points similar to that in domestic dogs. Critically, this pattern was totally different in a control population of foxes that was bred randomly. It is important to note that this population of foxes was bred solely based on their fear and aggression, and not on their responsiveness to human social cues. The heightened sensitivity to pointing thus appears to have evolved as a biproduct of selective pressure for different traits.

Communication of Emotion through the Hand

It is not only the movements of the body that we call "gestures" that communicate to other people. We can communicate our emotional states through facial expressions and by the posture of the body. Within psychology and neuroscience, research has focused overwhelmingly on facial expressions. There is, however, a long and fascinating tradition of research on bodily displays of emotion. One early contribution to this field came from Sir Charles Bell, whose influential book on the anatomy of the hand we discussed in chapter 1. In a fascinating book entitled *The Anatomy and Philosophy of Expression*, Bell (1847) explored this issue from the perspective of art. In addition to being an eminent anatomist, Bell was also a skilled artist. By combining these areas of expertise, Bell's book makes a remarkable contribution to both fields. Through a detailed discussion of anatomy, he identifies ways in which artists can enhance their depictions of emotion. At the same time, through an analysis of bodily postures in exceptional artworks (particularly classical Greek sculpture), Bell identifies specific patterns of posture in the body and face that skilled artists have used to communicate emotions effectively. Bell's book was highly influential on Charles Darwin, who in *The Expression of the Emotions in Man and Animals* (1872) describes a wide range of features of bodily movement that express many types of emotion, both in humans and in other animals.

In the twenty-first century, there has been a resurgence of interest in the communication of emotion through the body (de Gelder, 2009). Indeed, in some situations, our interpretation of the emotion a person is experiencing appears to be shaped even more heavily by the body than by the face (Aviezer et al., 2012). While many parts of the body can express emotion, there is evidence that the hand may play an especially important role. Ross and Flack (2020) took photos of actors expressing emotions through their body posture and deleted the arms and hands from the images. These deletions reduced the accuracy of emotion classification judgments. Even when only the hands were deleted, there were clear reductions in accuracy for recognizing fear and anger.

A recent study by Blythe et al. (2023) showed that emotions can be accurately recognized from even isolated body parts, and especially from hands. Images of actors displaying six different emotions were edited so that only individual body parts remained. When the full body was shown, participants were able to classify which emotion was shown on 65 percent of trials (chance performance is 17 percent). This level of performance was reduced, but still above chance, for isolated body parts, such as the hands (50 percent), arms (33 percent), and head (without face; 28 percent). Notably, performance was higher for the hands than for any of the other individual body parts, suggesting that the hand is particularly effective at communicating emotions.

Other studies have shown that emotion can also be communicated through the hand by touch. Hertenstein et al. (2006) brought pairs of strangers into the lab and asked one participant to communicate one of twelve emotions to the other participant by using their hand to touch the recipient's forearm. Recipients were well above chance performance for identifying most of the emotions. Since the recipient was not able to see the action of the encoder, these results show that the sense of touch alone can provide information about another's emotion. Interestingly, however, in another experiment, Hertenstein and colleagues showed that a third-person observer who saw—but did not feel—the touch could also decode the emotion being expressed at greater than chance levels. This shows that the information communicated about emotion through interpersonal touching behaviors can be carried through both the tactile and visual modalities separately. In many everyday situations, of course, information from both modalities will be present simultaneously.

A recent study by McIntyre et al. (2022) investigated the specific characteristics of touching movements that communicate specific emotions. They first brought pairs of participants into the lab who were in close relationships and asked them to communicate six different messages to each other. These messages concerned the sender's emotions toward the recipient (e.g., feeling love or gratitude toward them) or messages such as trying to calm the recipient or get their attention. As in the study by Hertenstein and colleagues, recipients could tell which message was being communicated at levels substantially higher than chance. By studying video recordings of these movements, McIntyre and colleagues identified regularities in the types of hand movements that senders used for different emotions. For example, getting the recipient's attention was communicated by taps of one or two fingers, whereas happiness was communicated by taps of larger numbers of fingers. Sadness was communicated by gripping the hand, while calming was communicated by stroking with the whole hand. These results show that people share intuitions about the signals communicating by touching other people with their hands. This work has interesting parallels with the findings of Lederman and Klatzky (1987), discussed in chapter 5, that people use a small number of exploratory procedures when haptically exploring objects.

App et al. (2011) asked participants to produce nonverbal expressions of eleven different emotions, using the "channels" of facial expression, body posture, and touch.

To prevent potential bias from the recipient's reactions, participants communicated the emotions to a life-sized mannequin as if it were a real person. The likelihood of using each channel varied systematically across emotions. For some emotions, such as happiness and sadness, the face was preferred. For other emotions, such as pride and shame, body posture was preferred. And finally, for love and sympathy, touch was preferred. These preferences were mirrored by confidence ratings in how effective the communication had been when participants were restricted to using just one of the channels. Facial expressions have dominated research on the expression of emotion. These results show, however, that it is not simply that visual displays of body posture and touch are able to communicate emotion when facial expressions are absent. Rather, for many emotions, these channels may be superior to the face.

The Hand as a Linguistic Channel

The hand can do much more than communicate emotional states. The hand can also be a channel for the full communicative power of language. The clearest example of this is sign language. Historically, sign languages were viewed as little more than iconic gestures, unable to express symbolic and abstract thought (Goldin-Meadow & Brentari, 2017). Linguistic research in the 1960s and 1970s, however, showed that sign languages such as American Sign Language have rich linguistic structure with features directly parallel to features of spoken languages in terms of phonology, morphology, and grammar (Klima & Bellugi, 1979; Stokoe, 1980). This view has been reinforced by studies of the development of sign language acquisition. Infants exposed to sign language show manual "babbling," producing rhythmic movements of the hands at frequencies characteristic of sign language, which is strikingly similar to that seen for speech in children exposed to spoken language (Petitto & Marentette, 1991). This phenomenon occurs as well in hearing children born to deaf parents (Petitto et al., 2001), showing that it is due to exposure to sign language and not to deafness per se. Similarly, in populations of deaf individuals placed together outside of an existing sign language community, sign languages emerge spontaneously and become increasingly rich across successive generations of speakers—a process of linguistic creation that has been described in populations in Nicaragua (Senghas et al., 2004) and a Bedouin group in Israel (Sandler et al., 2005). The emergence of sign systems has been observed even in individual deaf children being raised by hearing parents (Goldin-Meadow & Mylander, 1998).

The close link between sign language and spoken language is reinforced by neuropsychological and neuroimaging studies. Signers who suffer brain damage suffer from sign language aphasia, which, like aphasia for spoken language, occurs overwhelmingly following left-hemisphere lesions (Hickok et al., 1998). Hickok et al. (1996) tested twenty-three signers who had suffered unilateral brain lesions: thirteen with left-hemisphere damage and ten with right-hemisphere damage. They adapted a standard neurological examination of aphasia (the Boston Diagnostic Aphasia Examination) for use with American Sign Language. Across all measures of aphasia, patients with left-hemisphere lesions performed worse than those with right-hemisphere lesions, mirroring the pattern that has been well established in aphasia for spoken language since the classic studies of Dax and Broca in the mid-nineteenth century. Similarly, sign language aphasia resulting from more anterior lesions of the frontal lobes tends to affect production more than comprehension (analogous to Broca's aphasia), while aphasia due to more posterior lesions affects comprehension more than production (analogous to Wernicke's aphasia; Poizner et al., 1987). Such deficits appear to be specific to sign language and do not affect other types of manual gestures. For example, the patient known as "Charles" was a signer of British Sign Language who became aphasic following a stroke (Marshall et al., 2004). Although highly impaired at producing the conventional signs for concepts (e.g., for a hammer or toothbrush), he was nevertheless able to produce clear gestures pantomiming the use of the concepts. This dissociation is particularly striking given that for many such concepts, the British Sign Language signs are highly similar to pantomimes.

Similar conclusions have come from neuroimaging studies, which have consistently found that sign language activates similar regions as spoken language. For example, both spoken language and sign language produce clear activations in the inferior frontal gyrus—Broca's area—and in the posterior superior temporal lobe—Wernicke's area—of the left hemisphere (MacSweeney et al., 2002, 2004; Neville et al., 1998; Sakai et al., 2005). For example, MacSweeney et al. (2004) showed signers and non-signers visual stimuli of signs from British Sign Language or gestures from TicTac, a gestural system used by bookies at racetracks, which was unfamiliar to all participants but involves broadly similar types of hand movements. Activation in the classical left-hemisphere language regions was found in signers in response to British Sign Language signs, but critically not in response to TicTac gestures. Neither type of hand movement activated language regions in non-signers. While one early study suggested that activations in sign language may be less left lateralized than spoken language (Neville et al., 1998), this has not been found in later studies that compared sign language with control stimuli that also involved seen hand movements (MacSweeney et al., 2004).

Just as sign language provides a striking example of how the hand can substitute for spoken language in deaf people, so does braille provide an analogous example of the hand substituting for vision in reading. In the same way that viewing sign language involves the same linguistic neural network involved in perception of speech, braille reading by blind people activates a similar neural network to visual reading of text. As discussed in chapter 5, in blind people, braille reading produces clear activations of the primary visual cortex (Büchel, Price, Frackowiak, et al., 1998; Sadato et al., 1996, 1998), while disruption of this area using TMS impairs braille reading (Cohen et al., 1997). Similarly, in one blind patient, a stroke that damaged the bilateral occipital lobes produced an inability to read braille without any apparent deficits in other tactile abilities (Hamilton et al., 2000).

Such results have typically been discussed as examples of compensatory plasticity, in which the regions of the cortex that subserve vision in sighted individuals become co-opted for tactile functions in blindness. There is also evidence, however, that braille reading not only involves a generic repurposing of visual cortex but specifically co-opts networks involved in visual reading. For example, many studies of sighted individuals using fMRI have shown a specific region in the left fusiform gyrus in the ventral visual pathway, the so-called visual word form area, which shows stronger responses to visual presentation of words than other categories of stimuli (McCandliss et al., 2003). Several studies using fMRI have shown that a very similar area is activated when blind people read words in braille (Büchel, Price, & Friston, 1998; Reich et al., 2011), suggesting that the visual word form area is involved in reading regardless of the sensory modality involved. Some recent studies, while replicating these activations, have also identified some differences in how the visual word form area functions in visual reading and braille reading in the blind (Kim et al., 2017; Tian et al., 2023). For example, Kim and colleagues found that in blind people, the location of the visual word form area not only responded to braille words but also showed increased activation for spoken sentences of high grammatical complexity—an effect that was not seen in sighted people. This shows that while braille reading in the blind may involve similar networks to visual reading in the sighted, these networks nevertheless develop in different ways.

The similarity of networks involved in reading visual text and braille has also been shown in sighted people who have learned to read braille. In a heroic study, Siuda-Krzywicka et al. (2016) scanned sighted participants in the MRI scanner before and after an intensive, nine-month course in learning how to read braille. Before learning, visual words produced clear activations in the visual word form area, while feeling braille writing produced deactivations in this area. After learning, in contrast, clear activations were found in this region for words in both vision and braille. Across participants, the strength of these activations was correlated with the skill of braille reading, measured by the speed at which participants were able to read words. Moreover, the application of repetitive TMS to this area disrupted braille reading. The finding that involvement of the ventral visual pathway in braille reading is not specific to blind individuals shows that it is related to the development of expertise in braille reading and does not reflect generic compensatory plasticity as a result of blindness.

Together, this research on sign language and braille shows that the hand can function as a remarkable substitute for both speech and writing. In each case, the brain networks involved in sensorimotor control of the hand become closely linked with the networks involved in language. Sign languages and braille have had extraordinary benefits in helping deaf and blind people to lead full and productive lives. I will finish this section by briefly describing an emerging language that is doing the same for people who are both deaf and blind. Historically, DeafBlind individuals, such as Helen Keller, have communicated using braille finger spelling in which individual letters are drawn onto the skin, and tactile sign language in which the recipient places their hands on the hands of the signer. These methods, however, are both slow and error prone. For example, in one study, DeafBlind individuals asked to recognize individual American Sign Language signs by touch were successful on between 60 and 85 percent of trials (Reed et al., 1995).

The Protactile language first emerged in 2007 in the DeafBlind community in Seattle in the United States (Granda & Nuccio, 2018). While Protactile is in some ways based on American Sign Language, from which it draws much of its lexicon, it has been developed to rely entirely on tactile and proprioceptive cues. One intriguing feature of Protactile is that unlike sign languages, which involve the two hands of the signer, Protactile assigns linguistic roles to four hands: the signer's hands as well as the recipient's hands—what Edwards and Brentari (2020) call "proprioceptive constructions." The recipient receives linguistic information from their tactile perceptions of the signer's hand movements as well as tactile and proprioceptive information about the posture into which their own hands are placed. Similarly, several types of taps convey a range of linguistic signals (Edwards & Brentari, 2021). One intriguing feature of Protactile is the use of so-called backchannelling taps made by the recipient, which allow them to communicate reactions to ongoing signs without interrupting the signer. This allows Protactile conversations to operate as interactive dialogues rather than one-directional monologues. The Protactile language is continuing to develop, as are other tactile sign languages around the world, such as the emerging Tactile Italian Sign Language (Checchetto et al., 2018).

Gesture and Speech

Phenomena such as sign languages, braille, and Protactile show remarkable ways in which the hands can substitute for the typical vocal/auditory speech channel and the visual reading channel in deaf, blind, and DeafBlind individuals. However, it is not only in those with sensory loss that the hands are involved in language. All of us—not only Italians—gesture when we speak. In some cases, our gestures are deliberate efforts to increase the understanding of our speech, to substitute for speech, or to complement speech, as when pointing out a *gavagai* to a visiting foreigner. Many gestures, however, occur spontaneously, automatically, and without deliberate intent to produce them. We gesture even when speaking on the phone, when the person we are talking to obviously cannot see our hand movements. Remarkably, even congenitally blind people gesture when they speak, including when speaking to other blind people (Iverson & Goldin-Meadow, 1998), and so does patient I.W., described in chapter 2, who lacks touch and proprioception from the neck down. Despite having to perform everyday actions such as walking or holding a coffee cup under vigorous visual monitoring, I.W.'s gestures are spontaneous and natural, showing the same characteristic synchrony with his speech as in everyone else (Cole et al., 2002).

Classification of Manual Gestures
Over the years, numerous authors have provided classifications of the sorts of gestures people make with their hands while speaking (Ekman & Friesen, 1969, 1972; Krauss et al., 2000; McNeill, 1992; Wiener et al., 1972). In general, these classifications are in good agreement, although some make finer (Ekman & Friesen, 1969) and others courser (Krauss et al., 2000) distinctions. Here, I will follow McNeill (1992), who divides gestures into four categories: iconic gestures, metaphoric gestures, deictic gestures, and beat gestures. I will describe each of these in turn.

Iconic gestures are hand movements that have some literal physical resemblance to the idea or meaning being referred to. For example, when describing a Lego piece that you are looking for, you may position your thumb and index finger at a specific distance to indicate the size of the piece. The distance between your fingertips represents the size of the object in a transparent way. Ekman and Friesen (1969) make finer distinctions within this category. In a pictographic gesture, the hand movement indicates the shape of an object. Describing a seashell, we might move our finger upward in a narrowing spiral to indicate its shape. A kinetographic gesture represents the manner in which an action was performed. While describing our difficulty opening a jar, we might rotate our hand with fingers outstretched, pantomiming the act of turning the lid. Finally, spatial movement gestures represent the movement trajectory of a person or objects. For example, giving directions to another person we might say, "It's around the corner" while moving our finger along a line, then making a ninety-degree turn and moving in another direction. In each of these cases, the gestures represent in a literal and generally transparent way some physical property or motion of an object or action.

Metaphoric gestures, in contrast, involve using the hands to create a physical representation of an idea or concept that is itself more abstract. For example, McNeill (1985) gives an example of a discussion between two mathematicians discussing the concept of mathematical duality, in which the abstract relation between two mathematical objects can be inverted. Verbal references to the concept of duality were frequently accompanied by gestures in which the two hands alternated between spatial locations.

Deictic gestures direct a recipient's attention to a particular person, object, or location in the world. The most obvious deictic gesture is pointing—a gesture that is very important in human communication and which I have already discussed above. Broadly, deictic gestures such as points have two main communicative functions (Bates et al., 1975). Imperative gestures relate to a desire to obtain an object, and function as a request. For example, faced with a selection of choices at an ice cream parlor, you may point to indicate your desired flavor. Declarative gestures, in contrast, aim to draw the recipient's attention to something in the world, either to make it an object of joint-attention or to provide needed information. For example, if asked to give directions to the train station, you might point to indicate in which direction the person needs to go.

Finally, beat gestures are hand movements that follow, and emphasize, the rhythmic flow of speech. They tend to have a similar, stereotyped form, regardless of the specific content of the speech, usually consisting of short, staccato movements made in the periphery of the gesture space. Beat gestures can be used to give emphasis to specific words, highlighting their significance. They can also be used to help segment the flow of speech into discrete ideas or packages. For example, when describing a list of items, a beat gesture might be made at the onset of each item, highlighting the discreteness of the list.

Integration of Gesture and Speech

Based on an extensive series of studies of gestures, David McNeill (1985, 1992) has argued that gestures are integrated with speech into a single cognitive system. The overwhelming majority of gestures occur during speech, and there is a tight temporal synchrony between the onset of words and gestures. In general, the onset of gestures slightly precedes the onset of the corresponding word. Interestingly, the length of this delay is related to the frequency of the word being produced. Gestures linked to infrequent words start earlier and last longer than those linked to more familiar words (Morrel-Samuels & Krauss, 1992).

Developmental studies of the emergence of spoken language and gesture have also revealed striking correlations and tight developmental emergence of these behaviors (Bates et al., 1979, 1989; Bates & Dick, 2002; Sansavini et al., 2010). Bates and Dick lay out a compelling set of relations between linguistic milestones as shown in spoken language and in gesture.

Another line of evidence comes from situations in which speech is dysfluent. For example, both clinical stuttering (Mayberry et al., 1998) and artificial delays to auditory feedback of one's own speech (McNeill, 1992) produce delays and dysfluencies in speech. Yet, in both cases, the onset of gestures remains temporally locked to the onset of speech. Thus, when speech is delayed, gesture is delayed correspondingly, preserving the tight linkage between the two.

Consistent evidence comes from studies of brain-damaged patients with aphasia. Patients with dysfluent (Broca's) aphasia, for example, fail to produce beat gestures time locked to speech, as do healthy people and patients with semantic (Wernicke's) aphasia (Cicone et al., 1979). Notably, these same patients produced hand movements similar to beat gestures when trying to retrieve words, showing that their absence during speech is not due to motor impairment in producing the movements. Patients with Wernicke's aphasia produced more gestures at the boundaries between clauses when the clauses were semantically discontinuous (Delis et al., 1979). Finally, Glosser et al. (1986) quantified the complexity of gestures produced by aphasics, showing that gestural complexity was negatively correlated with the level of linguistic impairment.

Some researchers argue that gesturing enhances speech fluency (Krauss, 1998; Rauscher et al., 1996), although this is controversial. Rauscher and colleagues measured the speed of speech, as well as the number of dysfluencies, such as pauses and speech repairs, when speakers were allowed to gesture or were restricted from doing so. They reported less fluent speech, with more pauses, when participants were prevented from gesturing, although this effect appeared specific for spatial language. Other studies, however, have failed to find analogous dysfluencies in the absence of gesture (Graham & Heywood, 1975; Hoetjes et al., 2014; Kısa et al., 2022; Rimé et al., 1984). One recent study using a large dataset found no evidence for an effect of gesture on fluency for either literal or metaphorical spatial language (Kısa et al., 2022). Thus, gesture does not appear to have a specific effect on speech fluency or lexical access. Nevertheless, as we will see, gesture does have widespread effects on cognition in several domains.

Gesture as a Reflection of Thought Processes
Gesture may also provide information about cognitive processes beyond that available in spoken speech. Some studies show that developmental transitions in conceptual understanding may occur earlier in gesture than in speech, and that mismatches between speech and gesture may be a sign that children are in a position to benefit from instruction. For example, Church and Goldin-Meadow (1986) asked children between five and eight years of age to solve Piagetian conservation tasks, such as conservation of volume, length, and number. For example, in a conservation of number task, children were shown two identical arrays of checkers. After agreeing that the arrays had the same number of checkers, one array was spread apart, increasing the distance between adjacent checkers. Consistent with Piaget's (1952) classic results, many children failed to demonstrate

conservation, claiming there were more checkers in the row with the spaced checkers. These children often produced gestures consistent with this reasoning—for example, changing the distance between their hands to model the difference in length of the array after the checkers were spread apart. Other children, however, while making similar verbal statements, produced hand gestures that demonstrated a different—and correct—way of conceptualizing the problem. The child might point between the first item in each row, showing an implicit understanding of the relevance of the one-to-one correspondence between items in the two rows. A training study showed that children who produced mismatching speech and gestures showed greater improvement at posttest following verbal instruction. Other research has shown similar results in other domains—for example, older children's understanding of mathematical equivalence (Perry et al., 1988). Thus, mismatching gestures appear to indicate that children have burgeoning awareness and are in a receptive state for training.

A detailed microgenetic study of children's understanding of mathematical equivalence by Alibali and Goldin-Meadow (1993) showed that children go through three stages in their developing understanding. Children first produce matching verbal explanations and gestures, both demonstrating an incorrect understanding. They then go through a phase of mismatching speech and gestures, followed by a final phase of matching speech and gestures in which both modalities show correct understanding of the concept. Another study from the same lab assessed not only mismatches in individual utterances but also the full repertoire of concepts employed in speech and gesture (Goldin-Meadow et al., 1993). Intriguingly, correct concepts emerged earlier in gesture than in speech.

There is also evidence that the very act of gesturing may enhance reasoning. For example, Chu and Kita (2011) found that adult participants were better able to correctly solve visual mental rotation problems when they were allowed to gesture during the task than when they were prevented from gesturing. Intriguingly, when participants were actively encouraged to gesture, their performance was better still. Preventing people from gesturing also leads to poorer retention of route information after studying maps (So et al., 2014). Children were more accurate at counting a set of objects when allowed to gesture (Alibali & DiRusso, 1999). Similar results have also been reported for children's learning of novel mathematical concepts. When children were told to produce gestures while trying to solve mathematical problems, they benefited more from subsequent instruction than children told not to gesture (Broaders et al., 2007). Moreover, children who are told to gesture show retention of learning when tested four weeks later that children who only speak do not (Cook et al., 2008).

Gesture may also have the effect of generally lightening cognitive load. Goldin-Meadow et al. (2001) asked participants to remember lists of letters or words while simultaneously explaining how they had solved math problems. Participants in the "gesture" condition were allowed to gesture while they made either explanation, whereas those in the "no gesture" condition were asked to place their hands on the table. Both adults and children remembered more of the items on the list in the "gesture" than in the "no gesture" condition. This suggests that the act of gesturing while speaking may have lightened the cognitive load involved, freeing more working memory resources for the memory task.

Other evidence, however, suggests that gesture may produce systematic changes in the nature of thinking as well (Goldin-Meadow & Beilock, 2010; Kita et al., 2017). For example, participants prevented from gesturing produce less spatial imagery in their speech (Rimé et al., 1984). Wolff and Gutstein (1972) asked participants to make either circular or linear movements of their hands while making up short, two- to three-sentence stories. Although the hand movements had no relevance to the story task, raters judged the stories made up during circular gestures to have more "circular" themes and the stories made up during linear gestures to have more "linear" themes. Alibali et al. (2011) asked participants to perform a task involving predicting the movements of systems of gears. When allowed to gesture, participants tended to adopt a strategy of simulation of the movement, whereas when gestures were restricted, they adopted a more conceptual approach.

Communicating and Learning via Gesture
Many studies have shown that gestures contribute to the communicative function of speech (Goldin-Meadow, 2003; Kendon, 1994). For example, Berger and Popelka (1971) asked participants to write down the exact words spoken by a distant person who spoke softly while wearing a face mask. The accuracy of transcriptions was substantially higher when the speaker gestured while speaking compared to a condition in which they did not gesture. Graham and Argyle (1975) asked native English and Italian speakers to be "encoders" and to describe abstract line drawings to "decoders," who had to try to draw the shape from the description. Accuracy of these drawings was substantially higher when the speakers gestured than when they were asked to keep their arms folded.[2] In another study, decoders were asked to guess the object being described from videos including gesture or from hearing the audio alone (Riseborough, 1981). Performance was again higher when gestures were available. In contrast, however, no such advantage was found when a different hand movement was seen, indicating that the advantage is linked to the specific form of the gesture, not to the occurrence of hand movements as such.

Other studies have shown that seeing the iconic gestures produced by speaking adults helps children learn the meaning of verbs (Aussems & Kita, 2021; Mumford & Kita, 2014; Wakefield et al., 2018). For example, Mumford and Kita (2014) tested the ability of three-year-old children to generalize the meaning of novel verbs used by an adult model in short videos. For example, the model might arrange several black pieces of felt to make a cloud shape. Children showed higher performance when the actress produced gestures matching the manner of object manipulation (i.e., by modeling the hand movements used to position the objects) compared to conditions where the gesture matched the end state of the object (i.e., by tracing an outline of the cloud shape) or was absent.

Comparable results have also been found in classroom settings. Valenzeno et al. (2003) showed preschool children video-recorded lessons about the concept of symmetry. In one condition, the teacher gestured to indicate which shapes she was talking about and to emphasize the comparison of the two sides of the shape. In another condition, the teacher spoke without gesturing. Children's understanding was measured by asking them to indicate whether line drawings of shapes were symmetric and to give an explanation of why. Responses were scored as correct only when children both responded correctly and gave a cogent explanation. Children in the verbal-plus-gesture condition produced correct responses on 34.7 percent of trials—more than twice as many as children in the verbal-only condition (14.2 percent).

Another study applied a similar design to investigate learning of Piagetian conservation problems (Church et al., 2004). Again, children who watched an instructional video including gesture showed larger improvement from pretest to posttest than children who had watched a video without gesture. Another interesting aspect of this study was that while it was conducted in the United States, a group of predominantly Spanish-speaking children in a Spanish-only classroom were also tested. While these Spanish speakers showed little improvement after watching the (English) video without gesture, they did improve after seeing the video with gesture. Indeed, the learning showed by the Spanish-speaking children who saw gesture was similar to that of the English-speaking children who saw the video without gesture.

There is also evidence that seeing hand gestures during instruction can affect the consolidation of learning across time. Cook et al. (2013) showed videos explaining concepts related to mathematical equivalence relations, which were matched except for the inclusion of gesture. They tested the effects of instruction both immediately after the instruction and twenty-four hours later. Children who saw the video with gesture showed better performance both immediately and the next day, showing retention of this learning. Even more striking, children in the gesture condition showed an improvement across sessions, actually performing better the following day than immediately after the lesson. In contrast, children who saw the video without gesture showed no such improvement, performing similarly in the two sessions. In the second session, Cook and colleagues also tested for generalization of learning to different types of mathematical problems, finding that the beneficial effect of gesture was nearly of the same magnitude as on the trained problems.

As described above, children frequently produce mismatches in the concepts they express in speech and gesture when they are on the cusp of more advanced understanding and most receptive to instruction. There is also evidence that teachers recognize this and adapt their own instructional-style in response, producing more varied explanations and more gesture/speech mismatches of their own (Goldin-Meadow & Singer, 2003). Singer and Goldin-Meadow (2005) provided instruction to eight- to ten-year-old children on mathematical equivalence problems (e.g., "$6+4+3 = \_ + 3$"). They provided different concepts to children about how to solve the problem, focusing either on the conceptual nature of the equivalence relation or on procedural techniques such as removing the final "+3" from both sides. Interestingly, presenting both concepts in spoken instruction was less effective than presenting just one concept. In striking contrast, however, the best performance was found when children had received one concept in speech (e.g., "We want to make the other side of the equal sign the same amount") and the other concept in gesture (e.g., pointing to the left "3" and then producing a flicking gesture near the right "3").

Gesture and the Evolution of Language

The evidence described in this section shows that manual gestures have rich connections both to speech and to cognition more widely. What, then, is the link between gestures and language? Numerous writers over the past three centuries have argued that spoken language emerged from an earlier manual sign language (Condillac, 1756; Corballis, 2002; Hewes, 1973). While spoken language is unique to humans, many animals use various forms of manual signals to communicate. Jane Goodall (1986) described a range of gestures used by the chimpanzees at Gombe, which tended to be used in consistent ways to communicate specific things. For example, chimpanzees would stretch out their hand in the direction of another animal to beg for desired food to be shared. Other gestures have been described above in relation to pointing, such as "directed scratches" to initiate grooming (Pika & Mitani, 2006) and "beckoning" gestures to initiate sex (Genty & Zuberbühler, 2014). Across the great apes, there is evidence for the use of a substantial set of nearly a hundred communicative gestures, which show high levels of correspondence between chimpanzees, bonobos, gorillas, and orangutans (Byrne et al., 2017). The consistency of these gestures across individuals is such that research has been able to generate essentially a dictionary of chimpanzee gestures (Hobaiter & Byrne, 2014). Human observers have no difficulty in correctly inferring the meaning of these gestures (Graham & Hobaiter, 2023), which indeed are highly similar to those made by human children in the first two years of life (Kersken et al., 2019).

While attempts to teach great apes to speak have failed, several apes have learned to use sign language to communicate with humans, including Washoe the chimpanzee (Gardner & Gardner, 1969), Koko the gorilla (Patterson, 1978), Kanzi the bonobo (Savage-Rumbaugh et al., 1986), and Chantek the orangutan (Miles, 1990). While these animals' use of sign language is striking, it appears to lack the full grammatical richness of human language. For example, Kanzi's linguistic performance at eight years of age was compared to that of a two-year-old human girl named Alia on a variety of tasks (Savage-Rumbaugh et al., 1993). While Kanzi outperformed Alia on some tasks, he performed worse on others, and overall, the two performed at approximately the same level. The sign language abilities of great apes should perhaps be counted as what the linguist Derek Bickerton (1995) has called "protolanguage." Protolanguages, such as the simple pidgin languages that develop for trade purposes between two language communities, may have an extensive vocabulary but lack the grammatical complexity of true language. The protolinguistic abilities in the manual modality of all four nonhuman species of great ape, shown in both their gestures in the wild and sign language in captivity, suggest that it was shared by the common ancestor of us and the other great apes and likely formed the evolutionary basis for the emergence of true language during hominin evolution during the past few million years.

Other evidence for a close link between spoken language and manual abilities comes from common left-hemisphere lateralization for both domains. As mentioned in chapter 13, the association between aphasia and left-hemisphere strokes has been clear since the mid-nineteenth-century studies of Dax and Broca (cf. Jackson, 1864). The same pattern holds for cases of sign language aphasia (Kimura, 1981). Similarly, as discussed in chapter 9, the association between apraxia and the left hemisphere has been clear since Liepmann's (1905) pioneering studies at the start of the twentieth century. The links between the left hemisphere and aphasia and apraxia are among the most heavily replicated effects in neurology, and there is evidence that the two conditions themselves arise from damage to similar brain regions within the left hemisphere (Goldenberg & Randerath, 2015). This link between language and action is also seen in the association between language lateralization and handedness. Knecht et al. (2000) measured language laterality using a technique called functional transcranial Doppler sonography, which allowed them to measure blood flow in the arteries supplying the left and right hemispheres while participants performed a word-generation task. There was a strong association between handedness measured using a self-report questionnaire and language laterality. Among 107 strongly right-handed participants, 103 (96 percent) showed left-hemisphere lateralization for the word generation task. In striking contrast, among thirty-seven strongly left-handed participants, only twenty-seven (73 percent) showed left lateralization. Notably, left lateralization for language was the predominant pattern in both of these groups, although the rate of right lateralization increased by a factor of more than six in the left-handed group. However, self-reported handedness may not be a perfect measure of brain lateralization for praxis. A recent study by Vingerhoets et al. (2013) used fMRI to measure brain lateralization for both language (using a word-generation task) and praxis (using a tool-use pantomime task), finding a perfect correspondence between these, regardless of handedness.

As discussed in the previous chapter, there is little evidence for strong overall patterns of handedness in great apes. There are, however, two notable exceptions to this pattern: African great apes (i.e., chimpanzees, bonobos, gorillas) tend to show right-hand preferences for precision grips (Meguerditchian et al., 2015) and for communicative gestures (Meguerditchian et al., 2013). This suggests that the left lateralization for language and praxis in humans likely emerged from a weaker bias for both communicative signals and precise manual dexterity in our last common ancestor with chimpanzees. A brain region likely to be important to this process is Broca's area, which, as discussed in chapter 11, has undergone important changes in hominin evolution. While Broca's area is more than six times as large in humans as in chimpanzees (Schenker et al., 2010), a homologous region has been found in all great ape species (Schenker et al., 2008). Intriguingly, like in humans, Broca's area shows structural asymmetries in great apes, being larger in the left hemisphere in chimpanzees, bonobos, and gorillas (Cantalupo & Hopkins, 2001). This asymmetry is also linked to individual differences between chimpanzees in hand preferences for communicative gestures: chimpanzees who show a right-hand preference for communicative gestures show greater left-lateralization of Broca's area (Taglialatela et al., 2006). Research using neuroimaging has shown that Broca's area is also active when chimpanzees produce communicative gestures (Taglialatela et al., 2008).

In humans, there is strong evidence that Broca's area is involved in the hierarchical organization of manual actions into meaningful "chunks" (Koechlin & Jubault, 2006), as discussed in chapter 8, as well as in purely linguistic grammaticality judgments (Embick et al., 2000). It is plausible that both these abilities emerged from changes in the hominin lineage within the lateralized network for communicative gesture that existed in our last common ancestor with chimpanzees. It is tempting, although highly speculative, to link these changes to the emergence of hafted tools in the Middle Paleolithic (Ambrose, 2001), which, as discussed in chapter 10, is linked in Europe to the emergence of Neanderthals and in Africa to the emergence of anatomically modern humans. As with chunking of motor actions and linguistic grammar, the construction of hafted tools requires a hierarchical conception of multiple components being put together to a make a larger unit. This potential link between language and tool use is supported by a recent study using fMRI, which found common patterns of activation in the basal ganglia when participants used tools and made complex syntactic judgments about language (Thibault et al., 2021). As discussed in chapter 8, the basal ganglia have been widely implicated in the formation of motor chunks (Graybiel, 1998; Wymbs et al., 2012).

Given the apparent three-way association between the emergence of hierarchical structure in manual action, tools, and language, it is an intriguing possibility that language first emerged in the context of manual gestures. The first human languages may thus have been sign languages (Corballis, 2002). Sign languages do not exist only in deaf communities. Many traditional societies around the world have had sign languages as complements to spoken language—for example, in Aboriginal Australians (Kendon, 1988), in the !Kung San of southern Africa (Lee, 1979), in Anglo-Saxon England (Barley, 1974), in Native Americans of the Great Plains (Davis, 2010), and in monastic communities (Umiker-Sebeok & Sebeok, 1987). Similarly, modern Chinese has a conventionalized set of hand gestures to indicate the numbers one to ten (Nicoladis et al., 2019).

One interesting line of evidence for the antiquity of sign languages comes from the representation of hands in cave paintings, discussed in the previous chapter as evidence for the antiquity of right-handedness. In addition to the preponderance of left hands, there is a curious tendency for many hands to have missing fingers or fingertips. Traditionally, this had been interpreted as evidence that ancient people had lost fingers whether as a result of accidents or ritual disfigurement. The French anthropologist André Leroi-Gourhan (1967, 1986) challenged this interpretation, studying the twenty-seven-thousand-year-old hand paintings in Gargas cave in France. Leroi-Gourhan identified patterns in the color and spatial arrangement of hands with specific parts of fingers missing. Such patterns suggested to him that the fingers were deliberately left out of the painting. He proposed instead that they reflected manual gestures from sign language. On this interpretation, these hands were not just *art*, they were *writing*.

A recent study of the hand paintings in Gargas cave has provided intriguing support for Leroi-Gourhan's argument. Etxepare and Irurtzun (2021) analyzed the posture of the hand that would be required to produce stencils with the missing digits observed in Gargas by bending the hand into different configurations. They show that all of the hand shapes observed can be produced in mid-air, without using the cave wall to support the fingers. Moreover, the postures required to produce these stencils are within the range of hand postures actually used in known sign languages. While speculative, these findings have very interesting implications for the historical emergence of language as well as understanding the nature and significance of Paleolithic hand paintings.

Epilogue: The Future of the Invisible Hand

People have been fascinated by their hands for thousands of years and will continue to be as long as there are people. This book started with the seeming paradox that at the same time that the hand is held up as a paragon of perfection and responsible for uniquely human abilities and accomplishments, it is also described as primitive and undistinguished among the limbs of mammals. The answer to this paradox lies in the distinction between the visible and invisible hands. While the visible hand may be primitive in many respects, the invisible hand equips the hand with the full power of the human brain. Evolution does not stop, and the human hand and brain will continue to evolve. In closing, I wish to comment briefly on what the future has in store for the invisible hand and for our ability to understand it.

Aristotle, as mentioned in the Preface, called the hand the "tool of tools" because, by using a wide range of tools, the hand could achieve an equally wide range of tasks. The basic physiology of the human hand and brain have changed little since the ancient Greeks or, for that matter, since the Middle Paleolithic era. The things we can do with our hands, however, have changed dramatically due to the development of ever more technologically advanced tools. I am writing this sentence on my smartphone—an extraordinarily powerful tool that lets my hand become a flashlight, calendar, weather forecast, television, games console, and, by leveraging the power of an entire system of satellites orbiting the Earth, a communications system letting me interact instantly with people around the world. The hand has come a long way since its ability to become "talon, hoof, and horn" so impressed Aristotle. And, no doubt, the hand has a long way yet to go. Modern developments in prosthetics, exoskeletons, and tele-operation allow the functions of the hand to be extended in ways that were the domain of science fiction only a few decades ago.

Technology has revolutionized not only the things we can do with our hands but also how we understand how the hand and brain work. As mentioned in chapter 13, the ancient Greeks had come to understand some basic principles of how the brain controls the hand, such as that movements are predominantly controlled by the contralateral brain hemisphere. Every scientific field I have discussed in this book relies on methods and technologies that would have been unimaginable to the Hippocratic authors or, for that matter, to Wilder Penfield in the mid-twentieth century. These methods, and the ingenuity of the scientists and engineers who have developed and used them, have provided the rich insight into brain function that I have described throughout this book. But again, these technologies are ever developing and will lead to further breakthroughs in the coming decades. Cognitive neuroscience continues to be transformed by higher-resolution neuroimaging methods, and neurology by methods allowing systematic mapping of brain lesions in large groups of patients. Neurophysiology, in turn, is being revolutionized by methods such as optogenetics, which allow brain networks to be manipulated in unprecedented ways. Finally, paleontology and archaeology are being transformed by newly found fossils, such as the early tetrapod *Tiktaalik* and the hominin *H. naledi*, as well as by advances in paleogenetics, which allow us to understand prehistory in unprecedented ways.

Notes

Chapter 2

1. Interestingly, however, Head appeared to retain deep sensitivity to pressure. Head and colleagues interpreted this dissociation in terms of a putative distinction between two systems of touch: a primitive *protopathic* system, which allowed perception of diffuse pressure without precise localization (and which was spared by Head's injury), and a more advanced *epicritic* system, which allowed precise localization (and which Head had initially lost). This distinction has not been supported by subsequent research. For a detailed critique, see Walshe (1942).

2. It is noteworthy that the drawing of the homunculus depicts a male body. It may be that there are some differences between males and females in the organization of these maps for body regions with large sexual dimorphism such as the genitals and upper chest (Di Noto et al., 2013). There is little evidence, however, for sex differences in the large-scale organization of these maps or for body parts—like the hand—that are similar in men and women.

Chapter 6

1. Numerous sources give this quote, the earliest that I am aware of being Riddoch (1941). However, I have been unable to find any clear evidence that Nelson actually said (or wrote) it.

Chapter 8

1. https://twitter.com/miblogestublog/status/1560826362704838657.

2. https://www.nytimes.com/2021/12/28/sports/tetris-game.html.

Chapter 9

1. At the 1938 US Open, Ray Ainsley also took twenty-three strokes on the par-4 sixteenth hole at Cherry Hills in Colorado. Since Armour's twenty-three was on a par-5 and Ainsley's on a par-4, the latter took the dubious record with nineteen over par. Ainsley's performance, however, seems

to have been a result not of dystonia but rather of hitting the ball into a stream whose current carried it away.

2. Note that the *Regierungsrat*, although the first patient described by Liepmann, is an exception to this generalization, showing apraxia for his right hand but not his left hand.

Chapter 10

1. This point may not hold, however, if the poker is instead used to threaten a visiting lecturer.

Chapter 12

1. Note that Halverson did not use the terms "power grip" and "precision grip," which were only introduced more than two decades later by Napier (1956). Instead, he uses the general terms "palmar grasp" and "digital grasp" as well as more specific labels for a variety of individual grasp types.

2. This is in striking contrast to the profound hemiplegia induced by comparable lesions in adult animals. Indeed, the idea that there is an inverse relation between the functional consequences of brain damage and the age of injury is often known as "Kennard's principle," although, as Dennis (2010) argues, it "is neither Kennard's nor a principle" (p. 1043).

Chapter 15

1. This may relate to basic anatomical differences in hand structure, which make the index finger more distinct from the other fingers while in a resting state in humans than in other primates (Povinelli & Davis, 1994).

2. Consistent with stereotypes, there was evidence that Italian speakers used gestures more effectively than English speakers.

References

Aboitiz, F., Scheibel, A. B., Fisher, R. S., & Zaidel, E. (1992). Fiber composition of the human corpus callosum. *Brain Research, 598*(1–2), 143–153.

Abrams, R. A., Davoli, C. C., Du, F., Knapp, W. H., & Paull, D. (2008). Altered vision near the hands. *Cognition, 107*(3), 1035–1047.

Abrams, R. A., & Weidler, B. J. (2014). Trade-offs in visual processing for stimuli near the hands. *Attention, Perception, & Psychophysics, 76*(2), 383–390.

Adie, W. J., & Critchley, M. (1927). Forced grasping and groping. *Brain, 50*(2), 142–170.

Adler, C. H., Crews, D., Hentz, J. G., Smith, A. M., & Caviness, J. N. (2005). Abnormal co-contraction in yips-affected but not unaffected golfers: Evidence for focal dystonia. *Neurology, 64*(10), 1813–1814.

Adler, D. S., Wilkinson, K. N., Blockley, S., Mark, D. F., Pinhasi, R., Schmidt-Magee, B. A., Nahapetyan, S., Mallol, C., Berna, F., Glauberman, P. J., Raczynski-Henk, Y., Wales, N., Frahm, E., Jöris, O., MacLeod, A., Smith, V. C., Cullen, V. L., & Gasparian, B. (2014). Early Levallois technology and the Lower to Middle Paleolithic transition in the southern Caucasus. *Science, 345*(6204), 1609–1613.

Aflalo, T. N., & Graziano, M. S. A. (2006). Possible origins of the complex topographic organization of motor cortex: Reduction of a multidimensional space onto a two-dimensional array. *Journal of Neuroscience, 26*(23), 6288–6297.

Aglioti, S. M., Beltramello, A., Bonazzi, A., & Corbetta, M. (1996). Thumb-pointing in humans after damage to somatic sensory cortex. *Experimental Brain Research, 109*(1), 92–100.

Aglioti, S. M., Smania, N., Moro, V., & Peru, A. (1998). Tactile salience influences extinction. *Neurology, 50*(4), 1010–1014.

Ahlberg, P. E., Clack, J. A., & Lukševičs, E. (1996). Rapid braincase evolution between *Panderichthys* and the earliest tetrapods. *Nature, 381*, 61–64.

Aiello, B. R., Olsen, A. M., Mathis, C. E., Westneat, M. W., & Hale, M. E. (2020). Pectoral fin kinematics and motor patterns are shaped by fin ray mechanosensation during steady swimming in *Scarus quoyi*. *Journal of Experimental Biology, 223*(Pt 2), jeb211466.

Aiello, B. R., Stewart, T. A., & Hale, M. E. (2016). Mechanosensation in an adipose fin. *Proceedings of the Royal Society B*, *283*(1826), 20152794.

Aizawa, H., Mushiake, H., Inase, M., & Tanji, J. (1990). An output zone of the monkey primary motor cortex specialized for bilateral hand movement. *Experimental Brain Research*, *82*(1), 219–221.

Ajax, E. T. (1982). Trapshooter's cramp. *Archives of Neurology*, *39*(2), 131.

Akelaitis, A. J. (1945). Studies on the corpus callosum. IV. Diagnostic dyspraxia in epileptics following partial and complete section of the corpus callosum. *American Journal of Psychiatry*, *101*(5), 594–599.

Alba, D. M., Moyà-Solà, S., & Köhler, M. (2003). Morphological affinities of the *Australopithecus afarensis* hand on the basis of manual proportions and relative thumb length. *Journal of Human Evolution*, *44*(2), 225–254.

Albert, S. T., & Shadmehr, R. (2016). The neural feedback response to error as a teaching signal for the motor learning system. *Journal of Neuroscience*, *36*(17), 4832–4845.

Alexander, G. E., & Crutcher, M. D. (1990). Functional architecture of basal ganglia circuits: Neural substrates of parallel processing. *Trends in Neurosciences*, *13*(7), 266–271.

Alibali, M. W., & DiRusso, A. A. (1999). The function of gesture in learning to count: More than keeping track. *Cognitive Development*, *14*(1), 37–56.

Alibali, M. W., & Goldin-Meadow, S. (1993). Gesture-speech mismatch and mechanisms of learning: What the hands reveal about a child's state of mind. *Cognitive Psychology*, *25*(4), 468–523.

Alibali, M. W., Spencer, R. C., Knox, L., & Kita, S. (2011). Spontaneous gestures influence strategy choices in problem solving. *Psychological Science*, *22*(9), 1138–1144.

Allen, F. J. (1896). Mirror writing. *Brain*, *19*(2–3), 385–387.

Allen, H. A., & Humphreys, G. W. (2009). Direct tactile stimulation of dorsal occipito-temporal cortex in a visual agnosic. *Current Biology*, *19*(12), 1044–1049.

Allen, J. S., Bruss, J., & Damasio, H. (2006). Looking for the lunate sulcus: A magnetic resonance imaging study in modern humans. *Anatomical Record, Part A*, *288A*(8), 867–876.

Allievi, A. G., Arichi, T., Tusor, N., Kimpton, J., Arulkumaran, S., Counsell, S. J., Edwards, A. D., & Burdet, E. (2016). Maturation of sensori-motor functional responses in the preterm brain. *Cerebral Cortex*, *26*(1), 402–413.

Allison, T., McCarthy, G., Wood, C. C., Darcey, T. M., Spencer, D. D., & Williamson, P. D. (1989). Human cortical potentials evoked by stimulation of the median nerve. I. Cytoarchitectonic areas generating short-latency activity. *Journal of Neurophysiology*, *62*(3), 694–710.

Allison, T., McCarthy, G., Wood, C. C., & Jones, S. J. (1991). Potentials evoked in human and monkey cerebral cortex by stimulation of the median nerve. *Brain*, *114*(Pt 6), 2465–2503.

References

Allman, J. M., & Kaas, J. H. (1971). A representation of the visual field in the caudal third of the middle temporal gyrus of the owl monkey (*Aotus trivirgatus*). *Brain Research, 31*(1), 85–105.

Alloway, K. D. (2008). Information processing streams in rodent barrel cortex: The differential functions of barrel and septal circuits. *Cerebral Cortex, 18*(5), 979–989.

Almécija, S., & Alba, D. M. (2014). On manual proportions and pad-to-pad precision grasping in *Australopithecus afarensis*. *Journal of Human Evolution, 73*, 88–92.

Almécija, S., Alba, D. M., & Moyà-Solà, S. (2012). The thumb of Miocene apes: New insights from Castell de Barbera. *American Journal of Physical Anthropology, 148*(3), 436–450.

Almécija, S., Moyà-Solà, S., & Alba, D. M. (2010). Early origin for human-like precision grasping: A comparative study of pollical distal phalanges in fossil hominins. *PLOS One, 5*(7), e11727.

Almécija, S., Smaers, J. B., & Jungers, W. L. (2015). The evolution of human and ape hand proportions. *Nature Communications, 6*, 7717.

Alstermark, B., Pettersson, L. G., Nishimura, Y., Yoshino-Saito, K., Tsuboi, F., Takahashi, M., & Isa, T. (2011). Motor command for precision grip in the macaque monkey can be mediated by spinal interneurons. *Journal of Neurophysiology, 106*(1), 122–126.

Altenmüller, E. (2005). Robert Schumann's focal dystonia. In J. Bogousslavsky & F. Boller (Eds.), *Neurological disorders in famous artists* (pp. 179–188). Karger.

Altenmüller, E., Baur, V., Hofmann, A., Lim, V. K., & Jabusch, H.-C. (2012). Musician's cramp as manifestation of maladaptive brain plasticity: Arguments from instrumental differences. *Annals of the New York Academy of Sciences, 1252*, 259–265.

Ambrose, S. H. (2001). Paleolithic technology and human evolution. *Science, 291*(5509), 1748–1753.

Amedi, A., Jacobson, G., Hendler, T., Malach, R., & Zohary, E. (2002). Convergence of visual and tactile shape processing in the human lateral occipital complex. *Cerebral Cortex, 12*(11), 1202–1212.

Amedi, A., Malach, R., Hendler, T., Peled, S., & Zohary, E. (2001). Visuo-haptic object-related activation in the ventral visual pathway. *Nature Neuroscience, 4*(3), 324–330.

Ames, K. C., & Churchland, M. M. (2019). Motor cortex signals for each arm are mixed across hemispheres and neurons yet partitioned within the population response. *eLife, 8*, e46159.

Amunts, K., Schlaug, G., Schleicher, A., Steinmetz, H., Dabringhaus, A., Roland, P. E., & Zilles, K. (1996). Asymmetry in the human motor cortex and handedness. *NeuroImage, 4*(3), 216–222.

Andersen, R. A., Andersen, K. N., Hwang, E. J., & Hauschild, M. (2014). Optic ataxia: From Balint's syndrome to the parietal reach region. *Neuron, 81*(5), 967–983.

Andres, M., Pelgrims, B., Olivier, E., & Vannuscorps, G. (2017). The left supramarginal gyrus contributes to finger positioning for object use: A neuronavigated transcranial magnetic stimulation study. *European Journal of Neuroscience, 46*(12), 2835–2843.

Andrushko, J. W., Gould, L. A., Renshaw, D. W., Ekstrand, C., Hortobágyi, T., Borowsky, R., & Farthing, J. P. (2021). High force unimanual handgrip contractions increase ipsilateral sensorimotor activation and functional connectivity. *Neuroscience, 452*, 111–125.

Anema, H. A., van Zandvoort, M. J. E., de Haan, E. H. F., Kappelle, L. J., de Kort, P. L. M., Jansen, B. P. W., & Dijkerman, H. C. (2009). A double dissociation between somatosensory processing for perception and action. *Neuropsychologia, 47*(6), 1615–1620.

Angell, R. (1975, June 23). Down the drain. *The New Yorker.* https://www.newyorker.com/magazine/1975/06/23/down-the-drain

Angstmann, S., Madsen, K. S., Skimminge, A., Jernigan, T. L., Baaré, W. F. C., & Siebner, H. R. (2016). Microstructural asymmetry of the corticospinal tracts predicts right–left differences in circle drawing skill in right-handed adolescents. *Brain Structure and Function, 221*(9), 4475–4489.

Annett, M. (1970). A classification of hand preference by association analysis. *British Journal of Psychology, 61*(3), 303–321.

Annett, M. (2002). *Handedness and brain asymmetry: The right shift theory*. Psychology Press.

Anonymous. (1875a). A telegraphic malady. *The Lancet, 105*(2695), 585.

Anonymous. (1875b). Telegraph writers' cramp. *British Medical Journal, 1*(746), 515.

Antoniello, D., & Gottesman, R. (2013). Allochiria in acute right hemispheric dysfunction. *Neurology, 80*(7 Suppl), P01.009.

Antoniello, D., Kluger, B. M., Sahlein, D. H., & Heilman, K. M. (2010). Phantom limb after stroke: An underreported phenomenon. *Cortex, 46*(9), 1114–1122.

App, B., McIntosh, D. N., Reed, C. L., & Hertenstein, M. J. (2011). Nonverbal channel use in communication of emotion: How may depend on why. *Emotion, 11*(3), 603–617.

Arcaro, M. J., Schade, P. F., & Livingstone, M. S. (2019). Body map proto-organization in newborn macaques. *Proceedings of the National Academy of Sciences, 116*(49), 24861–24871.

Aristotle. (1882). *On the parts of animals* [W. Ogle, Trans.]. Kegan Paul.

Aristotle. (1908a). *Metaphysica* [W. D. Ross, Trans.]. Oxford University Press.

Aristotle. (1908b). *De somniis (On dreams)* [J. I. Beare, Trans.]. In *The Parva Naturalia* (pp. 458–462). Clarendon Press.

Aristotle. (1986). *De anima (On the soul)* [H. Lawson-Tancred, Trans.]. Penguin.

Armand, J., Olivier, E., Edgley, S. A., & Lemon, R. N. (1997). Postnatal development of corticospinal projections from motor cortex to the cervical enlargement in the macaque monkey. *Journal of Neuroscience, 17*(1), 251–266.

Armatas, C. A., Summers, J. J., & Bradshaw, J. L. (1994). Mirror movements in normal adult subjects. *Journal of Clinical and Experimental Neuropsychology, 16*(3), 405–413.

Armour, J. A., Davison, A., & McManus, I. C. (2014). Genome-wide association study of handedness excludes simple genetic models. *Heredity, 112*(3), 221–225.

Arroyo, A., & de la Torre, I. (2016). Assessing the function of pounding tools in the early stone age: A microscopic approach to the analysis of percussive artefacts from Beds I and II, Olduvai Gorge (Tanzania). *Journal of Archaeological Science, 74*, 23–34.

Arsuaga, J. L., Carretero, J.-M., Lorenzo, C., Gómez-Olivencia, A., Pablos, A., Rodríguez, L., García-González, R., Bonmatí, A., Quam, R. M., Pantoja-Pérez, A., Martínez, I., Aranburu, A., Gracia-Téllez, A., Poza-Rey, A., Sala, N., García, N., Alcázar de Velasco, A., Cuenca-Bescós, G., Bermúdez de Castro, J. M., & Carbonell, E. (2015). Postcranial morphology of the middle Pleistocene humans from Sima de los Huesos, Spain. *Proceedings of the National Academy of Sciences, 112*(37), 11524–11529.

Artieda, J., Pastor, M. A., Lacruz, F., & Obeso, J. A. (1992). Temporal discrimination is abnormal in Parkinson's disease. *Brain, 115*(Pt 1), 199–210.

Arthur, W. (1997). *The origin of animal body plans: A study in evolutionary developmental biology*. Cambridge University Press.

Arzy, S., Overney, L. S., Landis, T., & Blanke, O. (2006). Neural mechanisms of embodiment: Asomatognosia due to premotor cortex damage. *Archives of Neurology, 63*(7), 1022–1025.

Assimacopoulos, S., Kao, T., Issa, N. P., & Grove, E. A. (2012). Fibroblast growth factor 8 organizes the neocortical area map and regulates sensory map topography. *Journal of Neuroscience, 32*(21), 7191–7201.

Ausenda, C. D., Togni, G., Biffi, M., Morlacchi, S., Corrias, M., & Cristoforetti, G. (2014). A new idea for stroke rehabilitation: Bilateral transfer analysis from healthy hand to the paretic one with a randomized and controlled trial. *International Journal of Physical Medicine & Rehabilitation, S3*, 008.

Aussems, S., & Kita, S. (2021). Seeing iconic gesture promotes first- and second-order verb generalization in preschoolers. *Child Development, 92*(1), 124–141.

Aviezer, H., Trope, Y., & Todorov, A. (2012). Body cues, not facial expressions, discriminate between intense positive and negative emotions. *Science, 338*(6111), 1225–1229.

Azañón, E., Camacho, K., Morales, M., & Longo, M. R. (2018). The sensitive period for tactile remapping does not include early infancy. *Child Development, 89*(4), 1394–1404.

Azañón, E., Longo, M. R., Soto-Faraco, S., & Haggard, P. (2010). The posterior parietal cortex remaps touch into external space. *Current Biology, 20*(14), 1304–1309.

Azañón, E., & Soto-Faraco, S. (2008). Changing reference frames during the encoding of tactile events. *Current Biology, 18*(14), 1044–1049.

Babinski, J. (1914). Contribution à l'étude des troubles mentaux dans l'hemiplégie organique cérébrale (anosognosie) [Contribution to the study of the mental disorders in hemiplegia of organic cerebral origin (anosognosia)]. *Revue Neurologique, 27*, 845–848.

Babinski, J. (2014). Contribution to the study of the mental disorders in hemiplegia of organic cerebral origin (anosognosia) [K. G. Langer & D. N. Levine, Trans.]. *Cortex, 61*, 5–8.

Backwell, L. R., & d'Errico, F. (2001). Evidence of termite foraging by Swartkrans early hominids. *Proceedings of the National Academy of Sciences, 98*(4), 1358–1363.

Baier, B., & Karnath, H.-O. (2008). Tight link between our sense of limb ownership and self-awareness of actions. *Stroke, 39*(2), 486–488.

Baker, S. N., Olivier, E., & Lemon, R. N. (1994). Recording an identified pyramidal volley evoked by transcranial magnetic stimulation in a conscious macaque monkey. *Experimental Brain Research, 99*(3), 529–532.

Bakola, S., Passarelli, L., Huynh, T., Impieri, D., Worthy, K. H., Fattori, P., Galletti, C., Burman, K. J., & Rosa, M. G. P. (2017). Cortical afferents and myeloarchitecture distinguish the medial intraparietal area (MIP) from neighboring subdivisions of the macaque cortex. *eNeuro, 4*(6), e0344.

Baldwin, J. M. (1891). Suggestion in infancy. *Science, 17*(421), 113–117.

Baldwin, M. K. L., Cooke, D. F., Goldring, A. B., & Krubitzer, L. (2018). Representations of fine digit movements in posterior and anterior parietal cortex revealed using long-train intracortical microstimulation in macaque monkeys. *Cerebral Cortex, 28*(12), 4244–4263.

Balfour, S., Borthwick, S., Cubelli, R., & Della Sala, S. (2007). Mirror writing and reversing single letters in stroke patients and normal elderly. *Journal of Neurology, 254*(4), 436–441.

Bálint, R. (1909). Seelenlahmung des "schauens," optische ataxie, raumliche storung der aufmerksamkei [Psychic paralysis of gaze, optic ataxia, and spatial disorder of attention]. *Monatsschrift für Psychiatrie und Neurologie, 25*, 51–81.

Bálint, R. (1995). Psychic paralysis of gaze, optic ataxia, and spatial disorder of attention [M. Harvey, Trans.]. *Cognitive Neuropsychology, 12*(3), 265–281.

Balyaev, D. K. (1979). Destabilizing selection as a factor in domestication. *Journal of Heredity, 70*(5), 301–308.

Bara-Jimenez, W., Catalan, M. J., Hallett, M., & Gerloff, C. (1998). Abnormal somatosensory homunculus in dystonia of the hand. *Annals of Neurology, 44*(5), 1996–1999.

Barac-Cikoja, D., & Turvey, M. T. (1993). Haptically perceiving size at a distance. *Journal of Experimental Psychology: General, 122*(3), 347–370.

Barbieri, C., & De Renzi, E. (1988). The executive and ideational components of apraxia. *Cortex, 24*(4), 535–543.

Bardo, A., Moncel, M. H., Dunmore, C. J., Kivell, T. L., Pouydebat, E., & Cornette, R. (2020). The implications of thumb movements for Neanderthal and modern human manipulation. *Scientific Reports, 10*, 19323.

Bargalló, A., & Mosquera, M. (2014). Can hand laterality be identified through lithic technology? *Laterality, 19*(1), 37–41.

Barker, P. (1991). *Regeneration*. Viking Press.

Barley, N. F. (1974). Two Anglo-Saxon sign systems compared. *Semiotica, 12*(3), 227–237.

Baron-Cohen, S. (1989). Perceptual role taking and protodeclarative pointing in autism. *British Journal of Developmental Psychology, 7*(2), 113–127.

Barrett, P. H. (1974). Early writings of Charles Darwin. In H. E. Gruber (Ed.), *Darwin on man: A psychological study of scientific creativity*. Wildwood House.

Barrière, I., & Lorch, M. P. (2004). Premature thoughts on writing disorders. *Neurocase, 10*(2), 91–108.

Bartholow, R. (1874). Experimental investigations into the functions of the human brain. *American Journal of the Medical Sciences, 134*, 305–313.

Bartlett, F. C. (1932). *Remembering: A study in experimental and social psychology*. Cambridge University Press.

Barton, R. A., & Harvey, P. H. (2000). Mosaic evolution of brain structure in mammals. *Nature, 405*(6790), 1055–1058.

Bar-Yosef, O. (2002). The Upper Paleolithic revolution. *Annual Review of Anthropology, 31*, 363–393.

Bassett, D. S., Yang, M., Wymbs, N. F., & Grafton, S. T. (2015). Learning-induced autonomy of sensorimotor systems. *Nature Neuroscience, 18*(5), 744–751.

Bassolino, M., & Becchio, C. (2023). The "hand paradox": Distorted representations guide optimal actions. *Trends in Cognitive Sciences, 27*(1), 7–8.

Bates, E., Benigni, L., Bretherton, I., Camaioni, L., & Volterra, V. (1979). *The emergence of symbols: Cognition and communication in infancy*. Academic Press.

Bates, E., Camaioni, L., & Volterra, V. (1975). The acquisition of performatives prior to speech. *Merrill-Palmer Quarterly, 21*(3), 205–226.

Bates, E., & Dick, F. (2002). Language, gesture, and the developing brain. *Developmental Psychobiology, 40*(3), 293–310.

Bates, E., Thal, D., Whitesell, K., Fenson, L., & Oakes, L. (1989). Integrating language and gesture in infancy. *Developmental Psychology, 25*(6), 1004–1019.

Bawden, M., & Maynard, I. (2001). Towards an understanding of the personal experience of the 'yips' in cricketers. *Journal of Sports Sciences, 19*(12), 937–953.

Baxter, D. M., & Warrington, E. K. (1985). Category specific phonological dysgraphia. *Neuropsychologia, 23*(5), 653–666.

Baxter, D. M., & Warrington, E. K. (1986). Ideational agraphia: A single case study. *Journal of Neurology, Neurosurgery, and Psychiatry, 49*(4), 369–374.

Beaudet, A., Clarke, R. J., de Jager, E. J., Bruxelles, L., Carlson, K. J., Crompton, R. de Beer, F., Dhaene, J., Heaton, J. L., Jakata, K., Jashashvili, T., Kuman, K., McClymont, J., Pickering, T. R., & Stratford,

D. (2019). The endocast of StW 573 ("Little Foot") and hominin brain evolution. *Journal of Human Evolution, 126*, 112–123.

Beaudet, A., Dumoncel, J., de Beer, F., Durrleman, S., Gilissen, E., Oettlé, A., Subsol, G., Thackeray, J. F., & Braga, J. (2018). The endocranial shape of *Australopithecus africanus*: Surface analysis of the endocasts of Sts 5 and Sts 60. *Journal of Anatomy, 232*(2), 296–303.

Beauvois, M.-F., & Dérouesné, J. (1981). Lexical or orthographic agraphia. *Brain, 104*(Pt 1), 21–49.

Beck, M. M., Spedden, M. E., Dietz, M. J., Karabanov, A. N., Christensen, M. S., & Lundbye-Jensen, J. (2021). Cortical signatures of precision grip force control in children, adolescents, and adults. *eLife, 10*, e61018.

Beck, P. D., Pospichal, M. W., & Kaas, J. H. (1996). Topography, architecture, and connections of somatosensory cortex in opossums: Evidence for five somatosensory areas. *Journal of Comparative Neurology, 366*(1), 109–133.

Beck, S., & Hallett, M. (2011). Surround inhibition in the motor system. *Experimental Brain Research, 210*(2), 165–172.

Beck, S., Shamim, E. A., Richardson, S. P., Schubert, M., & Hallett, M. (2009). Inter-hemispheric inhibition is impaired in mirror dystonia. *European Journal of Neuroscience, 29*(8), 1634–1640.

Behne, T., Carpenter, M., & Tomasello, M. (2005). One-year-olds comprehend the communicative intentions behind gestures in a hiding game. *Developmental Science, 8*(6), 492–499.

Bell, C. (1833). *The hand: Its mechanism and vital endowments as evincing design.* William Pickering.

Bell, C. (1847). *The anatomy and philosophy of expression* (3r d ed.). John Murray.

Bello-Alonso, P., Rios-Garaizar, J., Panera, J., Rubio-Jara, S., Pérez-González, A., Rojas, R., Baquedano, E., Mabulla, A., Domínguez-Rodrigo, M., & Santonja, M. (2021). The first comprehensive micro use-wear analysis of an early Acheulean assemblage (Thiongo Korongo, Olduvai Gorge, Tanzania). *Quaternary Science Reviews, 263*, 106980.

Bender, M. B. (1952). *Disorders in perception: With particular reference to the phenomena of extinction and displacement.* Charles C. Thomas.

Bender, M. B., Shapiro, M. F., & Schappell, A. W. (1949). Extinction phenomenon in hemiplegia. *Archives of Neurology & Psychiatry, 62*(6), 717–724.

Benedetti, F. (1985). Processing of tactile spatial information with crossed fingers. *Journal of Experimental Psychology: Human Perception and Performance, 11*(4), 517–525.

Benedetti, F. (1986). Tactile diplopia (diplesthesia) on the human fingers. *Perception, 15*(1), 83–91.

Benke, T., Luzzatti, C., & Vallar, G. (2004). Hermann Zingerle's "Impaired perception of the own body due to organic brain disorders": An introductory comment, and an abridged translation. *Cortex, 40*(2), 265–274.

Bennati, F. (1831). Notice physiologique sur Paganini [Physiological note on Paganini]. *Revue de Paris, 26*, 54–55.

Bennett, E. A., Crevecoeur, I., Viola, B., Derevianko, A. P., Shunkov, M. V., Grange, T., Maureille, B., & Geigl, E.-M. (2019). Morphology of the Denisovan phalanx closer to modern humans than to Neanderthals. *Science Advances*, *5*(9), eaaw3950.

Bennike, P. (1985). *Paleopathology of Danish skeletons*. Akademisk Forlag.

Bensmaia, S. J., Denchev, P. V, Dammann, J. F., Craig, J. C., & Hsiao, S. S. (2008). The representation of stimulus orientation in the early stages of somatosensory processing. *Journal of Neuroscience*, *28*(3), 776–786.

Berardelli, A., Rothwell, J. C., Thompson, P. D., & Hallett, M. (2001). Pathophysiology of bradykinesia in Parkinson's disease. *Brain*, *124*(Pt 11), 2131–2146.

Berger, K. W., & Popelka, G. R. (1971). Extra-facial gestures in relation to speechreading. *Journal of Communication Disorders*, *3*(4), 302–308.

Bergman, H., Wichmann, T., & Delong, M. R. (1990). Reversal of experimental Parkinsonism by lesions of the subthalamic nucleus. *Science*, *249*(4975), 1436–1438.

Bermúdez de Castro, J.-M., Bromage, T. G., & Jalvo, Y. F. (1988). Buccal striations on fossil human anterior teeth: Evidence of handedness in the middle and early Upper Pleistocene. *Journal of Human Evolution*, *17*(4), 403–412.

Bernhard, C. G., Bohm, E., & Petersén, I. (1953). Investigations on the organization of the corticospinal system in monkeys. *Acta Physiologica Scandinavica*, *29*(S106), 79–105.

Bernstein, N. A. (1967). *Co-ordination and regulation of movements*. Pergamon Press.

Berret, B., Chiovetto, E., Nori, F., & Pozzo, T. (2011). Evidence for composite cost functions in arm movement planning: An inverse optimal control approach. *PLOS Computational Biology*, *7*(10), e1002183.

Bertenthal, B. I., Boyer, T. W., & Harding, S. (2014). When do infants begin to follow a point? *Developmental Psychology*, *50*(8), 2036–2048.

Berthier, N. E., Clifton, R. K., McCall, D. D., & Robin, D. J. (1999). Proximodistal structure of early reaching in human infants. *Experimental Brain Research*, *127*(3), 259–269.

Berthier, N. E., & Keen, R. (2006). Development of reaching in infancy. *Experimental Brain Research*, *169*(4), 507–518.

Berti, A., Bottini, G., Gandola, M., Pia, L., Smania, N., Stracciari, A., Castiglioni, I., Vallar, G., & Paulesu, E. (2005). Shared cortical anatomy for motor awareness and motor control. *Science*, *309*(5733), 488–491.

Berti, A., & Frassinetti, F. (2000). When far becomes near: Remapping of space by tool use. *Journal of Cognitive Neuroscience*, *12*(3), 415–420.

Berti, A., Oxbury, S., Oxbury, J., Affanni, P., Umilta, C., & Orlandi, L. (1999). Somatosensory extinction for meaningful objects in a patient with right hemispheric stroke. *Neuropsychologia*, *37*(3), 333–343.

Berti, A., Spinazzola, L., Pia, L., & Rabuffetti, M. (2007). Motor awareness and motor intention in anosognosia for hemiplegia. In P. Haggard, Y. Rossetti, & M. Kawato (Eds.), *Sensorimotor foundations of higher cognition series: Attention and performance XXI* (pp. 163–182). Oxford University Press.

Bestmann, S., & Krakauer, J. W. (2015). The uses and interpretations of the motor-evoked potential for understanding behaviour. *Experimental Brain Research, 233*(3), 679–689.

Beyene, Y., Katoh, S., WoldeGabriel, G., Hart, W. K., Uto, K., Sudo, M., Kondo, M., Hyodo, M., Renne, P. R., Suwa, G., & Asfaw, B. (2013). The characteristics and chronology of the earliest Acheulean at Konso, Ethiopia. *Proceedings of the National Academy of Sciences, 110*(5), 1584–1591.

Beznosov, P. A., Clack, J. A., Lukševičs, E., Ruta, M., & Ahlberg, P. E. (2019). Morphology of the earliest reconstructable tetrapod *Parmastega aelidae*. *Nature, 574*(7779), 527–531.

Bhat, A. N., & Galloway, J. C. (2006). Toy-oriented changes during early arm movements: Hand kinematics. *Infant Behavior and Development, 29*(3), 358–372.

Bickerton, D. (1995). *Language and human behavior*. University of Washington Press.

Binkofski, F., & Buxbaum, L. J. (2013). Two action systems in the human brain. *Brain and Language, 127*(2), 222–229.

Binkofski, F., Dohle, C., Posse, S., Stephan, K. M., Hefter, H., Seitz, R. J., & Freund, H. J. (1998). Human anterior intraparietal area subserves prehension. *Neurology, 50*(5), 1253–1259.

Binkofski, F., Kunesch, E., Classen, J., Seitz, R.J., & Freund, H.-J. (2001). Tactile apraxia: Unimodal apractic disorder of tactile object exploration associated with parietal lobe lesions. *Brain, 124*(Pt 1), 132–144.

Birznieks, I., Logina, I., & Wasner, G. (2012). Somatotopic mismatch following stroke: A pathophysiological condition escaping detection. *BMJ Case Reports, 2012*, 6304.

Birznieks, I., Logina, I., & Wasner, G. (2016). Somatotopic mismatch of hand representation following stroke: Is recovery possible? *Neurocase, 22*(1), 95–102.

Bishop, A. (1962). Control of the hand in lower primates. *Annals of the New York Academy of Sciences, 102*, 316–337.

Bisiach, E., Perani, D., Vallar, G., & Berti, A. (1986). Unilateral neglect: Personal and extra-personal. *Neuropsychologia, 24*(6), 759–767.

Bizley, J. K., & Cohen, Y. E. (2013). The what, where and how of auditory-object perception. *Nature Reviews Neuroscience, 14*(10), 693–707.

Bjoertomt, O., Cowey, A., & Walsh, V. (2002). Spatial neglect in near and far space investigated by repetitive transcranial magnetic stimulation. *Brain, 125*(Pt 9), 2012–2022.

Blangero, A., Ota, H., Delporte, L., Revol, P., Vindras, P., Rode, G., Boisson, D., Vighetto, A., Rossetti, Y., & Pisella, L. (2007). Optic ataxia is not only 'optic': Impaired spatial integration of proprioceptive information. *NeuroImage, 36*(Suppl 2), T61–T68.

Blankenburg, F., Ruben, J., Meyer, R., Schwiemann, J., & Villringer, A. (2003). Evidence for a rostral-to-caudal somatotopic organization in human primary somatosensory cortex with mirror-reversal in areas 3b and 1. *Cerebral Cortex, 13*(9), 987–993.

Blass, E. M., Fillion, T. J., Rochat, P., Hoffmeyer, L. B., & Metzger, M. A. (1989). Sensorimotor and motivational determinants of hand–mouth coordination in 1–3-day-old human infants. *Developmental Psychology, 25*(6), 963–975.

Bloch, J. I., & Boyer, D. M. (2002). Grasping primate origins. *Science, 298*(5598), 1606–1611.

Bloxham, C. A., Mindel, T. A., & Frith, C. D. (1984). Initiation and execution of predictable and unpredictable movements in Parkinson's disease. *Brain, 107*(Pt 2), 371–384.

Blythe, E., Garrido, L., & Longo, M. R. (2023). Emotion is perceived accurately from isolated body parts, especially hands. *Cognition, 230*, 105260.

Bodegård, A., Geyer, S., Grefkes, C., Zilles, K., & Roland, P. E. (2001). Hierarchical processing of tactile shape in the human brain. *Neuron, 31*(2), 317–328.

Boeckle, M., Schiestl, M., Frohnwieser, A., Gruber, R., Miller, R., Suddendorf, T., Gray, R. D., Taylor, A. H., & Clayton, N. S. (2020). New Caledonian crows plan for specific future tool use. *Proceedings of the Royal Society B, 287*(1938), 20201490.

Boëda, E., Connan, J., Dessort, D., Muhesen, S., Mercier, N., Valladas, H., & Tisnérat, N. (1996). Bitumen as a hafting material on Middle Palaeolithic artefacts. *Nature, 380*, 336–338.

Boesch, C., Head, J., & Robbins, M. M. (2009). Complex tool sets for honey extraction among chimpanzees in Loango National Park, Gabon. *Journal of Human Evolution, 56*(6), 560–569.

Boesch, C., Kalan, A. K., Mundry, R., Arandjelovic, M., Pika, S., Dieguez, P., Ayimisin, E. A., Barciela, A., Coupland, C., Egbe, V. E., Eno-Nku, M., Fay, J. M., Fine, D., Hernandez-Aguilar, R. A., Hermans, V., Kadam, P., Kambi, M., Llana, M., Maretti, G., & Kühl, H. S. (2020). Chimpanzee ethnography reveals unexpected cultural diversity. *Nature Human Behaviour, 4*(9), 910–916.

Bogen, J. E. (1993). The callosal syndromes. In K. M. Heilman & E. V. Valenstein (Eds.), *Clinical Neuropsychology* (3rd ed., pp. 337–407). Oxford University Press.

Bohlhalter, S., Fretz, C., & Weder, B. (2002). Hierarchical versus parallel processing in tactile object recognition: A behavioural-neuroanatomical study of aperceptive tactile agnosia. *Brain, 125*(Pt 11), 2537–2548.

Boisvert, C. A., Mark-Kurik, E., & Ahlberg, P. E. (2008). The pectoral fin of *Panderichthys* and the origin of digits. *Nature, 456*(7222), 636–638.

Boling, W., & Olivier, A. (2004). Localization of hand sensory function to the pli de passage moyen of Broca. *Journal of Neurosurgery, 101*(2), 278–283.

Boling, W., Olivier, A., Bittar, R. G., & Reutens, D. (1999). Localization of hand motor activation in Broca's pli de passage moyen. *Journal of Neurosurgery, 91*(6), 903–910.

Boling, W., Parsons, M., Kraszpulski, M., Cantrell, C., & Puce, A. (2008). Whole-hand sensorimotor area: Cortical stimulation localization and correlation with functional magnetic resonance imaging. *Journal of Neurosurgery, 108*(3), 491–500.

Bolognini, N., Convento, S., Banco, E., Mattioli, F., Tesio, L., & Vallar, G. (2015). Improving ideomotor limb apraxia by electrical stimulation of the left posterior parietal cortex. *Brain, 138*(Pt 2), 428–439.

Boraud, T., Bezard, E., Bioulac, B., & Gross, C. E. (2000). Ratio of inhibited-to-activated pallidal neurons decreases dramatically during passive limb movement in the MPTP-treated monkey. *Journal of Neurophysiology, 83*(3), 1760–1763.

Boroojerdi, B., Foltys, H., Krings, T., Spetzger, U., Thron, A., & Töpper, R. (1999). Localization of the motor hand area using transcranial magnetic stimulation and functional magnetic resonance imaging. *Clinical Neurophysiology, 110*(4), 699–704.

Borra, E., Belmalih, A., Calzavara, R., Gerbella, M., Murata, A., Rozzi, S., & Luppino, G. (2008). Cortical connections of the macaque anterior intraparietal (AIP) area. *Cerebral Cortex, 18*(5), 1094–1111.

Bortoff, G. A., & Strick, P. L. (1993). Corticospinal terminations in two New-World primates: Further evidence that corticomotoneuronal connections provide part of the neural substrate for manual dexterity. *Journal of Neuroscience, 13*(12), 5105–5118.

Bottini, G., Bisiach, E., Sterzi, R., & Vallar, G. (2002). Feeling touches in someone else's hand. *Neuroreport, 13*(2), 249–252.

Botvinick, M., & Cohen, J. (1998). Rubber hands 'feel' touch that eyes see. *Nature, 391*(6669), 756.

Bourke, J. (2016). Phantom suffering? *The Psychologist, 29*, 730–731.

Boyer, D. M., Yapuncich, G. S., Chester, S. G. B., Bloch, J. I., & Godinot, M. (2013). Hands of early primates. *American Journal of Physical Anthropology, 152*(S57), 33–78.

Boyer, E. O., Babayan, B. M., Bevilacqua, F., Noisternig, M., Warusfel, O., Roby-Brami, A., Hanneton, S., & Viaud-Delmon, I. (2013). From ear to hand: The role of the auditory-motor loop in pointing to an auditory source. *Frontiers in Computational Neuroscience, 7*, 26.

Bracci, S., Cavina-Pratesi, C., Ietswaart, M., Peelen, M. V., & Caramazza, A. (2012). Closely overlapping responses to tools and hands in left lateral occipitotemporal cortex. *Journal of Neurophysiology, 107*(5), 1443–1456.

Brain, C. K. (1970). New finds at the Swartkrans Australopithecine site. *Nature, 225*, 1112–1119.

Brain, W. R. (1941). Visual disorientation with special reference to lesions of the right cerebral hemisphere. *Brain, 64*(4), 244–272.

Braun, D. R., Aldeias, V., Archer, W., Arrowsmith, J. R., Baraki, N., Campisano, C. J., Deino, A. L., DiMaggio, E. N., Dupont-Nivet, G., Engda, B., Feary, D. A., Garello, D. I., Kerfelew, Z., McPherron, S. P., Patterson, D. B., Reeves, J. S., Thompson, J. C., & Reed, K. E. (2019). Earliest known Oldowan artifacts at >2.58 Ma from Ledi-Geraru, Ethiopia, highlight early technological diversity. *Proceedings of the National Academy of Sciences, 116*(24), 11712–11717.

Brazelton, T. B. (1956). Sucking in infancy. *Pediatrics, 17*(3), 400–404.

Bresciani, J.-P., Blouin, J., Sarlegna, F., Bourdin, C., Vercher, J.-L., & Gauthier, G. M. (2002). Online versus off-line vestibular-evoked control of goal-directed arm movements. *NeuroReport, 13*(12), 1563–1566.

Breuer, T., Ndoundou-Hockemba, M., & Fishlock, V. (2005). First observation of tool use in wild gorillas. *PLOS Biology, 3*(11), e380.

Brinkman, J., & Kuypers, H. G. J. M. (1972). Splitbrain monkeys: Cerebral control of ipsilateral and contralateral arm, hand, and finger movements. *Science, 176*(4034), 536–539.

Brion, S., & Jedynak, C. P. (1972). Trouble du transfert interhémisphérique à propos de trois observations se tumeurs du corps calleux: Le signe de la main étrangère [Interhemispheric transfer disorder apropos of three cases of tumors of the corpus callosum: The foreign hand sign]. *Revue Neurologique, 136*, 257–266.

Brisben, A. J., Hsiao, S. S., & Johnson, K. O. (1999). Detection of vibration transmitted through an object grasped in the hand. *Journal of Neurophysiology, 81*(4), 1548–1558.

Broaders, S. C., Cook, S. W., Mitchell, Z., & Goldin-Meadow, S. (2007). Making children gesture brings out implicit knowledge and leads to learning. *Journal of Experimental Psychology: General, 136*(4), 539–550.

Brochier, T., Habib, M., & Brouchon, M. (1994). Covert processing of information in hemianesthesia: A case report. *Cortex, 30*(1), 135–144.

Brodmann, K. (1909). *Vergleichende lokalisationslehre der grosshirnrinde in ihrem prinzipien dargestellt auf grund des zellenbaues [Comparative theory of localization of the cerebral cortex presented in its principles on the basis of the cell structure]*. Barth JA.

Bronowski, J. (1975). *The ascent of man*. Little, Brown and Company.

Brooks, R., & Meltzoff, A. N. (2008). Infant gaze following and pointing predict accelerated vocabulary growth through two years of age: A longitudinal, growth curve modeling study. *Journal of Child Language, 35*(1), 207–220.

Brooks, V. B., Rudomin, P., & Slayman, C. L. (1961). Peripheral receptive fields of neurons in the cat's cerebral cortex. *Journal of Neurophysiology, 24*(3), 302–325.

Brothwell, D. R., & Møller-Christensen, V. (1963). A possible case of amputation, dated to c. 2000 B.C. *Man, 63*, 192–194.

Brown, L. E., Kroliczak, G., Demonet, J.-F., & Goodale, M. A. (2008). A hand in blindsight: Hand placement near target improves size perception in the blind visual field. *Neuropsychologia, 46*(3), 786–802.

Brown, L. E., Morrissey, B. F., & Goodale, M. A. (2009). Vision in the palm of your hand. *Neuropsychologia*, *47*(6), 1621–1626.

Brown, P. B., Fuchs, J. L., & Tapper, D. N. (1975). Parametric studies of dorsal horn neurons responding to tactile stimulation. *Journal of Neurophysiology*, *38*(1), 19–25.

Brown, P. B., Koerber, H. R., & Millecchia, R. (2004). From innervation density to tactile acuity: 1. Spatial representation. *Brain Research*, *1011*(1), 14–32.

Brown, S., Massilani, D., Kozlikin, M. B., Shunkov, M. V., Derevianko, A. P., Stoessel, A., Jope-Street, B., Meyer, M., Kelso, J., Pääbo, S., Higham, T., & Douka, K. (2022). The earliest Denisovans and their cultural adaptation. *Nature Ecology & Evolution*, *6*, 28–35.

Brozzoli, C., Cardinali, L., Pavani, F., & Farnè, A. (2010). Action-specific remapping of peripersonal space. *Neuropsychologia*, *48*(3), 796–802.

Brozzoli, C., Gentile, G., & Ehrsson, H. H. (2012). That's near my hand! Parietal and premotor coding of hand-centered space contributes to localization and self-attribution of the hand. *Journal of Neuroscience*, *32*(42), 14573–14582.

Brozzoli, C., Gentile, G., Petkova, V. I., & Ehrsson, H. H. (2011). fMRI adaptation reveals a cortical mechanism for the coding of space near the hand. *Journal of Neuroscience*, *31*(24), 9023–9031.

Brozzoli, C., Pavani, F., Urquizar, C., Cardinali, L., & Farnè, A. (2009). Grasping actions remap peripersonal space. *NeuroReport*, *20*(10), 913–917.

Brugger, P. (2003). Supernumerary phantoms: A comment on Grossi et al.'s (2002) spare thoughts on spare limbs. *Perceptual and Motor Skills*, *97*(1), 3–10.

Brugger, P., Christen, M., Jellestad, L., & Hänggi, J. (2016). Limb amputation and other disability desires as a medical condition. *Lancet Psychiatry*, *3*(12), 1176–1186.

Brugger, P., Kollias, S. S., Müri, R. M., Crelier, G., Hepp-Reymond, M.-C., & Regard, M. (2000). Beyond re-membering: Phantom sensations of congenitally absent limbs. *Proceedings of the National Academy of Sciences*, *97*(11), 6167–6172.

Bruner, E. (2004). Geometric morphometrics and paleoneurology: Brain shape evolution in the genus Homo. *Journal of Human Evolution*, *47*(5), 279–303.

Bruner, E., & Lozano, M. (2014). Extended mind and visuo-spatial integration: Three hands for the Neandertal lineage. *Journal of Anthropological Sciences*, *92*, 273–280.

Bruner, E., Manzi, G., & Arsuaga, J. L. (2003). Encephalization and allometric trajectories in the genus *Homo*: Evidence from the Neandertal and modern lineages. *Proceedings of the National Academy of Sciences*, *100*(26), 15335–15340.

Bruner, J. S., & Koslowski, B. (1972). Visually preadapted constituents of manipulatory action. *Perception*, *1*(1), 3–14.

Bruno, N., & Bertamini, M. (2010). Haptic perception after a change in hand size. *Neuropsychologia*, *48*(6), 1853–1856.

Bruno, V., Carpinella, I., Rabuffetti, M., De Giuli, L., Sinigaglia, C., Garbarini, F., & Ferrarin, M. (2019). How tool-use shapes body metric representation: Evidence from motor training with and without robotic assistance. *Frontiers in Human Neuroscience, 13*, 299.

Brusatte, S. (2022). *The rise and reign of the mammals*. Picador.

Büchel, C., Price, C., Frackowiak, R. S. J., & Friston, K. (1998). Different activation patterns in the visual cortex of late and congenitally blind subjects. *Brain, 121*(Pt 3), 409–419.

Büchel, C., Price, C., & Friston, K. (1998). A multimodal language region in the ventral visual pathway. *Nature, 394*(6690), 274–277.

Buckner, R. L., Sepulcre, J., Talukdar, T., Krienen, F. M., Liu, H., Hedden, T., Andrews-Hanna, J. R., Sperling, R. A., & Johnson, K. A. (2009). Cortical hubs revealed by intrinsic functional connectivity: Mapping, assessment of stability, and relation to Alzheimer's disease. *Journal of Neuroscience, 29*(6), 1860–1873.

Buetefisch, C. M., Revill, K. P., Shuster, L., Hines, B., & Parsons, M. (2014). Motor demand-dependent activation of ipsilateral motor cortex. *Journal of Neurophysiology, 112*(4), 999–1009.

Buetti, S., Tamietto, M., Hervais-Adelman, A., Kerzel, D., de Gelder, B., & Pegna, A. J. (2013). Dissociation between goal-directed and discrete response localization in a patient with bilateral cortical blindness. *Journal of Cognitive Neuroscience, 25*(10), 1769–1775.

Bugbee, N. M., & Goldman-Rakic, P. S. (1983). Columnar organization of corticocortical projections in squirrel and rhesus monkeys: Similarity of column width in species differing in cortical volume. *Journal of Comparative Neurology, 220*(3), 355–364.

Buka, S. L., & Lipsitt, L. P. (1991). Newborn sucking behavior and its relation to grasping. *Infant Behavior and Development, 14*(1), 59–67.

Bundy, D. T., Szrama, N., Pahwa, M., & Leuthardt, E. C. (2018). Unilateral, 3D arm movement kinematics are encoded in ipsilateral human cortex. *Journal of Neuroscience, 38*(47), 10042–10056.

Bunn, H. T. (1981). Archaeological evidence for meat-eating by Plio-Pleistocene hominids from Koobi Fora and Olduvai Gorge. *Nature, 291*, 574–577.

Buquet-Marcon, C., Philippe, C., & Anaick, S. (2007). The oldest amputation on a Neolithic human skeleton in France. *Nature Precedings.* https://doi.org/10.1038/npre.2007.1278.1

Butler, B. C., Eskes, G. A., & Vandorpe, R. A. (2004). Gradients of detection in neglect: Comparison of peripersonal and extrapersonal space. *Neuropsychologia, 42*(3), 346–358.

Butterworth, G. (2003). Pointing is the royal road to language for babies. In S. Kita (Ed.), *Pointing: Where language, culture, and cognition meet* (pp. 9–33). Erlbaum.

Butterworth, G., & Hopkins, B. (1988). Hand–mouth coordination in the new-born baby. *British Journal of Developmental Psychology, 6*(4), 303–314.

Butterworth, G., Verweij, E., & Hopkins, B. (1997). The development of prehension in infants: Halverson revisited. *British Journal of Developmental Psychology, 15*(2), 223–236.

Butterworth, S., Francis, S., Kelly, E., McGlone, F., Bowtell, R., & Sawle, G. V. (2003). Abnormal cortical sensory activation in dystonia: An fMRI study. *Movement Disorders, 18*(6), 673–682.

Buxbaum, L. J. (2001). Ideomotor apraxia: A call to action. *Neurocase, 7*(6), 445–458.

Buxbaum, L. J., & Coslett, H. B. (1997). Subtypes of optic ataxia: Reframing the disconnection account. *Neurocase, 3*(3), 159–166.

Buxbaum, L. J., Kyle, K. M., & Menon, R. (2005). On beyond mirror neurons: Internal representations subserving imitation and recognition of skilled object-related actions in humans. *Cognitive Brain Research, 25*(1), 226–239.

Buxbaum, L. J., Kyle, K. M., Tang, K., & Detre, J. A. (2006). Neural substrates of knowledge of hand postures for object grasping and functional object use: Evidence from fMRI. *Brain Research, 1117*(1), 175–185.

Buxhoeveden, D. P., & Casanova, M. F. (2002). The minicolumn hypothesis in neuroscience. *Brain, 125*(Pt 5), 935–951.

Byl, N. N., Merzenich, M. M., & Jenkins, W. M. (1996). A primate genesis model of focal dystonia and repetitive strain injury: I. Learning-induced dedifferentiation of the representation of the hand in the primary somatosensory cortex in adult monkeys. *Neurology, 47*(2), 508–520.

Byrne, R. W., Cartmill, E., Genty, E., Graham, K. E., Hobaiter, C., & Tanner, J. (2017). Great ape gestures: Intentional communication with a rich set of innate signals. *Animal Cognition, 20*(4), 755–769.

Byrnit, J. T. (2004). Nonenculturated orangutans' (*Pongo pygmaeus*) use of experimenter-given manual and facial cues in an object-choice task. *Journal of Comparative Psychology, 118*(3), 309–315.

Byrnit, J. T. (2009). Gorillas' (*Gorilla gorilla*) use of experimenter-given manual and facial cues in an object-choice task. *Animal Cognition, 12*(2), 401–404.

Cabeza, R., Ciaramelli, E., Olson, I. R., & Moscovitch, M. (2008). The parietal cortex and episodic memory: An attentional account. *Nature Reviews Neuroscience, 9*(8), 613–625.

Cadete, D., Alsmith, A. J. T., & Longo, M. R. (2022). Curved sixth fingers: Flexible representation of the shape of supernumerary body parts. *Consciousness and Cognition, 105*, 103413.

Cadete, D., & Longo, M. R. (2020). A continuous illusion of having a sixth finger. *Perception, 49*(8), 807–821.

Cadete, D., & Longo, M. R. (2022). The long sixth finger illusion: The representation of the supernumerary finger is not a copy and can be felt with varying lengths. *Cognition, 218*, 104948.

Call, J., & Tomasello, M. (1994). Production and comprehension of referential pointing by orangutans (*Pongo pygmaeus*). *Journal of Comparative Psychology, 108*(4), 307–317.

Callier, V., Clack, J. A., & Ahlberg, P. E. (2009). Contrasting developmental trajectories in the earliest known tetrapod forelimbs. *Science, 324*(5925), 364–367.

Calvert, G. H. M., & Carson, R. G. (2022). Neural mechanisms mediating cross education: With additional considerations for the ageing brain. *Neuroscience and Biobehavioral Reviews, 132*, 260–288.

Calzolari, E., Azañón, E., Danvers, M., Vallar, G., & Longo, M. R. (2017). Adaptation aftereffects reveal that tactile distance is a basic somatosensory feature. *Proceedings of the National Academy of Sciences, 114*(17), 4555–4560.

Cameron, B. D., Cressman, E. K., Franks, I. M., & Chua, R. (2009). Cognitive constraint on the "automatic pilot" for the hand: Movement intention influences the hand's susceptibility to involuntary online corrections. *Consciousness and Cognition, 18*(3), 646–652.

Canavero, S., & Bonicalzi, V. (1998). The neurochemistry of central pain: Evidence from clinical studies, hypothesis and therapeutic implications. *Pain, 74*(2–3), 109–114.

Canavero, S., Bonicalzi, V., Castellano, G., Perozzo, P., & Massa-Micon, B. (1999). Painful supernumerary phantom arm following motor cortex stimulation for central poststroke pain. *Journal of Neurosurgery, 91*(1), 121–123.

Cantalupo, C., & Hopkins, W. D. (2001). Asymmetric Broca's area in great apes. *Nature, 414*, 505–505.

Canzoneri, E., Ubaldi, S., Rastelli, V., Finisguerra, A., Bassolino, M., & Serino, A. (2013). Tool-use reshapes the boundaries of body and peripersonal space representations. *Experimental Brain Research, 228*(1), 25–42.

Carbonell, E., Bermúdez de Castro, J. M., Parés, J. M., Pérez-González, A., Cuenca-Bescós, G., Ollé, A., Mosquera, M., Huguet, R., van der Made, J., Rosas, A., Sala, R., Vallverdú, J., García, N., Granger, D. E., Martinón-Torres, M., Rodríguez, X. P., Stock, G. M., Vergès, J. M., Allué, E., . . . Arsuaga, J. L. (2008). The first hominin of Europe. *Nature, 452*(7186), 465–469.

Cardinali, L., Brozzoli, C., Finos, L., Roy, A. C., & Farnè, A. (2016). The rules of tool incorporation: Tool morpho-functional and sensori-motor constraints. *Cognition, 149*, 1–5.

Cardinali, L., Frassinetti, F., Brozzoli, C., Urquizar, C., Roy, A. C., & Farnè, A. (2009). Tool-use induces morphological updating of the body schema. *Current Biology, 19*(12), R478–R479.

Cardini, F., Longo, M. R., & Haggard, P. (2011). Vision of the body modulates somatosensory intracortical inhibition. *Cerebral Cortex, 21*(9), 2014–2022.

Carello, C., Fitzpatrick, P., & Turvey, M. T. (1992). Haptic probing: Perceiving the length of a probe and the distance of a surface probed. *Perception & Psychophysics, 51*(6), 580–598.

Carlson, K. J., Stout, D., Jashashvili, T., de Ruiter, D. J., Tafforeau, P., Carlson, K., & Berger, L. R. (2011). The endocast of MH1, *Australopithecus sediba*. *Science, 333*(6048), 1402–1407.

Carpenter, M., Nagell, K., & Tomasello, M. (1998). Social cognition, joint attention, and communicative competence from 9 to 15 months of age. *Monographs of the Society for Research in Child Development, 63*, 1–143.

Carr, L. J., Harrison, L. M., Evans, A. L., & Stephens, J. A. (1993). Patterns of central motor reorganization in hemiplegic cerebral palsy. *Brain, 116*(Pt 5), 1223–1247.

Carson, R. G. (2020). Inter-hemispheric inhibition sculpts the output of neural circuits by co-opting the two cerebral hemispheres. *Journal of Physiology, 598*(21), 4781–4802.

Carter, C. S., Braver, T. S., Barch, D. M., Botvinick, M. M., Noll, D., & Cohen, J. D. (1998). Anterior cingulate cortex, error detection, and the online monitoring of performance. *Science, 280*(5364), 747–750.

Cartmill, E. A., & Byrne, R. W. (2007). Orangutans modify their gestural signaling according to their audience's comprehension. *Current Biology, 17*(15), 1345–1348.

Cartmill, M. (1974). Rethinking primate origins. *Science, 184*(4135), 436–443.

Carvalho, S., Biro, D., McGrew, W. C., & Matsuzawa, T. (2009). Tool-composite reuse in wild chimpanzees (*Pan troglodytes*): Archaeologically invisible steps in the technological evolution of early hominins? *Animal Cognition, 12*(Suppl 1), 103–114.

Carvalho, S., Cunha, E., Sousa, C., & Matsuzawa, T. (2008). Chaines opératoires and resource-exploitation strategies in chimpanzee (*Pan troglodytes*) nut cracking. *Journal of Human Evolution, 55*(1), 148–163.

Cascarano, G. D., Loconsole, C., Brunetti, A., Lattarulo, A., Buongiorno, D., Losavio, G., Di Sciascio, E., & Bevilacqua, V. (2019). Biometric handwriting analysis to support Parkinson's disease assessment and grading. *BMC Medical Informatics and Decision Making, 19*(Suppl 9), 252.

Case, L. K., Abrams, R. A., & Ramachandran, V. S. (2010). Immediate interpersonal and intermanual referral of sensations following anesthetic block of one arm. *Archives of Neurology, 67*(12), 1521–1523.

Caselli, R. J. (1991). Rediscovering tactile agnosia. *Mayo Clinic Proceedings, 66*(2), 129–142.

Caselli, R. J. (1993). Ventrolateral and dorsomedial somatosensory association cortex damage produces distinct somesthetic syndromes in humans. *Neurology, 43*(4), 762–762.

Castiello, U., Bennett, K. M., Egan, G. F., Tochon-Danguy, H. J., Kritikos, A., & Dunai, J. (1999). Human inferior parietal cortex "programs" the action class of grasping. *Journal of Cognitive Systems Research, 2*, 22–30.

Castiello, U., Paulignan, Y., & Jeannerod, M. (1991). Temporal dissociation of motor responses and subjective awareness. *Brain, 114*(Pt 6), 2639–2655.

Catani, M. (2017). A little man of some importance. *Brain, 140*(11), 3055–3061.

Catania, K. C., & Kaas, J. H. (1995). Organization of the somatosensory cortex of the star-nosed mole. *Journal of Comparative Neurology, 351*(4), 549–567.

Catania, K. C., & Kaas, J. H. (1997a). Somatosensory fovea in the star-nosed mole: Behavioral use of the star in relation to innervation patterns and cortical representation. *Journal of Comparative Neurology, 387*(2), 215–233.

Catania, K. C., & Kaas, J. H. (1997b). The mole nose instructs the brain. *Somatosensory & Motor Research, 14*(1), 56–58.

Catania, K. C., Northcutt, R. G., Kaas, J. H., & Beck, P. D. (1993). Nose stars and brain stripes. *Nature, 524,* 493.

Catania, K. C., & Remple, F. E. (2005). Asymptotic prey profitability drives star-nosed moles to the foraging speed limit. *Nature, 433,* 519–522.

Catania, K. C., & Remple, M. S. (2002). Somatosensory cortex dominated by the representation of teeth in the naked mole-rat brain. *Proceedings of the National Academy of Sciences, 99*(8), 5692–5697.

Chakraborty, M., & Jarvis, E. D. (2015). Brain evolution by brain pathway duplication. *Philosophical Transactions of the Royal Society B, 370*(1684), 20150056.

Chancel, M., & Ehrsson, H. H. (2020). Which hand is mine? Discriminating body ownership perception in a two-alternative forced-choice task. *Attention, Perception, & Psychophysics, 82*(8), 4058–4083.

Changizi, M. A., & Shimojo, S. (2005). Character complexity and redundancy in writing systems over human history. *Proceedings of the Royal Society B, 272*(1560), 267–275.

Chan, T.-C., & Turvey, M. T. (1991). Perceiving the vertical distances of surfaces by means of a hand-held probe. *Journal of Experimental Psychology: Human Perception and Performance, 17*(2), 347–358.

Chao, L. L., & Martin, A. (2000). Representation of manipulable man-made objects in the dorsal stream. *NeuroImage, 12*(4), 478–484.

Chaplin, T. A., Yu, H.-H., Soares, J. G. M., Gattass, R., & Rosa, M. G. P. (2013). A conserved pattern of differential expansion of cortical areas in simian primates. *Journal of Neuroscience, 33*(38), 15120–15125.

Charcot, J.-M., & Pitres, A. (1895). *Les centres moteurs corticaux chez l'homme [The cortical motor centers in man].* Rueff.

Checchetto, A., Geraci, C., Cecchetto, C., & Zucchi, S. (2018). The language instinct in extreme circumstances: The transition to tactile Italian Sign Language (LISt) by Deafblind signers. *Glossa, 3*(1), 66.

Chen, L. M., Friedman, R. M., Ramsden, B. M., LaMotte, R. H., & Roe, A. W. (2001). Fine-scale organization of SI (area 3b) in the squirrel monkey revealed with intrinsic optical imaging. *Journal of Neurophysiology, 86*(6), 3011–3029.

Cheney, P. D., & Fetz, E. E. (1984). Corticomotoneuronal cells contribute to long-latency stretch reflexes in the rhesus monkey. *Journal of Physiology, 349*(1), 249–272.

Cholewiak, R. W. (1999). The perception of tactile distance: Influences of body site, space, and time. *Perception, 28*(7), 851–876.

Chowdhury, R. H., Glaser, J. I., & Miller, L. E. (2020). Area 2 of primary somatosensory cortex encodes kinematics of the whole arm. *eLife, 9,* e48198.

Chowdhury, S. A., & Rasmusson, D. D. (2002). Comparison of receptive field expansion produced by GABA B and GABA A receptor antagonists in raccoon primary somatosensory cortex. *Experimental Brain Research, 144*(1), 114–121.

Chu, M., & Kita, S. (2011). The nature of gestures' beneficial role in spatial problem solving. *Journal of Experimental Psychology: General, 140*(1), 102–116.

Church, R. B., Ayman-Nolley, S., & Mahootian, S. (2004). The role of gesture in bilingual education: Does gesture enhance learning? *International Journal of Bilingual Education and Bilingualism, 7*(4), 303–319.

Church, R. B., & Goldin-Meadow, S. (1986). The mismatch between gesture and speech as an index of transitional knowledge. *Cognition, 23*(1), 43–71.

Churchill, S. E. (2001). Hand morphology, manipulation, and tool use in Neandertals and early modern humans of the Near East. *Proceedings of the National Academy of Sciences, 98*(6), 2953–2955.

Churchland, M. M., Cunningham, J. P., Kaufman, M. T., Foster, J. D., Nuyujukian, P., Ryu, S. I., & Shenoy, K. V. (2012). Neural population dynamics during reaching. *Nature, 487*(7405), 51–56.

Cicone, M., Wapner, W., Foldi, N., Zurif, E., & Gardner, H. (1979). The relation between gesture and language in aphasic communication. *Brain and Language, 8*(3), 324–349.

Cincotta, M., Borgheresi, A., Balestrieri, F., Giovannelli, F., Ragazzoni, A., Vanni, P., Benvenuti, F., Zaccara, G., & Ziemann, U. (2006). Mechanisms underlying mirror movements in Parkinson's disease: A transcranial magnetic stimulation study. *Movement Disorders, 21*(7), 1019–1039.

Cisek, P. (2007). Cortical mechanisms of action selection: The affordance competition hypothesis. *Philosophical Transactions of the Royal Society B, 362*(1485), 1585–1599.

Cisek, P., Crammond, D. J., & Kalaska, J. F. (2003). Neural activity in primary motor and dorsal premotor cortex in reaching tasks with the contralateral versus ipsilateral arm. *Journal of Neurophysiology, 89*(2), 922–942.

Cisek, P., & Kalaska, J. F. (2010). Neural mechanisms for interacting with a world full of action choices. *Annual Review of Neuroscience, 33*, 269–298.

Clack, J. A. (2009). The fish-tetrapod transition: New fossils and interpretations. *Evolution: Education and Outreach, 2*(Suppl 2), 213–223.

Clack, J. A., Ahlberg, P. E., Finney, S. M., Dominguez Alonso, P., Robinson, J., & Ketcham, R. A. (2003). A uniquely specialized ear in a very early tetrapod. *Nature, 425*(6953), 65–69.

Clark, G. (1977). *World prehistory in new perspective* (3rd ed). Cambridge University Press.

Clarke, P., Sheffield, D., & Akehurst, S. (2020). Personality predictors of yips and choking susceptibility. *Frontiers in Psychology, 10*, 2784.

Clarke, R. J. (1999). Discovery of complete arm and hand of the 3.3 million-year-old *Australopithecus* skeleton from Sterkfontein. *South African Journal of Science, 95*, 477–480.

Clifton, R. K., Muir, D. W., Ashmead, D. H., & Clarkson, M. G. (1993). Is visually guided reaching in early infancy a myth? *Child Development, 64*(4), 1099–1110.

Clifton, R. K., Rochat, P., Robin, D. J., & Berthier, N. E. (1994). Multimodal perception in the control of infant reaching. *Journal of Experimental Psychology: Human Perception and Performance, 20*(4), 876–886.

Clower, D. M., Dum, R. P., & Strick, P. L. (2005). Basal ganglia and cerebellar inputs to "AIP." *Cerebral Cortex, 15*(7), 913–920.

Coates, M. I., & Clack, J. A. (1990). Polydactyly in the earliest known tetrapod limbs. *Nature, 347*, 66–69.

Cocchini, G., Galligan, T., Mora, L., & Kuhn, G. (2018). The magic hand: Plasticity of mental hand representation. *Quarterly Journal of Experimental Psychology, 71*(11), 2314–2324.

Cody, F. W. J., Garside, R. A. D., Lloyd, D., & Poliakoff, E. (2008). Tactile spatial acuity varies with site and axis in the human upper limb. *Neuroscience Letters, 433*(2), 103–108.

Coelho, L. A., Schacher, J. P., Scammel, C., Doan, J. B., & Gonzalez, C. L. R. (2019). Long- but not short-term tool-use changes hand representation. *Experimental Brain Research, 237*(1), 137–146.

Cohen, D. A. D., Prud'Homme, M. J. L., & Kalaska, J. F. (1994). Tactile activity in primate primary somatosensory cortex during active arm movements: Correlation with receptive field properties. *Journal of Neurophysiology, 71*(1), 161–172.

Cohen, H. J., Taft, L. T., Mahadeviah, M. S., & Birch, H. G. (1967). Developmental changes in overflow in normal and aberrantly functioning children. *Journal of Pediatrics, 71*(1), 39–47.

Cohen, L. G., Celnik, P., Pascual-Leone, A., Corwell, B., Falz, L., Dambrosia, J., Honda, M., Sadato, N., Cerloff, C., Catalá, M. D., & Hallett, M. (1997). Functional relevance of cross-modal plasticity in blind humans. *Nature, 389*(6647), 180–183.

Cole, J. (1995). *Pride and a daily marathon*. MIT Press.

Cole, J. (2016). *Losing touch: A man without his body*. Oxford University Press.

Cole, J., Gallagher, S., & McNeill, D. (2002). Gesture following deafferentation: A phenomenologically informed experimental study. *Phenomenology and the Cognitive Sciences, 1*, 49–67.

Cole, K. J., & Abbs, J. H. (1986). Coordination of three-joint digit movements for rapid finger-thumb grasp. *Journal of Neurophysiology, 55*(6), 1407–1423.

Cole, M., Schutta, H. S., & Warrington, E. K. (1962). Visual disorientation in homonymous half-fields. *Neurology, 12*(4), 257–263.

Colonnesi, C., Stams, G. J. J. M., Koster, I., & Noom, M. J. (2010). The relation between pointing and language development: A meta-analysis. *Developmental Review, 30*(4), 352–366.

Condillac, E. (1756). *An essay on the origin of human knowledge* [T. Nugent, Trans.]. John Nourse.

Connolly, J. C. (1950). *External morphology of the primate brain*. Charles C. Thomas.

Connolly, J. D., Andersen, R. A., & Goodale, M. A. (2003). FMRI evidence for a parietal reach region in the human brain. *Experimental Brain Research*, *153*(2), 140–145.

Connolly, K. J., & Bishop, D. V. M. (1992). The measurement of handedness: A cross-cultural comparison of samples from England and Papua New Guinea. *Neuropsychologia*, *30*(1), 13–26.

Conte, A., Khan, N., Defazio, G., Rothwell, J. C., & Berardelli, A. (2013). Pathophysiology of somatosensory abnormalities in Parkinson disease. *Nature Reviews Neurology*, *9*(12), 687–697.

Conti, F., Fabri, M., & Manzoni, T. (1986). Bilateral receptive fields and callosal connectivity of the body midline representation in the first somatosensory area of primates. *Somatosensory Research*, *3*(4), 273–289.

Cook, S. W., Duffy, R. G., & Fenn, K. M. (2013). Consolidation and transfer of learning after observing hand gesture. *Child Development*, *84*(6), 1863–1871.

Cook, S. W., Mitchell, Z., & Goldin-Meadow, S. (2008). Gesturing makes learning last. *Cognition*, *106*(2), 1047–1058.

Cooper, H. M., Herbin, M., & Nevo, E. (1993). Ocular regression conceals adaptive progression of the visual system in a blind subterranean mammal. *Nature*, *361*, 156–159.

Corballis, M. C. (2002). *From hand to mouth: The origins of language.* Princeton University Press.

Coren, S., & Porac, C. (1977). Fifty centuries of right-handedness: The historical record. *Science*, *198*(4317), 631–632.

Corkin, S., Milner, B., & Rasmussen, T. (1970). Somatosensory thresholds: Contrasting effects of postcentral-gyrus and posterior parietal-lobe excisions. *Archives of Neurology*, *23*(1), 41–58.

Coslett, H. B., & Lie, E. (2004). Bare hands and attention: Evidence for a tactile representation of the human body. *Neuropsychologia*, *42*(14), 1865–1876.

Cosman, J. D., & Vecera, S. P. (2010). Attention affects visual perceptual processing near the hand. *Psychological Science*, *21*(9), 1254–1258.

Costantini, M., Robinson, J., Migliorati, D., Donno, B., Ferri, F., & Northoff, G. (2016). Temporal limits on rubber hand illusion reflect individuals' temporal resolution in multisensory perception. *Cognition*, *157*, 39–48.

Costantini, M., Urgesi, C., Galati, G., Romani, G. L., & Aglioti, S. M. (2011). Haptic perception and body representation in lateral and medial occipito-temporal cortices. *Neuropsychologia*, *49*(5), 821–829.

Costello, M. B., & Fragaszy, D. M. (1988). Prehension in *Cebus* and *Saimiri*: 1. Grip type and hand preference. *American Journal of Primatology*, *15*(3), 235–245.

Cowey, A., Small, M., & Ellis, S. (1994). Left visuo-spatial neglect can be worse in far than in near space. *Neuropsychologia*, *32*(9), 1059–1066.

Cowey, A., Small, M., & Ellis, S. (1999). No abrupt change in visual hemineglect from near to far space. *Neuropsychologia*, *37*(1), 1–6.

Craddock, M., & Lawson, R. (2009). Do left and right matter for haptic recognition of familiar objects? *Perception, 38*(9), 1355–1376.

Craig, J. C., & Johnson, K. O. (2000). The two-point threshold: Not a measure of tactile spatial resolution. *Current Directions in Psychological Science, 9*(1), 29–32.

Cressman, E. K., Franks, I. M., & Enns, J. T. (2006). No automatic pilot for visually guided aiming based on colour. *Experimental Brain Research, 171*(2), 174–183.

Crevecoeur, F., Munoz, D. P., & Scott, S. H. (2016). Dynamic multisensory integration: Somatosensory speed trumps visual accuracy during feedback control. *Journal of Neuroscience, 36*(33), 8598–8611.

Criscimagna-Hemminger, S. E., Donchin, O., Gazzaniga, M. S., & Shadmehr, R. (2003). Learned dynamics of reaching movements generalize from dominant to nondominant arm. *Journal of Neurophysiology, 89*(1), 168–176.

Critchley, M. (1928). *Mirror-writing*. Kegan Paul.

Critchley, M. (1953). *The parietal lobes*. Edward Arnold.

Critchley, M. (1955). Personification of paralysed limbs in hemiplegics. *British Medical Journal, 2*(4934), 284–286.

Crump, M. J. C., & Logan, G. D. (2010). Warning: This keyboard will deconstruct—The role of the keyboard in skilled typewriting. *Psychonomic Bulletin and Review, 17*(3), 394–399.

Cubelli, R. (1991). A selective deficit for writing vowels in acquired dysgraphia. *Nature, 353*, 258–260.

Cui, H., & Andersen, R. A. (2007). Posterior parietal cortex encodes autonomously selected motor plans. *Neuron, 56*(3), 552–559.

Cunningham, D. A., Machado, A., Yue, G. H., Carey, J. R., & Plow, E. B. (2013). Functional somatotopy revealed across multiple cortical regions using a model of complex motor task. *Brain Research, 1531*, 25–36.

Cushing, H. (1909). A note upon the faradic stimulation of the postcentral gyrus in conscious patients. *Brain, 32*(1), 44–53.

Dall'Orso, S., Arichi, T., Fitzgibbon, S. P., Edwards, A. D., Burdet, E., & Muceli, S. (2022). Development of functional organization within the sensorimotor network across the perinatal period. *Human Brain Mapping, 43*(7), 2249–2261.

Dall'Orso, S., Steinweg, J., Allievi, A. G., Edwards, A. D., Burdet, E., & Arichi, T. (2018). Somatotopic mapping of the developing sensorimotor cortex in the preterm human brain. *Cerebral Cortex, 28*(7), 2507–2515.

Damasio, A. R., & Benton, A. L. (1979). Impairment of hand movements under visual guidance. *Neurology, 29*(2), 170–178.

D'Amour, S., & Harris, L. R. (2014). Contralateral tactile masking between forearms. *Experimental Brain Research, 232*(3), 821–826.

Danforth, C. H. (1947). Heredity of polydactyly in the cat. *Journal of Heredity, 38*(4), 107–112.

Darian-Smith, I., Sugitani, M., Heywood, J., Karita, K., & Goodwin, A. (1982). Touching textured surfaces: Cells in somatosensory cortex respond both to finger movement and to surface features. *Science, 218*(4575), 906–909.

Dart, R. A. (1925). *Australopithecus africanus*: The man-ape of South Africa. *Nature, 115*, 195–199.

Darwin, C. (1859). *On the origin of species*. John Murray.

Darwin, C. (1868). *The variation of animals and plants under domestication*. John Murray.

Darwin, C. (1871). *The descent of man and selection in relation to sex*. John Murray.

Darwin, C. (1872). *The expression of the emotions in man and animals*. John Murray.

Daum, M. M., Ulber, J., & Gredebäck, G. (2013). The development of pointing perception in infancy: Effects of communicative signals on covert shifts of attention. *Developmental Psychology, 49*(10), 1898–1908.

Davare, M., Andres, M., Cosnard, G., Thonnard, J.-L., & Olivier, E. (2006). Dissociating the role of ventral and dorsal premotor cortex in precision grasping. *Journal of Neuroscience, 26*(8), 2260–2268.

Davare, M., Lemon, R., & Olivier, E. (2008). Selective modulation of interactions between ventral premotor cortex and primary motor cortex during precision grasping in humans. *Journal of Physiology, 586*(11), 2735–2742.

Davare, M., Rothwell, J. C., & Lemon, R. N. (2010). Causal connectivity between the human anterior intraparietal area and premotor cortex during grasp. *Current Biology, 20*(2), 176–181.

D'Avella, A., Portone, A., Fernandez, L., & Lacquaniti, F. (2006). Control of fast-reaching movements by muscle synergy combinations. *Journal of Neuroscience, 26*(30), 7791–7810.

Davis, J. E. (2010). *Hand talk: Sign language among American Indian nations*. Cambridge University Press.

Davis, W. W. (1898). Researches in cross-education. *Studies from the Yale Psychological Laboratory, 6*, 6–50.

Davoli, C. C., & Brockmole, J. R. (2012). The hands shield attention from visual interference. *Attention, Perception, & Psychophysics, 74*(7), 1386–1390.

Davoli, C. C., Brockmole, J. R., Du, F., & Abrams, R. A. (2014). Switching between global and local scopes of attention is resisted near the hands. *Visual Cognition, 20*(6), 659–668.

Davoli, C. C., Du, F., Montana, J., Garverick, S., & Abrams, R. A. (2010). When meaning matters, look but don't touch: The effects of posture on reading. *Memory & Cognition, 38*(5), 555–562.

Dawkins, R. (1986). *The blind watchmaker*. Norton.

Day, B. L., & Lyon, I. N. (2000). Voluntary modification of automatic arm movements evoked by motion of a visual target. *Experimental Brain Research, 130*(2), 159–168.

Day, L. (1929). *Paganini of Genoa*. Macaulay.

de Beaune, S. A. (2004). The invention of technology: Prehistory and cognition. *Current Anthropology, 45*(2), 139–162.

de Gelder, B. (2009). Why bodies? Twelve reasons for including bodily expressions in affective neuroscience. *Philosophical Transactions of the Royal Society B, 364*(1535), 3475–3484.

de Graaf, J. B., Jarrassé, N., Nicol, C., Touillet, A., Coyle, T., Maynard, L., Martinet, N., & Paysant, J. (2016). Phantom hand and wrist movements in upper limb amputees are slow but naturally controlled movements. *Neuroscience, 312*, 48–57.

de Haan, E. H. F., Fabri, M., Dijkerman, H. C., Foschi, N., Lattanzi, S., & Pinto, Y. (2020). Unified tactile detection and localisation in split-brain patients. *Cortex, 124*, 217–223.

de la Torre, I. (2004). Omo revisited: Evaluating the technological skill of Pliocene hominids. *Current Anthropology, 45*(4), 439–465.

de la Torre, I. (2016). The origins of the Acheulean: Past and present perspectives on a major transition in human evolution. *Philosophical Transactions of the Royal Society B, 371*(1698), 20150245.

De Renzi, E., Faglioni, P., & Sorgato, P. (1982). Modality-specific and supramodal mechanisms of apraxia. *Brain, 105*(Pt 2), 301–312.

De Renzi, E., Scotti, G., & Spinnler, H. (1969). Perceptual and associative disorders of visual recognition: Relationship to the side of the cerebral lesion. *Neurology, 19*(7), 634–642.

de Sousa, A. A., Sherwood, C. C., Mohlberg, H., Amunts, K., Schleicher, A., MacLeod, C. E., Hof, P. R., Frahm, H., & Zilles, K. (2010). Hominoid visual brain structure volumes and the position of the lunate sulcus. *Journal of Human Evolution, 58*(4), 281–292.

de Vignemont, F., Ehrsson, H. H., & Haggard, P. (2005). Bodily illusions modulate tactile perception. *Current Biology, 15*(14), 1286–1290.

de Vignemont, F., Majid, A., Jola, C., & Haggard, P. (2008). Segmenting the body into parts: Evidence from biases in tactile perception. *Quarterly Journal of Experimental Psychology, 62*(3), 500–512.

de Vries, J. I. P., Visser, G. H. A., & Prechtl, H. F. R. (1982). The emergence of fetal behaviour. I. Qualitative aspects. *Early Human Development, 7*(4), 301–322.

de Vries, J. I. P., Visser, G. H. A., & Prechtl, H. F. R. (1985). The emergence of fetal behaviour. II. Quantitative aspects. *Early Human Development, 12*(2), 99–120.

de Winter, W., & Oxnard, C. E. (2001). Evolutionary radiations and convergences in the structural organization of mammalian brains. *Nature, 409*, 710–714.

Dea, M., Hamadjida, A., Elgbeili, G., Quessy, S., & Dancause, N. (2016). Different patterns of cortical inputs to subregions of the primary motor cortex hand representation in *Cebus apella*. *Cerebral Cortex, 26*(4), 1747–1761.

DeCasien, A. R., Barton, R. A., & Higham, J. P. (2022). Understanding the human brain: Insights from comparative biology. *Trends in Cognitive Sciences, 26*(5), 432–445.

Decety, J., Jeannerod, M., & Preblanc, C. (1989). The timing of mentally represented actions. *Behavioural Brain Research, 34*(1–2), 35–42.

Dehno, N. S., Kamali, F., Shariat, A., & Jaberzadeh, S. (2021). Unilateral strength training of the less affected hand improves cortical excitability and clinical outcomes in patients with subacute stroke: A randomized controlled trial. *Archives of Physical Medicine and Rehabilitation, 102*(5), 914–924.

Delagnes, A., & Roche, H. (2005). Late Pliocene hominid knapping skills: The case of Lokalalei 2C, West Turkana, Kenya. *Journal of Human Evolution, 48*(5), 435–472.

Delis, D., Foldi, N. S., Hamby, S., Gardner, H., & Zurif, E. (1979). A note on temporal relations between language and gestures. *Brain and Language, 8*(3), 350–354.

Della Sala, S. (2005). The anarchic hand. *The Psychologist, 18*, 606–609.

Della Sala, S., Marchetti, C., & Spinnler, H. (1991). Right-sided anarchic (alien) hand: A longitudinal study. *Neuropsychologia, 29*(11), 1113–1127.

Della Sala, S., Marchetti, C., & Spinnler, H. (1994). The anarchic hand: A fronto-mesial sign. In F. Boller & J. Grafman (Eds.), *Handbook of Neuropsychology* (pp. 233–255). Elsevier.

Dellatolos, G., Annesi, I., Jallon, P., Chavance, M., & Lellouch, J. (1988). Mesure de la préférence manuelle par autoquestionnaire dans la population française adulte [Measurement of manual preference by self-questionnaire in the adult French population]. *Revue de Psychologie Appliquée, 38*, 117–136.

Dennis, M. (2010). Margaret Kennard (1899–1975): Not a "principle" of brain plasticity but a founding mother of developmental neuropsychology. *Cortex, 46*(8), 1043–1059.

Dennis, W. (1958). Early graphic evidence of dextrality in man. *Perceptual and Motor Skills, 8*(2), 147–149.

Denny-Brown, D. (1958). The nature of apraxia. *Journal of Nervous and Mental Disease, 126*(1), 9–32.

Denny-Brown, D., Meyer, J. S., & Horenstein, S. (1952). The significance of perceptual rivalry resulting from parietal lesion. *Brain, 75*(4), 433–471.

DeSilva, J. M. (2021). *First steps: How walking upright made us human*. William Collins.

Desmond, A. J. (1982). *Archetypes and ancestors: Paleontology in Victorian London, 1850–1875*. University of Chicago Press.

Desmurget, M., Epstein, C. M., Turner, R. S., Prablanc, C., Alexander, G. E., & Grafton, S.T. (1999). Role of the posterior parietal cortex in updating reaching movements to a visual target. *Nature Neuroscience, 2*(6), 563–567.

Desmurget, M., Reilly, K. T., Richard, N., Szathmari, A., Mottolese, C., & Sirigu, A. (2009). Movement intention after parietal cortex stimulation in humans. *Science, 324*(5928), 811–813.

Desmurget, M., & Sirigu, A. (2015). Revealing humans' sensorimotor functions with electrical cortical stimulation. *Philosophical Transactions of the Royal Society B, 370*(1677), 20140207.

Di Noto, P. M., Newman, L., Wall, S., & Einstein, G. (2013). The hermunculus: What is known about the representation of the female body in the brain? *Cerebral Cortex, 23*(5), 1005–1013.

di Pellegrino, G., & Frassinetti, F. (2000). Direct evidence from parietal extinction of enhancement of visual attention near a visible hand. *Current Biology, 10*(22), 1475–1477.

di Pellegrino, G., Làdavas, E., & Farnè, A. (1997). Seeing where your hands are. *Nature, 388*, 730.

DiCarlo, J. J., Johnson, K. O., & Hsiao, S. S. (1998). Structure of receptive fields in area 3b of primary somatosensory cortex in the alert monkey. *Journal of Neuroscience, 18*(7), 2626–2645.

Dick, F. K., Lehet, M. I., Callaghan, M. F., Keller, T. A., Sereno, M. I., & Holt, L. L. (2017). Extensive tonotopic mapping across auditory cortex is recapitulated by spectrally directed attention and systematically related to cortical myeloarchitecture. *Journal of Neuroscience, 37*(50), 12187–12201.

Dickens, C. (1863). Give me your hand. *All the Year Round, 10*, 345–349.

Diedrichsen, J., & Kornysheva, K. (2015). Motor skill learning between selection and execution. *Trends in Cognitive Sciences, 19*(4), 227–233.

Diedrichsen, J., Shadmehr, R., & Ivry, R. B. (2009). The coordination of movement: Optimal feedback control and beyond. *Trends in Cognitive Sciences, 14*(1), 31–39.

Diedrichsen, J., Wiestler, T., & Krakauer, J. W. (2013). Two distinct ipsilateral cortical representations for individuated finger movements. *Cerebral Cortex, 23*(6), 1362–1377.

Diers, M., Zieglgänsberger, W., Trojan, J., Drevensek, A. M., Erhardt-Raum, G., & Flor, H. (2013). Site-specific visual feedback reduces pain perception. *Pain, 154*(6), 890–896.

Diez-Martín, F., Sánchez, P., Domínguez-Rodrigo, M., Mabulla, A., & Barba, R. (2009). Were Olduvai hominins making butchering tools or battering tools? Analysis of a recently excavated lithic assemblage from BK (Bed II, Olduvai Gorge, Tanzania). *Journal of Anthropological Archaeology, 28*(3), 274–289.

Diez-Martín, F., Sánchez Yustos, P., Uribelarrea, D., Baquedano, E., Mark, D. F., Mabulla, A., Fraile, C., Duque, J., Díaz, I., Pérez-González, A., Yravedra, J., Egeland, C. P., Organista, E., & Domínguez-Rodrigo, M. (2015). The origin of the Acheulean: The 1.7 million-year-old site of FLK West, Olduvai Gorge (Tanzania). *Scientific Reports, 5*, 17839.

Dijkerman, H. C., & de Haan, E. H. F. (2007). Somatosensory processes subserving perception and action. *Behavioral and Brain Sciences, 30*(2), 189–201.

D'Imperio, D., Tomelleri, G., Moretto, G., & Moro, V. (2017). Modulation of somatoparaphrenia following left-hemisphere damage. *Neurocase, 23*(2), 162–170.

Diogo, R., Richmond, B. G., & Wood, B. (2012). Evolution and homologies of primate and modern human hand and forearm muscles, with notes on thumb movements and tool use. *Journal of Human Evolution, 63*(1), 64–78.

Disbrow, E., Roberts, T., & Krubitzer, L. (2000). Somatotopic organization of cortical fields in the lateral sulcus of *Homo sapiens*: Evidence for SII and PV. *Journal of Comparative Neurology, 418*(1), 1–21.

DiZio, P., & Lackner, J. R. (1995). Motor adaptation to Coriolis force perturbations of reaching movements: Endpoint but not trajectory adaptation transfers to the nonexposed arm. *Journal of Neurophysiology*, *74*(4), 1787–1792.

Dominey, P., Decety, J., Broussolle, E., Chazot, G., & Jeannerod, M. (1995). Motor imagery of a lateralized sequential task is asymmetrically slowed in hemi-Parkinson's patients. *Neuropsychologia*, *33*(6), 727–741.

Dominguez-Ballesteros, E., & Arrizabalaga, A. (2015). Flint knapping and determination of human handedness. Methodological proposal with quantifiable results. *Journal of Archaeological Science: Reports*, *3*, 313–320.

Domínguez-Rodrigo, M., Pickering, T. R., Almécija, S., Heaton, J. L., Baquedano, E., Mabulla, A., & Uribelarrea, D. (2015). Earliest modern human-like hand bone from a new >1.84-million-year-old site at Olduvai in Tanzania. *Nature Communications*, *6*, 7987.

Domínguez-Rodrigo, M., Pickering, T. R., Semaw, S., & Rogers, M. J. (2005). Cutmarked bones from Pliocene archaeological sites at Gona, Afar, Ethiopia: Implications for the function of the world's oldest stone tools. *Journal of Human Evolution*, *48*(2), 109–121.

Donchin, O., Gribova, A., Steinberg, O., Bergman, H., & Vaadia, E. (1998). Primary motor cortex is involved in bimanual coordination. *Nature*, *395*, 274–278.

Donchin, O., Gribova, A., Steinberg, O., Mitz, A. R., Bergman, H., & Vaadia, E. (2002). Single-unit activity related to bimanual arm movements in the primary and supplementary motor cortices. *Journal of Neurophysiology*, *88*(6), 3498–3517.

Downey, J. E., Quick, K. M., Schwed, N., Weiss, J. M., Wittenberg, G. F., Boninger, M. L., & Collinger, J. L. (2020). The motor cortex has independent representations for ipsilateral and contralateral arm movements but correlated representations for grasping. *Cerebral Cortex*, *30*(10), 5400–5409.

Downs, J. P., Daeschler, E. B., Jenkins, F. A., & Shubin, N. H. (2008). The cranial endoskeleton of *Tiktaalik roseae*. *Nature*, *455*, 925–929.

Du, F., Wang, X., Abrams, R. A., & Zhang, K. (2017). Emotional processing is enhanced in peri-hand space. *Cognition*, *165*, 39–44.

Duffy, F. H., & Burchfiel, J. L. (1971). Somatosensory system: Organizational hierarchy from single units in monkey area 5. *Science*, *172*(3980), 273–275.

Duhamel, J.-R., Colby, C. L., & Goldberg, M. E. (1998). Ventral intraparietal area of the macaque: Congruent visual and somatic response properties. *Journal of Neurophysiology*, *79*(1), 126–136.

Dum, R. P., & Strick, P. L. (2002). Motor areas in the frontal lobe of the primate. *Physiology & Behavior*, *77*(4–5), 677–682.

Dum, R. P., & Strick, P. L. (2005). Frontal lobe inputs to the digit representations of the motor areas on the lateral surface of the hemisphere. *Journal of Neuroscience*, *25*(6), 1375–1386.

References

Duncan, R. O., & Boynton, G. M. (2007). Tactile hyperacuity thresholds correlate with finger maps in primary somatosensory cortex (S1). *Cerebral Cortex, 17*(12), 2878–2891.

Duque, J., Mazzocchio, R., Dambrosia, J., Murase, N., Olivier, E., & Cohen, L.G. (2005). Kinematically specific interhemispheric inhibition operating in the process of generation of a voluntary movement. *Cerebral Cortex, 15*(5), 588–593.

Durbin, R., & Mitchison, G. (1990). A dimension reduction framework for understanding cortical maps. *Nature, 343*(6259), 644–647.

Dykes, R. W., & Gabor, A. (1981). Magnification functions and receptive field sequences for submodality-specific bands in SI cortex of cats. *Journal of Comparative Neurology, 202*(4), 597–620.

Dykes, R. W., Landry, P., Metherate, R., & Hicks, T. P. (1984). Functional role of GABA in cat primary somatosensory cortex: Shaping receptive fields of cortical neurons. *Journal of Neurophysiology, 52*(6), 1066–1093.

Easton, R. D., Greene, A. J., & Srinivas, K. (1997). Transfer between vision and haptics: Memory for 2-D patterns and 3-D objects. *Psychonomic Bulletin and Review, 4*(3), 403–410.

Ebner, F. F., & Myers, R. E. (1962). Corpus callosum and the interhemispheric transmission of tactual learning. *Journal of Neurophysiology, 25*(3), 380–391.

Eden, J., Bräcklein, M., Ibáñez, J., Barsakcioglu, D. Y., Di Pino, G., Farina, D., Burdet, E., & Mehring, C. (2022). Principles of human movement augmentation and the challenges in making it a reality. *Nature Communications, 13*(1), 1345.

Edgley, S. A., Eyret, J. A., Lemon, R. N., & Miller, S. (1990). Excitation of the corticospinal tract by electromagnetic and electrical stimulation of the scalp in the macaque monkey. *Journal of Physiology, 425*(1), 301–320.

Edwards, T., & Brentari, D. (2020). Feeling phonology: The conventionalization of phonology in Protactile communities in the United States. *Language, 96*(4), 819–840.

Edwards, T., & Brentari, D. (2021). The grammatical incorporation of demonstratives in an emerging tactile language. *Frontiers in Psychology, 11*, 579992.

Ehrenwald, H. (1930). Verändertes erleben des körperbildes mit konsekutiver wahnbildung bei linksseitiger hemiplegie [Altered experience of body image with resulting delusions in left-sided hemiplegia]. *Monatsschrift für Psychiatrie und Neurologie, 75*, 89–97.

Ehrsson, H. H., Holmes, N. P., & Passingham, R. E. (2005). Touching a rubber hand: Feeling of body ownership is associated with activity in multisensory brain areas. *Journal of Neuroscience, 25*(45), 10564–10573.

Ehrsson, H. H., Spence, C., & Passingham, R. E. (2004). That's my hand! Activity in premotor cortex reflects feeling of ownership of a limb. *Science, 305*(5685), 875–877.

Eidelberg, E., & Schwartz, A. S. (1971). Experimental analysis of the extinction phenomenon in monkeys. *Brain, 94*(1), 91–108.

Ejaz, N., Xu, J., Branscheidt, M., Hertler, B., Schambra, H., Widmer, M., Faria, A. V., Harran, M. D., Cortes, J. C., Kim, N., Celnik, P. A., Kitago, T., Luft, A. R., Krakauer, J. W., & Diedrichsen, J. (2018). Evidence for a subcortical origin of mirror movements after stroke: A longitudinal study. *Brain, 141*(3), 837–847.

Ekman, P., & Friesen, W. V. (1969). The repertoire of nonverbal behavior: Categories, origins, usage, and coding. *Semiotica, 1*(1), 49–98.

Ekman, P., & Friesen, W. V. (1972). Hand movements. *Journal of Communication, 22*(4), 353–374.

Elbert, T., Candia, V., Altenmüller, E., Rau, H., Sterr, A., Rockstroh, B., Pantev, C., & Taub, E. (1998). Alteration of digital representations in somatosensory cortex in focal hand dystonia. *NeuroReport, 9*(16), 3571–3575.

Elbert, T., Pantev, C., Wienbruch, C., Rockstroh, B., & Taub, E. (1995). Increased cortical representation of the fingers of the left hand in string players. *Science, 270*(5234), 305–307.

Eliasson, A.-C., Forssberg, H., Ikuta, K., Apel, I., Westling, G., & Johansson, R. S. (1995). Development of human precision grip. V. Anticipatory and triggered grip actions during sudden loading. *Experimental Brain Research, 106*(3), 425–433.

Ellis, A. W. (1979). Slips of the pen. *Visible Language, 13*(3), 265–282.

Embick, D., Marantz, A., Miyashita, Y., O'Neil, W., & Sakai, K. L. (2000). A syntactic specialization for Broca's area. *Proceedings of the National Academy of Sciences, 97*(11), 6150–6154.

Epstein, R. A., Patai, E. Z., Julian, J. B., & Spiers, H. J. (2017). The cognitive map in humans: Spatial navigation and beyond. *Nature Neuroscience, 20*(11), 1504–1513.

Ernst, M. O., & Banks, M. S. (2002). Humans integrate visual and haptic information in a statistically optimal fashion. *Nature, 415*, 429–433.

Eshel, N., Ruff, C. C., Spitzer, B., Blankenburg, F., & Driver, J. (2010). Effects of parietal TMS on somatosensory judgments challenge interhemispheric rivalry accounts. *Neuropsychologia, 48*(12), 3470–3481.

Eskes, G. A., Butler, B., McDonald, A., Harrison, E. R., & Phillips, S. J. (2003). Limb activation effects in hemispatial neglect. *Archives of Physical Medicine and Rehabilitation, 84*(3), 323–328.

Ettlinger, G., & Kalsbeck, J. E. (1962). Changes in tactile discrimination and in visual reaching after successive and simultaneous bilateral posterior parietal ablations in the monkey. *Journal Neurology, Neurosurgury, and Psychiatry, 25*(3), 256–268.

Etxepare, R., & Irurtzun, A. (2021). Gravettian hand stencils as sign language formatives. *Philosophical Transactions of the Royal Society B, 376*(1824), 20200205.

Evans, C., Milner, A. D., Humphreys, G. W., & Cavina-Pratesi, C. (2012). Optic ataxia affects the lower limbs: Evidence from a single case study. *Cortex, 49*(5), 1229–1240.

Evarts, E. V. (1966). Pyramidal tract activity associated with a conditioned hand movement in the monkey. *Journal of Neurophysiology, 29*(6), 1011–1027.

Evarts, E. V. (1968). Relation of pyramidal tract activity to force exerted during voluntary movement. *Journal of Neurophysiology, 31*(1), 14–27.

Evarts, E. V. (1973). Motor cortex reflexes associated with learned movement. *Science, 179*(4072), 501–503.

Evarts, E. V., & Tanji, J. (1976). Reflex and intended responses in motor cortex pyramidal tract neurons of monkey. *Journal of Neurophysiology, 39*(5), 1069–1080.

Eyre, J. A., Miller, S., Clowry, G. J., Conway, E. A., & Watts, C. (2000). Functional corticospinal projections are established prenatally in the human foetus permitting involvement in the development of spinal motor centres. *Brain, 123*(Pt 1), 51–64.

Eyre, J. A., Miller, S., & Ramesh, V. (1991). Constancy of central conduction delays during development in man: Investigation of motor and somatosensory pathways. *Journal of Physiology, 434*(1), 441–452.

Fabbri, S., Strnad, L., Caramazza, A., & Lingnau, A. (2014). Overlapping representations for grip type and reach direction. *NeuroImage, 94*, 138–146.

Fabri, M., Del Pesce, M., Paggi, A., Polonara, G., Bartolini, M., Salvolini, U., & Manzoni, T. (2005). Contribution of posterior corpus callosum to the interhemispheric transfer of tactile information. *Cognitive Brain Research, 24*(1), 73–80.

Fagard, J. (2000). Linked proximal and distal changes in the reaching behavior of 5- to 12-month-old human infants grasping objects of different sizes. *Infant Behavior and Development, 23*(3–4), 317–329.

Fagard, J., & Jacquet, A.-Y. (1989). Onset of bimanual coordination and symmetry versus asymmetry of movement. *Infant Behavior and Development, 12*(2), 229–235.

Fagot, J., Lacreuse, A., & Vauclair, J. (1993). Haptic discrimination of nonsense shapes: Hand exploratory strategies but not accuracy reveal laterality effects. *Brain and Cognition, 21*(2), 212–225.

Fagot, J., & Vauclair, J. (1991). Manual laterality in nonhuman primates: A distinction between handedness and manual specialization. *Psychological Bulletin, 109*(1), 76–89.

Falk, D. (1980). A reanalysis of the South African australopithecine natural endocasts. *American Journal of Physical Anthropology, 53*(4), 525–539.

Falk, D. (1983). Cerebral cortices of East African early hominids. *Science, 221*(4615), 1072–1074.

Falk, D. (1985). Hadar AL 162-28 endocast as evidence that brain enlargement preceded cortical reorganization in hominid evolution. *Nature, 313*, 45–47.

Falk, D. (2011). *The fossil chronicles*. University of California Press.

Falk, D. (2014). Interpreting sulci on hominin endocasts: Old hypotheses and new findings. *Frontiers in Human Neuroscience, 8*, 134.

Fan, C., Coppi, S., & Ehrsson, H. H. (2021). The supernumerary rubber hand illusion revisited: Perceived duplication of limbs and visuotactile events. *Journal of Experimental Psychology: Human Perception and Performance, 47*(6), 810–829.

Fang, F., & He, S. (2005). Cortical responses to invisible objects in the human dorsal and ventral pathways. *Nature Neuroscience, 8*(10), 1380–1385.

Fang, W., Li, J., Qi, G., Li, S., Sigman, M., & Wang, L. (2019). Statistical inference of body representation in the macaque brain. *Proceedings of the National Academy of Sciences, 116*(40), 20151–20157.

Farnè, A., Bonifazi, S., & Làdavas, E. (2005). The role played by tool-use and tool-length on the plastic elongation of peri-hand space: A single case study. *Cognitive Neuropsychology, 22*(3), 408–418.

Farnè, A., Dematté, M. L., & Làdavas, E. (2005). Neuropsychological evidence of modular organization of the near peripersonal space. *Neurology, 65*(11), 1754–1758.

Farnè, A., Iriki, A., & Làdavas, E. (2005). Shaping multisensory action-space with tools: Evidence from patients with cross-modal extinction. *Neuropsychologia, 43*(2), 238–248.

Farnè, A., & Làdavas, E. (2000). Dynamic size-change of hand peripersonal space following tool use. *NeuroReport, 11*(8), 1645–1649.

Farnè, A., Pavani, F., Meneghello, F., & Làdavas, E. (2000). Left tactile extinction following visual stimulation of a rubber hand. *Brain, 123*(Pt 11), 2350–2360.

Farnè, A., Serino, A., & Làdavas, E. (2007). Dynamic size-change of peri-hand space following tool-use: Determinants and spatial characteristics revealed through cross-modal extinction. *Cortex, 43*(3), 436–443.

Farrell, D. F., Burbank, N., Lettich, E., & Ojemann, G. A. (2007). Individual variation in human motor-sensory (Rolandic) cortex. *Journal of Clinical Neurophysiology, 24*(3), 286–293.

Farthing, J. P., Krentz, J. R., & Magnus, C. R. A. (2009). Strength training the free limb attenuates strength loss during unilateral immobilization. *Journal of Applied Physiology, 106*(3), 830–836.

Farthing, J. P., Krentz, J. R., Magnus, C. R. A., Barss, T.S., Lanovaz, J. L., Cummine, J., Esopenko, C., Sarty, G. E., & Borowsky, R. (2011). Changes in functional magnetic resonance imaging cortical activation with cross education to an immobilized limb. *Medicine and Science in Sports and Exercise, 43*(8), 1394–1405.

Farthing, J. P., & Zehr, E. P. (2014). Restoring symmetry: Clinical applications of cross-education. *Exercise and Sport Science Reviews, 42*(2), 70–75.

Fasano, A., Mazzoni, A., & Falotico, E. (2022). Reaching and grasping movements in Parkinson's disease: A review. *Journal of Parkinson's Disease, 12*(4), 1083–1113.

Fattori, P., Raos, V., Breveglieri, R., Bosco, A., Marzocchi, N., & Umana, F. (2010). The dorsomedial pathway is not just for reaching: Grasping neurons in the medial parieto-occipital cortex of the macaque monkey. *Journal of Neuroscience, 30*(1), 342–349.

Faugier-Grimaud, S., Frenois, C., & Peronnet, F. (1985). Effects of posterior parietal lesions on visually guided movements in monkeys. *Experimental Brain Research, 59*(1), 125–138.

Faurie, C., & Raymond, M. (2004). Handedness frequency over more than ten thousand years. *Proceedings of the Royal Society B*, *271*(Suppl.), S43–S45.

Faurie, C., Schiefenhövel, W., Le Bomin, S., Billiard, S., & Raymond, M. (2005). Variation in the frequency of left-handedness in traditional societies. *Current Anthropology*, *46*(1), 142–147.

Favorov, O. V, Diamond, M. E., & Whitsel, B. L. (1987). Evidence for a mosaic representation of the body surface in area 3b of the somatic cortex of cat. *Proceedings of the National Academy of Sciences*, *84*(18), 6606–6610.

Fechner, G. (1858). Beobachtingen, welche zu beweisen scheinen dass durch die unbung der glieder der einen seite die der anderen zugleich mit geübt werden. [Observations which seem to prove that by exercising the limbs on one side, those of the other are exercised at the same time]. *Berichte Über Der Königlich Sächsischen Gesellschaften Der Wissenschaften Zu Leipzig*, *10*, 70–76.

Feinberg, T. E., Gonzalez Rothi, L. J., & Heilman, K. M. (1986). Multimodal agnosia after unilateral left hemisphere lesion. *Neurology*, *36*(6), 864–867.

Feldman, J. F., & Brody, N. (1978). Non-elicited newborn behaviors in relation to state and prandial condition. *Merrill-Palmer Quarterly*, *24*(2), 79–84.

Felleman, D. J., Nelson, R. J., Sur, M., & Kaas, J. H. (1983). Representations of the body surface in areas 3b and 1 of postcentral parietal cortex of *Cebus* monkeys. *Brain Research*, *268*(1), 15–26.

Ferbert, A., Priori, A., Rothwell, J. C., Day, B. L., Colebatch, J. G., & Marsden, C. D. (1992). Interhemispheric inhibition of the human motor cortex. *Journal of Physiology*, *453*(1), 525–546.

Ferrier, D. (1873). Experimental researches in cerebral physiology and pathology. *Journal of Anatomy and Physiology*, *8*, 152–155.

Ferrier, D. (1876). *The functions of the brain*. Smith, Elder & Company.

Ferrier, D. (1882). Case of allochiria. *Brain*, *5*(3), 389–393.

Ferrier, D. (1890). The Croonian Lectures on cerebral localisation: Lecture III. *British Medical Journal*, *1*(1538), 1413–1418.

Ferrier, D., & Yeo, G. F. (1884). A record of experiments on the effects of lesion of different regions of the cerebral hemispheres. *Philosophical Transactions of the Royal Society*, *175*, 479–564.

Fessard, A., & Sand, A. (1937). Stretch receptors in the muscles of fishes. *Journal of Experimental Biology*, *14*(4), 383–404.

Filimon, F., Nelson, J. D., Huang, R.-S., & Sereno, M. I. (2009). Multiple parietal reach regions in humans: Cortical representations for visual and proprioceptive feedback during on-line reaching. *Journal of Neuroscience*, *29*(9), 2961–2971.

Finch, G. (1941). Chimpanzee handedness. *Science*, *94*, 117–118.

Finger, S., & Hustwit, M. P. (2003). Five early accounts of phantom limb in context: Paré, Descartes, Lemos, Bell, and Mitchell. *Neurosurgery*, *52*(3), 675–686.

Fink, G. R., Frackowiak, R. S. J., Pietrzyk, U., & Passingham, R. E. (1997). Multiple nonprimary motor areas in the human cortex. *Journal of Neurophysiology, 77*(4), 2164–2174.

Finlay, B. L., & Darlington, R. B. (1995). Linked regularities in the development and evolution of mammalian brains. *Science, 268*(5217), 1578–1584.

Finlay, B. L., Darlington, R. B., & Nicastro, N. (2001). Developmental structure in brain evolution. *Behavioral and Brain Sciences, 24*(2), 263–308.

Fiori, F., & Longo, M. R. (2018). Tactile distance illusions reflect a coherent stretch of tactile space. *Proceedings of the National Academy of Sciences, 115*(6), 1238–1243.

Fiorio, M., & Haggard, P. (2005). Viewing the body prepares the brain for touch: Effects of TMS over somatosensory cortex. *European Journal of Neuroscience, 22*(3), 773–777.

Flament, D., Hall, E. J., & Lemon, R. N. (1992). The development of cortico-motoneuronal projections investigated using magnetic brain stimulation in the infant macaque. *Journal of Physiology, 447*(1), 755–768.

Flash, T., & Hogan, N. (1985). The coordination of arm movements: An experimentally confirmed mathematical model. *Journal of Neuroscience, 5*(7), 1688–1703.

Flash, T., Inzelberg, R., Schechtman, E., & Korczyn, A. D. (1992). Kinematic analysis of upper limb trajectories in Parkinson's disease. *Experimental Neurology, 118*(2), 215–226.

Flechsig, P. (1920). *Anatomie des menschlichen gehirns und rückenmarks [Human brain and spinal cord anatomy]*. Verlag von Georg Thieme.

Fleisher, L., & Midgette, A. (2010). *My nine lives: A memoir of many careers in music*. Anchor Books.

Fleming, J. F. R., & Crosby, E. C. (1955). The parietal lobe as an additional motor area: The motor effects of electrical stimulation and ablation of cortical areas 5 and 7 in monkeys. *Journal of Comparative Neurology, 103*(3), 485–512.

Flesher, S. N., Collinger, J. L., Foldes, S. T., Weiss, J. M., Downey, J. E., Tyler-Kabara, E. C., Bensmaia, S. J., Schwarz, A. B., Boninger, M. L., & Gaunt, R. A. (2016). Intracortical microstimulation of human somatosensory cortex. *Science Translational Medicine, 8*(361), 361ra141.

Flor, H., Mühlnickel, W., Karl, A., Denke, C., Grüsser, S., Kurth, R., & Taub, E. (2000). A neural substrate for nonpainful phantom limb phenomena. *Neuroreport, 11*(7), 1407–1411.

Flowers, K. A. (1976). Visual "closed-loop" and "open-loop" characteristics of voluntary movement in patients with parkinsonism and intention tremor. *Brain, 99*(2), 269–310.

Floyer-Lea, A., & Matthews, P. M. (2005). Distinguishable brain activation networks for short- and long-term motor skill learning. *Journal of Neurophysiology, 94*(1), 512–518.

Foerster, O. (1926). Die pathogenese des epileptischen krampfanfalles [The pathogenesis of epileptic seizures]. *Deutsche Zeitschrift für Nervenheilkunde, 94*, 15–53.

References

Foerster, O. (1936). The motor cortex in man in the light of Hughlings Jackson's doctrines. *Brain, 59*(2), 135–159.

Fogassi, L., Gallese, V., Buccino, G., Craighero, L., Fadiga, L., & Rizzolatti, G. (2001). Cortical mechanism for the visual guidance of hand grasping movements in the monkey: A reversible inactivation study. *Brain, 124*(Pt 3), 571–586.

Fogassi, L., Gallese, V., Fadiga, L., Luppino, G., Matelli, M., & Rizzolatti, G. (1996). Coding of peripersonal space in inferior premotor cortex (area F4). *Journal of Neurophysiology, 76*(1), 141–157.

Fogel, A., & Hannan, T. E. (1985). Manual actions of nine- to fifteen-week-old human infants during face-to-face interaction with their mothers. *Child Development, 56*(5), 1271–1279.

Fogel, A., & Thelen, E. (1987). Development of early expressive and communicative action: Reinterpreting the evidence from a dynamic systems perspective. *Developmental Psychology, 23*(6), 747–761.

Foley, R., & Lahr, M. M. (1997). Mode 3 technologies and the evolution of modern humans. *Cambridge Archaeological Journal, 7*(1), 3–36.

Forss, N., Hietanen, M., Salonen, O., & Hari, R. (1999). Modified activation of somatosensory cortical network in patients with right-hemisphere stroke. *Brain, 122*(Pt 10), 1889–1899.

Forssberg, H., Eliasson, A. C., Kinoshita, H., Johansson, R. S., & Westling, G. (1991). Development of human precision grip I: Basic coordination of force. *Experimental Brain Research, 85*(2), 451–457.

Forster, B., & Eimer, M. (2005). Vision and gaze direction modulate tactile processing in somatosensory cortex: Evidence from event-related brain potentials. *Experimental Brain Research, 165*(1), 8–18.

Fossataro, C., Bruno, V., Bosso, E., Chiotti, V., Gindri, P., Farnè, A., & Garbarini, F. (2020). The sense of body-ownership gates cross-modal improvement of tactile extinction in brain-damaged patients. *Cortex, 127*, 94–107.

Fotopoulou, A., Tsakiris, M., Haggard, P., Vagopoulou, A., Rudd, A., & Kopelman, M. (2008). The role of motor intention in motor awareness: An experimental study on anosognosia for hemiplegia. *Brain, 131*(Pt 12), 3432–3442.

Fox, M. J. (2002). *Lucky man*. Ebury Press.

Frangeul, L., Pouchelon, G., Telley, L., Lefort, S., Luscher, C., & Jabaudon, D. (2016). A cross-modal genetic framework for the development and plasticity of sensory pathways. *Nature, 538*, 96–98.

Frank, S. M., Otto, A., Volberg, G., Tse, P. U., Watanabe, T., & Greenlee, M. W. (2022). Transfer of tactile learning from trained to untrained body parts supported by cortical coactivation in primary somatosensory cortex. *Journal of Neuroscience, 42*(31), 6131–6144.

Franz, E. A., & Ramachandran, V. S. (1998). Bimanual coupling in amputees with phantom limbs. *Nature Neuroscience, 1*(6), 443–444.

Frassinetti, F., Rossi, M., & Làdavas, E. (2001). Passive limb movements improve visual neglect. *Neuropsychologia, 39*(7), 725–733.

Frayer, D. W., Clarke, R. J., Fiore, I., Blumenschine, R. J., Pérez-Pérez, A., Martinez, L. M., Estebaranz, F., Holloway, R., & Bondioli, L. (2016). OH-65: The earliest evidence for right-handedness in the fossil record. *Journal of Human Evolution, 100*, 65–72.

Fried, I., Katz, A., McCarthy, G., Sass, K. J., Williamson, P., Spencer, S. S., & Spencer, D. D. (1991). Functional organization of human supplementary motor cortex studied by electrical stimulation. *Journal of Neuroscience, 11*(11), 3656–3666.

Friedman, R. M., Chen, L. M., & Roe, A. W. (2004). Modality maps within primate somatosensory cortex. *Proceedings of the National Academy of Sciences, 101*(34), 12724–12729.

Frisby, J. P., & Davies, I. R. L. (1971). Is the haptic Müller-Lyer a visual phenomenon? *Nature, 231*, 464–465.

Frith, C. D., Blakemore, S.-J., & Wolpert, D. M. (2000). Abnormalities in the awareness and control of action. *Philosophical Transactions of the Royal Society B, 355*(1404), 1771–1788.

Fritsch, G., & Hitzig, E. (1870). Über die elektrische erregbarkeit des grosshirns [Electric excitability of the cerebrum]. *Archive Für Anatomie, Physiologie Und Wissenschaftliche Medicin, 37*, 300–332.

Fritsch, G., & Hitzig, E. (2009). Electric excitability of the cerebrum (T. Crump & S. Lama, Trans.). *Epilepsy & Behavior, 15*(2), 123–130.

Frost, S. B., Milliken, G. W., Plautz, E. J., Masterton, R. B., & Nudo, R. J. (2000). Somatosensory and motor representations in cerebral cortex of a primitive mammal (*Monodelphis domestica*): A window into the early evolution of sensorimotor cortex. *Journal of Comparative Neurology, 421*(1), 29–51.

Froude, J. A. (1884). *Thomas Carlyle: A history of his life in London, 1834–1881* (Vol. 2). Longmans, Green.

Fujiwara, Y., Matsumoto, R., Nakae, T., Usami, K., Matsuhashi, M., Kikuchi, T., Yoshida, K., Kunieda, T., Miyamoto, S., Mima, T., Ikeda, A., & Osu, R. (2017). Neural pattern similarity between contra- and ipsilateral movements in high-frequency band of human electrocorticograms. *NeuroImage, 147*, 302–313.

Fukuchi-Shimogori, T., & Grove, E. A. (2001). Neocortex patterning by the secreted signaling molecule FGF8. *Science, 294*(5544), 1071–1074.

Fulton, J. F., Jacobsen, C. F., & Kennard, M. A. (1932). A note concerning the relation of the frontal lobes to posture and forced grasping in monkeys. *Brain, 55*(4), 524–536.

Furuya, S., & Altenmüller, E. (2013). Flexibility of movement organization in piano performance. *Frontiers in Human Neuroscience, 7*, 173.

Furuya, S., Flanders, M., & Soechting, J. F. (2011). Hand kinematics of piano playing. *Journal of Neurophysiology, 106*(6), 2849–2864.

Furuya, S., & Kinoshita, H. (2008). Expertise-dependent modulation of muscular and non-muscular torques in muti-joint arm movements during piano keystroke. *Neuroscience, 156*(2), 390–402.

Furuya, S., Osu, R., & Kinoshita, H. (2009). Effective utilization of gravity during arm downswing in keystrokes by expert pianists. *Neuroscience, 164*(2), 822–831.

Furuya, S., Tanibuchi, R., Nishioka, H., Kimoto, Y., Hirano, M., & Oku, T. (2023). Passive somatosensory training enhances piano skill in adolescent and adult pianists: A preliminary study. *Annals of the New York Academy of Sciences, 1519*(1), 167–172.

Furuya, S., Uehara, K., Sakamoto, T., & Hanakawa, T. (2018). Aberrant cortical excitability reflects the loss of hand dexterity in musician's dystonia. *Journal of Physiology, 596*(12), 2397–2411.

Gainotti, G., De Bonis, C., Daniele, A., & Caltagirone, C. (1989). Contralateral and ipsilateral tactile extinction in patients with right and left focal brain damage. *International Journal of Neuroscience, 45*(1–2), 81–89.

Galdikas, B. M. F. (1982). Orang-utantool-use at Tanjung Puting Reserve, Central Indonesian Borneo (Kalimantan Tengah). *Journal of Human Evolution, 11*(1), 19–33.

Galea, M. P., & Darian-Smith, I. (1995). Postnatal maturation of the direct corticospinal projections in the macaque monkey. *Cerebral Cortex, 5*(6), 518–540.

Gallagher, S., & Cole, J. (1995). Body image and body schema in a deafferented subject. *Journal of Mind and Behavior, 16*(4), 369–389.

Gallego, J. A., Perich, M. G., Naufel, S. N., Ethier, C., Solla, S. A., & Miller, L. E. (2018). Cortical population activity within a preserved neural manifold underlies multiple motor behaviors. *Nature Communications, 9*(1), 4233.

Gallese, V., Murata, A., Kaseda, M., Niki, N., & Sakata, H. (1994). Deficit of hand preshaping after muscimol injection in monkey parietal cortex. *NeuroReport, 5*(12), 1525–1529.

Galletta, L., Stephens, N. B., Bardo, A., Kivell, T. L., & Marchi, D. (2019). Three-dimensional geometric morphometric analysis of the first metacarpal distal articular surface in humans, great apes and fossil hominins. *Journal of Human Evolution, 132*, 119–136.

Gallivan, J. P., McLean, D. A., Flanagan, J. R., & Culham, J. C. (2013). Where one hand meets the other: Limb-specific and action-dependent movement plans decoded from preparatory signals in single human frontoparietal brain areas. *Journal of Neuroscience, 33*(5), 1991–2008.

Gallivan, J. P., McLean, D. A., Valyear, K. F., & Culham, J. C. (2013). Decoding the neural mechanisms of human tool use. *eLife, 2*, e00425.

Gandevia, S. C., Burke, D., & McKeon, B. B. (1983). Convergence in the somatosensory pathway between cutaneous afferents from the index and middle fingers in man. *Experimental Brain Research, 50*(2–3), 415–425.

Ganea, N., & Longo, M. R. (2017). Projecting the self outside the body: Body representations underlying proprioceptive imagery. *Cognition, 162*, 41–47.

Ganguly, K., Secundo, L., Ranade, G., Orsborn, A., Chang, E. F., Dimitrov, D. F., Wallis, J. D., Barbaro, N. M., Knight, R. T., & Carmena, J. M. (2009). Cortical representation of ipsilateral arm movements in monkey and man. *Journal of Neuroscience, 29*(41), 12948–12956.

Ganhos, F. N., & Ariyan, S. (1985). Hippocrates, the true father of hand surgery. *Surgery, Gynecology & Obstetrics, 160*(2), 178–184.

Garbarini, F., Fornia, L., Fossataro, C., Pia, L., Gindri, P., & Berti, A. (2014). Embodiment of others' hands elicits arousal responses similar to one's own hands. *Current Biology, 24*(16), R738–R739.

Garbarini, F., Fossataro, C., Berti, A., Gindri, P., Romano, D., Pia, L., della Gatta, F., Maravita, A., & Neppi-Modona, M. (2015). When your arm becomes mine: Pathological embodiment of alien limbs using tools modulates own body representation. *Neuropsychologia, 70*, 402–413.

Garbarini, F., Pia, L., Piedimonte, A., Rabuffetti, M., Gindri, P., & Berti, A. (2013). Embodiment of an alien hand interferes with intact-hand movements. *Current Biology, 23*(2), R57–R58.

Garbarini, F., Rabuffetti, M., Piedimonte, A., Pia, L., Ferrarin, M., Frassinetti, F., Gindri, P., Cantagallo, A., Driver, J., & Berti, A. (2012). "Moving" a paralysed hand: Bimanual coupling effect in patients with anosognosia for hemiplegia. *Brain, 135*(Pt 5), 1486–1497.

Garcha, H. S., & Ettlinger, G. (1978). The effects of unilateral or bilateral removals of the second somatosensory cortex (area SII): A profound tactile disorder in monkeys. *Cortex, 14*(3), 319–326.

Gardner, R. A., & Gardner, B. T. (1969). Teaching sign language to a chimpanzee. *Science, 165*(3894), 664–672.

Garraghty, P. E., Florence, S. L., & Kaas, J. H. (1990). Ablations of areas 3a and 3b of monkey somatosensory cortex abolish cutaneous responsivity in area 1. *Brain Research, 528*(1), 165–169.

Garraux, G., Bauer, A., Hanakawa, T., Wu, T., Kansaku, K., & Hallett, M. (2004). Changes in brain anatomy in focal hand dystonia. *Annals of Neurology, 55*(5), 736–739.

Gazzaniga, M. S., Bogen, J. E., & Sperry, R. W. (1962). Some functional effects of sectioning the cerebral commissures in man. *Proceedings of the National Academy of Sciences, 48*(10), 1765–1769.

Gazzaniga, M. S., Bogen, J. E., & Sperry, R. W. (1963). Laterality effects in somesthesis following cerebral commissurotomy in man. *Neuropsychologia, 1*(3), 209–215.

Gazzaniga, M. S., Bogen, J. E., & Sperry, R. W. (1967). Dyspraxia following division of the cerebral commissures. *Archives of Neurology, 16*(6), 606–612.

Gebo, D. L. (2011). Vertical clinging and leaping revisited: Vertical support use as the ancestral condition of strepsirrhine primates. *American Journal of Physical Anthropology, 146*(3), 323–335.

Gehrke, A. R., Schneider, I., de la Calle-Mustienes, E., Tena, J. J., Gomez-Marin, C., Chandran, M., Nakamura, T., Braasch, I., Postlethwait, J. H., Gómez-Skarmeta, J. L., & Shubin, N. H. (2015). Deep conservation of wrist and digit enhancers in fish. *Proceedings of the National Academy of Sciences, 112*(3), 803–808.

Gentaz, E., & Hatwell, Y. (2004). Geometrical haptic illusions: The role of exploration in the Müller-Lyer, vertical-horizontal, and Delboeuf illusions. *Psychonomic Bulletin and Review, 11*(1), 31–40.

Gentile, G., Guterstam, A., Brozzoli, C., & Ehrsson, H. H. (2013). Disintegration of multisensory signals from the real hand reduces default limb self-attribution: An fMRI study. *Journal of Neuroscience, 33*(33), 13350–13366.

Gentilucci, M. (2002). Object motor representation and reaching-grasping control. *Neuropsychologia, 40*(8), 1139–1153.

Gentner, R., & Classen, J. (2006). Modular organization of finger movements by the human central nervous system. *Neuron, 52*(4), 731–742.

Gentner, R., Gorges, S., Weise, D., Kampe, K., Buttmann, M., & Classen, J. (2010). Encoding of motor skill in the corticomuscular system of musicians. *Current Biology, 20*(20), 1869–1874.

Genty, E., & Zuberbühler, K. (2014). Spatial reference in a bonobo gesture. *Current Biology, 24*(14), 1601–1605.

Georgopoulos, A. P., Kalaska, J. F., Caminiti, R., & Massey, J. T. (1982). On the relations between the direction of two-dimensional arm movements and cell discharge in primate motor cortex. *Journal of Neuroscience, 2*(11), 1527–1537.

Georgopoulos, A. P., Kettner, R. E., & Schwartz, A. B. (1988). Primate motor cortex and free arm movements to visual targets in three-dimensional space. II. Coding of the direction of movement a neuronal population. *Journal of Neuroscience, 8*(8), 2928–2937.

Georgopoulos, A. P., Merchant, H., Naselaris, T., & Amirikian, B. (2007). Mapping of the preferred direction in the motor cortex. *Proceedings of the National Academy of Sciences, 104*(26), 11068–11072.

Georgopoulos, A. P., Schwartz, A. B., & Kettner, R. E. (1986). Neuronal population coding of movement direction. *Science, 233*(4771), 1416–1419.

Germann, J., Chakravarty, M. M., Collins, D. L., & Petrides, M. (2020). Tight coupling between morphological features of the central sulcus and somatomotor body representations: A combined anatomical and functional MRI study. *Cerebral Cortex, 30*(3), 1843–1854.

Gerstmann, J. (1918). Reine taktile agnosie [Pure tactile agnosia]. *Monatsschrift für Psychiatrie und Neurologie, 44*, 329–343.

Gerstmann, J. (1942). Problem of imperception of disease and of impaired body territories with organic lesions: Relation to body scheme and its disorders. *Archives of Neurology & Psychiatry, 48*(6), 890–913.

Gerstmann, J. (2001). Pure tactile agnosia [T. Benke, Trans.]. *Cognitive Neuropsychology, 18*(3), 267–274.

Geschwind, N. (1965). Disconnexion syndromes in animals and men. *Brain, 88*, 237–294, 585–644.

Geyer, S., Ledberg, A., Schleicher, A., Kinomura, S., Schormann, T., Bürgel, U., Klingberg, T., Larsson, J., Zilles, K., & Roland, P. E. (1996). Two different areas within the primary motor cortex of man. *Nature, 382*(6594), 805–807.

Geyer, S., Schleicher, A., & Zilles, K. (1997). The somatosensory cortex of human: Cytoarchitecture and regional distributions of receptor-binding sites. *NeuroImage, 6*(1), 27–45.

Geyer, S., Schleicher, A., & Zilles, K. (1999). Areas 3a, 3b, and 1 of human primary somatosensory cortex. *NeuroImage, 10*(1), 63–83.

Gharbawie, O. A., Stepniewska, I., & Kaas, J. H. (2011). Cortical connections of functional zones in posterior parietal cortex and frontal cortex motor regions in New World monkeys. *Cerebral Cortex*, *21*(9), 1981–2002.

Gibson, J. J. (1962). Observations on active touch. *Psychological Review*, *69*(6), 477–491.

Gibson, J. J. (1966). *The senses considered as perceptual systems*. Houghton Mifflin.

Gibson, J. J. (1979). *The ecological approach to visual perception*. Houghton Mifflin.

Gilbert, A. N., & Wysocki, C. J. (1992). Hand preference and age in the United States. *Neuropsychologia*, *30*(7), 601–608.

Gilliatt, R. W., & Pratt, R. T. C. (1952). Disorders of perception and performance in a case of right-sided cerebral thrombosis. *Journal of Neurology, Neurosurgery & Psychiatry*, *15*(4), 264–271.

Giummarra, M. J., Georgiou-Karistianis, N., Nicholls, M. E. R., & Bradshaw, J. L. (2011). The third hand: Ownership of a rubber hand in addition to the existing (phantom) hand. *Cortex*, *47*(8), 998–1000.

Giummarra, M. J., Georgiou-Karistianis, N., Nicholls, M. E. R., Gibson, S. J., . . . Bradshaw, J. L. (2010). Corporeal awareness and proprioceptive sense of the phantom. *British Journal of Psychology*, *101*(Pt 4), 791–808.

Glasser, M. F., Coalson, T. S., Robinson, E. C., Hacker, C. D., Harwell, J., Yacoub, E., Ugurbil, K., Andersson, J., Beckmann, C. F., Jenkinson, M., Smith, S. M., & Van Essen, D. C. (2016). A multimodal parcellation of human cerebral cortex. *Nature*, *536*(7615), 171–178.

Glosser, G., Wiener, M., & Kaplan, E. (1986). Communicative gestures in aphasia. *Brain and Language*, *27*(2), 345–359.

Glover, S., Miall, R. C., & Rushworth, M. F. S. (2005). Parietal rTMS disrupts the initiation but not the execution of on-line adjustments to a perturbation of object size. *Journal of Cognitive Neuroscience*, *17*(1), 124–136.

Goldberg, G. (1985). Supplementary motor area structure and function: Review and hypotheses. *Behavioral and Brain Sciences*, *8*(4), 567–588.

Goldberg, G., Mayer, N. H., & Toglia, J. U. (1981). Medial frontal cortex infarction and the alien hand sign. *Archives of Neurology*, *38*(11), 683–686.

Goldenberg, G., & Hagmann, S. (1998). Tool use and mechanical problem solving in apraxia. *Neuropsychologia*, *36*(7), 581–589.

Goldenberg, G., Hartmann, K., & Schlott, I. (2003). Defective pantomime of object use in left brain damage: Apraxia or asymbolia? *Neuropsychologia*, *41*(12), 1565–1573.

Goldenberg, G., & Randerath, J. (2015). Shared neural substrates of apraxia and aphasia. *Neuropsychologia*, *75*, 40–49.

Goldenberg, G., & Spatt, J. (2009). The neural basis of tool use. *Brain*, *132*(Pt 6), 1645–1655.

Goldin-Meadow, S. (2003). *Hearing gesture: How our hands help us think*. Harvard University Press.

Goldin-Meadow, S., Alibali, M. W., & Church, R. B. (1993). Transitions in concept acquisition: Using the hand to read the mind. *Psychological Review, 100*(2), 279–297.

Goldin-Meadow, S., & Beilock, S. L. (2010). Action's influence on thought: The case of gesture. *Perspectives on Psychological Science, 5*(6), 664–674.

Goldin-Meadow, S., & Brentari, D. (2017). Gesture, sign, and language: The coming of age of sign language and gesture studies. *Behavioral and Brain Sciences, 40*, e46.

Goldin-Meadow, S., & Mylander, C. (1998). Spontaneous sign systems created by deaf children in two cultures. *Nature, 391*(6664), 279–281.

Goldin-Meadow, S., Nusbaum, H., Kelly, S. D., & Wagner, S. (2001). Explaining math: Gesturing lightens the load. *Psychological Science, 12*(6), 516–522.

Goldin-Meadow, S., & Singer, M. A. (2003). From children's hands to adults' ears: Gesture's role in the learning process. *Developmental Psychology, 39*(3), 509–520.

Goldring, S., & Ratcheson, R. (1972). Human motor cortex: Sensory input data from single neuron recordings. *Science, 175*(4029), 1493–1495.

Goodale, M. A., & Milner, A. D. (2004). *Sight unseen*. Oxford University Press.

Goodale, M. A., Milner, A. D., Jakobson, L. S., & Carey, D. P. (1991). A neurological dissociation between perceiving objects and grasping them. *Nature, 349*, 154–156.

Goodale, M. A., Pelisson, D., & Preblanc, C. (1986). Large adjustments in visually guided reaching do not depend on vision of the hand or perception of target displacement. *Nature, 320*, 748–750.

Goodall, J. (1964). Tool-using and aimed throwing in a community of free-living chimpanzees. *Nature, 201*, 1264–1266.

Goodall, J. (1971). *In the shadow of man*. Houghton Mifflin.

Goodall, J. (1986). *The chimpanzees of Gombe*. Harvard University Press.

Goodall, J. (1998). Learning from the chimpanzees: A message humans can understand. *Science, 282*(5397), 2184–2185.

Gooddy, W. (1970). Admiral Lord Nelson's neurological illnesses. *Proceedings of the Royal Society of Medicine, 63*(3), 17–24.

Goodglass, H., & Kaplan, E. (1963). Disturbance of gesture and pantomime in aphasia. *Brain, 86*(4), 703–720.

Goodhew, S. C., Edwards, M., Ferber, S., & Pratt, J. (2015). Altered visual perception near the hands: A critical review of attentional and neurophysiological models. *Neuroscience and Biobehavioral Reviews, 55*, 223–233.

Goodman, J. M., Tabot, G. A., Lee, A. S., Suresh, A. K., Rajan, A. T., Hatsapoulos, N. G., & Bensmaia, S. (2019). Postural representations of the hand in the primate sensorimotor cortex. *Neuron, 104*(5), 1000–1009.

Goodwin, R. S., & Michel, G. F. (1981). Head orientation position during birth and in infant neonatal period, and hand preference at nineteen weeks. *Child Development, 52*(3), 819–826.

Gould, S. J. (1980). *The panda's thumb*. Norton.

Gowlett, J. A. J. (2015). Variability in an early hominin percussive tradition: The Acheulean versus cultural variation in modern chimpanzee artefacts. *Philosophical Transactions of the Royal Society B, 370*(1682), 20140358.

Gozli, D. G., West, G. L., & Pratt, J. (2012). Hand position alters vision by biasing processing through different visual pathways. *Cognition, 124*(2), 244–250.

Gradin, C. J. (1994). L'art rupestre dans la Patagonie argentine [Rock art in Argentine Patagonia]. *L'Anthropologie, 98*, 149–172.

Grafton, S. T., Hazeltine, E., & Ivry, R. B. (2002). Motor sequence learning with the nondominant left hand: A PET functional imaging study. *Experimental Brain Research, 146*(3), 369–378.

Graham, J. A., & Argyle, M. (1975). A cross-cultural study of the communication of extra-verbal meaning by gestures. *International Journal of Psychology, 10*(1), 57–67.

Graham, J. A., & Heywood, S. (1975). The effects of elimination of hand gestures and of verbal codability on speech performance. *European Journal of Social Psychology, 5*(2), 189–195.

Graham, K. E., & Hobaiter, C. (2023). Towards a great ape dictionary: Inexperienced humans understand common nonhuman ape gestures. *PLOS Biology, 21*(1), e3001939.

Granda, A. J., & Nuccio, J. (2018). *Protactile principles*. Tactile Communications. https://www.tactilecommunications.org/Documents/PTPrinciplesMoviesFinal.pdf

Graybiel, A. M. (1998). The basal ganglia and chunking of action repertoires. *Neurobiology of Learning and Memory, 70*(1–2), 119–136.

Graybiel, A. M. (2000). The basal ganglia. *Current Biology, 10*(14), R509–R511.

Graziano, M. S. A. (2009). *The intelligent movement machine*. Oxford University Press.

Graziano, M. S. A. (2016). Ethological action maps: A paradigm shift for the motor cortex. *Trends in Cognitive Sciences, 20*(2), 121–132.

Graziano, M. S. A., & Aflalo, T. N. (2007). Mapping behavioral repertoire onto the cortex. *Neuron, 56*(2), 239–251.

Graziano, M. S. A., Cooke, D. F., & Taylor, C. S. R. (2000). Coding the location of the arm by sight. *Science, 290*(5497), 1782–1786.

Graziano, M. S. A., Hu, X. T., & Gross, C. G. (1997). Visuospatial properties of ventral premotor cortex. *Journal of Neurophysiology, 77*(5), 2268–2292.

Graziano, M. S. A., Yap, G. S., & Gross, C. G. (1994). Coding of visual space by premotor neurons. *Science, 266*(5187), 1054–1057.

Gréa, H., Pisella, L., Rossetti, Y., Desmurget, M., Tilikete, C., Grafton, S., Preblanc, C., & Vighetto, A. (2002). A lesion of the posterior parietal cortex disrupts on-line adjustments during aiming movements. *Neuropsychologia, 40*(13), 2471–2480.

Gredebäck, G., Melinder, A., & Daum, M. (2010). The development and neural basis of pointing comprehension. *Social Neuroscience, 5*(5–6), 441–450.

Green, B. G. (1982). The perception of distance and location for dual tactile pressures. *Perception & Psychophysics, 31*(4), 315–323.

Green, D. J., & Gordon, A. D. (2008). Metacarpal proportions in *Australopithecus africanus*. *Journal of Human Evolution, 54*(5), 705–719.

Green, L. A., & Gabriel, D. A. (2018). The effect of unilateral training on contralateral limb strength in young, older, and patient populations: A meta-analysis of cross education. *Physical Therapy Reviews, 23*(4–5), 238–249.

Grefkes, C., Geyer, S., Schormann, T., Roland, P., & Zilles, K. (2001). Human somatosensory area 2: Observer-independent cytoarchitectonic mapping, interindividual variability, and population map. *NeuroImage, 14*(3), 617–631.

Griffin, D. M., Hoffman, D. S., & Strick, P. L. (2015). Corticomotoneuronal cells are "functionally tuned." *Science, 350*(6261), 667–670.

Grill-Spector, K., Kourtzi, Z., & Kanwisher, N. (2001). The lateral occipital complex and its role in object recognition. *Vision Research, 41*(10–11), 1409–1422.

Grivaz, P., Blanke, O., & Serino, A. (2017). Common and distinct brain regions processing multisensory bodily signals for peripersonal space and body ownership. *NeuroImage, 147*, 602–618.

Groenen, M. (2011). Images de mains dans la préhistoire [Hand images in prehistory]. *La Part de l'œil, 25*, 56–69.

Grove, E. A., & Fukuchi-Shimogori, T. (2003). Generating the cerebral cortical area map. *Annual Review of Neuroscience, 26*, 355–380.

Guadalupe, T., Willems, R. M., Zwiers, M. P., Vasquez, A. A., Hoogman, M., Hagoort, P., Fernandez, G., Buitelaar, J., Franke, B., Fisher, S. E., & Francks, C. (2014). Differences in cerebral cortical anatomy of left- and right-handers. *Frontiers in Psychology, 5*, 261.

Gumert, M. D., & Malaivijitnond, S. (2013). Long-tailed macaques select mass of stone tools according to food type. *Philosophical Transactions of the Royal Society B, 368*(1630), 20120413.

Gunz, P., Neubauer, S., Falk, D., Tafforeau, P., Le Cabec, A., Smith, T. M., Kimbel, W. H., Spoor, F., & Alemseged, Z. (2020). *Australopithecus afarensis* endocasts suggest ape-like brain organization and prolonged brain growth. *Science Advances, 6*(14), eaaz4729.

Gunz, P., Neubauer, S., & Maureille, B. (2010). Brain development after birth differs between Neanderthals and modern humans. *Current Biology, 20*(21), R921–R922.

Gunz, P., Tilot, A. K., Wittfeld, K., Teumer, A., Shapland, C. Y., van Erp, T. G. M., Dannemann, M., Vernot, B., Neubauer, S., Guadalupe, T., Fernández, G., Brunner, H. G., Enard, W., Fallon, J., Hosten, N., Völker, U., Profico, A., Di Vincenzo, F., Manzi, G., . . . Fisher, S. E. (2019). Neandertal introgression sheds light on modern human endocranial globularity. *Current Biology, 29*(1), 120–127.e5.

Guo, J., & Song, J.-H. (2019). Action fluency facilitates perceptual discrimination. *Psychological Science, 30*(10), 1434–1448.

Guterstam, A., Collins, K. L., Cronin, J. A., Zeberg, H., Darvas, F., Weaver, K. E., Ojemann, J. G., & Ehrsson, H. H. (2019). Direct electrophysiological correlates of body ownership in human cerebral cortex. *Cerebral Cortex, 29*(3), 1328–1341.

Guterstam, A., Petkova, V. I., & Ehrsson, H. H. (2011). The illusion of owning a third arm. *PLOS One, 6*(2), e17208.

Guterstam, A., Zeberg, H., Özçiftci, V. M., & Ehrsson, H. H. (2016). The magnetic touch illusion: A perceptual correlate of visuo-tactile integration in peripersonal space. *Cognition, 155*, 44–56.

Gutteling, T. P., Kenemans, J. L., & Neggers, S. F. W. (2011). Grasping preparation enhances orientation change detection. *PLOS One, 6*(3), 17675.

Gutteling, T. P., Park, S. Y., Kenemans, J. L., & Neggers, S. F. W. (2013). TMS of the anterior intraparietal area selectively modulates orientation change detection during action preparation. *Journal of Neurophysiology, 110*(1), 33–41.

Gutteling, T. P., Petridou, N., Dumoulin, S. O., Harvey, B. M., Aarnoutse, E. J., Kenemans, J. L., & Neggers, S. F. W. (2015). Action preparation shapes processing in early visual cortex. *Journal of Neuroscience, 35*(16), 6472–6480.

Haaland, K. Y., Harrington, D. L., & Knight, R. T. (2000). Neural representations of skilled movement. *Brain, 123*(Pt 11), 2306–2313.

Haar, S., Dinstein, I., Shelef, I., & Donchin, O. (2017). Effector-invariant movement encoding in the human motor system. *Journal of Neuroscience, 37*(37), 9054–9063.

Haerer, A. F., & Currier, R. D. (1966). Mirror movements. *Neurology, 16*(8), 757–760.

Haggard, P. (2009). The sources of human volition. *Science, 324*(5928), 731–732.

Haggard, P., Christakou, A., & Serino, A. (2007). Viewing the body modulates tactile receptive fields. *Experimental Brain Research, 180*(1), 187–193.

Hagmann, P., Cammoun, L., Gigandet, X., Meuli, R., Honey, C. J., Wedeen, V. J., & Sporns, O. (2008). Mapping the structural core of human cerebral cortex. *PLOS Biology, 6*(7), 1479–1493.

Hagoort, P. (2005). On Broca, brain, and binding: A new framework. *Trends in Cognitive Sciences, 9*(9), 416–423.

Haldane, J. B. S. (1927). On being the right size. In *Possible worlds and other essays* (pp. 18–26). Chatto and Windus.

Hale, M. E. (2021). Evolution of touch and proprioception of the limbs: Insights from fish and humans. *Current Opinion in Neurobiology, 71*, 37–43.

Hall, E. T. (1966). *The hidden dimension.* Anchor Books.

Hall, N. J., Udell, M. A. R., Dorey, N. R., Walsh, A. L., & Wynne, C. D. L. (2011). Megachiropteran bats (*Pteropus*) utilize human referential stimuli to locate hidden food. *Journal of Comparative Psychology, 125*(3), 341–346.

Hallett, M., & Khoshbin, S. (1980). A physiological mechanism of bradykinesia. *Brain, 103*(2), 301–314.

Halligan, P. W., Hunt, M., Marshall, J. C., & Wade, D. T. (1995). Sensory detection without localization. *Neurocase, 1*(3), 259–266.

Halligan, P. W., & Marshall, J. C. (1991). Left neglect for near but not far space in man. *Nature, 350*, 498–500.

Halligan, P. W., Marshall, J. C., Hunt, M., & Wade, D. T. (1997). Somatosensory assessment: Can seeing produce feeling? *Journal of Neurology, 244*(3), 199–203.

Halligan, P. W., Marshall, J. C., & Wade, D. T. (1993). Three arms: A case study of supernumerary phantom limb after right hemisphere stroke. *Journal of Neurology, Neurosurgery, and Psychiatry, 56*(2), 159–166.

Halligan, P. W., Marshall, J. C., Wade, D. T., Davey, J., & Morrison, D. (1993). Thumb in cheek? Sensory reorganization and perceptual plasticity after limb amputation. *Neuroreport, 4*(3), 233–236.

Halsband, U., & Passingham, R. E. (1985). Premotor cortex and the conditions for movement in monkeys (*Macaca fascicularis*). *Behavioural Brain Research, 18*(3), 269–277.

Halverson, H. M. (1931). An experimental study of prehension in infants by means of systematic cinema records. *Genetic Psychology Monographs, 10*, 107–285.

Halverson, H. M. (1932). A further study of grasping. *Journal of General Psychology, 7*(1), 34–64.

Halverson, H. M. (1937). Studies of the grasping responses of early infancy: I, II, and III. *Journal of Genetic Psychology, 51*(2), 371–449.

Hamilton, E. (1942). *Mythology.* Little, Brown and Company.

Hamilton, R., Keenan, J. P., Catala, M., & Pascual-Leone, A. (2000). Alexia for braille following bilateral occipital stroke in an early blind woman. *NeuroReport, 11*(2), 237–240.

Hammond, P. H. (1955). Involuntary activity in biceps following the sudden application of velocity to the abducted forearm. *Journal of Physiology, 127*(2), 23P–25P.

Hammond, P. H. (1956). The influence of prior instruction to the subject on an apparently involuntary neuro-muscular response. *Journal of Physiology, 132*(1), 17P–18P.

Hamrick, M. W. (2001). Primate origins: Evolutionary change in digital ray patterning and segmentation. *Journal of Human Evolution, 40*(4), 339–351.

Hamrick, M. W., Churchill, S. E., Schmitt, D., & Hylander, W. L. (1998). EMG of the human flexor pollicis longus muscle: Implications for the evolution of hominid tool use. *Journal of Human Evolution, 34*(2), 123–136.

Hamzei, F., Dettmers, C., Rzanny, R., Liepert, J., Büchel, C., & Weiller, C. (2002). Reduction of excitability ("inhibition") in the ipsilateral primary motor cortex is mirrored by fMRI signal decreases. *NeuroImage, 17*(1), 490–496.

Hanada, K., Yokoi, K., Futamura, A., Kinoshita, Y., Sakamoto, K., Ono, K., & Hirayama, K. (2021). Numbsense of shape, texture, and objects after left parietal infarction: A case report. *Journal of Neuropsychology, 15*(2), 204–214.

Hanajima, R., Ugawa, Y., Machii, K., Mochizuki, H., Terao, Y., Enomoto, H., Furubayashi, T., Shiio, Y., Uesugi, H., & Kanazawa, I. (2001). Interhemispheric facilitation of the hand motor area in humans. *Journal of Physiology, 531*(Pt 3), 849–859.

Handler, A., & Ginty, D. D. (2021). The mechanosensory neurons of touch and their mechanisms of activation. *Nature Reviews Neuroscience, 22*(9), 521–537.

Hannan, T. (1987). A cross-sequential assessment of the occurrences of pointing in 3- to 12-month-old human infants. *Infant Behavior and Development, 10*(1), 11–22.

Hanna-Pladdy, B., Mendoza, J. E., Apostolos, G. T., & Heilman, K. M. (2002). Lateralised motor control: Hemispheric damage and the loss of deftness. *Journal of Neurology, Neurosurgery, and Psychiatry, 73*(5), 574–577.

Hardy, A. R., & Hale, M. E. (2020). Sensing the structural characteristics of surfaces: Texture encoding by a bottom-dwelling fish. *Journal of Experimental Biology, 223*(Pt 21), jeb227280.

Hardy, A. R., Steinworth, B. M., & Hale, M. E. (2016). Touch sensation by pectoral fins of the catfish *Pimelodus pictus*. *Proceedings of the Royal Society B, 283*(1824), 20152652.

Hare, B., Brown, M., Williamson, C., & Tomasello, M. (2002). The domestication of social cognition in dogs. *Science, 298*(5598), 1634–1636.

Hare, B., Plyusnina, I., Ignacio, N., Schepina, O., Stepika, A., Wrangham, R., & Trut, L. (2005). Social cognitive evolution in captive foxes is a correlated by-product of experimental domestication. *Current Biology, 15*(3), 226–230.

Hare, B., & Tomasello, M. (2005). Human-like social skills in dogs? *Trends in Cognitive Sciences, 9*(9), 439–444.

Hari, R., Hänninen, R., Mäkinen, T., Jousmäki, V., Forss, N., Seppä, M., & Salonen, O. (1998). Three hands: Fragmentation of human bodily awareness. *Neuroscience Letters, 240*(3), 131–134.

Hari, R., & Jousmäki, V. (1996). Preference of personal to extrapersonal space in a visuomotor task. *Journal of Cognitive Neuroscience, 8*(3), 305–307.

Harmand, S., Lewis, J. E., Feibel, C. S., Lepre, C. J., Prat, S., Lenoble, A., Boës, X., Quinn, R. L., Brenet, M., Arroyo, A., Taylor, N., Clément, S., Daver, G., Brugal, J.-P., Leakey, L., Mortlock, R. A., Wright, J. D., Lokorodi, S., Kirwa, C., & Rocher, H. (2015). 3.3-million-year-old stone tools from Lomekwi 3, West Turkana, Kenya. *Nature, 521*(7552), 310–315.

Harrar, V., Spence, C., & Makin, T. R. (2014). Topographic generalization of tactile perceptual learning. *Journal of Experimental Psychology: Human Perception and Performance, 40*(1), 15–23.

Harris, J. A., Arabzadeh, E., Moore, C. A., & Clifford, C. W. G. (2007). Noninformative vision causes adaptive changes in tactile sensitivity. *Journal of Neuroscience, 27*(27), 7136–7140.

Harris, J. A., Harris, I. M., & Diamond, M. E. (2001). The topography of tactile learning in humans. *Journal of Neuroscience, 21*(3), 1056–1061.

Harris, J. A., Karlov, L., & Clifford, C. W. G. (2006). Localization of tactile stimuli depends on conscious detection. *Journal of Neuroscience, 26*(3), 948–952.

Harris, J. A., Thein, T., & Clifford, C. W. G. (2004). Dissociating detection from localization of tactile stimuli. *Journal of Neuroscience, 24*(14), 3683–3693.

Harvey, M. A., Saal, H. P., Dammann, J. F., & Bensmaia, S. J. (2013). Multiplexing stimulus information through rate and temporal codes in primate somatosensory cortex. *PLOS Biology, 11*(5), e1001558.

Haslinger, B., Erhard, P., Altenmüller, E., Hennenlotter, A., Schwaiger, M., Gräfin von Einsiedel, H., Rummeny, E., Conrad, B., & Ceballos-Baumann, A. O. (2004). Reduced recruitment of motor association areas during bimanual coordination in concert pianists. *Human Brain Mapping, 22*(3), 206–215.

Hauck, T. C., Connan, J., Charrié-Duhaut, A., & Le Tensorer, J.-M. (2013). Molecular evidence of bitumen in the Mousterian lithic assemblage of Hummal (central Syria). *Journal of Archaeological Science, 40*(8), 3252–3262.

Haward, B. M. (2019). *Telegraphists' cramp: The emergence and disappearance of an occupational disease between 1875 and 1930*. University of Edinburgh.

Hawkins, M. B., Henke, K., & Harris, M. P. (2021). Latent developmental potential to form limb-like skeletal structures in zebrafish. *Cell, 184*(4), 899–911.

Head, H. (1918). Sensation and the cerebral cortex. *Brain, 41*(2), 57–253.

Head, H., & Holmes, G. (1911). Sensory disturbances from cerebral lesions. *Brain, 34*(2–3), 102–254.

Head, H., Rivers, W. H. R., & Sherren, J. (1905). The afferent nervous system from a new aspect. *Brain, 28*(2), 99–115.

Heald, J. B., Ingram, J. N., Flanagan, J. R., & Wolpert, D. M. (2018). Multiple motor memories are learned to control different points on a tool. *Nature Human Behaviour, 2*(4), 300–311.

Hebb, D. O. (1949). *The organization of behavior*. Wiley.

Hecht, E. E., Gutman, D. A., Khreisheh, N., Taylor, S. V., Kilner, J., Faisal, A. A., Bradley, B. A., Chaminade, T., & Stout, D. (2015). Acquisition of paleolithic toolmaking abilities involves structural remodeling to inferior frontoparietal regions. *Brain Structure and Function, 220*(4), 2315–2331.

Hediger, H. (1955). *Studies of the psychology and behaviour of captive animals in zoos and circuses.* Butterworth.

Heed, T., Leone, F. T. M., Toni, I., & Medendorp, W. P. (2016). Functional versus effector-specific organization of the human posterior parietal cortex: Revisited. *Journal of Neurophysiology, 116*(4), 1885–1899.

Heffner, R. S., & Masterton, R. B. (1975). Variation in form of the pyramidal tract and its relationship to digital dexterity. *Brain, Behavior and Evolution, 12*(3), 161–200.

Heffner, R. S., & Masterton, R. B. (1983). The role of the corticospinal tract in the evolution of human digital dexterity. *Brain, Behavior and Evolution, 23*(3–4), 165–183.

Heilman, K. M., Maher, L. M., Greenwald, M. L., & Rothi, L. J. G. (1997). Conceptual apraxia from lateralized lesions. *Neurology, 49*(2), 457–464.

Heilman, K. M., Rothi, L. J., & Valenstein, E. (1982). Two forms of ideomotor apraxia. *Neurology, 32*(4), 342–346.

Helbig, H. B., Ernst, M. O., Ricciardi, E., Pietrini, P., Thielscher, A., Mayer, K. M., Schultz, J., & Noppeney, U. (2012). The neural mechanisms of reliability weighted integration of shape information from vision and touch. *NeuroImage, 60*(2), 1063–1072.

Held, R., & Bauer, J. A. (1967). Visually guided reaching in infant monkeys after restricted rearing. *Science, 155*(3763), 718–720.

Held, R., & Bauer, J. A. (1974). Development of sensorially-guided reaching in infant monkeys. *Brain Research, 71*(2–3), 265–271.

Held, R., Ostrovsky, Y., de Gelder, B., Gandhi, T., Ganesh, S., Mathur, U., & Sinha, P. (2011). The newly sighted fail to match seen with felt. *Nature Neuroscience, 14*(5), 551–553.

Heller, M. A. (1983). Haptic dominance in form perception with blurred vision. *Perception, 12*(5), 607–613.

Heller, M. A., McCarthy, M., & Clark, A. (2005). Pattern perception and pictures for the blind. *Psicológica, 26*(1), 161–171.

Heming, E. A., Cross, K. P., Takei, T., Cook, D. J., & Scott, S. H. (2019). Independent representations of ipsilateral and contralateral limbs in primary motor cortex. *eLife, 8*, e48190.

Hepper, P. G., McCartney, G. R., & Shannon, E. A. (1998). Lateralised behaviour in first trimester human foetuses. *Neuropsychologia, 36*(6), 531–534.

Hepper, P. G., Shahidullah, S., & White, R. (1990). Origins of fetal handedness. *Nature, 347*, 431.

Hepper, P. G., Shahidullah, S., & White, R. (1991). Handedness in the human fetus. *Neuropsychologia, 29*(11), 1107–1111.

References

Hepper, P. G., Wells, D. L., & Lynch, C. (2005). Prenatal thumb sucking is related to postnatal handedness. *Neuropsychologia, 43*(3), 313–315.

Herculano-Houzel, S. (2012). The remarkable, yet not extraordinary, human brain as a scaled-up primate brain and its associated cost. *Proceedings of the National Academy of Sciences, 109*(Suppl 1), 10661–10668.

Herron, P. (1978). Somatotopic organization of mechanosensory projections to SII cerebral neocortex in the raccoon (*Procyon lotor*). *Journal of Comparative Neurology, 181*(4), 717–728.

Hertenstein, M. J., Keltner, D., App, B., Bulleit, B. A., & Jaskolka, A. R. (2006). Touch communicates distinct emotions. *Emotion, 6*(3), 528–533.

Hess, C. W., Mills, K. R., & Murray, N. M. F. (1986). Magnetic stimulation of the human brain: Facilitation of motor responses by voluntary contraction of ipsilateral and contralateral muscles with additional observations on an amputee. *Neuroscience Letters, 71*(2), 235–240.

Heuer, H., Spijkers, W., Kleinsorge, T., & van der Loo, H. (1998). Period duration of physical and imaginary movement sequences affects contralateral amplitude modulation. *Quarterly Journal of Experimental Psychology, 51A*(4), 755–779.

Hewes, G. W. (1973). Primate communication and the gestural origin of language. *Current Anthropology, 14*(1–2), 5–24.

Hickok, G., Bellugi, U., & Klima, E. S. (1996). The neurobiology of sign language and its implications for the neural basis of language. *Nature, 381*, 699–702.

Hickok, G., Bellugi, U., & Klima, E. S. (1998). The neural organization of language: Evidence from sign language aphasia. *Trends in Cognitive Sciences, 2*(4), 129–136.

Hill, J., Inder, T., Neil, J., Dierker, D., Harwell, J., & Van Essen, D. (2010). Similar patterns of cortical expansion during human development and evolution. *Proceedings of the National Academy of Sciences, 107*(29), 13135–13140.

Hinde, R. A., Rowell, T. E., & Spencer-Booth, Y. (1964). Behaviour of socially living rhesus monkeys in their first six months. *Proceedings of the Zoological Society of London, 143*(4), 609–649.

Hippocrates. (1849a). On injuries of the head [F. Adams, Trans.]. In *The genuine works of Hippocrates* (Vol. 1, pp. 445–466). Sydenham Society.

Hippocrates. (1849b). On the articulations [F. Adams, Trans.]. In *The genuine works of Hippocrates* (Vol. 2, pp. 85–156). Syndham Society.

Hirano, M., Kimoto, Y., & Furuya, S. (2020). Specialized somatosensory–motor integration functions in musicians. *Cerebral Cortex, 30*(3), 1148–1158.

Hirano, M., Sakurada, M., & Furuya, S. (2020). Overcoming the ceiling effects of experts' motor expertise through active haptic training. *Science Advances, 6*(47), eabd2558.

Hirsh, I. J., & Sherrick, C. E. (1961). Perceived order in different sense modalities. *Journal of Experimental Psychology, 62*(5), 423–432.

Hobaiter, C., & Byrne, R. W. (2014). The meanings of chimpanzee gestures. *Current Biology, 24*(14), 1596–1600.

Hobaiter, C., Leavens, D. A., & Byrne, R. W. (2014). Deictic gesturing in wild chimpanzees (*Pan troglodytes*)? Some possible cases. *Journal of Comparative Psychology, 128*(1), 82–87.

Hoetjes, M., Krahmer, E., & Swerts, M. (2014). Does our speech change when we cannot gesture? *Speech Communication, 57*, 257–267.

Hoffecker, J. F. (2018). The complexity of Neanderthal technology. *Proceedings of the National Academy of Sciences, 115*(9), 1959–1961.

Hofman, M. A. (1983). Encephalization in hominids: Evidence for the model of punctuationalism. *Brain, Behavior and Evolution, 22*(2–3), 102–117.

Hofmann, A., Grossbach, M., Baur, V., Hermsdörfer, J., & Altenmüller, E. (2015). Musician's dystonia is highly task specific: No strong evidence for everyday fine motor deficits in patients. *Medical Problems of Performing Artists, 30*(1), 38–46.

Hohlstein, R. R. (1982). The development of prehension in normal infants. *American Journal of Occupational Therapy, 36*(3), 170–176.

Holloway, R. L. (1981a). Exploring the dorsal surface of hominoid brain endocasts by stereoplotter and discriminant analysis. *Philosophical Transactions of the Royal Society B, 292*(1057), 155–166.

Holloway, R. L. (1981b). Revisiting the South African Taung australopithecine endocast: The position of the lunate sulcus as determined by the stereoplotting technique. *American Journal of Physical Anthropology, 56*(1), 43–58.

Holloway, R. L., Hurst, S. D., Garvin, H. M., Schoenemann, P. T., Vanti, W. B., Berger, L. R., & Hawks, J. (2018). Endocast morphology of *Homo naledi* from the Dinaledi Chamber, South Africa. *Proceedings of the National Academy of Sciences, 115*(22), 5738–5743.

Holloway, V., Gadian, D. G., Vargha-Khadem, F., Porter, D. A., Boyd, S. G., & Connelly, A. (2000). The reorganization of sensorimotor function in children after hemispherectomy. *Brain, 123*(Pt 12), 2432–2444.

Holmes, N. P., Calvert, G. A., & Spence, C. (2004). Extending or projecting peripersonal space with tools? Multisensory interactions highlight only the distal and proximal ends of tools. *Neuroscience Letters, 372*(1–2), 62–67.

Hömke, L., Amunts, K., Bönig, L., Fretz, C., Binkofski, F., Zilles, K., & Weder, B. (2009). Analysis of lesions in patients with unilateral tactile agnosia using cytoarchitectonic probabilistic maps. *Human Brain Mapping, 30*(5), 1444–1456.

Honoré, J., Bourdeaud'Hui, M., & Sparrow, L. (1989). Reduction of cutaneous reaction time by directing eyes towards the source of stimulation. *Neuropsychologia, 27*(3), 367–371.

Hoosain, R. (1990). Left handedness and handedness switch amongst the Chinese. *Cortex, 26*(3), 451–454.

Hoover, J. E., & Strick, P. L. (1993). Multiple output channels in the basal ganglia. *Science, 259*(5096), 819–821.

Hopkins, W. D. (1995). Hand preferences for a coordinated bimanual task in 110 chimpanzees (*Pan troglodytes*): Cross-sectional analysis. *Journal of Comparative Psychology, 109*(3), 291–297.

Hopkins, W. D. (2006). Comparative and familial analysis of handedness in great apes. *Psychological Bulletin, 132*(4), 538–559.

Hopkins, W. D., & Cantalupo, C. (2004). Handedness in chimpanzees (*Pan troglodytes*) is associated with asymmetries of the primary motor cortex but not with homologous language areas. *Behavioral Neuroscience, 118*(6), 1176–1183.

Hopkins, W. D., Meguerditchian, A., Coulon, O., Bogart, S., Mangin, J.-F., Sherwood, C. C., Grabowski, M. W., Bennett, A. J., Pierre, P. J., Fears, S., Woods, R., Hof, P. R., & Vauclair, J. (2014). Evolution of the central sulcus morphology in primates. *Brain, Behavior, and Evolution, 84*(1), 19–30.

Hopkins, W. D., Phillips, K. A., Bania, A., Calcutt, S. E., Gardner, M., Russell, J., Schaeffer, J., Lonsdorf, E. V., Ross, S. R., & Schapiro, S. J. (2011). Hand preferences for coordinated bimanual actions in 777 great apes: Implications for the evolution of handedness in hominins. *Journal of Human Evolution, 60*(5), 605–611.

Hopkins, W. D., Taglialatela, J. P., Russell, J. L., Nir, T. M., & Schaeffer, J. (2010). Cortical representation of lateralized grasping in chimpanzees (*Pan troglodytes*): A combined MRI and PET study. *PLOS One, 5*(10), e13383.

Hopkins, W. D., & Vauclair, J. (2011). Evolution of behavioral and brain asymmetries in primates. In M. Tallerman & K. Gibson (Eds.), *Oxford handbook of language evolution* (pp. 184–197). Oxford University Press.

Hopkins, W. D., Wesley, M. J., Izard, M. K., Hook, M., & Schapiro, S. J. (2004). Chimpanzees (*Pan troglodytes*) are predominantly right-handed: Replication in three populations of apes. *Behavioral Neuroscience, 118*(3), 659–663.

Horsley, V., & Schäfer, E. A. (1888). A record of experiments upon the functions of the cerebral cortex. *Philosophical Transactions of the Royal Society, 179*, 1–45.

Hortobágyi, T., Richardson, S. P., Lomarev, M., Shamim, E., Meunier, S., Russman, H., Dang, N., & Hallett, M. (2011). Interhemispheric plasticity in humans. *Medicine and Science in Sports and Exercise, 43*(7), 1188–1199.

Hoyet, L., Argelaguet, F., Nicole, C., & Lécuyer, A. (2016). "Wow! I have six Fingers!": Would you accept structural changes of your hand in VR? *Frontiers in Robotics and AI, 3*, 27.

Hrbek, A., Karlberg, P., & Olsson, T. (1973). Development of visual and somatosensory evoked responses in pre-term newborn infants. *Electroencephalography and Clinical Neurophysiology, 34*(3), 225–232.

Hsieh, C.-L., Shima, F., Tobimatsu, S., Sun, S.-J., & Kato, M. (1995). The interaction of the somatosensory evoked potentials to simultaneous finger stimuli in the human central nervous system. A study using direct recordings. *Electroencephalography and Clinical Neurophysiology, 96*(2), 135–142.

Hu, Y., Marwick, B., Zhang, J.-F., Rui, X., Hou, Y.-M., Yue, J.-P., Chen, W.-R., Huang, W.-W., & Li, B. (2019). Late Middle Pleistocene Levallois stone-tool technology in southwest China. *Nature, 565*(7737), 82–85.

Huang, R.-S., Chen, C.-F., Tran, A. T., Holstein, K. L., & Sereno, M. I. (2012). Mapping multisensory parietal face and body areas in humans. *Proceedings of the National Academy of Sciences, 109*(44), 18114–18119.

Huang, R.-S., & Sereno, M. I. (2007). Dodecapus: An MR-compatible system for somatosensory stimulation. *NeuroImage, 34*(3), 1060–1073.

Huang, W., Ciochon, R., Gu, Y., Larick, R., Qiren, F., Schwarcz, H., Yonge, C., de Vos, J., & Rink, W. (1995). Early Homo and associated artefacts from Asia. *Nature, 378*(6554), 275–278.

Hubel, D. H., & Wiesel, T. N. (1974). Uniformity of monkey striate cortex: A parallel relationship between field size, scatter, and magnification factor. *Journal of Comparative Neurology, 158*(3), 295–306.

Hublin, J.-J., Sirakov, N., Aldeias, V., Bailey, S., Bard, E., Delvigne, V., Endarova, E., Fagault, Y., Fewlass, H., Hajdinjak, M., Kromer, B., Krumov, I., Marreiros, J., Martisius, N. L., Paskulin, L., Sinet-Mathiot, V., Meyer, M., Pääbo, S., Popov, V., & Tsanova, T. (2020). Initial Upper Palaeolithic *Homo sapiens* from Bacho Kiro cave, Bulgaria. *Nature, 581*, 299–302.

Huffman, K. J., & Krubitzer, L. A. (2001). Area 3a: Topographic organization and cortical connections in marmoset monkeys. *Cerebral Cortex, 11*(9), 849–867.

Hughes, E. (1908). Mirror-writing. *The Lancet, 171*(4403), 188.

Hunt, G. R. (1996). Manufacture and use of hook-tools by New Caledonian crows. *Nature, 379*, 249–251.

Hunt, G. R., & Gray, R. D. (2003). Diversification and cumulative evolution in New Caledonian crow tool manufacture. *Proceedings of the Royal Society B, 270*(1517), 867–874.

Huth, A. G., de Heer, W. A., Griffiths, T. L., Theunissen, F. E., & Gallant, J. L. (2016). Natural speech reveals the semantic maps that tile human cerebral cortex. *Nature, 532*(7600), 453–458.

Hwang, E. J., Hauschild, M., Wilke, M., & Andersen, R. A. (2012). Inactivation of the parietal reach region causes optic ataxia, impairing reaches but not saccades. *Neuron, 76*(5), 1021–1029.

Hyvärinen, J., & Poranen, A. (1978). Receptive field integration and submodality convergence in the hand area of the post-central gyrus of the alert monkey. *Journal of Physiology, 283*(1), 539–556.

Imamizu, H., Kuroda, T., Miyauchi, S., Yoshioka, T., & Kawato, M. (2003). Modular organization of internal models of tools in the human cerebellum. *Proceedings of the National Academy of Sciences, 100*(9), 5461–5466.

Imamizu, H., Miyauchi, S., Tamada, T., Sasaki, Y., Takino, R., Pütz, B., Yoshioka, T., & Kawato, M. (2000). Human cerebellar activity reflecting an acquired internal model of a new tool. *Nature, 403*(6766), 192–195.

Ingram, J. N., Howard, I. S., Flanagan, J. R., & Wolpert, D. M. (2010). Multiple grasp-specific representations of tool dynamics mediate skillful manipulation. *Current Biology, 20*(7), 618–623.

Inoue, Y., Inoue, E., & Itakura, S. (2004). Use of experimenter-given directional cues by a young white-handed gibbon (*Hylobates lar*). *Japanese Psychological Research, 46*(3), 262–267.

Ioannou, C. I., & Altenmüller, E. (2014). Psychological characteristics in musician's dystonia: A new diagnostic classification. *Neuropsychologia, 61*, 80–88.

Ireland, W. W. (1880). Notes on left-handedness. *Brain, 3*(2), 207–214.

Iriki, A., Tanaka, M., & Iwamura, Y. (1996). Coding of modified body schema during tool use by macaque postcentral neurones. *NeuroReport, 7*(14), 2325–2330.

Ivanco, T. L., Pellis, S. M., & Whishaw, I. Q. (1996). Skilled forelimb movements in prey catching and in reaching by rats (*Rattus norvegicus*) and opossums (*Monodelphis domestica*): Relations to anatomical differences in motor systems. *Behavioural Brain Research, 79*(1–2), 163–181.

Iverson, J. M., Capirci, O., Longobardi, E., & Caselli, M. C. (1999). Gesturing in mother–child interactions. *Cognitive Development, 14*(1), 57–75.

Iverson, J. M., & Goldin-Meadow, S. (1998). Why people gesture when they speak. *Nature, 396*, 228.

Iwamura, Y. (1998). Hierarchical somatosensory processing. *Current Opinion in Neurobiology, 8*(4), 522–528.

Iwamura, Y., Iriki, A., & Tanaka, M. (1994). Bilateral hand representation in the postcentral somatosensory cortex. *Nature, 369*, 554–556.

Iwamura, Y., Tanaka, M., & Hikosaka, O. (1980). Overlapping representation of fingers in the somatosensory cortex (area 2) of the conscious monkey. *Brain Research, 197*(2), 516–520.

Iwamura, Y., Tanaka, M., Sakamoto, M., & Hikosaka, O. (1983a). Converging patterns of finger representation and complex response properties of neurons in area 1 of the first somatosensory cortex of the conscious monkey. *Experimental Brain Research, 51*, 327–337.

Iwamura, Y., Tanaka, M., Sakamoto, M., & Hikosaka, O. (1983b). Functional subdivisions representing different finger regions in area 3 of the first somatosensory cortex of the conscious monkey. *Experimental Brain Research, 51*, 315–326.

Iwamura, Y., Tanaka, M., Sakamoto, M., & Hikosaka, O. (1985). Diversity in receptive field properties of vertical neuronal arrays in the crown of the postcentral gyrus of the conscious monkey. *Experimental Brain Research, 58*(2), 400–411.

Iwamura, Y., Tanaka, M., Sakamoto, M., & Hikosaka, O. (1993). Rostrocaudal gradients in the neuronal receptive field complexity in the finger region of the alert monkey's postcentral gyrus. *Experimental Brain Research, 92*(3), 360–368.

Iwamura, Y., Taoka, M., & Iriki, A. (2001). Bilateral activity and callosal connections in the somatosensory cortex. *The Neuroscientist, 7*(5), 419–429.

Iwaniuk, A. N., Pellis, S. M., & Whishaw, I. Q. (1999). Is digital dexterity really related to corticospinal projections?: A re-analysis of the Heffner and Masterton data set using modern comparative statistics. *Behavioural Brain Research, 101*(2), 173–187.

Iwaniuk, A. N., & Whishaw, I. Q. (1999). How skilled are the skilled limb movements of the raccoon (*Procyon lotor*)? *Behavioural Brain Research, 99*(1), 35–44.

Jackson, J. H. (1864). Hemiplegia on the right side, with loss of speech. *British Medical Journal, 1*(177), 572–573.

Jackson, S. R., Jackson, G. M., Harrison, J., Henderson, L., & Kennard, C. (1995). The internal control of action and Parkinson's disease: A kinematic analysis of visually-guided and memory-guided prehension movements. *Experimental Brain Research, 105*(1), 147–162.

Jackson, S. R., Newport, R., Husain, M., Fowlie, J. E., O'Donoghue, M., & Bajaj, N. (2009). There may be more to reaching than meets the eye: Re-thinking optic ataxia. *Neuropsychologia, 47*(6), 1397–1408.

Jackson, S. R., Newport, R., Mort, D., Husain, M., Jackson, G. M., Swainson, R., Pears, S., & Wilson, B. (2005). Action binding and the parietal lobes: Some new perspectives on optic ataxia. In G. Humphreys & M. Riddoch (Eds.), *Attention in action* (pp. 303–324). Psychology Press.

Jackson, C. V., & Zangwill, O. L. (1952). Experimental finger dyspraxia. *Quarterly Journal of Experimental Psychology, 4*(1), 1–10.

Jacobs, S., Danielmeier, C., & Frey, S. H. (2010). Human anterior intraparietal and ventral premotor cortices support representations of grasping with the hand or a novel tool. *Journal of Cognitive Neuroscience, 22*(11), 2594–2608.

Jain, N., Catania, K. C., & Kaas, J. H. (1998). A histologically visible representation of the fingers and palm in primate area 3b and its immutability following long-term deafferentations. *Cerebral Cortex, 8*(3), 227–236.

Jakobson, L. S., Archibald, Y. M., Carey, D. P., & Goodale, M. A. (1991). A kinematic analysis of reaching and grasping movements in a patient recovering from optic ataxia. *Neuropsychologia, 29*(8), 803–809.

Jakobson, L. S., Servos, P., Goodale, M. A., & Lassonde, M. (1994). Control of proximal and distal components of prehension in callosal agenesis. *Brain, 117*(Pt 5), 1107–1113.

James, T. W., Servos, P., Kilgour, A. R., Huh, E., & Lederman, S. (2006). The influence of familiarity on brain activation during haptic exploration of 3-D facemasks. *Neuroscience Letters, 397*(3), 269–273.

James, W. (1890). *The principles of psychology.* Dover.

Janecka, J. E., Miller, W., Pringle, T. H., Wiens, F., Zitzmann, A., Helgen, K. M., Springer, M. S., & Murphy, W. J. (2007). Molecular and genomic data identify the closest living relative of primates. *Science, 318*(5851), 792–794.

Jarrassé, N., Nicol, C., Richer, F., Touillet, A., Martinet, N., Paysant, J., & de Graaf, J. B. (2017). Voluntary phantom hand and finger movements in transhumeral amputees could be used to naturally control polydigital prostheses. *International Conference on Rehabilitation Robotics*, 1239–1245. https://ieeexplore.ieee.org/document/8009419

Jeannerod, M. (1981). Intersegmental coordination during reaching at natural visual objects. In J. Long & A. Baddeley (Eds.), *Attention and Performance IX* (pp. 153–168). Erlbaum.

Jeannerod, M. (1984). The timing of natural prehension movements. *Journal of Motor Behavior*, *16*(3), 235–254.

Jeannerod, M. (1986). The formation of finger grip during prehension: A cortically mediated visuomotor pattern. *Behavioural Brain Research*, *19*(2), 99–116.

Jeannerod, M. (1988). *The neural and behavioural organization of goal-directed movements*. Oxford University Press.

Jeannerod, M., Arbib, M. A., Rizzolatti, G., & Sakata, H. (1995). Grasping objects: The cortical mechanisms of visuomotor transformation. *Trends in Neurosciences*, *18*(7), 314–320.

Jeannerod, M., Decety, J., & Michel, F. (1994). Impairment of grasping movements following a bilateral posterior parietal lesion. *Neuropsychologia*, *32*(4), 369–380.

Jenkins, I. H., Fernandez, W., Playford, E. D., Lees, A. J., Frackowiak, R. S., Passingham, R. E., & Brooks, D. J. (1992). Impaired activation of the supplementary motor area in Parkinson's disease is reversed when akinesia is treated with apomorphine. *Annals of Neurology*, *32*(6), 749–757.

Jenkinson, P. M., Moro, V., & Fotopoulou, A. (2023). Disorders of body ownership. In A. J. T. Alsmith & M. R. Longo (Eds.), *The Routledge Handbook of Bodily Awareness* (pp. 398–410). Routledge.

Jenkinson, P. M., Papadaki, C., Besharati, S., Moro, V., Gobbetto, V., Crucianelli, L., Kirsch, L. P., Avesani, R., Ward, N. S., & Fotopoulou, A. (2020). Welcoming back my arm: Affective touch increases body ownership following right-hemisphere stroke. *Brain Communications*, *2*(1), fcaa034.

Jerde, T. E., Soechting, J. F., & Flanders, M. (2003). Biological constraints simplify the recognition of hand shapes. *IEEE Transactions on Biomedical Engineering*, *50*(2), 265–269.

Jerison, H. J. (1973). *Evolution of the brain and intelligence*. Academic Press.

Johansson, R. S., & Vallbo, Å. B. (1983). Tactile sensory coding in the glabrous skin of the human hand. *Trends in Neurosciences*, *6*(1), 27–32.

Johnson, K. O. (2001). The roles and functions of cutaneous mechanoreceptors. *Current Opinion in Neurobiology*, *11*(4), 455–461.

Jones, E. (1909). The dyschiric syndrome. *Journal of Abnormal Psychology*, *4*(5), 311–327.

Jones, E. G. (1993). GABAergic neurons and their role in cortical plasticity in primates. *Cerebral Cortex*, *3*(5), 361–372.

Jones, P. R. (1980). Experimental butchery with modern stone tools and its relevance for Palaeolithic archaeology. *World Archaeology*, *12*(2), 153–165.

Jüeptner, M., Stephan, K. M., Frith, C. D., Brooks, D. J., Frackowiak, R. S., & Passingham, R. E. (1997). Anatomy of motor learning. I. Frontal cortex and attention to action. *Journal of Neurophysiology, 77*(3), 1313–1324.

Jung, W. P., Kahrs, B. A., & Lockman, J. J. (2015). Manual action, fitting, and spatial planning: Relating objects by young children. *Cognition, 134*, 128–139.

Kaas, J. H. (1987). The organization of neocortex in mammals: Implications for theories of brain function. *Annual Review of Psychology, 38*, 129–151.

Kaas, J. H. (2013). The evolution of brains from early mammals to humans. *Wiley Interdisciplinary Reviews Cognitive Science, 4*(1), 33–45.

Kaas, J. H., & Catania, K. C. (2002). How do features of sensory representations develop? *BioEssays, 24*(4), 334–343.

Kaas, J. H., Nelson, R. J., Sur, M., Dykes, R. W., & Merzenich, M. M. (1984). The somatotopic organization of the ventroposterior thalamus of the squirrel monkey, *Saimiri sciureus*. *Journal of Comparative Neurology, 226*(1), 111–140.

Kaas, J. H., Nelson, R. J., Sur, M., Lin, C.-S., & Merzenich, M. M. (1979). Multiple representations of the body within the primary somatosensory cortex of primates. *Science, 204*(4392), 521–523.

Kaas, J. H., Qi, H.-X., & Stepniewska, I. (2018). The evolution of parietal cortex in primates. In G. Vallar & H. B. Coslett (Eds.), *Handbook of clinical neurology: The parietal lobe* (1st ed., Vol. 151, pp. 31–52). Elsevier.

Kaas, J. H., & Stepniewska, I. (2016). Evolution of posterior parietal cortex and parietal-frontal networks for specific actions in primates. *Journal of Comparative Neurology, 524*(3), 595–608.

Kadmon Harpaz, N., Flash, T., & Dinstein, I. (2013). Scale-invariant movement encoding in the human motor system. *Neuron, 81*(2), 452–462.

Kahn, D. M., & Krubitzer, L. (2002). Massive cross-modal cortical plasticity and the emergence of a new cortical area in developmentally blind mammals. *Proceedings of the National Academy of Sciences, 99*(17), 11429–11434.

Kakebeeke, T. H., Knaier, E., Chaouch, A., Caflisch, J., Rousson, V., Largo, R. H., & Jenni, O. G. (2018). Neuromotor development in children. Part 4: New norms from 3 to 18 years. *Developmental Medicine & Child Neurology, 60*(8), 810–819.

Kakei, S., Hoffman, D. S., & Strick, P. L. (1999). Muscle and movement representations in the primary motor cortex. *Science, 285*(5436), 2136–2139.

Kakei, S., Hoffman, D. S., & Strick, P. L. (2001). Direction of action is represented in the ventral premotor cortex. *Nature Neuroscience, 4*(10), 1020–1025.

Kalaska, J. F., Cohen, D. A. D., Hyde, M. L., & Prud'homme, M. (1989). A comparison of movement direction-related versus load direction-related activity in primate motor cortex, using a two-dimensional reaching task. *Journal of Neuroscience, 9*(6), 2080–2102.

Kalaska, J. F., Cohen, D. A. D., Prud'Homme, M., & Hyde, M. L. (1990). Parietal area 5 neuronal activity encodes movement kinematics, not movement dynamics. *Experimental Brain Research, 80*(2), 351–364.

Kaminski, J., Riedel, J., Call, J., & Tomasello, M. (2005). Domestic goats, *Capra hircus*, follow gaze direction and use social cues in an object choice task. *Animal Behaviour, 69*(1), 11–18.

Kammers, M. P. M., Longo, M. R., Tsakiris, M., Dijkerman, H. C., & Haggard, P. (2009). Specificity and coherence of body representations. *Perception, 38*(12), 1804–1820.

Kanwisher, N. (2010). Functional specificity in the human brain: A window into the functional architecture of the mind. *Proceedings of the National Academy of Sciences, 107*(25), 11163–11170.

Kapandji, A. I. (2019). *The physiology of the joints. 1: The upper limb* (7th ed.). Handspring.

Kaplan, G. (2011). Pointing gesture in a bird—merely instrumental or a cognitively complex behavior? *Current Zoology, 57*(4), 453–467.

Karakostis, F. A., Haeufle, D., Anastopoulou, I., Moraitis, K., Hotz, G., Tourloukis, V., & Harvati, K. (2021). Biomechanics of the human thumb and the evolution of dexterity. *Current Biology, 31*(6), 1317–1325.

Karakostis, F. A., Hotz, G., Tourloukis, V., & Harvati, K. (2018). Evidence for precision grasping in Neandertal daily activities. *Science Advances, 4*(9), eaat2369.

Karnath, H.-O., & Perenin, M.-T. (2005). Cortical control of visually guided reaching: Evidence from patients with optic ataxia. *Cerebral Cortex, 15*(10), 1561–1569.

Karni, A., Meyer, G., Jezzard, P., Adams, M. M., Turner, R., & Ungerleider, L. G. (1995). Functional MRI evidence for adult motor cortex plasticity during motor skill learning. *Nature, 377*(6545), 155–158.

Karni, A., Meyer, G., Rey-Hipolito, C., Jezzard, P., Adams, M. M., Turner, R., & Ungerleider, L. G. (1998). The acquisition of skilled motor performance: Fast and slow experience-driven changes in primary motor cortex. *Proceedings of the National Academy of Sciences, 95*(3), 861–868.

Kastner, S., Chen, Q., Jeong, S. K., & Mruczek, R. E. B. (2017). A brief comparative review of primate posterior parietal cortex: A novel hypothesis on the human toolmaker. *Neuropsychologia, 105*, 123–134.

Katz, D. (1925). *The world of touch* [L. E. Krueger, Trans.]. Erlbaum.

Kawai, R., Markman, T., Poddar, R., Ko, R., Fantana, A. L., Dhawale, A. K., Kampff, A. R., & Ölveczky, B. P. (2015). Motor cortex is required for learning but not for executing a motor skill. *Neuron, 86*(3), 800–812.

Kawamura, M., Hirayama, K., Shinohara, Y., Watanabe, Y., & Sugishita, M. (1987). Alloaesthesia. *Brain, 110*(Pt 1), 225–236.

Kay, B. A., Turvey, M. T., & Meijer, O. G. (2003). An early oscillator model: Studies on the biodynamics of the piano strike (Bernstein & Popova, 1930). *Motor Control, 7*(1), 1–45.

Kebschull, J. M., Richman, E. B., Ringach, N., Friedmann, D., Albarran, E., Kolluru, S. S., Jones, R. C., Allen, W. E., Wang, Y., Cho, S. W., Zhou, H., Ding, J. B., Chang, H. Y., Deisseroth, K., Quake, S. R., & Luo, L. (2020). Cerebellar nuclei evolved by repeatedly duplicating a conserved cell-type set. *Science, 370*(6523), eabd5059.

Keeley, L. H. (1977). The functions of paleolithic flint tools. *Scientific American, 237*(5), 108–127.

Keeley, L. H., & Toth, N. (1981). Microwear polishes on early stone tools from Koobi Fora, Kenya. *Nature, 293*, 464–465.

Kendon, A. (1988). *Sign languages of aboriginal Australia*. Cambridge University Press.

Kendon, A. (1994). Do gestures communicate? A review. *Research on Language and Social Interaction, 27*(3), 175–200.

Kennard, M. A. (1936). Age and other factors in motor recovery from precentral lesions in monkeys. *American Journal of Physiology, 115*(1), 138–146.

Kennard, M. A. (1938). Reorganization of motor function in the cerebral cortex of monkeys deprived of motor and premotor areas in infancy. *Journal of Neurophysiology, 1*(6), 477–496.

Kennard, M. A. (1940). Relation of age to motor impairment in man and in subhuman primates. *Archives of Neurology & Psychiatry, 44*(2), 377–397.

Kennett, S., Spence, C., & Driver, J. (2002). Visuo-tactile links in covert exogenous spatial attention remap across changes in unseen hand posture. *Perception and Psychophysics, 64*(7), 1083–1094.

Kennett, S., Taylor-Clarke, M., & Haggard, P. (2001). Noninformative vision improves the spatial resolution of touch in humans. *Current Biology, 11*(15), 1188–1191.

Kenward, B., Weir, A. A. S., Rutz, C., & KacelIA. (2005). Tool manufacture by naive juvenile crows. *Nature, 433*, 121.

Kersken, V., Gómez, J.-C., Liszkowski, U., Soldati, A., & Hobaiter, C. (2019). A gestural repertoire of 1- to 2-year-old human children: In search of the ape gestures. *Animal Cognition, 22*(4), 577–595.

Key, A. J. M., Dunmore, C. J., & Marzke, M. W. (2019). The unexpected importance of the fifth digit during stone tool production. *Scientific Reports, 9*, 16724.

Key, A. J. M., Jarić, I., & Roberts, D. L. (2021). Modelling the end of the Acheulean at global and continental levels suggests widespread persistence into the Middle Palaeolithic. *Humanities and Social Sciences Communications, 8*, 55.

Key, A. J. M., Merritt, S. R., & Kivell, T. L. (2018). Hand grip diversity and frequency during the use of Lower Palaeolithic. *Journal of Human Evolution, 125*, 137–158.

Khateb, A., Simon, S. R., Dieguez, S., Lazeyras, F., Momjian-Mayor, I., Blanke, O., Landis, T., Pegna, A. J., & Annoni, J.-M. (2009). Seeing the phantom: A functional magnetic resonance imaging study of a supernumerary phantom limb. *Annals of Neurology, 65*(6), 698–705.

Kibii, J. M., Clarke, R. J., & Tocheri, M. W. (2011). A hominin scaphoid from Sterkfontein, Member 4: Morphological description and first comparative phenetic 3D analyses. *Journal of Human Evolution*, *61*(4), 510–517.

Kieliba, P., Clode, D., Maimon-Mor, R. O., & Makin, T. R. (2021). Robotic hand augmentation drives changes in neural body representation. *Science Robotics*, *6*(54), eabd7935.

Kikkert, S., Kolasinski, J., Jbabdi, S., Tracey, I., Beckmann, C. F., Johansen-Berg, H., & Makin, T. R. (2016). Revealing the neural fingerprints of a missing hand. *eLife*, *5*, e15292.

Kilgour, A. R., Kitada, R., Servos, P., James, T. W., & Lederman, S. J. (2005). Haptic face identification activates ventral occipital and temporal areas: An fMRI study. *Brain and Cognition*, *59*(3), 246–257.

Kim, J. S., Kanjlia, S., Merabet, L. B., & Bedny, M. (2017). Development of the visual word form area requires visual experience: Evidence from blind braille readers. *Journal of Neuroscience*, *37*(47), 11495–11504.

Kim, S.-G., Ashe, J., Hendrich, K., Ellermann, J. M., Merkle, H., Uğurbil, K., & Georgopoulos, A. P. (1993). Functional magnetic resonance imaging of motor cortex: Hemispheric asymmetry and handedness. *Science*, *261*(5121), 615–617.

Kim, U., & Ebner, F. F. (1999). Barrels and septa: Separate circuits in rat barrel field cortex. *Journal of Comparative Neurology*, *408*(4), 489–505.

Kimura, D. (1981). Neural mechanisms in manual signing. *Sign Language Studies*, *33*, 291–312.

Kinoshita, M., Matsui, R., Kato, S., Hasegawa, T., Kasahara, H., Watakabe, A., Yamamori, T., Nishimura, Y., Alstermark, B., Watanabe, D., Kobayashi, K., & Isa, T. (2012). Genetic dissection of the circuit for hand dexterity in primates. *Nature*, *487*, 235–238.

Kinsbourne, M. (1987). Mechanisms of unilateral neglect. In M. Jeannerod (Ed.), *Neurophysiological and neuropsychological aspects of spatial neglect* (pp. 69–86). Elsevier.

Kinsbourne, M., & Rosenfield, D. B. (1974). Agraphia selective for written spelling: An experimental case study. *Brain and Language*, *1*(3), 215–225.

Kita, K., Furuya, S., Osu, R., Sakamoto, T., & Hanakawa, T. (2021). Aberrant cerebello-cortical connectivity in pianists with focal task-specific dystonia. *Cerebral Cortex*, *31*(10), 4853–4863.

Kita, S., Alibali, M. W., & Chu, M. (2017). How do gestures influence thinking and speaking? The gesture-for-conceptualization hypothesis. *Psychological Review*, *124*(3), 245–266.

Kitada, R., Johnsrude, I. S., Kochiyama, T., & Lederman, S. J. (2009). Functional specialization and convergence in the occipito-temporal cortex supporting haptic and visual identification of human faces and body parts. *Journal of Cognitive Neuroscience*, *21*(10), 2027–2045.

Kivell, T. L. (2015). Evidence in hand: Recent discoveries and the early evolution of human manual manipulation. *Philosophical Transactions of the Royal Society B*, *370*(1682), 20150105.

Kivell, T. L., Deane, A. S., Tocheri, M. W., Orr, C. M., Schmid, P., Hawks, J., Berger, L. R., & Churchill, S. E. (2015). The hand of *Homo naledi*. *Nature Communications*, *6*, 8431.

Kivell, T. L., Kibii, J. M., Churchill, S. E., Schmid, P., & Berger, L. R. (2011). *Australopithecus sediba* hand demonstrates mosaic evolution of locomotor and manipulative abilities. *Science, 333*(6048), 1411–1417.

Kivell, T. L., & Schmitt, D. (2009). Independent evolution of knuckle-walking in African apes shows that humans did not evolve from a knuckle-walking ancestor. *Proceedings of the National Academy of Sciences, 106*(34), 14241–14246.

Kısa, Y. D., Goldin-Meadow, S., & Casasanto, D. (2022). Do gestures really facilitate speech production? *Journal of Experimental Psychology: General, 151*(6), 1252–1271.

Klatzky, R. L., & Lederman, S. J. (1995). Identifying objects from a haptic glance. *Perception & Psychophysics, 57*(8), 1111–1123.

Klatzky, R. L., & Lederman, S. J. (2002). Perceiving texture through a probe. In M. L. McLaughlin, J. P. Hespanha, & G. S. Sukhatme (Eds.), *Touch in Virtual Environments* (pp. 180–193). Prentice Hall.

Klatzky, R. L., Lederman, S. J., & Metzger, V. A. (1985). Identifying objects by touch: An "expert system." *Perception & Psychophysics, 37*(4), 4–7.

Klatzky, R. L., Lederman, S. J., Hamilton, C., Grindley, M., & Swendsen, R. H. (2003). Feeling textures through a probe: Effects of probe and surface geometry and exploratory factors. *Perception & Psychophysics, 65*(4), 613–631.

Kleeman, L. T., & Shafritz, A. B. (2013). The Krukenberg procedure. *Journal of Hand Surgery, 38*(1), 173–175.

Klein, R. G. (2000). Archeology and the evolution of human behavior. *Evolutionary Anthropology, 9*(1), 17–36.

Klima, E., & Bellugi, U. (1979). *The signs of language*. Harvard University Press.

Kline, D. G. (2016). Silas Weir Mitchell and "The strange case of George Dedlow." *Neurosurgical Focus, 41*(1), E5.

Knecht, S., Dräger, B., Deppe, Bobe, L., Lohmann, H., Flöel, A., Ringelstein, E. B., & Henningsen, H. (2000). Handedness and hemispheric language dominance in healthy humans. *Brain, 123*(Pt 12), 2512–2518.

Knights, E., Mansfield, C., Tonin, D., Saada, J., Smith, F. W., & Rossit, S. (2021). Hand-selective visual regions represent how to grasp 3D tools: Brain decoding during real actions. *Journal of Neuroscience, 41*(24), 5263–5273.

Koechlin, E., & Jubault, T. (2006). Broca's area and the hierarchical organization of human behavior. *Neuron, 50*(6), 963–974.

Koffka, K. (1935). *Principles of gestalt psychology*. Lund Humphries.

Koh, T. H. H. G., & Eyre, J. A. (1988). Maturation of corticospinal tracts assessed by electromagnetic stimulation of the motor cortex. *Archives of Disease in Childhood, 63*, 1347–1352.

Köhler, W. (1925). *The mentality of apes* (2nd ed) [E. Winter, Trans.]. Kegan Paul.

Kohonen, T. (2006). Self-organizing neural projections. *Neural Networks, 19*(6–7), 723–733.

Kolasinski, J., Dima, D. C., Mehler, D. M. A., Stephenson, A., Valadan, S., Kusmia, S., & Rossiter, H. E. (2020). Spatially and temporally distinct encoding of muscle and kinematic information in rostral and caudal primary motor cortex. *Cerebral Cortex Communications, 1*(1), tgaa009.

Kolasinski, J., Makin, T. R., Jbabdi, S., Clare, S., Stagg, C. J., & Johansen-Berg, H. (2016). Investigating the stability of fine-grain digit somatotopy in individual human participants. *Journal of Neuroscience, 36*(4), 1113–1127.

Konczak, J., & Dichgans, J. (1997). The development toward stereotypic arm kinematics during reaching in the first 3 years of life. *Experimental Brain Research, 117*(2), 346–354.

Konczak, J., Sciutti, A., Avanzino, L., Squeri, V., Gori, M., Masia, L., Abbruzzese, G., & Sandini, G. (2012). Parkinson's disease accelerates age-related decline in haptic perception by altering somatosensory integration. *Brain, 135*(Pt 11), 3371–3379.

Konen, C. S., & Haggard, P. (2014). Multisensory parietal cortex contributes to visual enhancement of touch in humans. *Cerebral Cortex, 24*(2), 501–507.

Konen, C. S., & Kastner, S. (2008). Two hierarchically organized neural systems for object information in human visual cortex. *Nature Neuroscience, 11*(2), 224–231.

Konen, C. S., Mruczek, R. E. B., Montoya, J. L., & Kastner, S. (2013). Functional organization of human posterior parietal cortex: Grasping- and reaching-related activations relative to topographically organized cortex. *Journal of Neurophysiology, 109*(12), 2897–2908.

Krause, M. A., & Fouts, R. S. (1997). Chimpanzee (*Pan troglodytes*) pointing: Hand shapes, accuracy, and the role of eye gaze. *Journal of Comparative Psychology, 111*(4), 330–336.

Krause, M. A., Udell, M. A. R., Leavens, D. A., & Skopos, L. (2018). Animal pointing: Changing trends and findings from 30 years of research. *Journal of Comparative Psychology, 132*(3), 326–345.

Krauss, R. M. (1998). Why do we gesture when we speak? *Current Directions in Psychological Science, 7*(2), 54–60.

Krauss, R. M., Chen, Y., & Gottesman, R. F. (2000). Lexical gestures and lexical access: A process model. In D. McNeill (Ed.), *Language and gesture* (pp. 261–283). Cambridge University Press.

Kravitz, H., & Boehm, J. J. (1971). Rhythmic habit patterns in infancy: Their sequence, age of onset, and frequency. *Child Development, 42*(2), 399–413.

Kravitz, A. V, Freeze, B. S., Parker, P. R. L., Kay, K., Thwin, M. T., Deisseroth, K., &. Kreitzer, A. C. (2010). Regulation of parkinsonian motor behaviours by optogenetic control of basal ganglia circuitry. *Nature, 466*(7306), 622–626.

Krubitzer, L. A. (1995). The organization of neocortex in mammals: Are species differences really so different? *Trends in Neurosciences, 18*(9), 408–417.

Krubitzer, L. A. (2007). The magnificent compromise: Cortical field evolution in mammals. *Neuron*, *56*(2), 201–208.

Krubitzer, L. A., & Calford, M. B. (1992). Five topographically organized fields in the somatosensory cortex of the flying fox: Microelectrode maps, myeloarchitecture, and cortical modules. *Journal of Comparative Neurology*, *317*(1), 1–30.

Krubitzer, L. A., & Disbrow, E. A. (2008). The evolution of parietal areas involved in hand use in primates. In J. H. Kaas & E. P. Gardner (Eds.), *The senses: A comprehensive reference, somatosensation* (Vol. 6, pp. 183–214). Elsevier.

Krubitzer, L. A., & Huffman, K. J. (2000). Arealization of the neocortex in mammals: Genetic and epigenetic contributions to the phenotype. *Brain, Behavior and Evolution*, *55*(6), 322–335.

Krubitzer, L. A., Huffman, K. J., Disbrow, E., & Recanzone, G. (2004). Organization of area 3a in macaque monkeys: Contributions to the cortical phenotype. *Journal of Comparative Neurology*, *471*(1), 97–111.

Krubitzer, L. A., & Kaas, J. H. (1988). Responsiveness and somatotopic organization of anterior parietal field 3b and adjoining cortex in newborn and infant monkeys. *Somatosensory and Motor Research*, *6*(2), 179–205.

Krubitzer, L. A., & Kaas, J. H. (1990). The organization and connections of somatosensory cortex in marmosets. *Journal of Neuroscience*, *10*(3), 952–974.

Krubitzer, L. A., Manger, P., Pettigrew, J., & Calford, M. (1995). Organization of somatosensory cortex in monotremes: In search of the prototypical plan. *Journal of Comparative Neurology*, *351*(2), 261–306.

Kuehn, E., Dinse, J., Jakobsen, E., Long, X., Schäfer, A., Bazin, P.-L., Villringer, A., Sereno, M. I., & Margulies, D. S. (2017). Body topography parcellates human sensory and motor cortex. *Cerebral Cortex*, *27*(7), 3790–3805.

Kuhtz-Buschbeck, J. P., Stolze, H., Jöhnk, K., Boczek-Funcke, A., & Illert, M. (1998). Development of prehension movements in children: A kinematic study. *Experimental Brain Research*, *122*(4), 424–432.

Kuman, K., & Clarke, R. J. (2000). Stratigraphy, artefact industries and hominid associations for Sterkfontein, Member 5. *Journal of Human Evolution*, *38*(6), 827–847.

Kurjak, A., Vecek, N., Hafner, T., Bozek, T., Funduk-Kurjak, B., &. Ujevic, B. (2002). Prenatal diagnosis: What does four-dimensional ultrasound add? *Journal of Perinatal Medicine*, *30*(1), 57–62.

Kuypers, H. G. J. M. (1960). Central cortical projections to motor and somato-sensory cell groups: An experimental study in the rhesus monkey. *Brain*, *83*(1), 161–184.

Kuypers, H. G. J. M. (1962). Corticospinal connections: Postnatal development in the rhesus monkey. *Science*, *138*(3541), 678–680.

Kuypers, H. G. J. M. (1978). The motor system and the capacity to execute highly fractionated distal extremity movements. *Electroencephalography and Clinical Neurophysiology*, *34*, 429–431.

Lacoste, A. M. B., Schoppik, D., Robson, D. N., Haesemeyer, M., Portugues, R., Li, J. M., Randlett, O., Wee, C. L., Engert, F., & Schier, A. F. (2015). A convergent and essential interneuron pathway for Mauthner-cell-mediated escapes. *Current Biology, 25*(11), 1526–1534.

Làdavas, E., di Pellegrino, G., Farnè, A., & Zeloni, G. (1998). Neuropsychological evidence of an integrated visuotactile representation of peripersonal space in humans. *Journal of Cognitive Neuroscience, 10*(5), 581–589.

LaMotte, R. H. (2000). Softness discrimination with a tool. *Journal of Neurophysiology, 83*(4), 1777–1786.

LaMotte, R. H., & Acuña, C. (1978). Defects in accuracy of reaching after removal of posterior parietal cortex in monkeys. *Brain Research, 139*(2), 309–326.

Lashley, K. S. (1924). Studies of cerebral function in learning. V. The retention of motor habits after destruction of the so-called motor areas in primates. *Archives of Neurology & Psychiatry, 12*(3), 249–276.

Lashley, K. S. (1930). Basic neural mechanisms in behavior. *Psychological Review, 37*(1), 265–283.

Lashley, K. S. (1942). The problem of cerebral organization in vision. In H. Klüver (Ed.), *Visual mechanisms* (pp. 301–322). Jaques Cattell Press.

Lashley, K. S. (1951). The problem of serial order in behavior. In L. A. Jeffress (Ed.), *Cerebral mechanisms in behavior* (pp. 112–131). Wiley.

Laskin, S. E., & Spencer, W. A. (1979a). Cutaneous masking. II. Geometry of excitatory and inhibitory receptive fields of single units in somatosensory cortex of the cat. *Journal of Neurophysiology, 42*(4), 1061–1082.

Laskin, S. E., & Spencer, W. A. (1979b). Cutaneous masking. I. Psychophysical observations on interactions of multipoint stimuli in man. *Journal of Neurophysiology, 42*(4), 1048–1060.

Laszlo, J. I., Baguley, R. A., & Bairstow, P. J. (1970). Bilateral transfer in tapping skill in the absence of peripheral information. *Journal of Motor Behavior, 2*(4), 261–271.

Lawrence, D. G., & Hopkins, D. A. (1976). The development of motor control in the rhesus monkey: Evidence concerning the role of corticomotoneuronal connections. *Brain, 99*(2), 235–254.

Lawrence, D. G., & Kuypers, H. G. J. M. (1968). The functional organization of the motor system in the monkey. I. The effects of bilateral pyramidal lesions. *Brain, 91*(1), 1–14.

Lawson, R. (2014). Recognizing familiar objects by hand and foot: Haptic shape perception generalizes to inputs from unusual locations and untrained body parts. *Attention, Perception, & Psychophysics, 76*(2), 541–558.

Le Cornu Knight, F., Bremner, A. J., & Cowie, D. (2020). Does the language we use to segment the body, shape the way we perceive it? A study of tactile perceptual distortions. *Cognition, 197*, 104127.

Le Cornu Knight, F., Cowie, D., & Bremner, A. J. (2017). Part-based representations of the body in early childhood: Evidence from perceived distortions of tactile space across limb boundaries. *Developmental Science, 20*(6), e12439.

Le Cornu Knight, F., Longo, M. R., & Bremner, A. J. (2014). Categorical perception of tactile distance. *Cognition, 131*(2), 254–262.

Le Gros Clark, W. E. (1959). *The antecedents of man*. Edinburgh University Press.

Leakey, L. S. B. (1959). A new fossil skull from Olduvai. *Nature, 184*, 491–493.

Leakey, L. S. B., Hopwood, A. T., & Reck, H. (1931). New yields from the Oldoway bone beds, Tanganyika Territory. *Nature, 128*, 1075.

Leakey, L. S. B., Tobias, P. V., & Napier, J. R. (1964). A new species of the genus *Homo* from Olduvai Gorge. *Nature, 202*, 7–9.

Leakey, M. D. (1966). A review of the Oldowan culture form Olduvai Gorge, Tanzania. *Nature, 210*, 462–466.

Leakey, M. D. (1971). *Olduvai Gorge: Volume 3: Excavations in beds I and II, 1960–1963*. Cambridge University Press.

Leavens, D. A., & Hopkins, W. D. (1998). Intentional communication by chimpanzees: A cross-sectional study of the use of referential gestures. *Developmental Psychology, 34*(5), 813–822.

Leavens, D. A., & Hopkins, W. D. (1999). The whole-hand point: The structure and function of pointing from a comparative perspective. *Journal of Comparative Psychology, 113*(4), 417–425.

Leavens, D. A., Hopkins, W. D., & Bard, K. A. (2005). Understanding the point of chimpanzee pointing: Epigenesis and ecological validity. *Current Directions in Psychological Science, 14*(4), 185–189.

Leavens, D. A., Hopkins, W. D., & Thomas, R. K. (2004). Referential communication by chimpanzees (*Pan troglodytes*). *Journal of Comparative Psychology, 118*(1), 48–57.

Leavens, D. A., Russell, J. L., & Hopkins, W. D. (2005). Intentionality as measured in the persistence and elaboration of communication by chimpanzees (*Pan troglodytes*). *Child Development, 76*(1), 291–306.

Lebel, C., & Beaulieu, C. (2011). Longitudinal development of human brain wiring continues from childhood into adulthood. *Journal of Neuroscience, 31*(30), 10937–10947.

Lederman, S. J., & Klatzky, R. L. (1987). Hand movements: A window into haptic object recognition. *Cognitive Psychology, 19*(3), 342–368.

Lee, H., Devlin, J. T., Shakeshaft, C., Stewart, L. H., Brennan, A., Glensman, J., Pitcher, K., Crinion, J., Mechelli, A., Frackowiak, R. S. J., Green, D. W., & Price, C. J. (2007). Anatomical traces of vocabulary acquisition in the adolescent brain. *Journal of Neuroscience, 27*(5), 1184–1189.

Lee, M., Hinder, M. R., Gandevia, S. C., & Carroll, T. J. (2010). The ipsilateral motor cortex contributes to cross-limb transfer of performance gains after ballistic motor practice. *Journal of Physiology, 588*(Pt 1), 201–212.

Lee, M.-H., Ranganathan, R., & Newell, K. M. (2011). Changes in object-oriented arm movements that precede the transition to goal-directed reaching in infancy. *Developmental Psychobiology, 53*(7), 685–693.

References

Lee, R. B. (1979). *The !Kung San: Men, women, and work in a foraging society*. Cambridge University Press.

Legerstee, M., Corter, C., & Kienapple, K. (1990). Hand, arm, and facial actions of young infants to a social and nonsocial stimulus. *Child Development, 61*(3), 774–784.

Lehmann, S. J., & Scherberger, H. (2013). Reach and gaze representations in macaque parietal and premotor grasp areas. *Journal of Neuroscience, 33*(16), 7038–7049.

Leiguarda, C., & Marsden, C. D. (2000). Limb apraxias: Higher-order disorders of sensorimotor integration. *Brain, 123*(Pt 5), 860–879.

Leinonen, L., Hyvärinen, J., Nyman, G., & Linnankoski, I. (1979). Functional properties of neurons in lateral part of associative area 7 in awake monkeys. *Experimental Brain Research, 34*(2), 299–320.

Lemelin, P. (1999). Morphological correlates of substrate use in didelphid marsupials: Implications for primate origins. *Journal of Zoology, 247*(2), 165–175.

Lemelin, P., & Diogo, R. (2016). Anatomy, function, and evolution of the primate hand musculature. In T. L. Kivell, P. Lemelin, & B. G. Richmond (Eds.), *The evolution of the primate hand* (pp. 155–193). Springer.

Lemon, R. N. (1993). The G. L. Brown Prize Lecture: Cortical control of the primate hand. *Experimental Physiology, 78*(3), 263–301.

Lemon, R. N., Kirkwood, P. A., Maier, M. A., Nakajima, K., & Nathan, P. (2004). Direct and indirect pathways for corticospinal control of upper limb motoneurons in the primate. *Progress in Brain Research, 143*, 263–279.

Lende, R. A. (1963). Cerebral cortex: A sensorimotor amalgam in the marsupialia. *Science, 141*(3582), 730–732.

Lenka, A., & Jankovic, J. (2021). Sports-related dystonia. *Tremor and Other Hyperkinetic Movements, 11*, 54.

Leo, A., Handjaras, G., Bianchi, M., Marino, H., Gabiccini, M., Guidi, A., Scilingo, E. P., Pietrini, P., Bicchi, A., Santello, M., & Ricciardi, E. (2016). A synergy-based hand control is encoded in human motor cortical areas. *eLife, 5*, e13420.

Leont'ev, A. N., & Zaporozhets, A. V. (1960). *Rehabilitation of hand function* [B. Haigh, Trans.]. Pergamon Press.

Lepre, C. J., Roche, H., Kent, D. V, Harmand, S., Quinn, R. L., Brugal, J.-P., Texier, P.-J., Lenoble, A., & Feibel, C. S. (2011). An earlier origin for the Acheulian. *Nature, 477*(7362), 82–85.

Leroi-Gourhan, A. (1967). Les mains de Gargas. Essai pour une étude d'ensemble [The hands of Gargas. Toward a general study]. *Bulletin de la Société Préhistorique Française, 64*, 107–122.

Leroi-Gourhan, A. (1986). The hands of Gargas: Toward a general study [A. Michelson, Trans.]. *October, 37*, 18–34.

Leung, E. H. L., & Rheingold, H. L. (1981). Development of pointing as a social gesture. *Developmental Psychology, 17*(2), 215–220.

Lewis, J. E., & Harmand, S. (2016). An earlier origin for stone tool making: Implications for cognitive evolution and the transition to *Homo. Philosophical Transactions of the Royal Society B, 371*(1698), 20150233.

Lewis, J. W. (2006). Cortical networks related to human use of tools. *Neuroscientist, 12*(3), 211–231.

Lewis, O. J. (1977). Joint remodelling and the evolution of the human hand. *Journal of Anatomy, 123*(Pt 1), 157–201.

Ley, P., Bottari, D., Shenoy, B. H., Kekunnaya, R., & Röder, B. (2013). Partial recovery of visual-spatial remapping of touch after restoring vision in a congenitally blind man. *Neuropsychologia, 51*(6), 1119–1123.

Leyton, A. S. F., & Sherrington, C. S. (1917). Observations on the excitable cortex of the chimpanzee, orang-utan, and gorilla. *Quarterly Journal of Experimental Physiology, 11*(2), 135–222.

Lhermitte, F. (1983). "Utilization behaviour" and its relation to lesions of the frontal lobes. *Brain, 106*(Pt 2), 237–255.

Li, Q., & Ni, X. (2016). An early Oligocene fossil demonstrates treeshrews are slowly evolving "living fossils." *Scientific Reports, 6*, 18627.

Libertus, K., Joh, A. S., & Needham, A. W. (2016). Motor training at 3 months affects object exploration 12 months later. *Developmental Science, 19*(6), 1058–1066.

Libertus, K., & Needham, A. (2010). Teach to reach: The effects of active vs. passive reaching experiences on action and perception. *Vision Research, 50*(24), 2750–2757.

Liddell, E. G. T., & Sherrington, C. S. (1924). Reflexes in response to stretch (myotatic reflexes). *Proceedings of the Royal Society B, 96*(675), 212–242.

Lieberman, D. E., McBratney, B. M., & Krovitz, G. (2002). The evolution and development of cranial form in *Homo sapiens. Proceedings of the National Academy of Sciences, 99*(3), 1134–1139.

Liepmann, H. (1900). Das krankheitsbild der apraxie (motorische asymbolie): Auf grand eines falles von einseitiger apraxie [The syndrome of apraxia (motor asymboly): On the basis of a case of unilateral apraxia]. *Monatsschrift für Psychiatrie und Neurologie, 8*, 15–44.

Liepmann, H. (1905). Die linke hemisphäre und das handeln [The left hemisphere and action]. *Münchener Medizinische Wochenschrift, 52*, 2322–2326.

Liepmann, H. (1977). The syndrome of apraxia (motor asymboly) based on a case of unilateral apraxia [W. H. O. Bohne, K. Liepmann, & D. A. Rottenberg, Trans.]. In D. A. Rottenberg & F. H. Hochberg (Eds.), *Neurological Classics in Modern Translation* (pp. 155–183). Hafner Press.

Liszkowski, U., Carpenter, M., Henning, A., Striano, T., & Tomasello, M. (2004). Twelve-month-olds point to share attention and interest. *Developmental Science, 7*(3), 297–307.

Liszkowski, U., Carpenter, M., & Tomasello, M. (2007a). Pointing out new news, old news, and absent referents at 12 months of age. *Developmental Science, 10*(2), F1–F7.

Liszkowski, U., Carpenter, M., & Tomasello, M. (2007b). Reference and attitude in infant pointing. *Journal of Child Language, 34*(1), 1–20.

Liu, Y., Neal, A. O., Rafal, R. D., & Medina, J. (2020). Intact tactile detection yet biased tactile localization in a hand-centered frame of reference: Evidence from a dissociation. *Neuropsychologia, 147*, 107585.

Livingstone, M. S., & Hubel, D. H. (1987). Psychophysical evidence for separate channels for the perception of form, color, movement, and depth. *Journal of Neuroscience, 7*(11), 3416–3468.

Livingstone, M. S., & Hubel, D. H. (1988). Segregation of form, color, movements, and depth: Anatomy, physiology, and perception. *Science, 240*(4853), 740–749.

Llorente, M., Mosquera, M., & Fabré, M. (2009). Manual laterality for simple reaching and bimanual coordinated task in naturalistic housed *Pan troglodytes*. *International Journal of Primatology, 30*(1), 183–197.

Lloyd, D. M. (2007). Spatial limits on referred touch to an alien limb may reflect boundaries of visuo-tactile peripersonal space surrounding the hand. *Brain and Cognition, 64*(1), 104–109.

Locke, J. (1690). *An essay concerning human understanding*. Routledge.

Lockman, J. J., Ashmead, D. H., & Bushnell, E. W. (1984). The development of anticipatory hand orientation during infancy. *Journal of Experimental Child Psychology, 37*(1), 176–186.

Logan, G. D. (1988). Toward an instance theory of automatization. *Psychological Review, 95*(4), 492–527.

Logan, G. D., & Crump, M. J. C. (2009). The left hand doesn't know what the right hand is doing. *Psychological Science, 20*(10), 1296–1300.

Logan, G. D., & Crump, M. J. C. (2010). Cognitive illusions of authorship reveal hierarchical error detection in skilled typists. *Science, 330*(6004), 683–686.

Lombard, M., & Pargeter, J. (2008). Hunting with Howiesons Poort segments: Pilot experimental study and the functional interpretation of archaeological tools. *Journal of Archaeological Science, 35*(9), 2523–2531.

Long, C., Conrad, P. W., Hall, E. A., & Furler, S. L. (1970). Intrinsic-extrinsic muscle control of the hand in power grip and precision handling: An electromyographic study. *Journal of Bone and Joint Surgery, 52*(5), 853–867.

Long, X., Goltz, D., Margulies, D. S., Nierhaus, T., & Villringer, A. (2014). Functional connectivity-based parcellation of the human sensorimotor cortex. *European Journal of Neuroscience, 39*(8), 1332–1342.

Longo, M. R. (2014). The effects of immediate vision on implicit hand maps. *Experimental Brain Research, 232*(4), 1241–1247.

Longo, M. R. (2018). The effects of instrumental action on perceptual hand maps. *Experimental Brain Research, 236*(11), 3113–3119.

Longo, M. R. (2020). Tactile distance anisotropy on the palm: A meta-analysis. *Attention, Perception, & Psychophysics, 82*, 2137–2146.

Longo, M. R. (2023). Motor adaptation and distorted body representations. *Trends in Cognitive Sciences, 27*(1), 9.

Longo, M. R., Azañón, E., & Haggard, P. (2010). More than skin deep: Body representation beyond primary somatosensory cortex. *Neuropsychologia, 48*(3), 655–668.

Longo, M. R., Betti, V., Aglioti, S. M., & Haggard, P. (2009). Visually induced analgesia: Seeing the body reduces pain. *Journal of Neuroscience, 29*(39), 12125–12130.

Longo, M. R., & Golubova, O. (2017). Mapping the internal geometry of tactile space. *Journal of Experimental Psychology: Human Perception and Performance, 43*(10), 1815–1827.

Longo, M. R., & Haggard, P. (2010). An implicit body representation underlying human position sense. *Proceedings of the National Academy of Sciences, 107*(26), 11727–11732.

Longo, M. R., & Haggard, P. (2011). Weber's illusion and body shape: Anisotropy of tactile size perception on the hand. *Journal of Experimental Psychology: Human Perception and Performance, 37*(3), 720–726.

Longo, M. R., & Haggard, P. (2012). Implicit body representations and the conscious body image. *Acta Psychologica, 141*(2), 164–168.

Longo, M. R., Iannetti, G. D., Mancini, F., Driver, J., & Haggard, P. (2012). Linking pain and the body: Neural correlates of visually induced analgesia. *Journal of Neuroscience, 32*(8), 2601–2607.

Longo, M. R., Kammers, M. P. M., Gomi, H., Tsakiris, M., & Haggard, P. (2009). Contraction of body representation induced by proprioceptive conflict. *Current Biology, 19*(17), R727–R728.

Longo, M. R., Long, C., & Haggard, P. (2012). Mapping the invisible hand: A body model of a phantom limb. *Psychological Science, 23*(7), 740–742.

Longo, M. R., & Lourenco, S. F. (2006). On the nature of near space: Effects of tool use and the transition to far space. *Neuropsychologia, 44*(6), 977–981.

Longo, M. R., Mancini, F., & Haggard, P. (2015). Implicit body representations and tactile spatial remapping. *Acta Psychologica, 160*, 77–87.

Longo, M. R., Pernigo, S., & Haggard, P. (2011). Vision of the body modulates processing in primary somatosensory cortex. *Neuroscience Letters, 489*(3), 159–163.

Longo, M. R., Schüür, F., Kammers, M. P. M., Tsakiris, M., & Haggard, P. (2008). What is embodiment? A psychometric approach. *Cognition, 107*(3), 978–998.

Lorenzo, C., Arsuaga, J. L., & Carretero, J.-M. (1999). Hand and foot remains from the Gran Dolina Early Pleistocene site (Sierra de Atapuerca, Spain). *Journal of Human Evolution, 37*(3–4), 501–522.

Lorenzo, C., Pablos, A., Carretero, J. M., Huguet, R., Valverdú, J., Martinón-Torres, M., Arsuaga, J. L., Carbonell, E., & Bermúdez de Castro, J. M. (2015). Early Pleistocene human hand phalanx from the Sima del Elefante (TE) cave site in Sierra de Atapuerca (Spain). *Journal of Human Evolution, 78*, 114–121.

Lotze, H. (1885). *Microcosmus: An essay concerning man and his relation to the world* [E. Hamilton & E. E. Constance Jones, Trans.]. T&T Clark.

Louis, E. D. (1994). Contralateral control: Evolving concepts of the brain–body relationship from Hippocrates to Morgagni. *Neurology, 44*(12), 2398–2398.

Lovejoy, C. O., Simpson, S. W., White, T. D., Asfaw, B., & Suwa, G. (2009). Careful climbing in the Miocene: The forelimbs of *Ardipithecus ramidus* and humans are primitive. *Science, 326*(5949), 70e1–e8.

Lozano, M., Estalrrich, A., Bondioli, L., Fiore, I., Bermúdez de Castro, J.-M., Arsuaga, J. L., Carbonell, E., Rosas, A., & Frayer, D. W. (2017). Right-handed fossil humans. *Evolutionary Anthropology, 26*(6), 313–324.

Lukos, J. R., Lee, D., Poizner, H., & Santello, M. (2010). Anticipatory modulation of digit placement for grasp control is affected by Parkinson's disease. *PLOS One, 5*(2), e9184.

Luppino, G., Murata, A., Govoni, P., & Matelli, M. (1999). Largely segregated parietofrontal connections linking rostral intraparietal cortex (areas AIP and VIP) and the ventral premotor cortex (areas F5 and F4). *Experimental Brain Research, 128*(1–2), 181–187.

MacIver, M. A., Schmitz, L., Mugan, U., Murphey, T. D., & Mobley, C. D. (2017). Massive increase in visual range preceded the origin of terrestrial vertebrates. *Proceedings of the National Academy of Sciences, 114*(12), E2375–E2384.

MacNeilage, P. F. (2014). Evolution of the strongest vertebrate rightward action asymmetries: Marine mammal sidedness and human handedness. *Psychological Bulletin, 140*(2), 587–609.

MacNeilage, P. F., Studdert-Kennedy, M. G., & Lindblom, B. (1987). Primate handedness reconsidered. *Behavioral and Brain Sciences, 10*(2), 247–303.

MacSweeney, M., Campbell, R., Woll, B., Giampietro, V., David, A. S., McGuire, P. K., Calvert, G. A., & Brammer, M. J. (2004). Dissociating linguistic and nonlinguistic gestural communication in the brain. *NeuroImage, 22*(4), 1605–1618.

MacSweeney, M., Woll, B., Campbell, R., McGuire, P. K., David, A. S., Williams, S. C. R., Suckling, J., Calvert, G. A., & Brammer, M. J. (2002). Neural systems underlying British Sign Language and audio-visual English processing in native users. *Brain, 125*(Pt 7), 1583–1593.

Maeda, R. S., Gribble, P. L., & Pruszynski, J. A. (2020). Learning new feedforward motor commands based on feedback responses. *Current Biology, 30*(10), 1941–1948.

Magnus, C. R. A., Arnold, C. M., Johnston, G., Dal-Bello Haas, V., Basran, J., Krentz, J. R., & Farthing, J. P. (2013). Cross-education for improving strength and mobility after distal radius fractures. *Archives of Physical Medicine and Rehabilitation, 94*(7), 1247–1255.

Maier, M. A., Armand, J., Kirkwood, P. A., Yang, H., Davis, J. N., & Lemon, R. N. (2002). Differences in the corticospinal projection from primary motor cortex and supplementary motor area to macaque upper limb motoneurons: An anatomical and electrophysiological study. *Cerebral Cortex*, *12*(3), 281–296.

Maier, M. A., Illert, M., Kirkwood, P. A., Nielsen, J., & Lemon, R. N. (1998). Does a C3–C4 propriospinal system transmit corticospinal excitation in the primate? An investigation in the macaque monkey. *Journal of Physiology*, *511*(Pt 1), 191–212.

Mainka, T., Azañón, E., Zeuner, K. E., Knutzen, A., Bäumer, T., Neumann, W.-J., Borngräber, F., Kühn, A. A., Longo, M. R., & Ganos, C. (2021). Intact organization of tactile space in isolated focal dystonia. *Movement Disorders*, *36*(8), 1949–1955.

Maister, L., Sebanz, N., Knoblich, G., & Tsakiris, M. (2013). Experiencing ownership over a dark-skinned body reduces implicit racial bias. *Cognition*, *128*(2), 170–178.

Majsak, M. J., Kaminski, T., Gentile, A. M., & Flanagan, J. R. (1998). The reaching movements of patients with Parkinson's disease under self-determined maximal speed and visually cued conditions. *Brain*, *121*(Pt 4), 755–766.

Makin, T. R., Holmes, N. P., & Zohary, E. (2007). Is that near my hand? Multisensory representation of peripersonal space in human intraparietal sulcus. *Journal of Neuroscience*, *27*(4), 731–740.

Makin, T. R., Wilf, M., Schwartz, I., & Zohary, E. (2010). Amputees "neglect" the space near their missing hand. *Psychological Science*, *21*(1), 55–57.

Malfait, N., & Ostry, D. J. (2004). Is interlimb transfer of force-field adaptation a cognitive response to the sudden introduction of load? *Journal of Neuroscience*, *24*(37), 8084–8089.

Maloney, T. R., Dilkes-Hall, I. E., Vlok, M., Oktaviana, A. A., Setiawan, P., Priyatno, A. A. D., Ririmasse, M., Geria, I. M., Effendy, M. A. R., Istiawan, B., Atmoko, F. T., Adhityatama, S., Moffat, I., Joannes-Boyau, R., Brumm, A., & Aubert, M. (2022). Surgical amputation of a limb 31,000 years ago in Borneo. *Nature*, *609*(7927), 547–551.

Manca, A., Dragone, D., Dvir, Z., & Deriu, F. (2017). Cross-education of muscular strength following unilateral resistance training: A meta-analysis. *European Journal of Applied Physiology*, *117*(11), 2335–2354.

Mancini, F., Bricolo, E., & Vallar, G. (2010). Multisensory integration in the Müller-Lyer illusion: From vision to haptics. *Quarterly Journal of Experimental Psychology*, *63*(4), 818–830.

Mancini, F., Haggard, P., Iannetti, G. D., Longo, M. R., & Sereno, M. I. (2012). Fine-grained nociceptive maps in primary somatosensory cortex. *Journal of Neuroscience*, *32*(48), 17155–17162.

Mancini, F., Longo, M. R., Kammers, M. P. M., & Haggard, P. (2011). Visual distortion of body size modulates pain perception. *Psychological Science*, *22*(3), 325–330.

Manger, P., Sum, M., Szymanski, M., Ridgway, S., & Krubitzer, L. (1995). Modular subdivisions of dolphin insular cortex: Does evolutionary history repeat itself? *Journal of Cognitive Neuroscience*, *10*(2), 153–166.

Manni, E., & Petrosini, L. (2004). A century of cerebellar somatotopy: A debated representation. *Nature Reviews Neuroscience, 5*(3), 241–249.

Mannu, M., & Ottoni, E. B. (2009). The enhanced tool-kit of two groups of wild bearded capuchin monkeys in the Caatinga: Tool making, associative use, and secondary tools. *American Journal of Primatology, 71*(3), 242–251.

Manohar, S. G., Chong, T. T.-J., Apps, M. A. J., Batla, A., Stamelou, M., Jarman, P. R., Bhatia, K. P., & Husain, M. (2015). Reward pays the cost of noise reduction in motor and cognitive control. *Current Biology, 25*(13), 1707–1716.

Manzoni, T., Barbaresi, P., Bellardinelli, E., & Caminiti, R. (1980). Callosal projections from the two body midlines. *Experimental Brain Research, 39*(1), 1–9.

Manzoni, T., Barbaresi, P., Conti, F., & Fabri, M. (1989). The callosal connections of the primary somatosensory cortex and the neural bases of midline fusion. *Experimental Brain Research, 76*(2), 251–266.

Maravita, A. (1997). Implicit processing of somatosensory stimuli disclosed by a perceptual aftereffect. *Neuroreport, 8*(7), 1671–1674.

Maravita, A., Husain, M., Clarke, K., & Driver, J. (2001). Reaching with a tool extends visual-tactile interactions into far space: Evidence from cross-modal extinction. *Neuropsychologia, 39*(6), 580–585.

Maravita, A., Spence, C., Kennett, S., & Driver, J. (2002). Tool-use changes multimodal spatial interactions between vision and touch in normal humans. *Cognition, 83*(2), 25–34.

Marcel, A., Mackintosh, B., Postma, P., Cusack, R., Vuckovich, J., Nimmo-Smith, I., & Cox, S. M. L. (2006). Is susceptibility to perceptual migration and fusion modality-specific or multimodal? *Neuropsychologia, 44*(5), 693–710.

Marcel, A., Postma, P., Gillmeister, H., Cox, S., Rorden, C., Nimmo-Smith, I., & Mackintosh, B. (2004). Migration and fusion of tactile sensation—Premorbid susceptibility to allochiria, neglect and extinction? *Neuropsychologia, 42*(13), 1749–1767.

Marcel, A. J., Tegnér, R., & Nimmo-Smith, I. (2004). Anosognosia for plegia: Specificity, extension, partiality and disunity of bodily unawareness. *Cortex, 40*(1), 19–40.

Marchant, L. F., McGrew, W. C., & Eibl-Eibesfeldt, I. (1995). Is human handedness universal? Ethological analyses from three traditional cultures. *Ethology, 101*(3), 239–258.

Marchi, D., Proctor, D. J., Huston, E., Nicholas, C. L., & Fischer, F. (2017). Morphological correlates of the first metacarpal proximal articular surface with manipulative capabilities in apes, humans and South African hominins. *Comptes Rendus Palevol, 16*(5–6), 645–654.

Margiotoudi, K., Marie, D., Claidiére, N., Coulon, O., Roth, M., Nazarian, B., Lacoste, R., Hopkins, W. D., Molesti, S., Fresnais, P., Anton, J.-L., & Meguerditchian, A. (2019). Handedness in monkeys reflects hemispheric specialization within the central sulcus. An in vivo MRI study in right- and left-handed olive baboons. *Cortex, 118*, 203–211.

Margolin, D. I. (1984). The neuropsychology of writing and spelling: Semantic, phonological, motor, and perceptual processes. *Quarterly Journal of Experimental Psychology, 36A*(3), 459–489.

Margolis, A. N., & Longo, M. R. (2015). Visual detail about the body modulates tactile localisation biases. *Experimental Brain Research, 233*(2), 351–358.

Maros, K., Gácsi, M., & Miklósi, Á. (2008). Comprehension of human pointing gestures in horses (*Equus caballus*). *Animal Cognition, 11*(3), 457–466.

Marsden, C. D., Merton, P. A., & Morton, H. B. (1981). Human postural responses. *Brain, 104*(3), 513–534.

Marsden, C. D., & Sheehy, M. P. (1990). Writer's cramp. *Trends in Neurosciences, 13*(4), 148–153.

Marsh, A. P. L., Heron, D., Edwards, T. J., Quartier, A., Galea, C., Nava, C., Rastetter, A., Moutard, M.-L., Anderson, V., Bitoun, P., Bunt, J., Faudet, A., Garel, C., Gillies, G., Gobius, I., Guegan, J., Heide, S., Keren, B., Lesne, F., & Depienne, C. (2017). Mutations in DCC cause isolated agenesis of the corpus callosum with incomplete penetrance. *Nature Genetics, 49*(4), 511–514.

Marshall, J., Atkinson, J., Smulovitch, E., Thacker, A., & Woll, B. (2004). Aphasia in a user of British Sign Language: Dissociation between sign and gesture. *Cognitive Neuropsychology, 21*(5), 537–554.

Martin, W. L. B., & Freitas, M. B. (2002). Mean mortality among Brazilian left- and right- handers: Modification or selective elimination? *Laterality, 7*(1), 31–44.

Martinaud, O., Besharati, S., Jenkinson, P. M., & Fotopoulou, A. (2017). Ownership illusions in patients with body delusions: Different neural profiles of visual capture and disownership. *Cortex, 87,* 174–185.

Marzke, M. W. (1983). Joint functions and grips of the *Australopithecus afarensis* hand, with special reference to the region of the capitate. *Journal of Human Evolution, 12*(2), 197–211.

Marzke, M. W. (1997). Precision grips, hand morphology, and tools. *American Journal of Physical Anthropology, 102*(1), 91–110.

Marzke, M. W. (2013). Tool making, hand morphology and fossil hominins. *Philosophical Transactions of the Royal Society B, 368*(1630), 20120414.

Marzke, M. W., & Shackley, M. S. (1986). Hominid hand use in the Pliocene and Pleistocene: Evidence from experimental archaeology and comparative morphology. *Journal of Human Evolution, 15*(6), 439–460.

Marzke, M. W., Toth, N., Schick, K., Reece, S., Steinberg, B., Hunt, K., Linscheid, R. L., & An, K.-N. (1998). EMG study of hand muscle recruitment during hard hammer percussion manufacture of Oldowan tools. *American Journal of Physical Anthropology, 105*(3), 315–332.

Marzke, M. W., & Wullstein, K. L. (1996). Chimpanzee and human grips: A new classification with a focus on evolutionary morphology. *International Journal of Primatology, 17*(1), 117–139.

Masataka, N. (1995). The relation between index-finger extension and the acoustic quality of cooing in three-month-old infants. *Journal of Child Language, 22*(2), 247–257.

Maschke, M., Gomez, C. M., Tuite, P. J., & Konczak, J. (2003). Dysfunction of the basal ganglia, but not the cerebellum, impairs kinaesthesia. *Brain, 126*(Pt 10), 2312–2322.

Mason, J., Frazer, A. K., Horvath, D. M., Pearce, A. J., Avela, J., Howatson, G., & Kidgell, D. J. (2017). Ipsilateral corticomotor responses are confined to the homologous muscle following cross-education of muscular strength. *Applied Physiology, Nutrition, and Metabolism, 43*(1), 11–22.

Mason, W. A., Harlow, H. F., & Rueping, R. R. (1959). The development of manipulatory responsiveness in the infant rhesus monkey. *Journal of Comparative and Physiological Psychology, 52*(5), 555–558.

Matelli, M., Camarda, R., Glickstein, M., & Rizzolatti, G. (1986). Afferent and efferent projections of the inferior area 6 in the macaque monkey. *Journal of Comparative Neurology, 251*(3), 281–298.

Matelli, M., & Luppino, G. (2001). Parietofrontal circuits for action and space perception in the macaque monkey. *NeuroImage, 14*(1), S27–S32.

Matelli, M., Luppino, G., & Rizzolatti, G. (1985). Patterns of cytochrome oxidase activity in the frontal agranular cortex of the macaque monkey. *Behavioural Brain Research, 18*(2), 125–136.

Matiello, M., Zimmerman, E., Caplan, D., & Cohen, A. B. (2015). Reversible cursive agraphia. *Neurology, 85*(3), 295–296.

Matthews, P. B. C., Farmer, S. F., & Ingram, D. A. (1990). On the localization of the stretch reflex of intrinsic hand muscles in a patient with mirror movements. *Journal of Physiology, 428*(1), 561–577.

Mattingley, J. B., Driver, J., Beschin, N., & Robertson, I. H. (1997). Attentional competition between modalities: Extinction between touch and vision after right hemisphere damage. *Neuropsychologia, 35*(6), 867–880.

Mauguière, F., Desmedt, J. E., & Courjon, J. (1983). Astereognosis and dissociated loss of frontal or parietal components of somatosensory evoked potentials in hemispheric lesions. *Brain, 106*(Pt 2), 271–311.

Mayberry, R. I., Jaques, J., & Dede, G. (1998). What stuttering reveals about the development of the gesture–speech relationship. *New Directions in Child Development, 79*, 77–87.

Mayston, M. J., Harrison, L. M., & Stephens, J. A. (1999). A neurophysiological study of mirror movements in adults and children. *Annals of Neurology, 45*(5), 583–594.

Mazza, P. P. A., Martini, F., Sala, B., Magi, M., Colombini, M. P., Giachi, G., Landucci, F., Lemorini, C., Modugno, F., & Ribechini, E. (2006). A new palaeolithic discovery: Tar-hafted stone tools in a European Mid-Pleistocene bone-bearing bed. *Journal of Archaeological Science, 33*(9), 1310–1318.

Mazzoni, P., Hristova, A., & Krakauer, J. W. (2007). Why don't we move faster? Parkinson's disease, movement vigor, and implicit motivation. *Journal of Neuroscience, 27*(27), 7105–7116.

Mazzoni, P., & Krakauer, J. W. (2006). An implicit plan overrides an explicit strategy during visuomotor adaptation. *Journal of Neuroscience, 26*(14), 3642–3645.

McBride, J., Sumner, P., & Husain, M. (2012). Conflict in object affordance revealed by grip force. *Quarterly Journal of Experimental Psychology, 65*(1), 13–24.

McBride, J., Sumner, P., Jackson, S. R., Bajaj, N., & Husain, M. (2013). Exaggerated object affordance and absent automatic inhibition in alien hand syndrome. *Cortex, 49*(8), 2040–2054.

McCabe, C. S., Haigh, R. C., Halligan, P. W., & Blake, D. R. (2005). Simulating sensory-motor incongruence in healthy volunteers: Implications for a cortical model of pain. *Rheumatology, 44*(4), 509–516.

McCandliss, B. D., Cohen, L., & Dehaene, S. (2003). The visual word form area: Expertise for reading in the fusiform gyrus. *Trends in Cognitive Sciences, 7*(7), 293–299.

McCarroll, H. R. (2000). Congenital anomalies: A 25-year overview. *Journal of Hand Surgery, 25*(6), 1007–1037.

McDaniel, K. D., Cummings, J. L., & Shain, S. (1989). The "yips": A focal dystonia of golfers. *Neurology, 39*(2), 192–195.

McDonnell, P. M. (1979). Patterns of eye-hand coordination in the first year of life. *Canadian Journal of Psychology, 33*(4), 253–267.

McGonigle, D. J., Hänninen, R., Salenius, S., Hari, R., Frackowiak, R. S. J., & Frith, C. D. (2002). Whose arm is it anyway? An fMRI case study of supernumerary phantom limb. *Brain, 125*(Pt 6), 1265–1274.

McGraw, M. B. (1935). *Growth: A study of Johnny and Jimmy.* Appleton-Century.

McGraw, M. B. (1940). Suspension grasp behavior of the human infant. *American Journal of Disabled Children, 60*(4), 799–811.

McGraw, M. B. (1941). Neural maturation as exemplified in the reaching-prehensile behavior of the human infant. *Journal of Psychology, 11*(1), 127–141.

McGrew, W. C. (2010). Chimpanzee technology. *Science, 328*(5978), 579–580.

McGrew, W. C. (2013). Is primate tool use special? Chimpanzee and New Caledonian crow compared. *Philosophical Transactions of the Royal Society B, 368*(1630), 20120422.

McGrew, W. C., & Marchant, L. F. (1997). On the other hand: Current issues in and meta-analysis of the behavioural laterality of hand function in nonhuman primates. *Yearbook of Physical Anthropology, 40*, 201–232.

McIntyre, S., Hauser, S. C., Kusztor, A., Boehme, R., Moungou, A., Isager, P. M., Homman, L., Novembre, G., Nagi, S. S., Israr, A., Lumpkin, E. A., Abnousi, F., Gerling, G. J., & Olausson, H. (2022). The language of social touch is intuitive and quantifiable. *Psychological Science, 33*(9), 1477–1494.

McLennan, J. E., Nakano, K., Tyler, H. R., & Schwab, R. S. (1972). Micrographia in Parkinson's disease. *Journal of the Neurological Sciences, 15*(2), 141–152.

McManus, I. C. (1985). Handedness, language dominance and aphasia: A genetic model. *Psychological Medicine Monograph Supplement, 8*, 3–40.

McManus, I. C. (2002). *Right hand, left hand.* Phoenix.

McManus, I. C., & Hartigan, A. (2007). Declining left-handedness in Victorian England seen in the films of Mitchell and Kenyon. *Current Biology, 17*(18), R793–R794.

McManus, I. C., Van Horn, J. D., & Bryden, P. J. (2016). The Tapley and Bryden test of performance differences between the hands: The original data, newer data, and the relation to pegboard and other tasks. *Laterality, 21*(4–6), 371–396.

McNeill, D. (1985). So you think gestures are nonverbal? *Psychological Review, 92*(3), 350–371.

McNeill, D. (1992). *Hand and mind: What gestures reveal about thought.* University of Chicago Press.

McPherron, S. P., Alemseged, Z., Marean, C. W., Wynn, J. G., Reed, D., Geraads, D., Bobe, R., & Béarat, H. A. (2010). Evidence for stone-tool-assisted consumption of animal tissues before 3.39 million years ago at Dikika, Ethiopia. *Nature, 466*(7308), 857–860.

Meador, K. J., Loring, D. W., Lee, G. P., Brooks, B. S., Thompson, E. E., Thompson, W. O., & Heilman, K. M. (1988). Right cerebral specialization for tactile attention as evidenced by intracarotid sodium amytal. *Neurology, 38*(11), 1763–1766.

Mechelli, A., Crinion, J. T., Noppeney, U., O'Doherty, J., Ashburner, J., Frackowiak, R. S., & Price, C. J. (2004). Structural plasticity in the bilingual brain. *Nature, 431*(7010), 757.

Medina, J., & Coslett, H. B. (2010). From maps to form to space: Touch and the body schema. *Neuropsychologia, 48*(3), 645–654.

Medina, J., & Rapp, B. (2008). Phantom tactile sensations modulated by body position. *Current Biology, 18*(24), 1937–1942.

Medina, S., Tamè, L., & Longo, M. R. (2018). Tactile localization biases are modulated by gaze direction. *Experimental Brain Research, 236*(10), 31–42.

Medland, S. E., Duffy, D. L., Wright, M. J., Geffen, G. M., & Martin, N. G. (2006). Handedness in twins: Joint analysis of data from 35 samples. *Twin Research and Human Genetics, 9*(1), 46–53.

Meguerditchian, A., Gardner, M. J., Schapiro, S. J., & Hopkins, W. D. (2012). The sound of one-hand clapping: Handedness and perisylvian neural correlates of a communicative gesture in chimpanzees. *Proceedings of the Royal Society B, 279*(1735), 1959–1966.

Meguerditchian, A., Molesti, S., & Vauclair, J. (2011). Right-handedness predominance in 162 baboons (*Papio anubis*) for gestural communication: Consistency across time and groups. *Behavioral Neuroscience, 125*(4), 653–660.

Meguerditchian, A., Phillips, K. A., Chapelain, A., Mahovetz, L. M., Milne, S., Stoinski, T., Bania, A., Lonsdorf, E., Schaeffer, J., Russell, J., & Hopkins, W. D. (2015). Handedness for unimanual grasping in 564 great apes: The effect on grip morphology and a comparison with hand use for a bimanual coordinated task. *Frontiers in Psychology, 6*, 1794.

Meguerditchian, A., Vauclair, J., & Hopkins, W. D. (2013). On the origins of human handedness and language: A comparative review of hand preferences for bimanual coordinated actions and gestural communication in nonhuman primates. *Developmental Psychobiology, 55*(6), 637–650.

Mehring, C., Akselrod, M., Bashford, L., Mace, M., Choi, H., Blüher, M., Buschhoff, A.-S., Pistohl, T., Salomon, R., Cheah, A., Blanke, O., Serino, A., & Burdet, E. (2019). Augmented manipulation ability in humans with six-fingered hands. *Nature Communications, 10*, 2401.

Meltzoff, A. N., Saby, J. N., & Marshall, P. J. (2019). Neural representations of the body in 60-day-old human infants. *Developmental Science, 22*(1), e12698.

Melzack, R., Israel, R., Lacroix, R., & Schultz, G. (1997). Phantom limbs in people with congenital limb deficiency or amputation in early childhood. *Brain, 120*(Pt 9), 1603–1620.

Menichelli, A., Rapp, B., & Semenza, C. (2008). Allographic agraphia: A case study. *Cortex, 44*(7), 861–868.

Mercader, J., Barton, H., Gillespie, J., Harris, J., Kuhn, S., Tyler, R., & Boesch, C. (2007). 4,300-year-old chimpanzee sites and the origins of percussive stone technology. *Proceedings of the National Academy of Sciences, 104*(9), 3043–3048.

Mercader, J., Panger, M., & Boesch, C. (2002). Excavation of a chimpanzee stone tool site in the African rainforest. *Science, 296*(5572), 1452–1456.

Merchant, S. H. I., Frangos, E., Parker, J., Bradson, M., Wu, T., Vial-Underraga, F., Leodori, G., Bushnell, M. C., Horovitz, S. G., Hallett, M., & Popa, T. (2020). The role of the inferior parietal lobule in writer's cramp. *Brain, 143*(6), 1766–1779.

Mercier, C., Reilly, K. T., Vargas, C. D., Aballea, A., & Sirigu, A. (2006). Mapping phantom movement representations in the motor cortex of amputees. *Brain, 129*(Pt 8), 2202–2210.

Merrick, C. M., Dixon, T. C., Breska, A., Lin, J., Chang, E. F., King-Stephens, D., Laxer, K. D., Weber, P. B., Carmena, J., Knight, R. T., & Ivry, R. B. (2022). Left hemisphere dominance for bilateral kinematic encoding in the human brain. *eLife, 11*, e69977.

Merton, P. A. (1972). How we control the contraction of our muscles. *Scientific American, 226*(5), 30–37.

Merton, P. A., & Morton, H. B. (1980). Stimulation of the cerebral cortex in the intact human subject. *Nature, 285*, 227.

Merzenich, M. M., Kaas, J. H., Sur, M., & Lin, C.-S. (1978). Double representation of the body surface within cytoarchitectonic areas 3b and 1 in "SI" in the owl monkey (*Aotus trivirgatus*). *Journal of Comparative Neurology, 181*(1), 41–73.

Merzenich, M. M., Nelson, R. J., Kaas, J. H., Stryker, M. P., Jenkins, W. M., Zook, J. M., Cynader, M. S., & Schoppmann, A. (1987). Variability in hand surface representations in areas 3b and 1 in adult owl and squirrel monkeys. *Journal of Comparative Neurology, 258*(2), 281–296.

Merzenich, M. M., Nelson, R. J., Stryker, M. P., Cynader, M. S., Schoppmann, A., & Zook, J. M. (1984). Somatosensory cortical map changes following digit amputation in adult monkeys. *Journal of Comparative Neurology, 224*(4), 591–605.

Messina, C., El-Moneim, S. M. A., Pozzi, M., Tomaino, A., Biehler-Gomez, L., Cummaudo, M., Cattaneo, C., & Piacentini, P. (2022). Evidence of possible lower limb amputation in a tomb in an ancient Egyptian necropolis: The case report of an on-site radiographic analysis. *BJR Case Reports*, *8*(6), 20220090.

Miall, R. C., Afanasyeva, D., Cole, J. D., & Mason, P. (2021). Perception of body shape and size without touch or proprioception: Evidence from individuals with congenital and acquired neuropathy. *Experimental Brain Research*, *239*(4), 1203–1221.

Michaels, J. A., & Scherberger, H. (2018). Population coding of grasp and laterality-related information in the macaque fronto-parietal network. *Scientific Reports*, *8*, 1710.

Michel, G. F. (2021). Handedness development: A model for investigating the development of hemispheric specialization and interhemispheric coordination. *Symmetry*, *13*(6), 992.

Miklósi, Á., Kubinyi, E., Topál, J., Gácsi, M., Virányi, Z., & Csányi, V. (2003). A simple reason for a big difference: Wolves do not look back at humans, but dogs do. *Current Biology*, *13*(9), 763–766.

Miklósi, Á., Polgárdi, R., Tomál, J., & Csányi, V. (1998). Use of experimenter-given cues in dogs. *Animal Cognition*, *1*(2), 113–121.

Miles, L. (1990). The cognitive foundations for reference in a signing orangutan. In S. Parker & K. Gibson (Eds.), *Language and intelligence in monkeys and apes* (pp. 511–539). Cambridge University Press.

Miller, G. A., Galanter, E., & Pribram, K. H. (1960). *Plans and the structure of behavior*. Henry Holt.

Miller, L. E., Fabio, C., Ravenda, V., Bahmad, S., Koun, E., Salemme, R., Luauté, J., Bolognini, N., Hayward, B., & Farnè, A. (2019). Somatosensory cortex efficiently processes touch located beyond the body. *Current Biology*, *29*(24), 4276–4283.

Miller, L. E., Longo, M. R., & Saygin, A. P. (2014). Tool morphology constrains the effects of tool use on body representations. *Journal of Experimental Psychology: Human Perception and Performance*, *40*(6), 2143–2153.

Miller, L. E., Longo, M. R., & Saygin, A. P. (2016). Mental body representations retain homuncular shape distortions: Evidence from Weber's illusion. *Consciousness and Cognition*, *40*, 17–25.

Miller, L. E., Montroni, L., Koun, E., Salemme, R., Hayward, V., & Farnè, A. (2018). Sensing with tools extends somatosensory processing beyond the body. *Nature*, *561*(7722), 239–242.

Milner, A. D., Dijkerman, H. C., Pisella, L., McIntosh, R. D., Tilikete, C., Vighetto, A., & Rossetti, Y. (2001). Grasping the past: Delay can improve visuomotor performance. *Current Biology*, *11*(23), 1896–1901.

Milner, A. D., & Goodale, M. A. (2006). *The visual brain in action* (2nd ed.). Oxford University Press.

Milner, A. D., Paulignan, Y., Dijkerman, H. C., Michel, F., & Jeannerod, M. (1999). A paradoxical improvement of misreaching in optic ataxia: New evidence for two separate neural systems for visual localization. *Proceedings of the Royal Society B*, *266*(1434), 2225–2229.

Milner, B. (1962). Les troubles de la mémoire accompagnant des lésions hippocampiques bilatérales [Memory problems accompanying bilateral hippocampal lesions]. In *Physiologic de l'Hippocampe* (pp. 257–272). Centre National de la Recherche Scientifique.

Mishkin, M. (1979). Analogous neural models for tactual and visual learning. *Neuropsychologia, 17*(2), 139–151.

Mitchell, S. W. (1871). Phantom limbs. *Lippincott's Magazine, 8*(48), 563–569.

Mitchell, S. W. (1872). *Injuries of nerves and their consequences*. Lippincott.

Mithen, S. (1996). *The prehistory of mind*. Thames and Hudson.

Mittra, E. S., Smith, H. F., Lemelin, P., & Jungers, W. L. (2007). Comparative morphometrics of the primate apical tuft. *American Journal of Physical Anthropology, 134*(4), 449–459.

Moffett, A., Ettlinger, G., Morton, H. B., & Piercy, M. F. (1967). Tactile discrimination performance in the monkey: The effect of ablation of various subdivisions of posterior parietal cortex. *Cortex, 3*(1), 59–96.

Monaco, S., Gallivan, J. P., Figley, T. D., Singhal, A., & Culham, J. C. (2017). Recruitment of foveal retinotopic cortex during haptic exploration of shapes and actions in the dark. *Journal of Neuroscience, 37*(48), 11572–11591.

Moore, C., & Corkum, V. (1994). Social understanding at the end of the first year of life. *Developmental Review, 14*(4), 349–372.

Moore, C., & D'Entremont, B. (2001). Developmental changes in pointing as a function of attentional focus. *Journal of Cognition and Development, 2*(2), 109–129.

Mooshagian, E., Holmes, C. D., & Snyder, L. H. (2021). Local field potentials in the parietal reach region reveal mechanisms of bimanual coordination. *Nature Communications, 12*, 2514.

Mora, L., Sedda, A., Esteban, T., & Cocchini, G. (2021). The signing body: Extensive sign language practice shapes the size of hands and face. *Experimental Brain Research, 239*(7), 2233–2249.

Mora, R., & de la Torre, I. (2005). Percussion tools in Olduvai Beds I and II (Tanzania): Implications for early human activities. *Journal of Anthropological Archaeology, 24*(2), 179–192.

Moro, V., Pacella, V., Scandola, M., Besharati, S., Rossato, E., Jenkinson, P. M., & Fotopoulou, A. (2023). A fronto-insular-parietal network for the sense of body ownership. *Cerebral Cortex, 33*(3), 512–522.

Morrel-Samuels, P., & Krauss, R. M. (1992). Word familiarity predicts temporal asynchrony of hand gestures and speech. *Journal of Experimental Psychology: Learning, Memory, and Cognition, 18*(3), 615–622.

Morton, S. M., Lang, C. E., & Bastian, A. J. (2001). Inter- and intra-limb generalization of adaptation during catching. *Experimental Brain Research, 141*(4), 438–445.

Moscovitch, M., & Behrmann, M. (1994). Coding of spatial information in the somatosensory system: Evidence from patients with neglect following parietal lobe damage. *Journal of Cognitive Neuroscience, 6*(2), 151–155.

Moseley, G. L., McCormick, K., Hudson, M., & Zalucki, N. (2006). Disrupted cortical proprioceptive representation evokes symptoms of peculiarity, foreignness and swelling, but not pain. *Rheumatology, 45*(2), 196–200.

Moss, A. D., & Turnbull, O. H. (1996). Hatred of the hemiparetic limbs (misoplegia) in a 10 year old child. *Journal of Neurology, Neurosurgery & Psychiatry, 61*(2), 210–211.

Mott, F. W. (1892). Results of hemisection of the spinal cord in monkeys. *Proceedings of the Royal Society, 50*, 1–59.

Mountcastle, V. B. (1957). Modality and topographic properties of single neurons of cat's somatic sensory cortex. *Journal of Neurophysiology, 20*(4), 408–434.

Mountcastle, V. B. (1997). The columnar organization of the neocortex. *Brain, 120*(Pt 4), 701–722.

Mountcastle, V. B. (2005). *The sensory hand: Neural mechanisms of somatic sensation.* Harvard University Press.

Mountcastle, V. B., Lynch, J. C., Georgopoulos, A., Sakata, H., & Acuna, C. (1975). Posterior parietal association cortex of the monkey: Command functions for operations within extrapersonal space. *Journal of Neurophysiology, 38*(4), 871–908.

Moyà-Solà, S., Köhler, M., & Rook, L. (1999). Evidence of hominid-like precision grip capability in the hand of the Miocene ape *Oreopithecus*. *Proceedings of the National Academy of Sciences, 96*(1), 313–317.

Mruczek, R. E. B., von Loga, I. S., & Kastner, S. (2013). The representation of tool and non-tool object information in the human intraparietal sulcus. *Journal of Neurophysiology, 109*(12), 2883–2896.

Muir, R. B., & Lemon, R. N. (1983). Corticospinal neurons with a special role in precision grip. *Brain Research, 261*(2), 312–316.

Müller, F., Kunesch, E., Binkofski, F., & Freund, H.-J. (1991). Residual sensorimotor functions in a patient after right-sided hemispherectomy. *Neuropsychologia, 29*(2), 125–145.

Müller, K., Ebner, B., & Hömberg, V. (1994). Maturation of fastest afferent and efferent central and peripheral pathways: No evidence for a constancy of central conduction delays. *Neuroscience Letters, 166*(1), 9–12.

Mulliken, G. H., Musallam, S., & Andersen, R. A. (2008). Forward estimation of movement state in posterior parietal cortex. *Proceedings of the National Academy of Sciences, 105*(24), 8170–8177.

Mumford, K. H., & Kita, S. (2014). Children use gesture to interpret novel verb meanings. *Child Development, 85*(3), 1181–1189.

Murata, A., Fadiga, L., Fogassi, L., Gallese, V., Raos, V., & Rizzolatti, G. (1997). Object representation in the ventral premotor cortex (area F5) of the monkey. *Journal of Neurophysiology, 78*(4), 2226–2230.

Murata, A., Gallese, V., Luppino, G., Kaseda, M., & Sakata, H. (2000). Selectivity for the shape, size, and orientation of objects for grasping in neurons of monkey parietal area AIP. *Journal of Neurophysiology, 83*(5), 2580–2601.

Muret, D., Dinse, H. R., Macchione, S., Urquizar, C., Farnè, A., & Reilly, K.T. (2014). Touch improvement at the hand transfers to the face. *Current Biology, 24*(16), R736–R737.

Murphy, W. F., & Langley, A. L. (1963). Common bullous lesions—presumably self-inflicted—occurring in utero in the newborn infant. *Pediatrics, 32*, 1099–1101.

Musgrave, J. H. (1971). How dextrous was Neanderthal man? *Nature, 233*, 538.

Mushiake, H., Inase, M., & Tanji, J. (1991). Neuronal activity in the primate premotor, supplementary, and precentral motor cortex during visually guided and internally determined sequential movements. *Journal of Neurophysiology, 66*(3), 705–718.

Myers, R. E., & Henson, C. O. (1960). Role of corpus callosum in transfer of tactuokinesthetic learning in chimpanzee. *Archives of Neurology, 3*(4), 404–409.

Myowa-Yamakoshi, M., & Takeshita, H. (2006). Do human fetuses anticipate self-oriented actions? A study by four-dimensional (4D) ultrasonography. *Infancy, 10*(3), 289–301.

Nagarajan, S. S., Blake, D. T., Wright, B. A., Byl, N., & Merzenich, M. M. (1998). Practice-related improvements in somatosensory interval discrimination are temporally specific but generalize across skin location, hemisphere, and modality. *Journal of Neuroscience, 18*(4), 1559–1570.

Nakajima, K., Maier, M. A., Kirkwood, P. A., & Lemon, R. N. (2000). Striking differences in transmission of corticospinal excitation to upper limb motoneurons in two primate species. *Journal of Neurophysiology, 84*(2), 698–709.

Nakamichi, M., & Yamada, K. (2009). Distribution of dorsal carriage among simians. *Primates, 50*(2), 153–168.

Nakamura, T., Gehrke, A. R., Lemberg, J., Szymaszek, J., & Shubin, N. H. (2016). Digits and fin rays share common developmental histories. *Nature, 537*, 225–228.

Napier, J. R. (1956). The prehensile movements of the human hand. *Journal of Bone and Joint Surgery, 38*, 902–913.

Napier, J. R. (1961). Prehensility and opposability in the hands of primates. *Symposia of the Zoological Society of London, 5*, 115–132.

Napier, J. R. (1962a). Fossil hand bones from Olduvai Gorge. *Nature, 196*, 409–411.

Napier, J. R. (1962b). The evolution of the hand. *Scientific American, 207*(6), 56–62.

Napier, J. R. (1970). *The roots of mankind*. Smithsonian Institution Press.

Napier, J. R. (1993). *Hands* (Rev. ed.). Princeton University Press.

Napier, J. R., & Davis, P. R. (1959). The forelimb skeleton and associated remains of *Proconsul africanus*. *Fossil Mammals of Africa*, *16*, 1–69.

Napier, J. R., & Walker, A. C. (1967). Vertical clinging and leaping—A newly recognized category of locomotor behaviour of primates. *Folia Primatologica*, *6*(3), 204–219.

Nashed, J. Y., Crevecoeur, F., & Scott, S. H. (2012). Influence of the behavioral goal and environmental obstacles on rapid feedback responses. *Journal of Neurophysiology*, *108*(4), 999–1009.

Nathan, P. W. (1956). Reference of sensation at the spinal level. *Journal of Neurology, Neurosurgery & Psychiatry*, *19*(2), 88–100.

Needham, A., Barrett, T., & Peterman, K. (2002). A pick-me-up for infants' exploratory skills: Early simulated experiences reaching for objects using "sticky mittens" enhances young infants' object exploration skills. *Infant Behavior and Development*, *25*(3), 279–295.

Negrotti, A., Secchi, C., & Gentilucci, M. (2005). Effects of disease progression and L-dopa therapy on the control of reaching-grasping in Parkinson's disease. *Neuropsychologia*, *43*(3), 450–459.

Nelson, E. L. (2022). Insights into human and nonhuman primate handedness from measuring both hands. *Current Directions in Psychological Science*, *31*(2), 154–161.

Nelson, R. J., Sur, M., Felleman, D. J., & Kaas, J. H. (1980). Representations of the body surface in postcentral parietal cortex of *Macaca fascicularis*. *Journal of Comparative Neurology*, *192*(4), 611–643.

Neubauer, S., Gunz, P., & Hublin, J.-J. (2009). The pattern of endocranial ontogenetic shape changes in humans. *Journal of Anatomy*, *215*(3), 240–255.

Neubauer, S., Hublin, J.-J., & Gunz, P. (2018). The evolution of modern human brain shape. *Science Advances*, *4*(1), eaao5961.

Neville, H. J., Bavelier, D., Corina, D., Rauschecker, J., Karna, A., Lalwani, A., Braun, A., Clark, V., Jezzard, P., & Turner, R. (1998). Cerebral organization for language in deaf and hearing subjects: Biological constraints and effects of experience. *Proceedings of the National Academy of Sciences*, *95*(3), 922–929.

Newell, K. M., McDonald, P. V, & Baillargeon, R. (1993). Body scale and infant grip configurations. *Developmental Psychobiology*, *26*(4), 195–205.

Newell, K. M., Scully, D. M., McDonald, P. V, & Baillargeon, R. (1989). Task constraints and infant grip configurations. *Developmental Psychobiology*, *22*(8), 817–832.

Newport, R., & Gilpin, H. R. (2011). Multisensory disintegration and the disappearing hand trick. *Current Biology*, *21*(19), R804–R805.

Newport, R., Pearce, R., & Preston, C. (2010). Fake hands in action: Embodiment and control of supernumerary limbs. *Experimental Brain Research*, *204*(3), 385–395.

Newport, R., & Preston, C. (2010). Pulling the finger off disrupts agency, embodiment and peripersonal space. *Perception, 39*(9), 1296–1298.

Newport, R., Wong, D. Y., Howard, E. M., & Silver, E. (2016). The Anne Boleyn illusion is a six-fingered salute to sensory remapping. *I-Perception, 7*(5), 2041669516669732.

Nicoladis, E., Marentette, P., & Pika, S. (2019). How many fingers am I holding up? The answer depends on children's language background. *Developmental Science, 22*(4), e12781.

Niekus, M. J. L. Th., Kozowyk, P. R. B., Langejans, G. H. J., Ngan-Tillard, D., van Keulen, H., van der Plicht, J., Cohen, K. M., van Wingerden, W., van Os, B., Smit, B. I., Amkreutz, L. W. S. W., Johansen, L., Verbaas, A., & Dusseldorp, G. L. (2019). Middle Paleolithic complex technology and a Neandertal tar-backed tool from the Dutch North Sea. *Proceedings of the National Academy of Sciences, 116*(44), 22081–22087.

Nielsen, J. M. (1938). Gerstmann syndrome: Finger agnosia, agraphia, confusion of right and left and acalculia. *Archives of Neurology & Psychiatry, 39*(3), 536–560.

Niewoehner, W. A. (2001). Behavioral inferences from the Skhul/Qafzeh early modern human hand remains. *Proceedings of the National Academy of Sciences, 98*(6), 2979–2984.

Niewoehner, W. A., Bergstrom, A., Eichele, D., Zuroff, M., & Clark, J. T. (2003). Manual dexterity in Neanderthals. *Nature, 422*, 395.

Nightingale, S. (1982). Somatoparaphrenia: A case report. *Cortex, 18*(3), 463–467.

Nirkko, A. C., Ozdoba, C., Redmond, S. M., Bürki, M., Schroth, G., Hess, C. W., & Wiesendanger, M. (2001). Different ipsilateral representations for distal and proximal movements in the sensorimotor cortex: Activation and deactivation patterns. *NeuroImage, 13*(5), 825–835.

Nishimura, Y., Morichika, Y., & Isa, T. (2009). A subcortical oscillatory network contributes to recovery of hand dexterity after spinal cord injury. *Brain, 132*(3), 709–721.

Nishimura, Y., Onoe, H., Morichika, Y., Perfiliev, S., Tsukada, H., & Isa, T. (2007). Time-dependent central compensatory mechanisms of finger dexterity after spinal cord injury. *Science, 318*(5853), 1150–1155.

Noachtar, S., Lüders, H. O., Dinner, D. S., & Klem, G. (1997). Ipsilateral median somatosensory evoked potentials recorded from human somatosensory cortex. *Electroencephalography and Clinical Neurophysiology, 104*(3), 189–198.

Nonaka, T., Bril, B., & Rein, R. (2010). How do stone knappers predict and control the outcome of flaking? Implications for understanding early stone tool technology. *Journal of Human Evolution, 59*(2), 155–167.

Nudo, R. J., & Masterton, R. B. (1990). Descending pathways to the spinal cord, IV: Some factors related to the amount of cortex devoted to the corticospinal tract. *Journal of Comparative Neurology, 296*(4), 584–597.

Obayashi, S., Suhara, T., Kawabe, K., Okauchi, T., Maeda, J., Akine, Y., Onoe, H., & Iriki, A. (2001). Functional brain mapping of monkey tool use. *NeuroImage, 14*(4), 853–861.

Obersteiner, H. (1882). On allochiria: A peculiar sensory disorder. *Brain, 4*(2), 153–163.

Ochoa, J., & Torebjörk, E. (1983). Sensations evoked by intraneural microstimulation of single mechanoreceptor units innervating the human hand. *Journal of Physiology, 342*(1), 633–654.

Ocklenburg, S., Schmitz, J., Moinfar, Z., Moser, D., Klose, R., Lor, S., Kunz, G., Tegenthoff, M., Faustmann, P., Francks, C., Epplen, J. T., Kumsta, R., & Güntürkün, O. (2017). Epigenetic regulation of lateralized fetal spinal gene expression underlies hemispheric asymmetries. *eLife, 6*, e22784.

Ogawa, K., Mitsui, K., Imai, F., & Nishida, S. (2019). Long-term training-dependent representation of individual finger movements in the primary motor cortex. *NeuroImage, 202*, 116051.

Ohsaki, K., Osumi, N., & Nakamura, S. (2002). Altered whisker patterns induced by ectopic expression of *Shh* are topographically represented by barrels. *Developmental Brain Research, 137*(2), 159–170.

Ohtake, H., Fujii, T., Yamadori, A., Fujimori, M., Hayakawa, Y., & Suzuki, K. (2001). The influence of misnaming on object recognition: A case of multimodal agnosia. *Cortex, 37*(2), 175–186.

Oliveri, M., Rossini, P. M., Filippi, M. M., Traversa, R., Cicinelli, P., Palmieri, M. G., Pasqualetti, P. & Caltagirone, C. (2000). Time-dependent activation of parieto-frontal networks for directing attention to tactile space: A study with paired transcranial magnetic stimulation pulses in right-brain-damaged patients with extinction. *Brain, 123*(Pt 9), 1939–1947.

Oliveri, M., Rossini, P. M., Pasqualetti, P., Traversa, R., Cicinelli, P., Palmieri, M. G., Tomaiuolo, F., & Caltagirone, C. (1999). Interhemispheric asymmetries in the perception of unimanual and bimanual cutaneous stimuli: A study using transcranial magnetic stimulation. *Brain, 122*(Pt 9), 1721–1729.

Olivier, E., Edgley, S. A., Armand, J., & Lemon, R. N. (1997). An electrophysiological study of the postnatal development of the corticospinal system in the macaque monkey. *Journal of Neuroscience, 17*(1), 267–276.

Omrani, M., Murnaghan, C. D., Pruszynski, J. A., & Scott, S. H. (2016). Distributed task-specific processing of somatosensory feedback for voluntary motor control. *eLife, 5*, e13141.

Orban, G. A., & Caruana, F. (2014). The neural basis of human tool use. *Frontiers in Psychology, 5*, 310.

Orban, G. A., Claeys, K., Nelissen, K., Smans, R., Sunaert, S., Todd, J. T., Wardak, C., Durand, J.-B., & Vanduffel, W. (2006). Mapping the parietal cortex of human and non-human primates. *Neuropsychologia, 44*(13), 2647–2667.

Osiurak, F., Lesourd, M., Navarro, J., & Reynaud, E. (2020). Technition: When tools come out of the closet. *Perspectives on Psychological Science, 15*(4), 880–897.

Osiurak, F., Reynaud, E., Baumard, J., Rossetti, Y., Bartolo, A., & Lesourd, M. (2021). Pantomime of tool use: Looking beyond apraxia. *Brain Communications, 3*(4), fcab263.

Ossmy, O., Han, D., Cheng, M., Kaplan, B. E., & Adolph, K. E. (2020). Look before you fit: The real-time planning cascade in children and adults. *Journal of Experimental Child Psychology, 189*, 104696.

Ossmy, O., Kaplan, B. E., Han, D., Xu, M., Bianco, C., Mukamel, R., & Adolph, K. E. (2022). Real-time processes in the development of action planning. *Current Biology, 32*(1), 190–199.

Ossmy, O., & Mukamel, R. (2016). Neural network underlying intermanual skill transfer in humans. *Cell Reports, 17*(11), 2891–2900.

Osumi, M., Nobusako, S., Zama, T., Taniguchi, M., Shimada, S., & Morioka, S. (2018). Sensorimotor incongruence alters limb perception and movement. *Human Movement Science, 57*, 251–257.

Otsuru, N., Hashizume, A., Nakamura, D., Endo, Y., Inui, K., Kakigi, R., & Yuge, L. (2014). Sensory incongruence leading to hand disownership modulates somatosensory cortical processing. *Cortex, 58*, 1–8.

Over, R. (1966). A comparison of haptic and visual judgments of some illusions. *American Journal of Psychology, 79*(4), 590–595.

Overduin, S. A., D'Avella, A., Carmena, J. M., & Bizzi, E. (2012). Microstimulation activates a handful of muscle synergies. *Neuron, 76*(6), 1071–1077.

Owen, D. (2014, May 19). The yips. *The New Yorker*. https://www.newyorker.com/magazine/2014/05/26/the-yips

Owen, R. (1849). *On the nature of limbs*. John Van Voorst.

Padberg, J., Disbrow, E., & Krubitzer, L. (2005). The organization and connections of anterior and posterior parietal cortex in Titi monkeys: Do New World monkeys have an area 2? *Cerebral Cortex, 15*(12), 1938–1963.

Padberg, J., Franca, J. G., Cooke, D. F., Soares, J. G. M., Rosa, M. G. P., Fiorani, M., Gattass, R., & Krubitzer, L. (2007). Parallel evolution of cortical areas involved in skilled hand use. *Journal of Neuroscience, 27*(38), 10106–10115.

Paillard, J., Michel, F., & Stelmach, G. (1983). Localization without content: A tactile analogue of "blind sight." *Archives of Neurology, 40*(9), 548–551.

Pammi, V. S. C., Miyapuram, K. P., Ahmed, Samejima, K., Bapi, R. S., & Doya, K. (2012). Changing the structure of complex visuo-motor sequences selectively activates the fronto-parietal network. *NeuroImage, 59*(2), 1180–1189.

Pan, B. A., Rowe, M. L., Singer, J. D., & Snow, C. E. (2005). Maternal correlates of growth in toddler vocabulary production in low-income families. *Child Development, 76*(4), 763–782.

Papadatou-Pastou, M., Ntolka, E., Schmitz, J., Martin, M., Munafò, M. R., Ocklenburg, S., & Paracchini, S. (2020). Human handedness: A meta-analysis. *Psychological Bulletin, 146*(6), 481–524.

Paré, A. (1634). *The workes of that famous chirurgion Ambrose Parey* [T. Johnson, Trans.]. Th. Cotes and R. Young.

Pargeter, J., Khreisheh, N., Shea, J. J., & Stout, D. (2020). Knowledge vs. know-how? Dissecting the foundations of stone knapping skill. *Journal of Human Evolution, 145*, 102807.

Parins-Fukuchi, C., Greiner, E., MacLatchy, L. M., & Fisher, D. C. (2019). Phylogeny, ancestors, and anagenesis in the hominin fossil record. *Paleobiology, 45*(2), 378–393.

Parkinson, J. (1817). *An essay on the shaking palsy*. Sherwood, Neely & Jones.

Parlitz, D., Peschel, T., & Altenmüller, E. (1998). Assessment of dynamic finger forces in pianists: Effects of training and expertise. *Journal of Biomechanics, 31*(11), 1063–1067.

Parlow, S. E., & Kinsbourne, M. (1989). Asymmetrical transfer of training between hands: Implications for interhemispheric communication in normal brain. *Brain and Cognition, 11*(1), 98–113.

Pascual-Leone, A., Grafman, J., & Hallett, M. (1994). Modulation of cortical motor output maps during development of implicit and explicit knowledge. *Science, 263*(5151), 1287–1289.

Pascual-Leone, A., & Torres, F. (1993). Plasticity of the sensorimotor cortex representation of the reading finger in braille readers. *Brain, 116*(Pt 1), 39–52.

Passingham, R. E., Perry, H., & Wilkinson, F. (1978). Failure to develop a precision grip in monkeys with unilateral neocortical lesions made in infancy. *Brain Research, 145*(2), 410–414.

Passingham, R. E. (1975). Changes in the size and organisation of the brain in man and his ancestors. *Brain, Behavior, and Evolution, 11*(2), 73–90.

Passingham, R. E. (1993). *The frontal lobes and voluntary action*. Oxford University Press.

Passingham, R. E., & Wise, S. P. (2012). *The neurobiology of the prefrontal cortex*. Oxford University Press.

Patané, I., Cardinali, L., Salemme, R., Pavani, F., Farnè, A., & Brozzoli, C. (2019). Action planning modulates peripersonal space. *Journal of Cognitive Neuroscience, 31*(8), 1141–1154.

Paton, R. L., Smithson, T. R., & Clack, J. A. (1999). An amniote-like skeleton from the Early Carboniferous of Scotland. *Nature, 398*, 508–513.

Patterson, F. G. (1978). The gestures of a gorilla: Language acquisition in another pongid. *Brain and Language, 5*(1), 72–97.

Patterson, K., & Wing, A. M. (1989). Processes in handwriting: A case for case. *Cognitive Neuropsychology, 6*(1), 1–23.

Patterson, N., Richter, D. J., Gnerre, S., Lander, E. S., & Reich, D. (2006). Genetic evidence for complex speciation of humans and chimpanzees. *Nature, 441*, 1103–1108.

Paul, R. L., Merzenich, M. M., & Goodman, H. (1972). Representation of slowly and rapidly adapting cutaneous mechanoreceptors of the hand in Brodmann's areas 3 and 1 of *Macaca mulatta*. *Brain Research, 36*(2), 229–249.

Paulignan, Y., MacKenzie, C., Marteniuk, R., & Jeannerod, M. (1991). Selective perturbation of visual input during prehension movements 1. The effects of changing object position. *Experimental Brain Research, 83*(3), 502–512.

Pause, M., Kunesch, E., Binkofski, F., & Freund, H.-J. (1989). Sensorimotor disturbances in patients with lesions of the parietal cortex. *Brain, 112*(Pt 6), 1599–1625.

Pavani, F., & Castiello, U. (2004). Binding personal and extrapersonal space through body shadows. *Nature Neuroscience, 7*(1), 13–14.

Pavani, F., Spence, C., & Driver, J. (2000). Visual capture of touch: Out-of-the-body experiences with rubber gloves. *Psychological Science, 11*(5), 353–359.

Pearce, J. M. S. (2005). A note on scrivener's palsy. *Journal of Neurology, Neurosurgery & Psychiatry, 76*(4), 513.

Pearcey, G. E. P., Smith, L. A., Sun, Y., & Zehr, E. P. (2022). 1894 revisited: Cross-education of skilled muscular control in women and the importance of representation. *PLOS One, 17*(3), e0264686.

Peckre, L., Fabre, A.-C., Wall, C. E., Brewer, D., Ehmke, E., Haring, D., Shaw, E., Welser, K., & Pouydebat, E. (2016). Holding-on: Co-evolution between infant carrying and grasping behaviour in strepsirrhines. *Scientific Reports, 6*, 37729.

Pedrazzini, A., Martelli, A., & Tocco, S. (2015). Niccolò Paganini: The hands of a genius. *Acta Biomedica, 86*(1), 27–31.

Peeters, R. R., Rizzolatti, G., & Orban, G. A. (2013). Functional properties of the left parietal tool use region. *NeuroImage, 78*, 83–93.

Peeters, R. R., Simone, L., Nelissen, K., Vanduffel, W., Rizzolatti, G., & Orban, G. A. (2009). The representation of tool use in humans and monkeys: Common and uniquely human features. *Journal of Neuroscience, 29*(37), 11523–11539.

Pei, Y.-C., Denchev, P. V, Hsiao, S. S., Craig, J. C., & Bensmaia, S. J. (2009). Convergence of submodality-specific input onto neurons in primary somatosensory cortex. *Journal of Neurophysiology, 102*(3), 1843–1853.

Peignot, P., & Anderson, J. R. (1999). Use of experimenter-given manual and facial cues by gorillas (*Gorilla gorilla*) in an object-choice task. *Journal of Comparative Psychology, 113*(3), 253–260.

Pelgrims, B., Olivier, E., & Andres, M. (2011). Dissociation between manipulation and conceptual knowledge of object use in the supramarginalis gyrus. *Human Brain Mapping, 32*(11), 1802–1810.

Pellijeff, A., Bonilha, L., Morgan, P. S., McKenzie, K., & Jackson, S. R. (2006). Parietal updating of limb posture: An event-related fMRI study. *Neuropsychologia, 44*(13), 2685–2690.

Penfield, W., & Boldrey, E. (1937). Somatic motor and sensory representation in the cerebral cortex of man as studied by electrical stimulation. *Brain, 60*(4), 389–443.

Penfield, W., & Jasper, H. (1954). *Epilepsy and the functional anatomy of the human brain*. Little, Brown and Company.

Penfield, W., & Rasmussen, T. (1950). *The cerebral cortex of man*. Macmillan.

Penfield, W., & Welch, K. (1951). The supplementary motor area of the cerebral cortex: A clinical and experimental study. *Archives of Neurology & Psychiatry, 66*(3), 289–317.

Pereira-Pedro, A. S., Bruner, E., Gunz, P., & Neubauer, S. (2020). A morphometric comparison of the parietal lobe in modern humans and Neanderthals. *Journal of Human Evolution, 142*, 102770.

Perelle, I. B., & Ehrman, L. (1994). An international study of human handedness: The data. *Behavior Genetics, 24*(3), 217–227.

Perenin, M.-T., & Jeannerod, M. (1978). Visual function within the hemianopic field following early cerebral hemidecortication in man—I. Spatial localization. *Neuropsychologia, 16*(1), 1–13.

Perenin, M.-T., & Rossetti, Y. (1996). Grasping without form discrimination in a hemianopic field. *NeuroReport, 7*(3), 793–797.

Perenin, M.-T., & Vighetto, A. (1988). Optic ataxia: A specific disruption in visuomotor mechanisms. I. Different aspects of the deficit in reaching for objects. *Brain, 111*(Pt 3), 643–674.

Perez, M. A., & Cohen, L. G. (2008). Mechanisms underlying functional changes in the primary motor cortex ipsilateral to an active hand. *Journal of Neuroscience, 28*(22), 5631–5640.

Perez, M. A., Tanaka, S., Wise, S. P., Sadato, N., Tanabe, H. C., Willingham, D. T., & Cohen, L. G. (2007). Neural substrates of intermanual transfer of a newly acquired motor skill. *Current Biology, 17*(21), 1896–1902.

Perez, M. A., Wise, S. P., Willingham, D. T., & Cohen, L. G. (2007). Neurophysiological mechanisms involved in transfer of procedural knowledge. *Journal of Neuroscience, 27*(5), 1045–1053.

Perris, E. E., & Clifton, R. K. (1988). Reaching in the dark toward sound as a measure of auditory localization in infants. *Infant Behavior and Development, 11*(4), 473–491.

Perry, C. J., Sergio, L. E., Crawford, J. D., & Fallah, M. (2015). Hand placement near the visual stimulus improves orientation selectivity in V2 neurons. *Journal of Neurophysiology, 113*(7), 2859–2870.

Perry, M., Church, R. B., & Goldin-Meadow, S. (1988). Transitional knowledge in the acquisition of concepts. *Cognitive Development, 3*(4), 359–400.

Pesaran, B., Nelson, M. J., & Andersen, R. A. (2006). Dorsal premotor neurons encode the relative position of the hand, eye, and goal during reach planning. *Neuron, 51*(1), 125–134.

Petitto, L. A., Holowka, S., Sergio, L. E., & Ostry, D. (2001). Language rhythms in baby hand movements. *Nature, 413*, 35–36.

Petitto, L. A., & Marentette, P. F. (1991). Babbling in the manual mode: Evidence for the ontogeny of language. *Science, 251*(5000), 1493–1496.

Petkova, V. I., & Ehrsson, H. H. (2009). When right feels left: Referral of touch and ownership between the hands. *PLOS One, 4*(9), e6933.

Petrides, M., Caroret, G., & Mackey, S. (2005). Orofacial somatomotor responses in the macaque monkey homologue of Broca's area. *Nature, 435*, 1235–1238.

Peviani, V., & Bottini, G. (2018). The distorted hand metric representation serves both perception and action. *Journal of Cognitive Psychology, 30*(8), 880–893.

Peviani, V., Liotta, J., & Bottini, G. (2020). The motor system (partially) deceives body representation biases in absence of visual correcting cues. *Acta Psychologica, 203,* 103003.

Phillips, K. A., & Sherwood, C. C. (2005). Primary motor cortex asymmetry is correlated with handedness in capuchin monkeys (*Cebus apella*). *Behavioral Neuroscience, 119*(6), 1701–1704.

Pia, L., Fossataro, C., Burin, D., Bruno, V., Spinazzola, L., Gindri, P., Fotopoulou, K., Berti, A., & Garbarini, F. (2020). The anatomo-clinical picture of the pathological embodiment over someone else's body part after stroke. *Cortex, 130,* 203–219.

Pia, L., Garbarini, F., Fossataro, C., Fornia, L., & Berti, A. (2013). Pain and body awareness: Evidence from brain-damaged patients with delusional body ownership. *Frontiers in Human Neuroscience, 7,* 298.

Piaget, J. (1936). *The origins of intelligence in children* [M. Cook, Trans.]. International Universities Press.

Piaget, J. (1952). *The child's conception of number* [G. Gattegno & F. M. Hodgson, Trans.]. Routledge.

Piazza, M., Pinel, P., Le Bihan, D., & Dehaene, S. (2007). A magnitude code common to numerosities and number symbols in human intraparietal cortex. *Neuron, 53*(2), 293–305.

Picard, N., Matsuzaka, Y., & Strick, P. L. (2013). Extended practice of a motor skill is associated with reduced metabolic activity in M1. *Nature Neuroscience, 16*(9), 1340–1347.

Picard, N., & Strick, P. L. (2001). Imaging the premotor areas. *Current Opinion in Neurobiology, 11*(6), 663–672.

Pickering, T. R., Heaton, J. L., Clarke, R. J., & Stratford, D. (2018). Hominin hand bone fossils from Sterkfontein Caves, South Africa (1998–2003 excavations). *Journal of Human Evolution, 118,* 89–102.

Piedimonte, A., Garbarini, F., Rabuffetti, M., Pia, L., Montesano, A., Ferrarin, M., & Berti, A. (2015). Invisible grasps: Grip interference in anosognosia for hemiplegia. *Neuropsychology, 29*(5), 776–781.

Pietrini, P., Furey, M. L., Ricciardi, E., Gobbini, M. I., Wu, W.-H. C., Cohen, L., Guazzelli, M., & Haxby, J. V. (2004). Beyond sensory images: Object-based representation in the human ventral pathway. *Proceedings of the National Academy of Sciences, 101*(15), 5658–5663.

Pika, S., & Mitani, J. (2006). Referential gestural communication in wild chimpanzees (*Pan troglodytes*). *Current Biology, 16*(6), R191–R192.

Pike, A. A., Marlow, N., & Dawson, C. (1997). Posterior tibial somatosensory evoked potentials in very preterm infants. *Early Human Development, 47*(1), 71–84.

Pilbeam, D., & Gould, S. J. (1974). Size and scaling in human evolution. *Science, 186*(4167), 892–901.

Pisella, L., Binkofski, F., Lasek, K., Toni, I., & Rossetti, Y. (2006). No double-dissociation between optic ataxia and visual agnosia: Multiple sub-streams for multiple visuo-manual integrations. *Neuropsychologia, 44*(13), 2734–2748.

Pisella, L., Gréa, H., Tilikete, C., Vighetto, A., Desmurget, M., Rode, G., Boisson, D., & Rossetti, Y. (2000). An "automatic pilot" for the hand in human posterior parietal cortex: Toward reinterpreting optic ataxia. *Nature Neuroscience, 3*(7), 729–736.

Pitzalis, S., Galletti, C., Huang, R.-S., Patria, F., Committeri, G., Galati, G, Fattori, P., & Sereno, M. I. (2006). Wide-field retinotopy defines human cortical visual area V6. *Journal of Neuroscience, 26*(30), 7962–7973.

Planton, S., Jucla, M., Roux, F.-E., & Démonet, J.-F. (2013). The "handwriting brain": A meta-analysis of neuroimaging studies of motor versus orthographic processes. *Cortex, 49*(10), 2772–2787.

Plato. (1875). *Laws* [B. Jowett, Trans.]. Clarendon Press.

Platz, T. (1996). Tactile agnosia: Casuistic evidence and theoretical remarks on modality-specific meaning representations and sensorimotor integration. *Brain, 119*(Pt 5), 1565–1574.

Playford, E. D., Jenkins, I. H., Passingham, R. E., Nutt, J., Frackowiak, R. S. J., & Brooks, D. J. (1992). Impaired mesial frontal and putamen activation in Parkinson's disease. *Annals of Neurology, 32*(2), 151–161.

Pliny the Elder. (1855). *The natural history* [J. Bostock & H. T. Riley, Trans.]. Henry Bohn.

Plummer, T. W., Oliver, J. S., Finestone, E. M., Ditchfield, P. W., Bishop, L. C., Blumenthal, S. A., Lemorini, C., Caricola, I., Bailey, S. E., Herries, A. I. R., Parkinson, J. A., Whitfield, E., Hertel, F., Kinyanjui, R. N., Vincent, T. H., Li, Y., Louys, J., Frost, S. R., Braun, D. R., . . . Potts, R. (2023). Expanded geographic distribution and dietary strategies of the earliest Oldowan hominins and *Paranthropus*. *Science, 379*(6632), 561–566.

Poeck, K. (1964). Phantoms following amputation in early childhood and in congenital absence of limbs. *Cortex, 1*(3), 269–275.

Poeck, K., & Orgass, B. (1971). The concept of the body schema: A critical review and some experimental results. *Cortex, 7*(3), 254–277.

Poizner, H., Klima, E. S., & Bellugi, U. (1987). *What the hands reveal about the brain*. MIT Press.

Ponce de León, M. S., Bienvenu, T., Marom, A., Engel, S., Tafforeau, P., Warren, J. L. A., Lordkipanidze, D., Kurniawan, I., Murti, D. B., Suriyanto, R. A., Koesbardiati, T., & Zollikofer, C. P. E. (2021). The primitive brain of early Homo. *Science, 372*(6538), 165–171.

Ponce de León, M. S., & Zollikofer, C. P. E. (2001). Neanderthal cranial ontogeny and its implications for late hominid diversity. *Nature, 412*, 534–538.

Pons, T. P., Garraghty, P. E., Cusick, C. G., & Kaas, J. H. (1985). The somatotopic organization of area 2 in macaque monkeys. *Journal of Comparative Neurology, 241*(4), 445–466.

Pons, T. P., Garraghty, P. E., Friedman, D. P., & Mishkin, M. (1987). Physiological evidence for serial processing in somatosensory cortex. *Science, 237*(4813), 417–420.

Pons, T. P., Garraghty, P. E., & Mishkin, M. (1992). Serial and parallel processing of tactual information in somatosensory cortex of rhesus monkeys. *Journal of Neurophysiology*, *68*(2), 518–527.

Pons, T. P., Garraghty, P. E., Ommaya, A. K., Kaas, J. H., Taub, E., & Mishkin, M. (1991). Massive cortical reorganization after sensory deafferentation in adult macaques. *Science*, *252*(5014), 1857–1860.

Pontzer, H., Holloway, J. H., Raichlen, D. A., & Lieberman, D. E. (2009). Control and function of arm swing in human walking and running. *Journal of Experimental Biology*, *212*(Pt 4), 523–534.

Pool, E.-M., Rehme, A. K., Fink, G. R., Eickhoff, S. B., & Grefkes, C. (2014). Handedness and effective connectivity of the motor system. *NeuroImage*, *99*, 451–460.

Porro, C. A., Martinig, M., Facchin, P., Maieron, M., Jones, A. K. P., & Fadiga, L. (2007). Parietal cortex involvement in the localization of tactile and noxious mechanical stimuli: A transcranial magnetic stimulation study. *Behavioural Brain Research*, *178*(2), 183–189.

Porter, R., & Lemon, R. N. (1993). *Corticospinal function and voluntary movement*. Oxford University Press.

Posnansky, M. (1959). Some functional considerations on the handaxe. *Man*, *59*, 42–44.

Potts, R., & Shipman, P. (1981). Cutmarks made by stone tools on bones from Olduvai Gorge, Tanzania. *Nature*, *291*, 577–580.

Pourrier, S. D., Nieuwstraten, W., Van Cranenburgh, B., Schreuders, T. A. R., Stam, H. J., & Selles, R. W. (2010). Three cases of referred sensation in traumatic nerve injury of the hand: Implications for understanding central nervous system reorganization. *Journal of Rehabilitation Medicine*, *42*(4), 357–361.

Povinelli, D. J., & Davis, D. R. (1994). Differences between chimpanzees (*Pan troglodytes*) and humans (*Homo sapiens*) in the resting state of the index finger: Implications for pointing. *Journal of Comparative Psychology*, *108*(2), 134–139.

Povinelli, D. J., Reaux, J. E., & Frey, S. H. (2010). Chimpanzees' context-dependent tool use provides evidence for separable representations of hand and tool even during active use within peripersonal space. *Neuropsychologia*, *48*(1), 243–247.

Poza-Rey, E. M., Lozano, M., & Arsuaga, J. L. (2017). Brain asymmetries and handedness in the specimens from the Sima de los Huesos site (Atapuerca, Spain). *Quaternary International*, *433*(Part A), 32–44.

Prado, J., Clavagnier, S., Otzenberger, H., Schieber, C., Kennedy, H., & Perenin, M.-T. (2005). Two cortical systems for reaching in central and peripheral vision. *Neuron*, *48*(5), 849–858.

Prang, T. C., Ramirez, K., Grabowski, M., & Williams, S. A. (2021). Ardipithecus hand provides evidence that humans and chimpanzees evolved from an ancestor with suspensory adaptations. *Science Advances*, *7*(9), eabf2474.

Prechtl, H. F. R. (1977). *The neurological examination of the full-term newborn infant* (2nd ed.). Spastics International Medical Publications.

References

Prechtl, H. F. R. (1985). Ultrasound studies of human fetal behaviour. *Early Human Development, 12*(2), 91–98.

Press, C., Taylor-Clarke, M., Kennett, S., & Haggard, P. (2004). Visual enhancement of touch in spatial body representation. *Experimental Brain Research, 154*(2), 238–245.

Preston, C., & Newport, R. (2011). Differential effects of perceived hand location on the disruption of embodiment by apparent physical encroachment of the limb. *Cognitive Neuroscience, 2*(3–4), 163–170.

Preuss, T. M., & Goldman-Rakic, P. S. (1991). Myelo- and cytoarchitecture of the granular frontal cortex and surrounding regions in the strepsirhine primate galago and the anthropoid primate *Macaca*. *Journal of Comparative Neurology, 310*(4), 429–474.

Preuss, T. M., Stepniewska, I., & Kaas, J. H. (1996). Movement representation in the dorsal and ventral premotor areas of owl monkeys: A microstimulation study. *Journal of Comparative Neurology, 371*(4), 649–676.

Previc, F. H. (1991). A general theory concerning the prenatal origins of cerebral lateralization in humans. *Psychological Review, 98*(3), 299–334.

Proffitt, T., Luncz, L. V, Falótico, T., Ottoni, E. B., de la Torre, I., & Haslam, M. (2016). Wild monkeys flake stone tools. *Nature, 539*(7627), 85–88.

Proske, U., & Gandevia, S. C. (2012). The proprioceptive senses: Their roles in signaling body shape, body position and movement, and muscle force. *Physiological Reviews, 92*(4), 1651–1697.

Pruszynski, J. A., Johansson, R. S., & Flanagan, J. R. (2016). A rapid tactile-motor reflex automatically guides reaching toward handheld objects. *Current Biology, 26*(6), 788–792.

Pruszynski, J. A., Kurtzer, I., Nashed, J. Y., Omrani, M., Brouwer, B., & Scott, S. H. (2011). Primary motor cortex underlies multi-joint integration for fast feedback control. *Nature, 478*, 387–390.

Pruszynski, J. A., Kurtzer, I., & Scott, S. H. (2008). Rapid motor responses are appropriately tuned to the metrics of a visuospatial task. *Journal of Neurophysiology, 100*(1), 224–238.

Pruszynski, J. A., Omrani, M., & Scott, S. H. (2014). Goal-dependent modulation of fast feedback responses in primary motor cortex. *Journal of Neuroscience, 34*(13), 4608–4617.

Pubols, B. H., & Pubols, L. M. (1971). Somatotopic organization of spider monkey somatic sensory cerebral cortex. *Journal of Comparative Neurology, 141*(1), 63–75.

Puts, N. A. J., Edden, R. A. E., Evans, C. J., McGlone, F., & McGonigle, D. J. (2011). Regionally specific human GABA concentration correlates with tactile discrimination thresholds. *Journal of Neuroscience, 31*(46), 16556–16560.

Putt, S. S., Wijeakumar, S., Franciscus, R. G., & Spencer, J. P. (2017). The functional brain networks that underlie early stone age tool manufacture. *Nature Human Behaviour, 1*, 0102.

Quallo, M. M., Kraskov, A., & Lemon, R. N. (2012). The activity of primary motor cortex corticospinal neurons during tool use by macaque monkeys. *Journal of Neuroscience, 32*(48), 17351–17364.

Quallo, M. M., Price, C. J., Ueno, K., Asamizuya, T., Cheng, K., Lemon, R. N., & Iriki, A. (2009). Gray and white matter changes associated with tool-use learning in macaque monkeys. *Proceedings of the National Academy of Sciences, 106*(43), 18379–18384.

Quartarone, A., Siebner, H. R., & Rothwell, J. C. (2006). Task-specific hand dystonia: Can too much plasticity be bad for you? *Trends in Neurosciences, 29*(4), 192–199.

Quine, W. V. O. (1960). *Word and object*. MIT Press.

Rabbitt, P., & Rodgers, B. (1977). What does a man do after he makes an error? An analysis of response programming. *Quarterly Journal of Experimental Psychology, 29*(4), 727–743.

Rabin, E., & Gordon, A. M. (2004). Tactile feedback contributes to consistency of finger movements during typing. *Experimental Brain Research, 155*(3), 362–369.

Rader, N., & Stern, J. D. (1982). Visually elicited reaching in neonates. *Child Development, 53*(4), 1004–1007.

Radinsky, L. (1973). *Aegyptopithecus* endocasts: Oldest record of a pongid brain. *American Journal of Physical Anthropology, 39*(2), 239–248.

Radinsky, L. (1975). Primate brain evolution. *American Scientist, 63*(6), 656–663.

Raffin, E., Giraux, P., & Reilly, K. T. (2012). The moving phantom: Motor execution or motor imagery? *Cortex, 48*(6), 746–757.

Raffin, E., Mattout, J., Reilly, K. T., & Giraux, P. (2012). Disentangling motor execution from motor imagery with the phantom limb. *Brain, 135*(Pt 2), 582–595.

Ragert, P., Schmidt, A., Altenmüller, E., & Dinse, H. R. (2004). Superior tactile performance and learning in professional pianists: Evidence for meta-plasticity in musicians. *European Journal of Neuroscience, 19*(2), 473–478.

Rakic, P. (1988). Specification of cerebral cortical areas. *Science, 241*(4862), 170–176.

Rakic, P., Suñer, I., & Williams, R. W. (1991). A novel cytoarchitectonic area induced experimentally within the primate visual cortex. *Proceedings of the National Academy of Sciences, 88*(6), 2083–2087.

Ramachandran, V. S., & Blakeslee, S. (1998). *Phantoms in the brain*. William Morrow.

Ramachandran, V. S., Rogers-Ramachandran, D., & Cobb, S. (1995). Touching the phantom limb. *Nature, 377*, 489–490.

Ramachandran, V. S., Rogers-Ramachandran, D., & Stewart, M. (1992). Perceptual correlates of massive cortical reorganization. *Science, 258*(5085), 1159–1160.

Randolph, M., & Semmes, J. (1974). Behavioral consequences of selective subtotal ablations in the postcentral gyrus of *Macaca mulatta*. *Brain Research, 70*(1), 55–70.

Ranhorn, K., & Tryon, C. A. (2018). New radiocarbon dates from Nasera rockshelter (Tanzania): Implications for studying spatial patterns in Late Pleistocene technology. *Journal of African Archaeology, 16*(2), 211–222.

Raos, V., Franchi, G., Gallese, V., & Fogassi, L. (2003). Somatotopic organization of the lateral part of area F2 (dorsal premotor cortex) of the macaque monkey. *Journal of Neurophysiology, 89*(3), 1503–1518.

Rapp, B., & Caramazza, A. (1997). From graphemes to abstract letter shapes: Levels of representation in written spelling. *Journal of Experimental Psychology: Human Perception and Performance, 23*(4), 1130–1152.

Rapp, B., Hendel, S. K., & Medina, J. (2002). Remodeling of somatasensory hand representations following cerebral lesions in humans. *Neuroreport, 13*(2), 207–211.

Ratcliff, G., & Davies-Jones, G. A. B. (1972). Defective visual localization in focal brain wounds. *Brain, 95*(1), 49–60.

Rathelot, J.-A., Dum, R. P., & Strick, P. L. (2017). Posterior parietal cortex contains a command apparatus for hand movements. *Proceedings of the National Academy of Sciences, 114*(16), 4255–4260.

Rathelot, J.-A., & Strick, P. L. (2006). Muscle representation in the macaque motor cortex: An anatomical perspective. *Proceedings of the National Academy of Sciences, 103*(21), 8257–8262.

Rathelot, J.-A., & Strick, P. L. (2009). Subdivisions of primary motor cortex based on corticomotoneuronal cells. *Proceedings of the National Academy of Sciences, 106*(3), 918–923.

Rauscher, F. H., Krauss, R. M., & Chen, Y. (1996). Gesture, speech, and lexical access: The role of lexical movements in speech production. *Psychological Science, 7*(4), 226–232.

Raymond, M., & Pontier, D. (2004). Is there geographical variation in human handedness? *Laterality, 9*(1), 35–51.

Reales, J. M., & Ballesteros, S. (1999). Implicit and explicit memory for visual and haptic objects: Cross-modal priming depends on structural descriptions. *Journal of Experimental Psychology: Learning, Memory, and Cognition, 25*(3), 644–663.

Reed, C. L., & Caselli, R. J. (1994). The nature of tactile agnosia: A case study. *Neuropsychologia, 32*(5), 527–539.

Reed, C. L., Caselli, R. J., & Farah, M. J. (1996). Tactile agnosia: Underlying impairment and implications for normal tactile object recognition. *Brain, 119*(Pt 3), 875–888.

Reed, C. L., Grubb, J. D., & Steele, C. (2006). Hands up: Attentional prioritization of space near the hand. *Journal of Experimental Psychology: Human Perception and Performance, 32*(1), 166–177.

Reed, C. M., Delhorne, L. A., Durlach, N. I., & Fischer, S. D. (1995). A study of the tactual reception of sign language. *Journal of Speech, Language, and Hearing Research, 38*(2), 477–489.

Reed, J. L., Pouget, P., Qi, H.-X., Zhou, Z., Bernard, M. R., Burish, M. J., Haitas, J., Bonds, A. B., & Kaas, J. H. (2008). Widespread spatial integration in primary somatosensory cortex. *Proceedings of the National Academy of Sciences, 105*(29), 10233–10237.

Reed, J. L., Qi, H.-X., Pouget, P., Burish, M. J., Bonds, A. B., & Kaas, J. H. (2010). Modular processing in the hand representation of primate primary somatosensory cortex coexists with widespread activation. *Journal of Neurophysiology, 104*(6), 3136–3145.

Regaiolli, B., Spiezio, C., & Hopkins, W. D. (2018). Hand preference on unimanual and bimanual tasks in Barbary macaques (*Macaca sylvanus*). *American Journal of Primatology, 80*(3), e22745.

Reich, L., Szwed, M., Cohen, L., & Amedi, A. (2011). A ventral visual stream reading center independent of visual experience. *Current Biology, 21*(5), 363–368.

Reichenbach, A., Bresciani, J.-P., Peer, A., Bülthoff, H. H., & Thielscher, A. (2011). Contributions of the PPC to online control of visually guided reaching movements assessed with fMRI-guided TMS. *Cerebral Cortex, 21*(7), 1602–1612.

Reilly, K. T., Mercier, C., Schieber, M. H., & Sirigu, A. (2006). Persistent hand motor commands in the amputees' brain. *Brain, 129*(Pt 8), 2211–2223.

Reissland, N., Francis, B., Aydin, E., Mason, J., & Schaal, B. (2013). The development of anticipation in the fetus: A longitudinal account of human fetal mouth movements in reaction to and anticipation of touch. *Developmental Psychobiology, 56*(5), 955–963.

Remy, P., Zilbovicius, M., Degos, J.-D., Bachoud-Levi, A.-C., Rancurel, G., Cesaro, P., & Samson, Y. (1999). Somatosensory cortical activations are suppressed in patients with tactile extinction. *Neurology, 52*(3), 571–577.

Révész, G. (1934). System der optischen und haptischen Raumtäuschungen [A system of optic and haptic space illusions]. *Zeitschrift Für Psychologie, 131*, 296–375.

Ricci, R., Salatino, A., Caldano, M., Perozzo, P., Cerrato, P., Pyasik, M., Pia, L., & Berti, A. (2019). Phantom touch: How to unmask sensory unawareness after stroke. *Cortex, 121*, 253–263.

Richmond, B. G., Begun, D. R., & Strait, D. S. (2001). Origin of human bipedalism: The knuckle-walking hypothesis revisited. *Yearbook of Physical Anthropology, 44*, 70–105.

Richmond, B. G., Roach, N. T., & Ostrovsky, K. R. (2016). Evolution of the early hominin hand. In T. L. Kivell, P. Lemelin, B. G. Richmond, & D. Schmitt (Eds.), *The evolution of the primate hand* (pp. 515–543). Springer.

Richmond, B. G., & Strait, D. S. (2000). Evidence that humans evolved from a knuckle-walking ancestor. *Nature, 404*, 382–385.

Richter, C. P. (1931). The grasping reflex in the new-born monkey. *Archives of Neurology and Psychiatry, 26*(4), 784–790.

Ricklan, D. E. (1988). Functional anatomy of the hand of *Australopithecus africanus*. *Journal of Human Evolution, 16*(7–8), 643–664.

Riddoch, G. (1935). Visual disorientation in homonymous half-fields. *Brain, 58*(3), 376–382.

Riddoch, G. (1941). Phantom limbs and body shape. *Brain, 64*(4), 197–222.

Riddoch, M. J., Edwards, M. G., Humphreys, G. W., West, R., & Heafield, T. (1998). Visual affordances direct action: Neuropsychological evidence from manual interference. *Cognitive Neuropsychology, 15*(6–8), 645–683.

Riddoch, M. J., & Humphreys, G. W. (1987). A case of integrative visual agnosia. *Brain, 110*(Pt 6), 1431–1462.

Riedel, J., Schumann, K., Kaminski, J., Call, J., & Tomasello, M. (2008). The early ontogeny of human-dog communication. *Animal Behaviour, 75*(3), 1003–1014.

Rijntjes, M., Dettmers, C., Büchel, C., Kiebel, S., Frackowiak, R. S. J., & Weiller, C. (1999). A blueprint for movement: Functional and anatomical representations in the human motor system. *Journal of Neuroscience, 19*(18), 8043–8048.

Rilling, J. K., & Insel, T. R. (1999). The primate neocortex in comparative perspective using magnetic resonance imaging. *Journal of Human Evolution, 37*(2), 191–223.

Rimé, B., Schiaratura, L., Hupet, M., & Ghysselinckx, A. (1984). Effects of relative immobilization on the speaker's nonverbal behavior and on the dialogue imagery level. *Motivation and Emotion, 8*(4), 311–325.

Riseborough, M. G. (1981). Physiographic gestures as decoding facilitators: Three experiments exploring a neglected facet of communication. *Journal of Nonverbal Behavior, 5*(3), 172–183.

Rivers, W. H. R., & Head, H. (1908). A human experiment in nerve division. *Brain, 31*(3), 323–450.

Rizzolatti, G., & Arbib, M. A. (1998). Language within our grasp. *Trends in Neurosciences, 21*(5), 188–194.

Rizzolatti, G., Camarda, R., Fogassi, L., Gentilucci, M., Luppino, G., & Matelli, M. (1988). Functional organization of inferior area 6 in the macaque monkey. II. Area F5 and the control of distal movements. *Experimental Brain Research, 71*(3), 491–507.

Rizzolatti, G., Matelli, M., & Pavesi, G. (1983). Deficits in attention and movement following the removal of postarcuate (area 6) and prearcuate (area 8) cortex in macaque monkeys. *Brain, 106*(Pt 3), 655–673.

Rizzolatti, G., Scandolara, C., Matelli, M., & Gentilucci, M. (1981). Afferent properties of periarcuate neurons in macaque monkeys. II. Visual responses. *Behavioural Brain Research, 2*(2), 147–163.

Ro, T., & Koenig, L. (2021). Unconscious touch perception after disruption of the primary somatosensory cortex. *Psychological Science, 32*(4), 549–557.

Robertson, I. H., & North, N. (1993). Active and passive activation of left limbs: Influence on visual and sensory neglect. *Neuropsychologia, 31*(3), 293–300.

Robinson, L. (1891). Darwinism in the nursery. *Nineteenth Century, 30*, 831–842.

Rochat, P., Blass, E. M., & Hoffmeyer, L. B. (1988). Oropharyngeal control of hand–mouth coordination in newborn infants. *Developmental Psychology, 24*(4), 459–463.

Roche, H., Delagnes, A., Brugal, J.-P., Feibel, C., Kibunjia, M., Mourre, V., & Texier, P.-J. (1999). Early hominid stone tool production and technical skill 2.34 Myr ago in West Turkana, Kenya. *Nature, 399*, 57–60.

Rock, I., & Victor, J. (1964). Vision and touch: An experimentally created conflict between the two senses. *Science, 143*(3606), 594–596.

Röder, B., Rösler, F., & Spence, C. (2004). Early vision impairs tactile perception in the blind. *Current Biology, 14*(2), 121–124.

Roediger, H. L., & Blaxton, T. A. (1987). Effects of varying modality, surface features, and retention interval on priming in word-fragment completion. *Memory & Cognition, 15*(5), 379–388.

Roel Lesur, M., Weijs, M. L., Simon, C., Kannape, O. A., & Lenggenhager, B. (2020). Psychometrics of disembodiment and its differential modulation by visuomotor and visuotactile mismatches. *iScience, 23*(3), 100901.

Roffman, I., Savage-Rumbaugh, S., Rubert-Pugh, E., Ronen, A., & Nevo, E. (2012). Stone tool production and utilization by bonobo-chimpanzees (*Pan paniscus*). *Proceedings of the National Academy of Sciences, 109*(36), 14500–14503.

Rogers, M. J., & Semaw, S. (2009). From nothing to something: The appearance and context of the earliest archaeological record. In M. Camps & P. Chauhan (Eds.), *Sourcebook of paleolithic transitions* (pp. 155–171). Springer.

Rohde, M., Di Luca, M., & Ernst, M. O. (2011). The rubber hand illusion: Feeling of ownership and proprioceptive drift do not go hand in hand. *PLOS One, 6*(6), e21659.

Rohlfing, K. J., Longo, M. R., & Bertenthal, B. I. (2012). Dynamic pointing triggers shifts of visual attention in young infants. *Developmental Science, 15*(3), 426–435.

Roland, P. E. (1976). Astereognosis: Tactile discrimination after localized hemispheric lesions in man. *Archives of Neurology, 33*(8), 543–550.

Roland, P. E. (1987). Somatosensory detection of microgeometry, macrogeometry and kinesthesia after localized lesions of the cerebral hemispheres in man. *Brain Research Reviews, 12*(1), 43–94.

Roland, P. E., O'Sullivan, B. T., & Kawashima, R. (1998). Shape and roughness activate different somatosensory areas in the human brain. *Proceedings of the National Academy of Sciences, 95*(6), 3295–3300.

Rolian, C., & Gordon, A. D. (2013). Reassessing manual proportions in *Australopithecus afarensis*. *American Journal of Physical Anthropology, 152*(3), 393–406.

Rolian, C., Lieberman, D. E., & Hallgrímsson, B. (2010). The coevolution of human hands and feet. *Evolution, 64*(6), 1558–1568.

Rolian, C., Lieberman, D. E., & Paul, J. (2011). Hand biomechanics during simulated stone tool use. *Journal of Human Evolution, 61*(1), 26–41.

Romano, D., & Maravita, A. (2019). The dynamic nature of the sense of ownership after brain injury. Clues from asomatognosia and somatoparaphrenia. *Neuropsychologia, 132*, 107119.

Romano, D., Uberti, E., Caggiano, P., Cocchini, G., & Maravita, A. (2019). Different tool training induces specific effects on body metric representation. *Experimental Brain Research, 237*(2), 493–501.

Romo, R., Hernández, A., Zainos, A., Brody, C. D., & Lemus, L. (2000). Sensing without touching: Psychophysical performance based on cortical microstimulation. *Neuron, 26*(1), 273–278.

Romo, R., Hernández, A., Zainos, A., & Salinas, E. (1998). Somatosensory discrimination based on cortical microstimulation. *Nature, 392*, 387–390.

Ronchi, R., Bassolino, M., Viceic, D., Bellmann, A., Vuadens, P., Blanke, O., & Vallar, G. (2020). Disownership of body parts as revealed by a visual scale evaluation. *Neuropsychologia, 138*, 107337.

Rondot, P., De Recondo, J., & Ribadeau Dumas, J. L. (1977). Visuomotor ataxia. *Brain, 100*(2), 355–376.

Rosenblum, S., Samuel, M., Zlotnik, S., Erikh, I., & Schlesinger, I. (2013). Handwriting as an objective tool for Parkinson's disease diagnosis. *Journal of Neurology, 260*(9), 2357–2361.

Rosenzopf, H., Wiesen, D., Basilakos, A., Yourganov, G., Bonilha, L., Rorden, C., Fridriksson, J., Karnath, H.-O., & Sperber, C. (2022). Mapping the human praxis network: An investigation of white matter disconnection in limb apraxia of gesture production. *Brain Communications, 4*(1), fcac004.

Ross, C. (2001). Park or ride? Evolution of infant carrying. *International Journal of Primatology, 22*(5), 749–771.

Ross, P., & Flack, T. (2020). Removing hand form information specifically impairs emotion recognition for fearful and angry body stimuli. *Perception, 49*(1), 98–112.

Ross Russell, R. W., & Bharucha, N. (1984). Visual localisation in patients with occipital infarction. *Journal of Neurology, Neurosurgery, and Psychiatry, 47*(2), 153–158.

Rossetti, Y., Revol, P., McIntosh, R., Pisella, L., Rode, G., Danckert, J., Tilikete, C., Dijkerman, H. C., Boisson, D., Vighetto, A., Michel, F., & Milner, A. D. (2005). Visually guided reaching: Bilateral posterior parietal lesions cause a switch from fast visuomotor to slow cognitive control. *Neuropsychologia, 43*(2), 162–177.

Rossetti, Y., Rode, G., & Boisson, D. (1995). Implicit processing of somaesthetic information: A dissociation between where and how? *NeuroReport, 6*(3), 506–510.

Rothi, L. J. G., Ochipa, C., & Heilman, K. M. (1991). A cognitive neuropsychological model of limb praxis. *Cognitive Neuropsychology, 8*(6), 443–458.

Rothwell, J. C., Traub, M. M., Day, B. L., Obeso, J.A., Thomas, P. K., & Marsden, C. D. (1982). Manual motor performance in a deafferented man. *Brain, 105*(Pt 3), 515–542.

Rothwell, J. C., Traub, M. M., & Marsden, C. D. (1980). Influence of voluntary intent on the human long-latency stretch reflex. *Nature, 286*, 496–498.

Roux, F.-E., Djidjeli, I., & Durand, J.-B. (2018). Functional architecture of the somatosensory homunculus detected by electrostimulation. *Journal of Physiology, 596*(5), 941–956.

Roux, F.-E., Dufor, O., Giussani, C., Wamain, Y., Draper, L., Longcamp, M., & Démonet, J.-F. (2009). The graphemic/motor frontal area Exner's area revisited. *Annals of Neurology, 66*(4), 537–545.

Roux, F.-E., Niare, M., Charni, S., Giussani, C., & Durand, J.-B. (2020). Functional architecture of the motor homunculus detected by electrostimulation. *Journal of Physiology, 598*(23), 5487–5504.

Rowe, M. L., Özçalışkan, Ş., & Goldin-Meadow, S. (2008). Learning words by hand: Gesture's role in predicting vocabulary development. *First Language, 28*(2), 182–199.

Rowe, T. B., Macrini, T. E., & Luo, Z.-X. (2011). Fossil evidence on origin of the mammalian brain. *Science, 332*(6032), 955–957.

Rubel, E. W. (1971). A comparison of somatotopic organization in sensory neocortex of newborn kittens and adult cats. *Journal of Comparative Neurology, 143*(4), 447–480.

Rumelhart, D. E., & Norman, D. A. (1982). Simulating a skilled typist: A study of skilled cognitive-motor performance. *Cognitive Science, 6*(1), 1–36.

Russo, A. A., Bittner, S. R., Perkins, S. M., Seely, J. S., London, B. M., Lara, A. H., Miri, A., Marshall, N. J., Kohn, A., Jessell, T. M., Abbott, L. F., Cunningham, J. P., & Churchland, M. M. (2018). Motor cortex embeds muscle-like commands in an untangled population response. *Neuron, 97*(4), 953–966.

Russell, J. R., & Reitan, R. M. (1955). Psychological abnormalities in agenesis of the corpus callosum. *Journal of Nervous and Mental Disease, 121*(3), 205–214.

Saadah, E. S. M., & Melzack, R. (1994). Phantom limb experiences in congenital limb-deficient adults. *Cortex, 30*(3), 479–485.

Saadon-Grosman, N., Loewenstein, Y., & Arzy, S. (2020). The "creatures" of the human cortical somatosensory system. *Brain Communications, 2*(1), fcaa003.

Saal, H. P., & Bensmaia, S. J. (2014). Touch is a team effort: Interplay of submodalities in cutaneous sensibility. *Trends in Neurosciences, 37*(12), 689–697.

Saal, H. P., Delhaye, B. P., Rayhaun, B. C., & Bensmaia, S. J. (2017). Simulating tactile signals from the whole hand with millisecond precision. *Proceedings of the National Academy of Sciences, 114*(28), E5693–E5702.

Saby, J. N., Meltzoff, A. N., & Marshall, P. J. (2015). Neural body maps in human infants: Somatotopic responses to tactile stimulation in 7-month-olds. *NeuroImage, 118*, 74–78.

Sacrey, L. R., Karl, J. M., & Whishaw, I. Q. (2012). Development of rotational movements, hand shaping, and accuracy in advance and withdrawal for the reach-to-eat movement in human infants aged 6–12 months. *Infant Behavior and Development, 35*(3), 543–560.

Sadato, N., Pascual-Leone, A., Grafman, J., Deiber, M.-P., Ibañez, V., & Hallett, M. (1998). Neural networks for braille reading by the blind. *Brain, 121*(Pt 7), 1213–1229.

Sadato, N., Pascual-Leone, A., Grafman, J., Ibañez, V., Deiber, M. P., Gold, G., & Hallett, M. (1996). Activation of the primary visual cortex by braille reading in blind subjects. *Nature, 380*, 526–528.

Sadnicka, A., Kornysheva, K., Rothwell, J. C., & Edwards, M. J. (2017). A unifying motor control framework for task-specific dystonia. *Nature Reviews Neurology, 14*(2), 116–124.

Sadnicka, A., Wiestler, T., Butler, K., Altenmüller, E., Edwards, M. J., Ejaz, N., & Diedrichsen, J. (2023). Intact finger representation within primary sensorimotor cortex of musician's dystonia. *Brain*, *146*(4), 1511–1522.

Saetta, G., Cognolato, M., Atzori, M., Faccio, D., Giacomino, K., Hager, A.-G. M., Tiengo, C., Bassetto, F., Müller, H., & Brugger, P. (2020). Gaze, behavioral, and clinical data for phantom limbs after hand amputation from 15 amputees and 29 controls. *Scientific Data*, *7*(1), 60.

Saetta, G., Hänggi, J., Gandola, M., Zapparoli, L., Salvato, G., Berlingeri, M., Sberna, M., Paulesu, E., Bottini, G., & Brugger, P. (2020). Neural correlates of body integrity dysphoria. *Current Biology*, *30*(11), 2191–2195.

Sahnouni, M., Parés, J. M., Duval, M., Cáceres, I., Harichane, Z., van der Made, J., Pérez-González, A., Abdessadok, S., Kandi, N., Derradji, A., Medig, M., Boulaghraif, K., & Semaw, S. (2018). 1.9-million- and 2.4-million-year-old artifacts and stone tool-cutmarked bones from Ain Boucherit, Algeria. *Science*, *362*(6420), 1297–1301.

Sakai, K. L., Tatsuno, Y., Suzuki, K., Kimura, H., & Ichida, Y. (2005). Sign and speech: Amodal commonality in left hemisphere dominance for comprehension of sentences. *Brain*, *128*(Pt 6), 1407–1417.

Sakata, H., Taira, M., Murata, A., & Mine, S. (1995). Neural mechanisms of visual guidance of hand action in the parietal cortex of the monkey. *Cerebral Cortex*, *5*(5), 429–438.

Sakata, H., Takaoka, Y., Kawarasaki, A., & Shibutani, H. (1973). Somatosensory properties of neurons in the superior parietal cortex (area 5) of the rhesus monkey. *Brain Research*, *64*, 85–102.

Saleh, M., Takahashi, K., & Hatsopoulos, N. G. (2012). Encoding of coordinated reach and grasp trajectories in primary motor cortex. *Journal of Neuroscience*, *32*(4), 1220–1232.

Salmaso, D., & Longoni, A. M. (1985). Problems in the assessment of hand preference. *Cortex*, *21*(4), 533–549.

Sambo, C. F., Gillmeister, H., & Forster, B. (2009). Viewing the body modulates neural mechanisms underlying sustained spatial attention in touch. *European Journal of Neuroscience*, *30*(1), 143–150.

Sanchez-Panchuelo, R. M., Besle, J., Beckett, A., Bowtell, R., Schluppeck, D., & Francis, S. (2012). Within-digit functional parcellation of Brodmann areas of the human primary somatosensory cortex using functional magnetic resonance imaging at 7 tesla. *Journal of Neuroscience*, *32*(45), 15815–15822.

Sánchez-Panchuelo, R.-M., Besle, J., Mougin, O., Gowland, P., Bowtell, R., Schluppeck, D., & Francis, S. (2014). Regional structural differences across functionally parcellated Brodmann areas of human primary somatosensory cortex. *NeuroImage*, *93*(Pt 2), 221–230.

Sánchez-Panchuelo, R.-M., Francis, S., Bowtell, R., & Schluppeck, D. (2010). Mapping human somatosensory cortex in individual subjects with 7T functional MRI. *Journal of Neurophysiology*, *103*(5), 2544–2556.

Sandler, W., Meir, I., Padden, C., & Aronoff, M. (2005). The emergence of grammar: Systematic structure in a new language. *Proceedings of the National Academy of Sciences*, *102*(7), 2661–2665.

Sanes, J. N., Donoghue, J. P., Thangaraj, V., Edelman, R. R., & Warach, S. (1995). Shared neural substrates controlling hand movements in human motor cortex. *Science, 268*(5218), 1775–1777.

Sansavini, A., Bello, A., Guarini, A., Savini, S. S., & Caselli, M. C. (2010). Early development of gestures, object-related-actions, word comprehension and word production, and their relationships in Italian infants: A longitudinal study. *Gesture, 10*(1), 52–85.

Santello, M., Flanders, M., & Soechting, J. F. (1998). Postural hand synergies for tool use. *Journal of Neuroscience, 18*(23), 10105–10115.

Sasaki, S., Isa, T., Pettersson, L. G., Alstermark, B., Naito, K., Yoshimura, K., Seki, K., & Ohki, Y. (2004). Dexterous finger movements in primate without monosynaptic corticomotoneuronal excitation. *Journal of Neurophysiology, 92*(5), 3142–3147.

Sathian, K. (2000). Intermanual referral of sensation to anesthetic hands. *Neurology, 54*(9), 1866–1868.

Sathian, K. (2016). Analysis of haptic information in the cerebral cortex. *Journal of Neurophysiology, 116*(4), 1795–1806.

Sathian, K., & Zangaladze, A. (1998). Perceptual learning in tactile hyperacuity: Complete intermanual transfer but limited retention. *Experimental Brain Research, 118*(1), 131–134.

Sathian, K., Zangaladze, A., Green, J., Vitek, J. L., & Delong, M. R. (1997). Tactile spatial acuity and roughness discrimination: Impairments due to aging and Parkinson's disease. *Neurology, 49*(1), 168–177.

Savage-Rumbaugh, S., McDonald, K., Sevcik, R. A., Hopkins, W. D., & Rubert, E. (1986). Spontaneous symbol acquisition and communicative use by pygmy chimpanzees (*Pan paniscus*). *Journal of Experimental Psychology: General, 115*(3), 211–235.

Savage-Rumbaugh, S., Murphy, J., Sevcik, R. A., Brakke, K. E., Williams, S. L., & Rumbaugh, Duane M. (1993). Language comprehension in ape and child. *Monographs of the Society for Research in Child Development, 58*, 233.

Saxe, R., & Kanwisher, N. (2003). People thinking about people. The role of the temporo-parietal junction in "theory of mind". *Neuroimage, 19*(4), 1835–1842.

Scaliti, E., Gruppioni, E., & Becchio, C. (2020). And yet it moves: What we currently know about phantom arm movements. *The Neuroscientist, 26*(4), 328–342.

Schady, W. J. L., Torebjörk, H. E., & Ochoa, J. L. (1983). Cerebral localisation function from the input of single mechanoreceptive units in man. *Acta Physiologica Scandinavica, 119*(3), 277–285.

Schaffelhofer, S., Agudelo-Toro, A., & Scherberger, H. (2015). Decoding a wide range of hand configurations from macaque motor, premotor, and parietal cortices. *Journal of Neuroscience, 35*(3), 1068–1081.

Schellekens, W., Thio, M., Badde, S., Winawer, J., Ramsey, N., & Petridou, N. (2021). A touch of hierarchy: Population receptive fields reveal fingertip integration in Brodmann areas in human primary somatosensory cortex. *Brain Structure and Function, 226*(7), 2099–2112.

Schendel, K., & Robertson, L. C. (2004). Reaching out to see: Arm position can attenuate human visual loss. *Journal of Cognitive Neuroscience, 16*(6), 935–943.

Schenker, N. M., Buxhoeveden, D. P., Blackmon, W. L., Amunts, K., Zilles, K., & Semendeferi, K. (2008). A comparative quantitative analysis of cytoarchitecture and minicolumnar organization in Broca's area in humans and great apes. *Journal of Comparative Neurology, 510*(1), 117–128.

Schenker, N. M., Hopkins, W. D., Spocter, M. A., Garrison, A. R., Stimpson, C. D., Erwin, J. M., Hof, P. R., & Sherwood, C. C. (2010). Broca's area homologue in chimpanzees (*Pan troglodytes*): Probabilistic mapping, asymmetry, and comparison to humans. *Cerebral Cortex, 20*(3), 730–742.

Scherer, R., Zanos, S. P., Miller, K. J., Rao, R. P. N., & Ojemann, J. G. (2009). Classification of contralateral and ipsilateral finger movements for electrocorticographic brain-computer interfaces. *Neurosurgury Focus, 27*(1), E12.

Schettino, L.F., Rajaraman, V., Jack, D., Adamovich, S. V, Sage, J., & Poizner, H. (2003). Deficits in the evolution of hand preshaping in Parkinson's disease. *Neuropsychologia, 42*(1), 82–94.

Schick, K. D., & Toth, N. (1993). *Making silent stones speak*. Simon & Schuster.

Schick, K. D., Toth, N., Garufi, G., Savage-Rumbaugh, E.S., Rumbaugh, D., & Sevcik, R. (1999). Continuing investigations into the stone tool-making and tool-using capabilities of a bonobo (*Pan paniscus*). *Journal of Archaeological Science, 26*(7), 821–832.

Schieber, M. H., & Hibbard, L. S. (1993). How somatotopic is the motor cortex hand area? *Science, 261*(5120), 22–25.

Schilder, P. (1935). *The image and appearance of the human body*. Kegan Paul.

Schlereth, T., Magerl, W., & Treede, R.-D. (2001). Spatial discrimination thresholds for pain and touch in human hairy skin. *Pain, 92*(1–2), 187–194.

Schluter, N. D., Krams, M., Rushworth, M. F. S., & Passingham, R. E. (2001). Cerebral dominance for action in the human brain: The selection of actions. *Neuropsychologia, 39*(2), 105–113.

Schmidt, P., Koch, T. J., & February, E. (2022). Archaeological adhesives made from *Podocarpus* document innovative potential in the African Middle Stone Age. *Proceedings of the National Academy of Sciences, 119*(40), e2209592119.

Schneiberg, S., Sveistrup, H., McFadyen, B., McKinley, P., & Levin, M. F. (2002). The development of coordination for reach-to-grasp movements in children. *Experimental Brain Research, 146*(2), 142–154.

Schneider, I., & Shubin, N. H. (2013). The origin of the tetrapod limb: From expeditions to enhancers. *Trends in Genetics, 29*(7), 419–426.

Schoenemann, P. T. (2006). Evolution of the size and functional areas of the human brain. *Annual Review of Anthropology, 35*, 379–406.

Schoenfeld, M. R. (1978). Nicolo Paganini: Musical magician and Marfan mutant? *JAMA, 239*(1), 40–42.

Schott, G. D. (1980). Mirror movements of the left arm following peripheral damage to the preferred right arm. *Journal of Neurology, Neurosurgery, and Psychiatry, 43*(9), 768–773.

Schott, G. D. (1999). Mirror writing: Allen's self observations, Lewis Carroll's "looking-glass" letters, and Leonardo da Vinci's maps. *The Lancet, 354*(9196), 2158–2161.

Schott, G. D., & Wyke, M. A. (1981). Congenital mirror movements. *Journal of Neurology, Neurosurgery & Psychiatry, 44*(7), 586–599.

Schwarz, B. (1983). *Great masters of the violin*. Simon & Schuster.

Schweisfurth, M. A., Frahm, J., Farina, D., & Schweizer, R. (2018). Comparison of fMRI digit representations of the dominant and non-dominant hand in the human primary somatosensory cortex. *Frontiers in Human Neuroscience, 12,* 492.

Schweizer, R., & Braun, C. (2001). The distribution of mislocalizations across fingers demonstrates training-induced neuroplastic changes in somatosensory cortex. *Experimental Brain Research, 139*(4), 435–442.

Schwenkreis, P., Janssen, F., Rommel, O., Pleger, B., Völker, B., Hosbach, I., Dertwinkel, R., Maier, C., & Tegenthoff, M. (2003). Bilateral motor cortex disinhibition in complex regional pain syndrome (CRPS) type I of the hand. *Neurology, 61*(4), 515–519.

Schwoebel, J., & Coslett, H. B. (2005). Evidence for multiple, distinct representations of the human body. *Journal of Cognitive Neuroscience, 17*(4), 543–553.

Scocchia, L., Stucchi, N., & Loomis, J. M. (2009). The influence of facing direction on the haptic identification of two-dimensional raised pictures. *Perception, 38*(4), 606–612.

Scott, S. H. (2004). Optimal feedback control and the neural basis of volitional motor control. *Nature Reviews Neuroscience, 5*(7), 534–546.

Scott, S. H. (2016). A functional taxonomy of bottom-up sensory feedback processing for motor actions. *Trends in Neurosciences, 39*(8), 512–526.

Scripture, E. W., Smith, T. L., & Brown, E. M. (1894). On the education of muscular control and power. *Studies from the Yale Psychological Laboratory, 2,* 114–119.

Sechzer, J. A. (1970). Prolonged learning and split-brain cats. *Science, 169*(3948), 889–892.

Seelke, A. M. H., Dooley, J. C., & Krubitzer, L. A. (2012). The emergence of somatotopic maps of the body in S1 in rats: The correspondence between functional and anatomical organization. *PLOS One, 7*(2), e32322.

Seelke, A. M. H., Padberg, J. J., Disbrow, E., Purnell, S.M., Recanzone, G., & Krubitzer, L. (2012). Topographic maps within Brodmann's area 5 of macaque monkeys. *Cerebral Cortex, 22*(8), 1834–1850.

Semaw, S., Renne, P., Harris, J. W. K., Feibel, C. S., Bernor, R. L., Fesseha, N., & Mowbray, K. (1997). 2.5-million-year-old stone tools from Gona, Ethiopia. *Nature, 385*(6614), 333–336.

References

Semaw, S., Rogers, M. J., Quade, J., Renne, P. R., Butler, R. F., Dominguez-Rodrigo, M., Stout, D., Hart, W. S., Pickering, T., & Simpson, S.W. (2003). 2.6-million-year-old stone tools and associated bones from OGS-6 and OGS-7, Gona, Afar, Ethiopia. *Journal of Human Evolution, 45*(2), 169–177.

Semmes, J. (1965). A non-tactual factor in astereognosis. *Neuropsychologia, 3*(4), 295–315.

Senghas, A., Kita, S., & Özyürek, A. (2004). Children creating core properties of language: Evidence from an emerging sign language in Nicaragua. *Science, 305*(5691), 1779–1782.

Senut, B., Pickford, M., Gommery, D., Mein, P., Cheboi, K., & Coppens, Y. (2001). First hominid from the Miocene (Lukeino Formation, Kenya). *Comptes Rendus de l'Académie des Sciences, 332*(2), 137–144.

Sereno, M. I., Dale, A. M., Reppas, J. B., Kwong, K. K., Belliveau, J. W., Brady, T. J., Rosen, B. R., & Tootell, R. B. H. (1995). Borders of multiple visual areas in humans revealed by functional magnetic resonance imaging. *Science, 268*(5212), 889–893.

Sereno, M. I., & Huang, R.-S. (2006). A human parietal face area contains aligned head-centered visual and tactile maps. *Nature Neuroscience, 9*(10), 1337–1343.

Serino, A., Farnè, A., Rinaldesi, M. L., Haggard, P., & Làdavas, E. (2007). Can vision of the body ameliorate impaired somatosensory function? *Neuropsychologia, 45*(5), 1101–1107.

Seyal, M., Ro, T., & Rafal, R. (1995). Increased sensitivity to ipsilateral cutaneous stimuli following transcranial magnetic stimulation of the parietal lobe. *Annals of Neurology, 38*(2), 264–267.

Seyal, M., Siddiqui, I., & Hundal, N. S. (1997). Suppression of spatial localization of a cutaneous stimulus following transcranial magnetic pulse stimulation of the sensorimotor cortex. *Electroencephalography and Clinical Neurophysiology, 105*(1), 24–28.

Seyffarth, H., & Denny-Brown, D. (1948). The grasp reflex and the instinctive grasp reaction. *Brain, 71*(2), 109–183.

Shafer, D. D. (1993). Patterns of handedness: Comparative study of nursery school children and captive gorillas. In J. P. Ward & W. D. Hopkins (Eds.), *Primate laterality* (pp. 267–283). Springer-Verlag.

Shallice, T. (1981). Phonological agraphia and the lexical route in writing. *Brain, 104*(3), 413–429.

Shallice, T., Burgess, P. W., Schon, F., & Baxter, D. M. (1989). The origins of utilization behaviour. *Brain, 112*(Pt 6), 1587–1598.

Shamim, E. A., Chu, J., Scheider, L.H., Savitt, J., Jinnah, H. A., & Hallett, M. (2011). Extreme task specificity in writer's cramp. *Movement Disorders, 26*(11), 2107–2109.

Shea, J. J. (1988). Spear points from the Middle Paleolithic of the Levant. *Journal of Field Archaeology, 15*(4), 441–450.

Shea, J. J. (2017). *Stone tools in human evolution*. Cambridge University Press.

Shea, J. J., Davis, Z., & Brown, K. (2001). Experimental tests of Middle Palaeolithic spear points using a calibrated crossbow. *Journal of Archaeological Science, 28*(8), 807–816.

Sheehy, M. P., & Marsden, C. D. (1982). Writers' cramp—A focal dystonia. *Brain, 105*(Pt 3), 461–480.

Sherrick, C. E. (1964). Effects of double simultaneous stimulation of the skin. *American Journal of Psychology, 77*(1), 42–53.

Sherrington, C. S. (1906). *The integrative action of the nervous system*. Oxford University Press.

Sherwood, C. C., Bauernfeind, A. L., Bianchi, S., Raghanti, M. A., & Hof, P. R. (2012). Human brain evolution writ large and small. *Progress in Brain Research, 195*, 237–254.

Shimada, S., Suzuki, T., Yoda, N., & Hayashi, T. (2014). Relationship between sensitivity to visuotactile temporal discrepancy and the rubber hand illusion. *Neuroscience Research, 85*, 33–38.

Shoham, D., & Grinvald, A. (2001). The cortical representation of the hand in macaque and human area S-I: High resolution optical imaging. *Journal of Neuroscience, 21*(17), 6820–6835.

Shrewsbury, M. M., Marzke, M. W., Linscheid, R. L., & Reece, S. P. (2003). Comparative morphology of the pollical distal phalanx. *American Journal of Physical Anthropology, 121*(1), 30–47.

Shubin, N. (2009). *Your inner fish*. Penguin.

Shubin, N. H., Daeschler, E. B., & Jenkins, F. A. (2006). The pectoral fin of *Tiktaalik roseae* and the origin of the tetrapod limb. *Nature, 440*, 764–771.

Shumaker, R. W., Walkup, K. R., & Beck, B. B. (2011). *Animal tool behavior*. Johns Hopkins University Press.

Silver, M. A, & Kastner, S. (2009). Topographic maps in human frontal and parietal cortex. *Trends in Cognitive Sciences, 13*(11), 488–495.

Simmel, M. L. (1961). The absence of phantoms for congenitally missing limbs. *American Journal of Psychology, 74*(3), 467–470.

Simonyan, K., Cho, H., Sichani, A. H., Rubien-Thomas, E., & Hallett, M. (2017). The direct basal ganglia pathway is hyperfunctional in focal dystonia. *Brain, 140*(12), 3179–3190.

Singer, M. A., & Goldin-Meadow, S. (2005). Children learn when their teacher's gestures and speech differ. *Psychological Science, 16*(2), 85–90.

Sinha, N., Manohar, S., & Husain, M. (2013). Impulsivity and apathy in Parkinson's disease. *Journal of Neuropsychology, 7*(2), 255–283.

Sirianni, G., Mundry, R., & Boesch, C. (2015). When to choose which tool: Multidimensional and conditional selection of nut-cracking hammers in wild chimpanzees. *Animal Behaviour, 100*, 152–165.

Sirigu, A., Cohen, L., Duhamel, J.-R., Pillon, B., Dubois, B., Agid, Y., & Pierrot-Deseilligny, C. (1995). Congruent unilateral impairments for real and imagined hand movements. *NeuroReport, 6*(7), 997–1001.

Sirigu, A., Duhamel, J.-R., Cohen, L., Pillon, B., Dubois, B., & Agid, Y. (1996). The mental representation of hand movements after parietal cortex damage. *Science, 273*(5281), 1564–1568.

Sirigu, A., Duhamel, J.-R., & Poncet, M. (1991). The role of sensorimotor experience in object recognition: A case of multimodal agnosia. *Brain, 114*(Pt 6), 2555–2573.

Siuda-Krzywicka, K., Bola, Ł., Paplińska, M., Sumera, E., Jednoróg, K., Marchewska, A., Śliwińska, M. W., Amedi, A., & Szwed, M. (2016). Massive cortical reorganization in sighted braille readers. *eLife, 5*, e10762.

Skinner, M. M., Stephens, N. B., Tsegai, Z. J., Foote, A. C., Nguyen, N. H., Gross, T., Pahr, D. H., Hublin, J.-J., & Kivell, T. L. (2015). Human-like hand use in *Australopithecus africanus. Science, 347*(6220), 395–399.

Skoyles, J. R. (1990). Is there a genetic component to body schema? *Trends in Neurosciences, 13*(10), 409.

Sloan, T. B., Fugina, M. L., & Toleikis, J. R. (1990). Effects of midazolam on median nerve somatosensory evoked potentials. *British Journal of Anaesthesia, 64*(5), 590–593.

Smaers, J. B., & Soligo, C. (2013). Brain reorganization, not relative brain size, primarily characterizes anthropoid brain evolution. *Proceedings of the Royal Society B, 280*(1759), 20130269.

Smetacek, V. (1992). Mirror-script and left-handedness. *Nature, 355*, 118–119.

Smit, M., Van Stralen, H. E., Van den Munckhof, B., Snijders, T. J., & Dijkerman, H. C. (2019). The man who lost his body: Suboptimal multisensory integration yields body awareness problems after a right temporoparietal brain tumour. *Journal of Neuropsychology, 13*(3), 603–612.

Smith, G. E. (1904). The morphology of the retrocalcarine region of the cortex cerebri. *Proceedings of the Royal Society, 73*, 59–65.

Smith, K. U. (1951). Learning and the associative pathways of the human cerebral cortex. *Science, 114*(2953), 117–120.

Snow, J. C., Goodale, M. A., & Culham, J. C. (2015). Preserved haptic shape processing after bilateral LOC lesions. *Journal of Neuroscience, 35*(40), 13745–13760.

Snow, J. C., Strother, L., & Humphreys, G. W. (2014). Haptic shape processing in visual cortex. *Journal of Cognitive Neuroscience, 26*(5), 1154–1167.

Snyder, L. H., Batista, A. P., & Andersen, R. A. (1997). Coding of intention in the posterior parietal cortex. *Nature, 386*, 167–170.

So, E. L., & Schauble, B. S. (2004). Ictal asomatognosia as a cause of epileptic falls: Simultaneous video, EMG, and invasive EEG. *Neurology, 63*(11), 2153–2154.

So, W. C., Ching, T. H.-W., Lim, P. E., Cheng, X., & Ip, K. Y. (2014). Producing gestures facilitates route learning. *PLOS One, 9*(11), e112543.

Soechting, J. F., & Flanders, M. (1992). Organization of sequential typing movements. *Journal of Neurophysiology, 67*(5), 1275–1290.

Sohn, Y. H., & Hallett, M. (2004). Disturbed surround inhibition in focal hand dystonia. *Annals of Neurology, 56*(4), 595–599.

Sohn, Y. H., Jung, H. Y., Kaelin-Lang, A., & Hallett, M. (2003). Excitability of the ipsilateral motor cortex during phasic voluntary hand movement. *Experimental Brain Research, 148*(2), 176–185.

Solly, S. (1864). Scrivener's palsy, or the paralysis of writers. *The Lancet, 84*(2156), 709–711.

Solvi, C., Gutierrez Al-Khudhairy, S., & Chittka, L. (2020). Bumble bees display cross-modal object recognition between visual and tactile senses. *Science, 367*(6480), 910–912.

Sommer, R. (1969). *Personal space*. Prentice-Hall.

Southey, R. (1813). *The life of Nelson*. Harper.

Sparling, J. W., Van Tol, J., & Chescheir, N. C. (1999). Fetal and neonatal hand movement. *Physical Therapy, 79*(1), 24–39.

Spelke, E. S., Breinlinger, K., Macomber, J., & Jacobson, K. (1992). Origins of knowledge. *Psychological Review, 99*(4), 605–632.

Spence, C., Kingstone, A., Shore, D. I., & Gazzaniga, M. S. (2001). Representation of visuotactile space in the split brain. *Psychological Science, 12*(1), 90–93.

Spence, C., Pavani, F., & Driver, J. (2000). Crossmodal links between vision and touch in covert endogenous spatial attention. *Journal of Experimental Psychology: Human Perception and Performance, 26*(4), 1298–1319.

Spence, C., Pavani, F., & Driver, J. (2004). Spatial constraints on visual-tactile cross-modal distractor congruency effects. *Cognitive, Affective, and Behavioral Neuroscience, 4*(2), 148–169.

Spencer, J. P., & Thelen, E. (2000). Spatially specific changes in infants' muscle coactivity as they learn to reach. *Infancy, 1*(3), 275–302.

Sperry, R. W. (1963). Chemoaffinity in the orderly growth of nerve fiber patterns and connections. *Proceedings of the National Academy of Sciences, 50*(4), 703–710.

Sperry, R. W., Gazzaniga, M. S., & Bogen, J. E. (1969). Interhemispheric relationships: The neocortical commissures; syndromes of hemisphere disconnection. In P. Vinken & G. Bruyn (Eds.), *Handbook of clinical neurology* (pp. 273–290). Wiley.

Spinazzola, L., Pia, L., Folegatti, A., Marchetti, C., & Berti, A. (2008). Modular structure of awareness for sensorimotor disorders: Evidence from anosognosia for hemiplegia and anosognosia for hemianaesthesia. *Neuropsychologia, 46*(3), 915–926.

Spinozzi, G., Truppa, V., & Laganà, T. (2004). Grasping behavior in tufted capuchin monkeys (*Cebus apella*): Grip types and manual laterality for picking up a small food item. *American Journal of Physical Anthropology, 125*(1), 30–41.

Sposito, A., Bolognini, N., Vallar, G., & Maravita, A. (2012). Extension of perceived arm length following tool-use: Clues to plasticity of body metrics. *Neuropsychologia, 50*(9), 2187–2194.

Sretavan, D., & Dykes, R. W. (1983). The organization of two cutaneous submodalities in the forearm region of area 3b of cat somatosensory cortex. *Journal of Comparative Neurology, 213*(4), 381–398.

Srour, M., Rivière, J.-B., Pham, J. M. T., Dubé, M.-P., Girard, S., Morin, S., Dion, P. A., Asselin, G., Rochefort, D., Hince, P., Diab, S., Sharafaddinzadeh, N., Chouinard, S., Théoret, H., Charron, F., & Rouleau, G. A. (2010). Mutations in DCC cause congenital mirror movements. *Science*, *328*(5978), 592.

St. Clair, J. J. H., & Rutz, C. (2013). New Caledonian crows attend to multiple functional properties of complex tools. *Philosophical Transactions of the Royal Society B*, *368*(1630), 20120415.

Stack, D. M., Muir, D. W., Sherriff, F., & Roman, J. (1989). Development of infant reaching in the dark to luminous objects and "invisible sounds." *Perception*, *18*(1), 69–82.

Stamelou, M., Edwards, M. J., Hallett, M., & Bhatia, K. P. (2012). The non-motor syndrome of primary dystonia: Clinical and pathophysiological implications. *Brain*, *135*(Pt 6), 1668–1681.

Stamm, J. S., & Sperry, R. W. (1957). Function of corpus callosum in contralateral transfer of somesthetic discrimination in cats. *Journal of Comparative and Physiological Psychology*, *50*(2), 138–143.

Stankevicius, A., Wallwork, S. B., Summers, S. J., Hordacre, B., & Stanton, T. R. (2021). Prevalence and incidence of phantom limb pain, phantom limb sensations and telescoping in amputees. *European Journal of Pain*, *25*(1), 23–38.

Starkovich, B. M., Cuthbertson, P., Kitagawa, K., Thompson, N., Konidaris, G. E., Rots, V., Münzel, S. C., Giusti, D., Schmid, V. C., Blanco-Lapaz, A., Lepers, C., & Tourloukis, V. (2020). Minimal tools, maximum meat: A pilot experiment to butcher an elephant foot and make elephant bone tools using lower paleolithic stone tool technology. *Ethnoarchaeology*, *12*(2), 118–147.

Starkstein, S. E., Berthier, M. L., Fedoroff, P., Price, T. R., & Robinson, R. G. (1990). Anosognosia and major depression in 2 patients with cerebrovascular lesions. *Neurology*, *40*(9), 1380–1382.

Staub, F., Bogousslavsky, J., Maeder, P., Maedor-Ingvar, M., Fornari, E., Ghika, J., Vingerhoets, F., & Assal, G. (2006). Intentional motor phantom limb syndrome. *Neurology*, *67*(12), 2140–2146.

Stepniewska, I., Cerkevich, C. M., & Kaas, J. H. (2016). Cortical connections of the caudal portion of posterior parietal cortex in prosimian galagos. *Cerebral Cortex*, *26*(6), 2753–2777.

Stepniewska, I., Friedman, R. M., Gharbawie, O. A., Cerkevich, C. M., Roe, A. W., & Kaas, J. H. (2011). Optical imaging in galagos reveals parietal—frontal circuits underlying motor behavior. *Proceedings of the National Academy of Sciences*, *108*(37), 725–732.

Stepniewska, I., Gharbawie, O. A., Burish, M. J., & Kaas, J. H. (2014). Effects of muscimol inactivations of functional domains in motor, premotor, and posterior parietal cortex on complex movements evoked by electrical stimulation. *Journal of Neurophysiology*, *111*(5), 1100–1119.

Stepniewska, I., Preuss, T. M., & Kaas, J. H. (1993). Architectonics, somatoptopic organization, and ipsilateral cortical connections of the primary motor area (M1) of owl monkeys. *Journal of Comparative Neurology*, *330*(2), 238–271.

Sterr, A., Müller, M. M., Elbert, T., Rockstroh, B., Pantev, C., & Taub, E. (1998). Changed perceptions in braille readers. *Nature*, *391*(6663), 134–135.

Sterzi, R., Bottini, G., Celani, M. G., Righetti, E., Lamassa, M., Ricci, S., & Vallar, G. (1993). Hemianopia, hemianaesthesia, and hemiplegia after right and left hemisphere damage. A hemispheric difference. *Journal of Neurology, Neurosurgery, and Psychiatry, 56*(3), 308–310.

Stokoe, W. C. (1980). Sign language structure. *Annual Review of Anthropology, 9*, 365–390.

Stoll, S. E. M., Finkel, L., Buchmann, I., Hassa, T., Spiteri, S., Liepert, J., & Randerath, J. (2022). 100 years after Liepmann–Lesion correlates of diminished selection and application of familiar versus novel tools. *Cortex, 146*, 1–23.

Stout, D., & Chaminade, T. (2007). The evolutionary neuroscience of tool making. *Neuropsychologia, 45*(5), 1091–1100.

Stout, D., Hecht, E., Khreisheh, N., Bradley, B., & Chaminade, T. (2015). Cognitive demands of lower paleolithic toolmaking. *PLOS One, 10*(4), e0121804.

Stout, D., Quade, J., Semaw, S., Rogers, M. J., & Levin, N. E. (2005). Raw material selectivity of the earliest stone toolmakers at Gona, Afar, Ethiopia. *Journal of Human Evolution, 48*(4), 365–380.

Stout, D., Toth, N., Schick, K., & Chaminade, T. (2008). Neural correlates of Early Stone Age toolmaking: Technology, language and cognition in human evolution. *Philosophical Transactions of the Royal Society B, 363*(1499), 1939–1949.

Straus, W. L. (1949). The riddle of man's ancestry. *Quarterly Review of Biology, 24*(3), 200–223.

Strick, P. L. (1983). The influence of motor preparation on the response of cerebellar neurons to limb displacements. *Journal of Neuroscience, 3*(10), 2007–2020.

Strick, P. L., Dum, R. P., & Rathelot, J.-A. (2021). The cortical motor areas and the emergence of motor skills: A neuroanatomical perspective. *Annual Review of Neuroscience, 44*, 425–447.

Strick, P. L., & Preston, J. B. (1982). Two representations of the hand in area 4 of a primate. I. Motor output organization. *Journal of Neurophysiology, 48*(1), 139–149.

Striedter, G. F. (2005). *Principles of brain evolution*. Sinauer Associates.

Striemer, C. L., Yukovsky, J., & Goodale, M. A. (2010). Can intention override the "automatic pilot"? *Experimental Brain Research, 202*(3), 623–632.

Ströckens, F., Güntürkün, O., & Ocklenburg, S. (2013). Limb preferences in non-human vertebrates. *Laterality, 18*(5), 536–575.

Summers, D. C., & Lederman, S. J. (1990). Perceptual asymmetries in the somatosensory system: A dichhaptic experiment and critical review of the literature from 1929 to 1986. *Cortex, 26*(2), 201–226.

Sun, T., Patoine, C., Abu-Khalil, A., Visvader, J., Sum, E., Cherry, T. J., Orkin, S. H., Geschwind, D. H., & Walsh, C. A. (2005). Early asymmetry of gene transcription in embryonic human left and right cerebral cortex. *Science, 308*(5729), 1794–1798.

Sur, M., Merzenich, M. M., & Kaas, J. H. (1980). Magnification, receptive-field area, and size in areas 3b and 1 of somatosensory cortex in owl monkeys. *Journal of Neurophysiology, 44*(2), 295–311.

References

Sur, M., Nelson, R. J., & Kaas, J. H. (1982). Representations of the body surface in cortical areas 3b and 1 of squirrel monkeys: Comparisons with other primates. *Journal of Comparative Neurology, 211*(2), 177–192.

Sur, M., Wall, J. T., & Kaas, J. H. (1981). Modular segregation of functional cell classes within the postcentral somatosensory cortex of monkeys. *Science, 212*(4498), 1059–1061.

Sur, M., Wall, J. T., & Kaas, J. H. (1984). Modular distribution of neurons with slowly adapting and rapidly adapting responses in area 3b of somatosensory cortex in monkeys. *Journal of Neurophysiology, 51*(4), 724–744.

Suresh, A. K., Goodman, J. M., Okorokova, E. V., Kaufman, M., Hatsopoulos, N. G., & Bensmaia, S. J. (2020). Neural population dynamics in motor cortex are different for reach and grasp. *eLife, 9*, e58848.

Susman, R. L. (1979). Comparative and functional morphology of hominoid fingers. *American Journal of Physical Anthropology, 50*(2), 215–236.

Susman, R. L. (1988). Hand of *Paranthropus robustus* from Member 1, Swartkrans: Fossil evidence for tool behavior. *Science, 240*(4853), 781–784.

Susman, R. L. (1994). Fossil evidence for early hominid tool use. *Science, 265*(5178), 1570–1573.

Suzuki, K., & Arashida, R. (1992). Geometrical haptic illusions revisited: Haptic illusions compared with visual illusions. *Perception & Psychophysics, 52*(3), 329–335.

Szalay, F. S., & Dagosto, M. (1988). Evolution of hallucial grasping in the primates. *Journal of Human Evolution, 17*(1–2), 1–33.

Taglialatela, J. P., Cantalupo, C., & Hopkins, W. D. (2006). Gesture handedness predicts asymmetry in the chimpanzee inferior frontal gyrus. *NeuroReport, 17*(9), 923–927.

Taglialatela, J. P., Russell, J. L., Schaeffer, J. A., & Hopkins, W. D. (2008). Communicative signaling activates "Broca's" homolog in chimpanzees. *Current Biology, 18*(5), 343–348.

Tajadura-Jiménez, A., Väljamäe, A., Toshima, I., Kimura, T., Tsakiris, M., & Kitagawa, N. (2012). Action sounds recalibrate perceived tactile distance. *Current Biology, 22*(13), R516–R517.

Takahashi, K., Best, M. D., Huh, N., Brown, K. A., Tobaa, A. A., & Hatsopoulos, N. G. (2017). Encoding of both reaching and grasping kinematics in dorsal and ventral premotor cortices. *Journal of Neuroscience, 37*(7), 1733–1746.

Takei, T., Lomber, S. G., Cook, D. J., & Scott, S. H. (2021). Transient deactivation of dorsal premotor cortex or parietal area 5 impairs feedback control of the limb in macaques. *Current Biology, 31*(7), 1476–1487.

Tal, Z., Geva, R., & Amedi, A. (2016). The origins of metamodality in visual object area LO: Bodily topographical biases and increased functional connectivity to S1. *NeuroImage, 127*, 363–375.

Tamburin, S., Manganotti, P., Marzi, C. A., Fiaschi, A., & Zanette, G. (2002). Abnormal somatotopic arrangement of sensorimotor interactions in dystonic patients. *Brain, 125*(Pt 12), 2719–2730.

Tamè, L., Azañón, E., & Longo, M. R. (2019). A conceptual model of tactile processing across body features of size, shape, side, and spatial location. *Frontiers in Psychology*, *10*, 291.

Tamè, L., Braun, C., Lingnau, A., Schwarzbach, J., Demarchi, G., Hegner, Y. L., Farnè, A., & Pavani, F. (2012). The contribution of primary and secondary somatosensory cortices to the representation of body parts and body sides: An fMRI adaptation study. *Journal of Cognitive Neuroscience*, *24*(12), 2306–2320.

Tamè, L., Farnè, A., & Pavani, F. (2011). Spatial coding of touch at the fingers: Insights from double simultaneous stimulation within and between hands. *Neuroscience Letters*, *487*(1), 78–82.

Tamè, L., & Longo, M. R. (2023). Emerging principles in functional representations of touch. *Nature Reviews Psychology*, *2*, 459–471.

Tamè, L., Pavani, F., Papadelis, C., Farnè, A., & Braun, C. (2015). Early integration of bilateral touch in the primary somatosensory cortex. *Human Brain Mapping*, *36*(4), 1506–1523.

Tamè, L., Tucciarelli, R., Sadibolova, R., Sereno, M. I., & Longo, M. R. (2021). Reconstructing neural representations of tactile space. *NeuroImage*, *229*, 117730.

Tanaka, Y., Yoshida, A., Kawahata, N., Hashimoto, R., & Obayashi, T. (1996). Diagonistic dyspraxia: Clinical characteristics, responsible lesion and possible underlying mechanism. *Brain*, *119*(Pt 3), 859–873.

Tang, R., Whitwell, R. L., & Goodale, M. A. (2015). The influence of visual feedback from the recent past on the programming of grip aperture is grasp-specific, shared between hands, and mediated by sensorimotor memory not task set. *Cognition*, *138*, 49–63.

Tang, R., Whitwell, R. L., & Goodale, M. A. (2016). Unusual hand postures but not familiar tools show motor equivalence with precision grasping. *Cognition*, *151*, 28–36.

Tanji, J., Okano, K., & Sato, K. C. (1987). Relation of neurons in the nonprimary motor cortex to bilateral hand movement. *Nature*, *327*, 618–620.

Tanji, J., Okano, K., & Sato, K. C. (1988). Neuronal activity in cortical motor areas related to ipsilateral, contralateral, and bilateral digit movements of the monkey. *Journal of Neurophysiology*, *60*(1), 325–343.

Tanji, J., & Wise, S. P. (1981). Submodality distribution in sensorimotor cortex of the unanesthetized monkey. *Journal of Neurophysiology*, *45*(3), 467–481.

Tapley, S. M., & Bryden, M. P. (1985). A group test for the assessment of performance between the hands. *Neuropsychologia*, *23*(2), 215–221.

Tastevin, J. (1937). En partant de l'expérience d'Aristote [On Aristotle's illusion]. *Encéphale*, *32*, 57–84, 140–158.

Taub, E., & Goldberg, I. A. (1973). Prism adaptation: Control of intermanual transfer by distribution of practice. *Science*, *180*(4087), 755–757.

Tavare, S., Marshall, C. R., Will, O., Soligo, C., & Martin, R. D. (2002). Using the fossil record to estimate the age of the last common ancestor of extant primates. *Nature, 416*, 726–729.

Taylor, H. G., & Heilman, K. M. (1980). Left-hemisphere motor dominance in righthanders. *Cortex, 16*(4), 587–603.

Taylor, J. A., Wojaczynski, G. J., & Ivry, R. B. (2011). Trial-by-trial analysis of intermanual transfer during visuomotor adaptation. *Journal of Neurophysiology, 106*(6), 3157–3172.

Taylor-Clarke, M., Jacobsen, P., & Haggard, P. (2004). Keeping the world a constant size: Object constancy in human touch. *Nature Neuroscience, 7*(3), 219–220.

Taylor-Clarke, M., Kennett, S., & Haggard, P. (2002). Vision modulates somatosensory cortical processing. *Current Biology, 12*(3), 233–236.

Tazoe, T., & Perez, M. A. (2013). Speed-dependent contribution of callosal pathways to ipsilateral movements. *Journal of Neuroscience, 33*(41), 16178–16188.

Teng, E. L., Lee, P.-H., Yang, K.-S., & Chang, P. C. (1976). Handedness in a Chinese population: Biological, social, and pathological factors. *Science, 193*(4258), 1148–1150.

Teuber, H.-L. (1965). Preface: Disorders of higher tactile and visual functions. *Neuropsychologia, 3*(4), 287–294.

Thaler, D., Chen, Y.-C., Nixon, P. D., Stern, C. E., & Passingham, R. E. (1995). The functions of the medial premotor cortex. I. Simple learned movements. *Experimental Brain Research, 102*(3), 445–460.

Thelen, E., Corbetta, D., Kamm, K., Spencer, J. P., Schneider, K., & Zernicke, R. F. (1993). The transition to reaching: Mapping intention and intrinsic dynamics. *Child Development, 64*(4), 1058–1098.

Thelen, E., Fisher, D. M., & Ridley-Johnson, R. (1984). The relationship between physical growth and a newborn reflex. *Infant Behavior and Development, 7*, 479–493.

Thibault, S., Py, R., Gervasi, A. M., Salemme, R., Koun, E., Lövden, M., Boulenger, V., Roy, A. C., & Brozzoli, C. (2021). Tool use and language share syntactic processes and neural patterns in the basal ganglia. *Science, 374*(6569), eabe0874.

Thieme, H. (1997). Lower Palaeolithic hunting spears from Germany. *Nature, 385*, 807–810.

Thomas, L. E. (2013). Grasp posture modulates attentional prioritization of space near the hands. *Frontiers in Psychology, 4*, 312.

Thomas, L. E. (2015). Grasp posture alters visual processing biases near the hands. *Psychological Science, 26*(5), 625–632.

Thomas, L. E. (2017). Action experience drives visual-processing biases near the hands. *Psychological Science, 28*(1), 124–131.

Tian, M., Saccone, E. J., Kim, J. S., Kanjlia, S., & Bedny, M. (2023). Sensory modality and spoken language shape reading network in blind readers of braille. *Cerebral Cortex, 33*(6), 2426–2440.

Tinazzi, M., Priori, A., Bertolasi, L., Frasson, E., Mauguière, F., & Fiaschi, A. (2000). Abnormal central integration of a dual somatosensory input in dystonia. *Brain, 123*(Pt 1), 42–50.

Tinazzi, M., & Zanette, G. (1998). Modulation of ipsilateral motor cortex in man during unimanual finger movements of different complexities. *Neuroscience Letters, 244*(3), 121–124.

Tinazzi, M., Zanette, G., Volpato, D., Testoni, R., Bonato, C., Manganotti, P., Miniussi, C., & Fiaschi, A. (1998). Neurophysiological evidence of neuroplasticity at multiple levels of the somatosensory system in patients with carpal tunnel syndrome. *Brain, 121*(Pt 9), 1785–1794.

Tipper, S. P., Lloyd, D., Shorland, B., Dancer, C., Howard, L. A., & McGlone, F. (1998). Vision influences tactile perception without proprioceptive orienting. *NeuroReport, 9*(8), 1741–1744.

Tobias, P. V. (1987). The brain of *Homo habilis*: A new level of organization in cerebral evolution. *Journal of Human Evolution, 16*(7–8), 741–761.

Tocheri, M. W., Orr, C. M., Jacofsky, M. C., & Marzke, M. W. (2008). The evolutionary history of the hominin hand since the last common ancestor of *Pan* and *Homo*. *Journal of Anatomy, 212*(4), 544–562.

Todd, J. J., & Marois, R. (2004). Capacity limit of visual short-term memory in human posterior parietal cortex. *Nature, 428*(6984), 751–754.

Todorov, E. (2004). Optimality principles in sensorimotor control. *Nature Neuroscience, 7*(9), 907–915.

Todorov, E., & Jordan, M. I. (2002). Optimal feedback control as a theory of motor coordination. *Nature Neuroscience, 5*(11), 1226–1235.

Tohyama, T., Kinoshita, M., Kobayashi, K., Isa, K., Watanabe, D., Kobayashi, K., Liu, M., & Isa, T. (2017). Contribution of propriospinal neurons to recovery of hand dexterity after corticospinal tract lesions in monkeys. *Proceedings of the National Academy of Sciences, 114*(3), 604–609.

Tomasello, M., Carpenter, M., & Liszkowski, U. (2007). A new look at infant pointing. *Child Development, 78*(3), 705–722.

Topham, J. R. (1998). Beyond the "common context": The production and reading of the Bridgewater Treatises. *Isis, 89*(2), 233–262.

Torigoe, T. (1985). Comparison of object manipulation among 74 species of non-human primates. *Primates, 26*(2), 182–194.

Tosi, G., Mentesana, B., & Romano, D. (2023). The correlation between proprioceptive drift and subjective embodiment during the rubber hand illusion: A meta-analytic approach. *Quarterly Journal of Experimental Psychology, 76*(10), 2197–2207.

Toth, N. (1985). Archaeological evidence for preferential right-handedness in the Lower and Middle Pleistocene, and its possible implications. *Journal of Human Evolution, 14*(6), 607–614.

Toth, N., Clark, D., & Ligabue, G. (1992). The last stone ax makers. *Scientific American, 267*(1), 88–93.

Toth, N., Schick, K., & Semaw, S. (2006). A comparative study of the stone tool-making skills of *Pan, Australopithecus*, and *Homo sapiens*. In N. Toth & K. Schick (Eds.), *The Oldowan*. Stone Age Institute Press.

Toth, N., Schick, K. D., Savage-Rumbaugh, E. S., Sevcik, R. A., & Rumbaugh, D. (1993). Pan the tool-maker: Investigations into the stone tool-making and tool-using capabilities of a bonobo (*Pan paniscus*). *Journal of Archaeological Science, 20*(1), 81–91.

Touwen, B. (1976). *Neurological development in infancy*. Heinemann.

Traub, M. M., Rothwell, J. C., & Marsden, C. D. (1980). A grab reflex in the human hand. *Brain, 103*(4), 869–884.

Trevarthen, C. (1977). Descriptive analyses of infant communicative behavior. In H. R. Schaffer (Ed.), *Studies in mother-infant interaction* (pp. 227–270). Academic Press.

Trillenberg, P., Sprenger, A., Petersen, D., Kömpf, D., Heide, W., & Helmchen, C. (2007). Functional dissociation of saccade and hand reaching control with bilateral lesions of the medial wall of the intraparietal sulcus: Implications for optic ataxia. *NeuroImage, 36*(Suppl 2), T69–T76.

Trinkaus, E., & Villemeur, I. (1991). Mechanical advantages of the Neandertal thumb in flexion: A test of an hypothesis. *American Journal of Physical Anthropology, 84*(3), 249–260.

Trinkaus, E., & Zimmerman, M. R. (1982). Trauma among the Shanidar Neandertals. *American Journal of Physical Anthropology, 57*(1), 61–76.

Trut, L. N. (1999). Early canid domestication: The farm-fox experiment. *American Scientist, 87*(2), 160–169.

Tsakiris, M., Carpenter, L., James, D., & Fotopoulou, A. (2010). Hands only illusion: Multisensory integration elicits sense of ownership for body parts but not for non-corporeal objects. *Experimental Brain Research, 204*(3), 343–352.

Tsakiris, M., & Haggard, P. (2005). The rubber hand illusion revisited: Visuotactile integration and self-attribution. *Journal of Experimental Psychology: Human Perception and Performance, 31*(1), 80–91.

Tsakiris, M., Hesse, M. D., Boy, C., Haggard, P., & Fink, G. R. (2007). Neural signatures of body ownership: A sensory network for bodily self-consciousness. *Cerebral Cortex, 17*(10), 2235–2244.

Tsakiris, M., Prabhu, G., & Haggard, P. (2006). Having a body versus moving your body: How agency structures body-ownership. *Consciousness and Cognition, 15*(2), 423–432.

Tseng, P., & Bridgeman, B. (2011). Improved change detection with nearby hands. *Experimental Brain Research, 209*(2), 257–269.

Tucker, M., & Ellis, R. (1998). On the relations between seen objects and components of potential actions. *Journal of Experimental Psychology: Human Perception and Performance, 24*(3), 830–846.

Tucker, M., & Ellis, R. (2001). The potentiation of grasp types during visual object categorization. *Visual Cognition, 8*(6), 769–800.

Tulving, E., & Schacter, D. L. (1990). Priming and human memory systems. *Science, 247*(4940), 301–306.

Tunik, E., Frey, S. H., & Grafton, S. T. (2005). Virtual lesions of the anterior intraparietal area disrupt goal-dependent on-line adjustments of grasp. *Nature Neuroscience, 8*(4), 505–511.

Turella, L., Rumiati, R., & Lingnau, A. (2020). Hierarchical action encoding within the human brain. *Cerebral Cortex, 30*(5), 2924–2938.

Turvey, M. T., Burton, G., Amazeen, E. L., Butwill, M., & Carello, C. (1998). Perceiving the width and height of a hend-held object by dynamic touch. *Journal of Experimental Psychology: Human Perception and Performance, 24*(1), 35–48.

Tuttle, R. H. (1969a). Knuckle-walking and the problem of human origins. *Science, 166*(3908), 953–961.

Tuttle, R. H. (1969b). Quantitative and functional studies on the hands of the *Anthropoidea*: I. The *Hominoidea*. *Journal of Morphology, 128*(3), 309–364.

Twitchell, T. E. (1965). The automatic grasping responses of infants. *Neuropsychologia, 3*(3), 247–259.

Tzourio-Mazoyer, N., Petit, L., Zago, L., Crivello, F., Vinuesa, N., Joliot, M., Jobard, G., Mellet, E., & Mazoyer, B. (2015). Between-hand difference in ipsilateral deactivation is associated with hand lateralization: fMRI mapping of 284 volunteers balanced for handedness. *Frontiers in Human Neuroscience, 9*, 5.

Umiker-Sebeok, J., & Sebeok, T. A. (Eds.). (1987). *Monastic sign languages*. Mouton de Gruyter.

Umiltà, M. A., Escola, L., Intskirveli, I., Grammont, F., Rochat, M., Caruana, F., Jezzini, A., Gallese, V., & Rizzolatti, G. (2008). When pliers become fingers in the monkey motor system. *Proceedings of the National Academy of Sciences, 105*(6), 2209–2213.

Ungerleider, L., & Mishkin, M. (1982). Two cortical visual systems. In D. J. Ingle, M. A. Goodale, & R. J. W. Mansfield (Eds.), *Analysis of motor behavior* (pp. 549–586). MIT Press.

Uno, Y., Kawato, M., & Suzuki, R. (1989). Formation and control of optimal trajectory in human multijoint arm movement: Minimum torque-change model. *Biological Cybernetics, 61*(2), 89–101.

Uomini, N. T., & Ruck, L. (2018). Manual laterality and cognition through evolution: An archeological perspective. *Progress in Brain Research, 238*, 295–323.

Vail, A. L., Manica, A., & Bshary, R. (2013). Referential gestures in fish collaborative hunting. *Nature Communications, 4*, 1765.

Vaina, L. M., Goodglass, H., & Daltroy, L. (1995). Inference of object use from pantomimed actions by aphasics and patients with right hemisphere lesions. *Synthese, 104*(1), 43–57.

Valenza, N., Ptak, R., Zimine, I., Badan, M., Lazeyras, F., & Schnider, A. (2001). Dissociated active and passive tactile shape recognition: A case study of pure tactile apraxia. *Brain, 124*(Pt 11), 2287–2298.

Valenzeno, L., Alibali, M. W., & Klatzky, R. (2003). Teachers' gestures facilitate students' learning: A lesson in symmetry. *Contemporary Educational Psychology, 28*(2), 187–204.

Vallar, G., Antonucci, G., Guariglia, C., & Pizzamiglio, L. (1993). Deficits of position sense, unilateral neglect and optokinetic stimulation. *Neuropsychologia, 31*(11), 1191–1200.

Vallar, G., & Ronchi, R. (2009). Somatoparaphrenia: A body delusion. A review of the neuropsychological literature. *Experimental Brain Research, 192*(3), 533–551.

Vallar, G., Rusconi, M. L., Bignamini, L., Geminiani, G., & Perani, D. (1994). Anatomical correlates of visual and tactile extinction in humans. *Journal of Neurology, Neurosurgery & Psychiatry, 57*(4), 464–470.

Valyear, K. F., Cavina-Pratesi, C., Stiglick, A. J., & Culham, J. C. (2007). Does tool-related fMRI activity within the intraparietal sulcus reflect the plan to grasp? *NeuroImage, 36*(Suppl 2), 94–108.

Van Boven, R. W., Ingeholm, J. E., Beauchamp, M. S., Bikle, P. C., & Ungerleider, L. G. (2005). Tactile form and location processing in the human brain. *Proceedings of the National Academy of Sciences, 102*(35), 12601–12605.

Van den Heuvel, M. P., Kersbergen, K. J., de Reus, M. A., Keunen, K., Kahn, R. S., Groenendaal, F., de Vries, L. S., & Benders, M. J. N. L. (2015). The neonatal connectome during preterm brain development. *Cerebral Cortex, 25*(9), 3000–3013.

Van der Loos, H., & Woolsey, T. A. (1973). Somatosensory cortex: Structural alterations following early injury to sense organs. *Science, 179*(4071), 395–398.

van der Meer, A. L. H., van der Weel, F. R., & Lee, D. N. (1995). The functional significance of arm movements in neonates. *Science, 267*(5198), 693–695.

van der Meer, A. L. H., van der Weel, F. R., & Lee, D. N. (1996). Lifting weights in neonates: Developing visual control of reaching. *Scandinavian Journal of Psychology, 37*(4), 424–436.

Van Essen, D. C., & Dierker, D. L. (2007). Surface-based and probabilistic atlases of primate Cerebral cortex. *Neuron, 56*(2), 209–225.

van Galen, G. P. (1991). Handwriting: Issues for a psychomotor theory. *Human Movement Science, 10*(2–3), 165–191.

Vanderpool, D., Minh, B. Q., Lanfear, R., Hughes, D., Murali, S., Harris, R. A., Raveendran, M., Muzny, D. M., Hibbins, M. S., Williamson, R. J., Gibbs, R. A., Worley, K. C., Rogers, J., & Hahn, M. W. (2020). Primate phylogenomics uncovers multiple rapid radiations and ancient interspecific introgression. *PLOS Biology, 18*(12), e3000954.

Vasari, G. (1568). *The lives of the artists*. Oxford University Press.

Vauclair, J., Meguerditchian, A., & Hopkins, W. D. (2005). Hand preferences for unimanual and coordinated bimanual tasks in baboons (*Papio anubis*). *Cognitive Brain Research, 25*(1), 210–216.

Vaught, G. M., Simpson, W. E., & Ryder, R. (1968). 'Feeling' with a stick. *Perceptual and Motor Skills, 26*, 848.

Veà, J. J., & Sabater-Pi, J. (1998). Spontaneous pointing behaviour in the wild pygmy chimpanzee (*Pan paniscus*). *Folia Primatologica, 69*(5), 289–290.

Vega-Bermudez, F., & Johnson, K. O. (2001). Differences in spatial acuity between digits. *Neurology, 56*(10), 1389–1391.

Venditti, F., Agam, A., Tirillò, J., Nunziante-Cesaro, S., & Barkai, R. (2021). An integrated study discloses chopping tools use from late Acheulean Revadim (Israel). *PLOS One, 16*(1), e0245595.

Verhagen, L., Dijkerman, H. C., Grol, M. J., & Toni, I. (2008). Perceptuo-motor interactions during prehension movements. *Journal of Neuroscience, 28*(18), 4726–4735.

Veronelli, L., Ginex, V., Dinacci, D., Cappa, S. F., & Corbo, M. (2014). Pure associative tactile agnosia for the left hand: Clinical and anatomo-functional correlations. *Cortex, 58*, 206–216.

Verstynen, T., Diedrichsen, J., Albert, N., Aparicio, P., & Ivry, R. B. (2005). Ipsilateral motor cortex activity during unimanual hand movements relates to task complexity. *Journal of Neurophysiology, 93*(3), 1209–1222.

Vetter, R. J., & Weinstein, S. (1967). The history of the phantom in congenitally absent limbs. *Neuropsychologia, 5*(4), 335–338.

Vingerhoets, G., Alderweireldt, A.-S., Vandemaele, P., Cai, Q., Van der Haegen, L., Brysbaert, M., & Achten, E. (2013). Praxis and language are linked: Evidence from co-lateralization in individuals with atypical language dominance. *Cortex, 49*(1), 172–183.

Virányi, Z., Gácsi, M., Kubinyi, E., Topál, J., Belényi, B., Ujfalussy, D., & Miklósi, Á. (2008). Comprehension of human pointing gestures in young human-reared wolves (*Canis lupus*) and dogs (*Canis familiaris*). *Animal Cognition, 11*(3), 373.

Vocat, R., Staub, F., Stroppini, T., & Vuilleumier, P. (2010). Anosognosia for hemiplegia: A clinical-anatomical prospective study. *Brain, 133*(Pt 12), 3578–3597.

Volkmann, A. W. (1858). Über den einfluss der uebung auf das erkennen räumlicher distanzen [On the effect of practice on the recognition of spatial distances]. *Berichte Über Die Verhandlungen Der Königlich Sächsischen Gesellschaft Der Wissenschaften Zu Leipzig, 10*, 38–69.

Von Bayern, A. M. P., Danel, S., Auersperg, A. M. I., Mioduszewska, B., & Kacelnik, A. (2018). Compound tool construction by New Caledonian crows. *Scientific Reports, 8*(1), 15676.

von Hofsten, C. (1980). Predictive reaching for moving objects by human infants. *Journal of Experimental Child Psychology, 30*(3), 369–382.

von Hofsten, C. (1982). Eye-hand coordination in the newborn. *Developmental Psychology, 18*(3), 450–461.

von Hofsten, C. (1983). Catching skills in infancy. *Journal of Experimental Psychology: Human Perception and Performance, 9*(1), 75–85.

von Hofsten, C. (1984). Developmental changes in the organization of prereaching movements. *Developmental Psychology, 20*(3), 378–388.

von Hofsten, C. (1991). Structuring of early reaching movements: A longitudinal study. *Journal of Motor Behavior, 23*(4), 280–292.

von Hofsten, C., & Fazel-Zandy, S. (1984). Development of visually guided hand orientation in reaching. *Journal of Experimental Child Psychology, 38*(2), 208–219.

von Hofsten, C., & Lindhagen, K. (1979). Observations on the development of reaching for moving objects. *Journal of Experimental Child Psychology, 28*(1), 158–173.

von Hofsten, C., & Rönnqvist, L. (1988). Preparation for grasping an object: A developmental study. *Journal of Experimental Psychology: Human Perception and Performance, 14*(4), 610–621.

von Hofsten, C., Vishton, P., Spelke, E. S., Feng, Q., & Rosander, K. (1998). Predictive action in infancy: Tracking and reaching for moving objects. *Cognition, 67*(3), 255–285.

von Reis, G. (1954). Electromyographic studies in writer's cramp. *Acta Medica Scandinavica, 149*(4), 253–260.

Vuilleumier, P., Reverdin, A., & Landis, T. (1997). Four legs: Illusory reduplication of the lower limbs after bilateral parietal lobe damage. *Archives of Neurology, 54*(12), 1543–1547.

Vuilleumier, P., Valenza, N., Mayer, E., Reverdin, A., & Landis, T. (1998). Near and far visual space in unilateral neglect. *Annals of Neurology, 43*(3), 406–410.

Wada, J., & Rasmussen, T. (1960). Intracarotid injection of sodium amytal for the lateralization of cerebral speech dominance: Experimental and clinical observations. *Journal of Neurosurgery, 17*(2), 266–282.

Wadley, L., Hodgskiss, T., & Grant, M. (2009). Implications for complex cognition from the hafting of tools with compound adhesives in the middle stone age, South Africa. *Proceedings of the National Academy of Sciences, 106*(24), 9590–9594.

Wagstyl, K., Ronan, L., Goodyer, I. M., & Fletcher, P. C. (2015). Cortical thickness gradients in structural hierarchies. *NeuroImage, 111*, 241–250.

Wakefield, E. M., Hall, C., James, K. H., & Goldin-Meadow, S. (2018). Gesture for generalization: Gesture facilitates flexible learning of words for actions on objects. *Developmental Science, 21*(5), e12656.

Wallace, P. S., & Whishaw, I. Q. (2003). Independent digit movements and precision grip patterns in 1–5-month-old human infants: Hand-babbling, including vacuous then self-directed hand and digit movements, precedes targeted reaching. *Neuropsychologia, 41*(14), 1912–1918.

Walshe, F. M. R. (1942). The anatomy and physiology of cutaneous sensibility: A critical review. *Brain, 65*(1), 48–112.

Wang, J., & Sainburg, R. L. (2004). Limitations in interlimb transfer of visuomotor rotations. *Experimental Brain Research, 155*(1), 1–8.

Ward, J. P., Milliken, G. W., Dodson, D. L., Stafford, D. K., & Wallace, M. (1990). Handedness as a function of sex and age in a large population of lemur. *Journal of Comparative Psychology, 104*(2), 167–173.

Ward, C. V, Kimbel, W. H., Harmon, E. H., & Johanson, D. C. (2012). New postcranial fossils of *Australopithecus afarensis* from Hadar, Ethiopia (1990–2007). *Journal of Human Evolution, 63*(1), 1–51.

Ward, C. V, Tocheri, M. W., Plavcan, J. M., Brown, F. H., & Kyalo, F. (2014). Early Pleistocene third metacarpal from Kenya and the evolution of modern human-like hand morphology. *Proceedings of the National Academy of Sciences, 111*(1), 121–124.

Warren, W. C., Hillier, L.W., Marshall Graves, J. A., Birney, E., Ponting, C. P., Grützner, F., Belov, K., Miller, W., Clarke, L., Chinwalla, A. T., Yang, S.-P., Heger, A., Locke, D. P., Miethke, P., Waters, P. D., Veyrunes, F., Fulton, L., Fulton, B., Graves, T., & Wilson, R. K. (2008). Genome analysis of the platypus reveals unique signatures of evolution. *Nature, 453*, 175–183.

Weaver, D. S., Perry, G. H., Macchiarelli, R., & Bondioli, L. (2000). A surgical amputation in 2nd century Rome. *The Lancet, 356*(9230), 686.

Weber, E. H. (1834). De subtilitate tactus. In H. E. Ross & D. J. Murray (Eds.), *E. H. Weber on the tactile senses* (pp. 21–128). Academic Press.

Weder, B. J., Leenders, K. L., Vontobel, P., Nienhusmeier, M., Keel, A., Zaunbauer, W., Vonesch, T., & Ludin, H.-P. (1999). Impaired somatosensory discrimination of shape in Parkinson's disease: Association with caudate nucleus dopaminergic function. *Human Brain Mapping, 8*(1), 1–12.

Weidenreich, F. (1936). Observations on the form and proportions of the endocranial casts of *Sinanthropus pekinensis*, other hominids and the great apes: A comparative study of brain size. *Paleontologica Sinica, Series D7*, 1–50.

Weinstein, E. A., Kahn, R. L., Malitz, S., & Rozanski, J. (1954). Delusional reduplication of parts of the body. *Brain, 77*(1), 45–60.

Weir, A. A. S., Chappell, J., & Kacelnik, A. (2002). Shaping of hooks in New Caledonian crows. *Science, 297*(5583), 981.

Weiskrantz, L. (2009). *Blindsight: A case study spanning 35 years and new developments* (3rd ed.). Oxford University Press.

Weiskrantz, L., Warrington, E. K., Sanders, M. D., & Marshall, J. (1974). Visual capacity in the hemianopic field following a restricted occipital ablation. *Brain, 97*(4), 709–728.

Weiss, E. J., & Flanders, M. (2004). Muscular and postural synergies of the human hand. *Journal of Neurophysiology, 92*(1), 523–535.

Welker, E., & Van der Loos, H. (1986). Is areal extent in sensory cerebral cortex determined by peripheral innervation density? *Experimental Brain Research, 63*(3), 650–654.

Welker, W. I., & Seidenstein, S. (1959). Somatic sensory representation in the cerebral cortex of the racoon (*Procyon lotor*). *Journal of Comparative Neurology, 111*(3), 469–501.

Wenger, E., Kühn, S., Verrel, J., Mårtensson, J., Bodammer, N. C., Lindenberger, U., & Lövdén, M. (2017). Repeated structural imaging reveals nonlinear progression of experience-dependent volume changes in human motor cortex. *Cerebral Cortex, 27*(5), 2911–2925.

Wernicke, K. (1895). Zwei Fälle von Rindenläsion [Two cases of cortical lesion]. *Arbeiten Aus Der Psychiatrischen Klinik Breslau, 2*, 33–53.

Wesselink, D. B., van den Heiligenberg, F. M. Z., Ejaz, N., Dempsey-Jones, H., Cardinali, L., Tarall-Jozwiak, A., Diedrichsen, J., & Makin, T. R. (2019). Obtaining and maintaining cortical hand representation as evidenced from acquired and congenital handlessness. *eLife, 8*, e37227.

White, B. L., Castle, P., & Held, R. (1964). Observations on the development of visually-directed reaching. *Child Development, 35*(2), 349–364.

White, R. C., Aimola Davies, A. M., & Kischka, U. (2010). Errors of somatosensory localisation in a patient with right-hemisphere stroke. *Neurocase, 16*(3), 238–258.

White, T. D., Asfaw, B., Beyene, Y., Haile-Selassie, Y., Lovejoy, C. O., Suwa, G., & WoldeGabriel, G. (2009). *Ardipithecus ramidus* and the paleobiology of early hominids. *Science, 326*(5949), 75–86.

Whitehead, K., Papadelis, C., Laudiano-Dray, M. P., Meek, J., & Fabrizi, L. (2019). The emergence of hierarchical somatosensory processing in late prematurity. *Cerebral Cortex, 29*(5), 2245–2260.

Whiten, A., Goodall, J., McGrew, W. C., Nishida, T., Reynolds, V., Sugiyama, Y., Tutin, C. E. G., Wrangham, R. W., & Boesch, C. (1999). Cultures in chimpanzees. *Nature, 399*, 682–685.

Whiten, A., Goodall, J., McGrew, W. C., Nishida, T., Reynolds, V., Sugiyama, Y., Tutin, C. E. G., Wrangham, R. W., & Boesch, C. (2001). Charting cultural variation in chimpanzees. *Behaviour, 138*(11–12), 1481–1516.

Whitwell, R. L., Sperandio, I., Buckingham, G., Chouinard, P. A., & Goodale, M. A. (2020). Grip constancy but not perceptual size constancy survives lesions of early visual cortex. *Current Biology, 30*(18), 3680–3686.

Whitwell, R. L., Striemer, C. L., Nicolle, D. A., & Goodale, M. A. (2011). Grasping the non-conscious: Preserved grip scaling to unseen objects for immediate but not delayed grasping following a unilateral lesion to primary visual cortex. *Vision Research, 51*(8), 908–924.

Wiener, M., Devoe, S., Rubinow, S., & Geller, J. (1972). Nonverbal behavior and non-verbal communication. *Psychological Review, 79*(3), 185–214.

Wiesel, T. N., & Hubel, D. H. (1963). Single-cell responses in striate cortex of kittens deprived of vision in one eye. *Journal of Neurophysiology, 26*(6), 1003–1017.

Wiestler, T., & Diedrichsen, J. (2013). Skill learning strengthens cortical representations of motor sequences. *eLife, 2*, e00801.

Wilkins, J., Schoville, B. J., Brown, K. S., & Chazan, M. (2012). Evidence for early hafted hunting technology. *Science, 338*(6109), 942–946.

Willett, F. R., Deo, D. R., Avansino, D. T., Rezaii, P., Hochberg, L. R., Henderson, J. M., & Shenoy, K. V. (2020). Hand knob area of premotor cortex represents the whole body in a compositional way. *Cell, 181*(2), 396–409.

Williams, R., & Hale, M. E. (2015). Fin ray sensation participates in the generation of normal fin movement in the hovering behavior of the bluegill sunfish (*Lepomis macrochirus*). *Journal of Experimental Biology, 218*(Pt 21), 3435–3447.

Williams, R., Neubarth, N., & Hale, M. E. (2013). The function of fin rays as proprioceptive sensors in fish. *Nature Communications, 4*, 1729.

Williams-Hatala, E. M., Hatala, K. G., Gordon, M., Key, A., Kasper, M., & Kivell, T. L. (2018). The manual pressures of stone tool behaviors and their implications for the evolution of the human hand. *Journal of Human Evolution, 119*, 14–26.

Willingham, D. B. (1998). A neuropsychological theory of motor skill learning. *Psychological Review, 105*(3), 558–584.

Wimmers, R. H., Savelsbergh, G. J. P., Beek, P. J., & Hopkins, B. (1998). Evidence for a phase transition in the early development of prehension. *Developmental Psychobiology, 32*(3), 235–248.

Wing, A. M. (1980). The height of handwriting. *Acta Psychologica, 46*(2), 141–151.

Witham, C. L., Fisher, K. M., Edgley, S. A., & Baker, S. N. (2016). Corticospinal inputs to primate motoneurons innervating the forelimb from two divisions of primary motor cortex and area 3a. *Journal of Neuroscience, 36*(9), 2605–2616.

Witherington, D. C. (2005). The development of prospective grasping control between 5 and 7 months: A longitudinal study. *Infancy, 7*(2), 143–161.

Wittgenstein, L. (1953). *Philosophical investigations* [G. E. M. Anscombe, Trans.]. Blackwell.

Wolbers, T., Klatzky, R. L., Loomis, J. M., Wutte, M. G., & Giudice, N. A. (2011). Modality-independent coding of spatial layout in the human brain. *Current Biology, 21*(11), 984–989.

Wolff, P., & Gutstein, J. (1972). Effects of induced motor gestures on vocal output. *Journal of Communication, 22*(3), 277–288.

Wolpert, D. M., Ghahramani, Z., & Jordan, M. I. (1995). An internal model for sensorimotor integration. *Science, 269*(5232), 1880–1882.

Wolpert, D. M., Goodbody, S. J., & Husain, M. (1998). Maintaining internal representations: The role of the human superior parietal lobe. *Nature Neuroscience, 1*(6), 529–533.

Wong, P., & Kaas, J. H. (2010). Architectonic subdivisions of neocortex in the galago (*Otolemur garnetti*). *Anatomical Record, 293*(6), 1033–1069.

Wood Jones, F. (1916). *Arboreal man*. Arnold.

Wood Jones, F. (1920). *The principles of anatomy as seen in the hand*. Churchill.

Woodward, A. L., & Guajardo, J. J. (2002). Infants' understanding of the point gesture as an object-directed action. *Cognitive Development, 17*(1), 1061–1084.

Woodworth, R. S. (1899). The accuracy of voluntary movement. *Psychological Review, 3*(Supp. 3), 1–119.

Woolsey, T. A., & Van der Loos, H. (1970). The structural organization of layer IV in the somatosensory region (SI) of mouse cerebral cortex. *Brain Research, 17*(2), 205–242.

Wright, R. V. S. (1972). Imitative learning of a flaked stone technology—The case of an orangutan. *Mankind, 8*(4), 296–306.

Wu, C. W., Bichot, N. P., & Kaas, J. H. (2000). Converging evidence from microstimulation, architecture, and connections for multiple motor areas in the frontal and cingulate cortex of prosimian primates. *Journal of Comparative Neurology, 423*(1), 140–177.

Wu, T., Zhang, J., Hallett, M., Feng, T., Hou, Y., & Chan, P. (2016). Neural correlates underlying micrographia in Parkinson's disease. *Brain, 139*(Pt 1), 144–160.

Wymbs, N. F., Bassett, D. S., Mucha, P. J., Porter, M. A., & Grafton, S. T. (2012). Differential recruitment of the sensorimotor putamen and frontoparietal cortex during motor chunking in humans. *Neuron, 74*(5), 936–946.

Wymbs, N. F., & Grafton, S. T. (2015). The human motor system supports sequence-specific representations over multiple training-dependent timescales. *Cerebral Cortex, 25*(11), 4213–4225.

Wynn, T., & Gowlett, J. (2018). The handaxe reconsidered. *Evolutionary Anthropology, 27*(1), 21–29.

Wynn, T., Hernandez-Aguilar, R. A., Marchant, L. F., & McGrew, W. C. (2011). "An ape's view of the Oldowan" revisited. *Evolutionary Anthropology, 20*(5), 181–197.

Wynn, T., & McGrew, W. C. (1989). An ape's view of the Oldowan. *Man, 24*(3), 383.

Xitco, M. J., Gory, J. D., & Kuczaj, S. A. (2001). Spontaneous pointing by bottlenose dolphins (*Tursiops truncatus*). *Animal Cognition, 4*(2), 115–123.

Yamamoto, S., & Kitazawa, S. (2001a). Reversal of subjective temporal order due to arm crossing. *Nature Neuroscience, 4*(7), 759–765.

Yamamoto, S., & Kitazawa, S. (2001b). Sensation at the tips of invisible tools. *Nature Neuroscience, 4*(10), 979–980.

Yan, Y., Goodman, J. M., Moore, D. D., Solla, S. A., & Bensmaia, S. J. (2020). Unexpected complexity of everyday manual behaviors. *Nature Communications, 11*, 3564.

Yeo, S. S., Jang, S. H., & Son, S. M. (2014). The different maturation of the corticospinal tract and corticoreticular pathway in normal brain development: Diffusion tensor imaging study. *Frontiers in Human Neuroscience, 8*, 573.

Young, R. R., & Benson, D. F. (1992). Where is the lesion in allochiria? *Archives of Neurology, 49*(4), 348–349.

Yousry, T. A., Schmid, U. D., Alkadhi, H., Schmidt, D., Peraud, A., Buettner, A., & Winkler, P. (1997). Localization of the motor hand area to a knob on the precentral gyrus. *Brain, 120*(Pt 1), 141–157.

Zaidel, E. (1998). Stereognosis in the chronic split brain: Hemispheric differences, ipsilateral control and sensory integration across the midline. *Neuropsychologia, 36*(10), 1033–1047.

Zaki, M. E., El-Din, A. M. S., Soliman, M. A.-T., Mahmoud, N. H., & Basha, W. A. B. (2010). Limb amputation in ancient Egyptians from Old Kingdom. *Journal of Applied Sciences Research, 6,* 913–917.

Zangaladze, A., Epstein, C. M., Grafton, S. T., & Sathian, K. (1999). Involvement of visual cortex in tactile discrimination of orientation. *Nature, 401,* 587–590.

Zeharia, N., Hertz, U., Flash, T., & Amedi, A. (2015). New whole-body sensory-motor gradients revealed using phase-locked analysis and verified using multivoxel pattern analysis and functional connectivity. *Journal of Neuroscience, 35*(7), 2845–2859.

Zeller, D., Gross, C., Bartsch, A., Johansen-Berg, H., & Classen, J. (2011). Ventral premotor cortex may be required for dynamic changes in the feeling of limb ownership. *Journal of Neuroscience, 31*(13), 4852–4857.

Zhang, X., Zhan, X., Ding, Y., Li, Y., Yeh, H.-Y., & Chen, L. (2022). A case of well-healed foot amputation in early China (8th-5th centuries BCE). *International Journal of Osteoarchaeology, 32*(1), 132–141.

Zhao, D., Hopkins, W. D., & Li, B. (2012). Handedness in nature: First evidence on manual laterality on bimanual coordinated tube task in wild primates. *American Journal of Physical Anthropology, 148*(1), 36–44.

Zingerle, H. (1913). Über störungen der wahrnehmung des eigenen körpers bei organischen gehirnerkrankungen [Impaired perception of one's own body due to organic brain disorders]. *Monatsschrift für Psychiatrie und Neurologie, 34,* 13–36.

Zoia, S., Blason, L., D'Ottavio, G., Bulgheroni, M., Pezzetta, E., Scabar, A., & Castiello, U. (2007). Evidence of early development of action planning in the human foetus: A kinematic study. *Experimental Brain Research, 176*(2), 217–226.

Zopf, R., Harris, J. A., & Williams, M. A. (2011). The influence of body-ownership cues on tactile sensitivity. *Cognitive Neuroscience, 2*(3–4), 147–154.

Author Index

Aboitiz, F., 295
Abrams, R. A., 139, 140
Acuña, C., 80
Adie, W. J., 72, 260
Adler, C. H., 171
Adler, D. S., 200
Aflalo, T. N., 59, 273
Aglioti, S. M., 32
Ahlberg, P. E., 223
Aiello, B. R., 223
Aizawa, H., 283
Ajax, E. T., 171
Akelaitis, A. J., 73
Alba, D. M., 234, 239
Albert, S. T., 87
Alexander, G. E., 179
Alibali, M. W., 336, 337
Allen, F. J., 288–289
Allen, H. A., 103
Allen, J. S., 252
Allievi, A. G., 271
Allison, T., 143, 272
Allman, J. M., 229
Alloway, K. D.
Almècija, S., 236–239
Alstermark, B., 245
Altenmüller, E., 170–172
Ambrose, S. H., 200, 207, 341
Amedi, A., 102
Ames, K. C., 285
Amunts, K., 311

Anaxagoras, ix, 1
Andersen, R. A., 67–68
Anderson, J. R., 325
Andres, M., 211
Andrushko, J. W., 284
Anema, H. A., 31
Angell, R., 172
Angstmann, S., 310–311
Annett, M., 310, 318
Anonymous, 167
Antoniello, D., 122, 299
App, B., 328–329
Arashida, R., 101
Arbib, M. A., 255
Arcaro, M. J., 270–271
Argyle, M., 337
Aristotle, ix, 37, 307, 343
Ariyan, S., 106
Armand, J., 274
Armatas, C. A., 287
Armour, J., 318
Armour, T., 171, 345–346
Arrizabalaga, A., 316
Arroyo, A., 198
Arsuaga, J. L., 241
Arthur, W., 222
Artieda, J., 185
Arzy, S., 114–115, 120
Assimacopoulos, S., 225, 229, 272
Ausenda, C. D., 293
Aussems, S., 337

Aviezer, H., 327
Azañón, E., 40–41

Babinski, J., 113
Backwell, L. R., 195, 240
Baier, B., 113, 114, 116, 120
Baker, S. N., 276
Bakola, S., 68
Baldwin, J. M., 72
Baldwin, M. K. L., 62
Balfour, S., 288
Bálint, R., 78
Ballesteros, S., 102
Balyaev, D. K., 326
Banks, M. S., 99–100
Barac-Cikoja, D., 214
Bara-Jimenez, W., 173
Barbieri, C., 211
Bardo, A., 241
Bargalló, A., 316
Barker, P., 11
Barley, N. F., 342
Baron-Cohen, S., 322
Barrett, P. H., 152
Barrière, I., 160
Bartholow, R., 13, 281
Bartlett, F. C., 28
Barton, R. A., 243
Bar-Yosef, O., 201
Bassett, D. S., 153
Bassolino, M., 37
Bates, E., 334
Bauer, J. A., 264
Bawden, M., 171–172
Baxter, D. M., 161–162
Beaudet, A., 252, 254, 255
Beaulieu, C., 276
Beauvois, M.-F., 161
Becchio, C., 37
Beck, M. M., 67, 268
Beck, P. D., 226
Beck, S., 287
Behne, T., 324

Behrmann, M., 39
Beilock, S. L., 337
Bell, C., xii, 1–2, 9, 327
Bello-Alonso, P., 199
Bellugi, U., 329
Bender, M. B., 132, 296–297, 299
Benedetti, F., 37–38
Benke, T., 114
Bennati, F., 145–146
Bennett, E. A., 241
Bennike, P., 307
Bensmaia, S., 24
Benson, D. F., 299
Benton, A. L., 78
Berardelli, A., 178, 183
Berger, K. W., 337
Bergman, H., 180
Bermúdez de Castro, J. M., 317–318
Bernhard, C. G., 51, 226
Bernstein, N. A., 47, 49, 89, 156
Berret, B., 88
Bertamini, M., 112
Bertenthal, B. I., 323
Berthier, N. E., 263
Berti, A., 32, 113, 120, 130, 213
Bestmann, S., 63, 281
Beyene, Y., 198–199, 242
Beznosov, P. A., 223
Bharucha, N., 78
Bhat, A. N., 262
Bickerton, D., 340
Binkofski, F., 64, 66, 76, 83, 98
Birznieks, I., 29, 31
Bishop, A., 260
Bishop, D. V. M., 308–309
Bisiach, E., 130
Bizley, J. K., 83
Bjoertomt, O., 131
Blakeslee, S., 107
Blangero, A., 79
Blankenburg, F., 16, 230, 250
Blass, E. M., 262
Blaxton, T. A., 101

Bloch, J. I., 232, 314
Bloxham, C. A., 183
Blythe, E., 327–328
Bodegård, A., 27
Boeckle, M., 194
Boëda, E., 201
Boehm, J. J., 261
Boesch, C., 194
Bogen, J. E., 73
Bohlhalter, S., 28
Boisvert, C. A., 223
Boldrey, E., 13, 15, 58, 281
Boling, W., 52–53
Bolognini, N., 192
Bonicalzi, V., 144
Bonifazi, S., 212
Boraud, T., 186
Boroojerdi, B., 52
Borra, E., 64, 102
Bortoff, G. A., 51, 234, 244–245
Bottini, G., 37, 116
Botvinick, M., 110
Bourke, J., 107
Boyer, D. M., 232, 314
Boyer, E. O., 85
Boynton, G. M., 20, 41, 249
Bracci, S., 209
Brain, C. K., 196
Brain, W. R., 130
Braun, C., 30
Braun, D. R., 196
Brazelton, T. B., 261
Brentari, D., 329, 332
Bresciani, J.-P., 85
Breuer, T., 194
Bridgeman, B., 139
Brinkman, J., 283
Brion, S., 116
Brisben, A. J., 214
Broaders, S. C., 336
Broca, P., 52, 160, 281, 330, 340
Brochier, T., 32
Brockmole, J. R., 137

Brodmann, K., 24, 228
Brody, N., 261
Bronowski, J., ix
Brooks, R., 322
Brooks, V. B., 43
Brothwell, D. R., 106
Brown, E. M., 289–290
Brown, L. E., 136–138
Brown, P. B., 20, 142
Brown, S., 195
Brozzoli, C., 133–134, 135
Brugger, P., 110, 117, 123
Bruner, E., 254, 317
Bruner, J., 262
Bruno, N., 112
Bruno, V., 219
Brusatte, S., 224
Bryden, P. J., 310
Büchel, C., 100, 331
Buckner, R. L., 252
Buetefisch, C. M., 284
Buetti, S., 81
Bugbee, N. M., 22
Buka, S. L., 261
Bundy, D. T., 285
Bunn, H. T., 197–198
Buquet-Marcon, C., 106
Burchfiel, J. L., 25, 34
Butler, B. C., 131
Butterworth, G., 261, 265, 267, 322, 324
Butterworth, S., 173
Buxbaum, L. J., 78, 83, 188, 209–210, 211
Buxhoeveden, D. P., 22
Byl, N. N., 172–173
Byrne, R. W., 325, 339
Byrnit, J. T., 325

Cabeza, R., 251
Cadete, D., 124
Call, J., 325–326
Callier, V., 223
Calvert, G. H. M., 291
Calzolari, E., 42

Cameron, B. D., 85
Canavero, S., 122, 144
Cantalupo, C., 315, 341
Canzoneri, E., 42, 220
Caramazza, A., 162–163
Carbonell, E., 196
Cardinali, L., 219–220
Cardini, F., 141, 143–144
Carello, C., 214
Carlson, K. J., 255
Carlyle, T., 313
Carpenter, M., 322
Carr, L. J., 287
Carson, R. G., 287, 291, 311
Carter, C. S., 87
Cartmill, E. A., 325
Cartmill, M., 232, 314
Caruana, F., 209, 251
Carvalho, S., 194, 205
Casanova, M. F., 22
Cascarano, G. D., 181
Case, L. K., 301
Caselli, R. J., 94–97
Castiello, U., 48, 84, 133
Catani, M., 15
Catania, K. C., 225, 227–228, 272, 273
Chakraborty, M., 229
Chaminade, T., 207
Chan, T.-C., 214
Chancel, M., 111
Changizi, M. A., 228
Chao, L. L., 209
Chaplin, T. A., 234, 251
Charcot, J.-M., 281
Checchetto, A., 332
Chen, L. M., 24
Cheney, P. D., 90
Cholewiak, R. W., 41
Chowdhury, R. H., 25
Chowdhury, S. A., 143
Chu, M., 336
Church, R. B., 335–336
Churchill, S. E., 241

Churchland, M. M., 59, 285
Cicone, M., 335
Cincotta, M., 287
Cisek, P., 76, 284, 285
Clack, J. A., 3, 222–224
Clark, G., 195–196
Clarke, P., 172
Clarke, R. J., 239, 240
Classen, J., 50
Clifton, R. K., 264
Clower, D. M., 64, 179
Coates, M. I., 3, 222
Cocchini, G., 36
Cody, F. W. J., 20, 42–43
Coelho, L. A., 36
Cognolato, M., 108
Cohen, D. A. D., 24
Cohen, H. J., 268
Cohen, J., 110
Cohen, L. G., 100, 286, 331
Cohen, Y. E., 83
Cole, J., 33–34, 333
Cole, K. J., 47
Cole, M., 78
Colonnesi, C., 322
Condillac, E., 339
Connolly, J. C., 255
Connolly, J. D., 68
Connolly, K. J., 308–309
Conte, A., 184
Conti, F., 293
Cook, S. W., 336, 338–339
Cooper, H. M., 225
Corballis, M. C., 339, 342
Coren, S., 307
Corkin, S., 95
Corkum, V., 324
Coslett, H. B., 28, 30, 78, 136
Cosman, J. D., 139
Costantini, M., 103, 111
Costello, M. B., 233, 244
Cowey, A., 130–131
Craddock, M., 313

Author Index

Craig, J. C., 141
Cressman, E. K., 85
Crevecoeur, F., 85
Criscimagna-Hemminger, S. E., 290–291
Critchley, M., 29, 72, 95, 114, 123, 260, 288
Crosby, E. C., 62
Crump, M. J. C., 164–165
Crutcher, M. D., 179
Cubelli, R., 162
Cui, H., 68
Cunningham, D. A., 48
Currier, R. D., 287
Cushing, H., 13, 114

Dagosto, M., 232
Dall'Orso, S., 271
Damasio, A. R., 78
D'Amour, S., 298
Danforth, C. H., 3
Darian-Smith, I., 24, 51, 274
Darlington, R. B., 225, 243
Dart, R. A., 252–253
Darwin, C., ix, 2, 152, 221, 241, 258, 318, 327
Daum, M. M., 323
Davare, M., 66
D'Avella, A., 49
Davies, I. R. L., 101
Davies-Jones, G. A. B., 78
Davis, D. R., 346
Davis, J. E., 342
Davis, P. R., 236
Davis, W. W., 290
Davoli, C. C., 137, 139
Dawkins, R., 1
Dax, M., 281, 330, 340
Day, B. L., 85
Day, L., 145
Dea, M., 55
de Beaune, S. A., 198, 206
DeCasien, A. R., 243
Decety, J., 63
Dehno, N. S., 293
de Gelder, B., 327

de Graaf, J. B., 108
de Haan, E. H. F., 27, 97, 296
Delagnes, A., 205
de la Torre, I., 198, 205, 206
Delis, D., 335
Della Sala, S., 73, 76
Dellatolos, G., 309
Dennis, M., 346
Dennis, W., 307
Denny-Brown, D., 29, 72, 189, 260
D'Entremont, B., 324
De Renzi, E., 189, 211
Dérouesné, J., 161
d'Errico, F., 195, 240
DeSilva, J. M., 234
Desmond, A. J., 2, 221
Desmurget, M., 15, 63, 89
de Sousa, A. A., 252
de Vignemont, F., 41–42, 220
de Vries, J. I. P., 261, 319
de Winter, W., 243–244
DiCarlo, J. J., 25
Dichgans, J., 263
Dick, F., 12, 334
Dickens, C., ix
Diedrichsen, J., 87–88, 147–148, 150, 285
Dierker, D. L., 251
Diers, M., 144
Diez-Martín, F., 195, 197, 198, 242
Dijkerman, H. C., 27, 97
D'Imperio, D., 116
Di Noto, P. M., 345
Diogo, R., 4–5
di Pellegrino, G., 132, 212
DiRusso, A. A., 336
Disbrow, E., 21, 250
DiZio, P., 290–291
Dominey, P., 63, 178
Dominguez-Ballesteros, E., 316
Domínguez-Rodrigo, M., 197, 242
Donchin, O., 284
Downey, J. E., 285
Downs, J. P., 223

Du, F., 139
Duffy, F. H., 25, 34
Duhamel, J.-R., 128, 134
Dum, R. P., 69, 254–255
Duncan, R. O., 20, 41, 249
Duque, J., 268, 287, 311
Durbin, R., 273
Dykes, R. W., 23, 142

Easton, R. D., 102
Ebner, F. F., 21, 303
Eden, J., 124
Edgley, S. A., 276
Edwards, T., 332
Eidelberg, E., 16, 301
Ehrenwald, H., 121
Ehrman, L., 306, 308
Ehrsson, H. H., 111, 112, 120, 301–302
Eimer, M., 141
Ejaz, N., 287–288
Ekman, P., 333
Elbert, T., 30, 147, 173
Eliasson, A.-C., 267
Ellis, A. W., 163
Ellis, R., 74–75
Embick, D., 341
Epstein, R. A., 12
Ernst, M. O., 99–100
Eshel, N., 298
Eskes, G. A., 135
Ettlinger, G., 80, 97
Etxepare, R., 342
Evans, C., 78
Evarts, E. V., 56–57, 90, 283
Exner, S., 163–164
Eyre, J. A., 276

Fabbri, S., 69
Fabri, M., 296
Fagard, J., 267–268
Fagot, J., 313, 314
Falk, D., 244, 252–253, 255
Fan, C., 124

Fang, F., 64, 76
Fang, W., 120
Farnè, A., 132, 212–213, 297
Farrell, D. F., 15
Farthing, J. P., 292
Fasano, A., 184
Fattori, P., 69
Faugier-Grimaud, S., 80
Faurie, C., 306, 307–308
Favorov, O., 22–23
Fazel-Zandy, S., 267
Fechner, G., 279
Feinberg, T. E., 103
Feldman, J. F., 261
Felleman, D. J., 249
Ferbert, A., 286
Ferrier, D., 13, 53, 56, 58, 80, 186, 281, 298
Fessard, A., 223
Fetz, E. E., 90
Filimon, F., 34, 68
Finch, G., 314
Finger, S., 106
Fink, G. R., 254
Finlay, B. L., 225, 243
Fiori, F., 41, 43, 175
Fiorio, M., 144
Flack, T., 327
Flament, D., 276
Flanders, M., 49–50, 165
Flash, T., 88, 183
Flechsig, P., 21–22
Fleisher, L., 170, 176
Fleming, J. F. R., 62
Flesher, S. N., 15
Flor, H., 299
Flowers, K. A., 181
Floyer-Lea, A., 149
Foerster, O., 13, 281
Fogassi, L., 64, 134
Fogel, A., 323–324
Foley, R., 200
Forss, N., 295–296
Forssberg, H., 267

Forster, B., 141
Fossataro, C., 123
Fotopoulou, A., 113–114
Fouts, R. S., 325
Fox, M. J., 177–178
Frackowiak, R. S. J., 331
Fragaszy, D. M., 233, 244
Frangeul, L., 230
Frank, S. M., 303
Franz, E. A., 109, 283
Frassinetti, F., 130, 132, 135, 213
Frayer, D. W., 317–318
Freitas, M. B., 309
Fried, I., 21, 77
Friedman, R. M., 24
Friesen, W. V., 333
Frisby, J. P., 101
Frith, C. D., 113
Fritsch, G., 13, 56, 58, 281
Frost, S. B., 226
Froude, J. A., 313
Fujiwara, Y., 285
Fukuchi-Shimogori, T., 225, 229–230
Fulton, J. F., 73
Furuya, S., 4, 49, 157, 159, 170, 174

Gabor, A., 23
Gabriel, D. A., 289
Gainotti, G., 297
Galdikas, B. M. F., 193, 194
Galea, M. P., 51, 274
Gallagher, S., 34
Gallego, J. A., 50
Gallese, V., 64
Galletta, L., 240, 241
Gallivan, J. P., 48, 210, 286
Galloway, J. C., 262
Gandevia, S. C., 35, 144
Ganea, N., 36
Ganguly, K., 283
Ganhos, F. N., 106
Garbarini, F., 113, 123, 283
Garcha, H. S., 97

Gardner, B. T., 340
Gardner, R. A., 340
Garraghty, P. E., 27
Garraux, G., 173
Gazzaniga, M. S., 191–192, 282–283, 296, 312
Gebo, D. L., 233
Gehrke, A. R., 222
Gentaz, E., 101
Gentile, G., 112, 118, 120
Gentilucci, M., 75
Gentner, R., 50, 158
Genty, E., 325, 339
Georgopoulos, A. P., 56–57
Germann, J., 52
Gerstmann, J., 95–96, 115, 123
Geschwind, N., 186, 188, 189, 182, 312
Geyer, S., 24, 53
Gharbawie, O. A., 250
Gibson, J. J., 37, 71, 92, 93, 214
Gilbert, A. N., 309, 311
Gilliatt, R. W., 114
Gilpin, H. R., 119
Ginty, D. D., 6
Giraux, P., 108
Giummarra, M. J., 107, 124
Glasser, M. F., 228
Glosser, G., 335
Glover, S., 89
Goldberg, G., 73, 76, 184
Goldberg, I. A., 290
Goldenberg, G., 189–190, 211, 340
Goldin-Meadow, S., 329, 333, 335–339
Goldman-Rakic, P. S., 22, 254–255
Goldring, S., 283
Golubova, O., 44
Goodale, M. A., 27, 61, 82–83, 94, 140
Goodall, J., 193, 339
Gooddy, W., 105
Goodglass, H., 188–189
Goodhew, S. C., 140
Goodman, J. M., 35
Goodwin, R. S., 319
Gordon, A. D., 239, 240

Gordon, A. M., 165
Gould, S. J., 3, 243
Gowlett, J. A. J., 198–199
Gozli, D. G., 140
Gradin, C., 305
Grafton, S. T., 149–150, 290
Graham, J. A., 335, 337
Graham, K. E., 340
Granda, A. J., 332
Gray, R. D., 194
Graybiel, A. M., 152, 179, 342
Graziano, M. S. A., 34, 50, 58–59, 129, 134, 261, 273
Gréa, H., 89
Gredebäck, G., 323
Green, B. G., 41
Green, D. J., 240
Green, L. A., 289
Grefkes, C., 24, 250
Griffin, D. M., 55
Grill-Spector, K., 102
Grinvald, A., 25
Grivaz, P., 120, 135
Groenen, M., 316
Grove, E. A., 225, 229–230, 272
Guadalupe, T., 311
Guajardo, J. J., 322–323
Gumert, M. D., 194, 249
Gunz, P., 254–255
Guo, J., 137
Guterstam, A., 111–112, 120, 124, 134
Gutstein, J., 337
Gutteling, T. P., 137

Haaland, K. Y., 192
Haar, S., 285
Haerer, A. F., 287
Haggard, P., 35–36, 41, 43, 45, 111–112, 143–144, 249
Hagmann, P., 252
Hagmann, S., 189–190, 211
Hagoort, P., 255
Haldane, J. B. S., 71

Hale, M. E., 223
Hall, E. T., 127
Hall, N. J., 326
Hallett, M., 174, 181–183, 287
Halligan, P. W., 29–31, 123, 130–131, 141
Halsband, U., 77
Halverson, H. M., 258–259, 263, 265–267, 274, 346
Hamilton, E., 145
Hamilton, R., 331
Hammond, P. H., 85, 90
Hamrick, M. W., 5, 204, 232
Hamzei, F., 284
Hanada, K., 32
Hanajima, R., 286–287
Handler, A., 6
Hänggi, J., 117
Hannan, T. E., 324
Hanna-Pladdy, B., 188, 192, 312
Hardy, A. R., 223
Hare, B., 326–327
Hari, R., 114, 121–122, 136
Harmand, S., 195–196, 204, 238–239
Harrar, V., 303
Harris, J. A., 32–33, 141–142, 303
Harris, L. R., 298
Hartigan, A., 309
Harvey, M. A., 24
Harvey, P. H., 243
Haslinger, B., 148
Hatwell, Y., 101
Hauck, T. C., 201
Haward, B. M., 167, 172
Hawkins, M. B., 222
He, S., 76
Head, H., 11–12, 28–29, 34
Heald, J. B., 217
Hebb, D. O., 46
Hecht, E. E., 207
Hediger, H., 127
Heed, T., 48
Heffner, R. S., 51, 226, 244
Heilman, K. M., 189–191, 211, 291

Helbig, H. B., 100
Held, R., 99, 264
Heller, M. A., 99, 141
Heming, E. A., 285
Henson, C. O., 303
Hepper, P. G., 261, 319
Herculano-Houzel, S., 244
Herron, P., 228
Hertenstein, M. J., 328
Hess, C. W., 286
Heuer, H., 283
Hewes, G. W., 339
Heywood, S., 335
Hibbard, L. S., 53
Hickok, G., 329–330
Hill, J., 251
Hinde, R. A., 274
Hippocrates, 106, 280
Hirano, M., 159
Hirsh, I. J., 298
Hitzig, E., 13, 56, 58, 281
Hobaiter, C., 325, 339, 340
Hoetjes, M., 335
Hoffecker, J. F., 195
Hofman, M. A., 243
Hofmann, A., 170
Hogan, N., 88
Hohlstein, R. R., 265
Holmes, G., 28–29, 34
Holmes, N. P., 133
Holloway, R. L., 244, 252, 254, 255
Holloway, V., 282
Hömke, L., 28, 97
Honoré, J., 141
Hoosain, R., 309
Hoover, J. E., 179
Hopkins, B., 261
Hopkins, D. A., 51, 267, 274–275
Hopkins, W. D., 52, 248, 314–316, 325–326, 341
Horsley, V., 14, 53–54
Hortobágyi, T., 291
Hoyet, L., 124

Hrbek, A., 272
Hsieh, C.-L., 144
Hu, Y., 200
Huang, R.-S., 20
Huang, W., 196
Hubel, D. H., 20, 139–140, 272
Hublin, J.-J., 201
Huffman, K. J., 229, 249
Hughes, E., 288
Humphreys, G. W., 103
Hunt, G. R., 194
Hustwit, M. P., 106
Huth, A. G., 12
Hwang, E. J., 81
Hyvärinen, J., 24–25

Iannetti, G. D., 144
Imamizu, H., 218
Ingram, J. N., 217
Inoue, Y, 325
Insel, T. R., 234
Ioannou, C. I., 172
Ireland, W. W., 309
Iriki, A., 129, 211–212
Irurtzun, A., 342
Ivanco, T. L., 226
Iverson, J. M., 322, 333
Ivry, R. B., 281
Iwamura, Y., 25, 61, 293–294, 303
Iwaniuk, A. N., 226–227

Jackson, C. V., 117
Jackson, J. H., 281, 340
Jackson, S. R., 78, 80, 183
Jacobs, S., 216–217
Jacquet, A.-Y., 268
Jain, N., 21
Jakobson, L. S., 78, 283
James, T. W., 103
James, W., 71, 74, 78, 111, 214, 260
Janecka, J. E., 231
Jankovic, J., 171
Jarrassé, N., 108

Jarvis, E. D., 229
Jasper, H., 281
Jeannerod, M., 60–61, 64, 66, 78–79, 81
Jedynak, C. P., 116
Jenkins, I. H., 184
Jenkinson, P. M., 114, 116
Jerde, T. E., 49
Jerison, H. J., 243
Johansson, R. S., 5
Johnson, K. O., 6, 20, 141, 249, 312
Jones, E., 299
Jones, E. G., 142
Jones, P. R., 197, 199
Jordan, M. I., 87
Jousmäki, V., 136
Jubault, T., 151–152, 341
Jueptner, M., 153
Jung, W. P., 267

Kaas, J. H., 16, 18, 21, 225, 227–231, 242, 248–250, 269–270, 272–273
Kadmon Harpaz, N., 47, 163
Kahn, D. M., 229
Kakebeeke, T. H., 268, 287
Kakei, S., 57–58, 63, 216
Kalaska, J. F., 57, 76
Kalsbeck, J. E., 80
Kaminski, J., 326
Kammers, M. P. M., 112, 118
Kant, I., ix
Kanwisher, N., 103, 251
Kapandji, A. I., 3–4
Kaplan, E., 188–189
Kaplan, G., 326
Karakostis, F. A., 204, 241
Karnath, H.-O., 81, 113–114, 116, 120
Karni, A., 149
Kastner, S., 12, 64, 209, 251
Katz, D., ix, xii, 92–93, 214
Kawai, R., 154
Kawamura, M., 299–300
Kay, B. A., 4, 156–157
Kebschull, J. M., 230

Keeley, L. H., 197, 316
Keen, R., 263
Kendon, A., 337, 342
Kennard, M., 186, 274–275, 281, 346
Kennett, S., 133, 141–142
Kenward, B., 194
Kersken, V., 340
Key, A. J. M., 198, 202–204
Khateb, A., 121–123
Khoshbin, S., 181
Kibii, J. M., 240
Kieliba, P., 124–125
Kikkert, S., 108–109
Kilgour, A. R., 103
Kim, J. S., 331
Kim, S.-G., 285, 311
Kim, U., 21
Kimoto, Y., 159
Kimura, D., 340
Kinoshita, H., 4, 157
Kinoshita, M., 245–246
Kinsbourne, M., 161, 291, 297
Kısa, Y. D., 335
Kita, K., 175
Kita, S., 336–337
Kitada, R., 103
Kitazawa, S., 39–40, 214
Kivell, T. L., 236–237, 239–240
Klatzky, R. L., 92–95, 214, 328
Kleeman, L. T., 91
Klein, R. G., 200–201
Klima, E., 329
Kline, D. G., 107
Knecht, S., 340–341
Knights, E., 210–211
Koechlin, E., 151–152, 341
Koenig, L., 33
Koh, T. H. H. G., 276
Köhler, W., 204
Kohonen, T., 273
Koffka, K., 72
Kolasinski, J., 57
Konczak, J., 185, 263

Author Index

Konen, C. S., 64, 69, 144, 251
Kornysheva, K., 147, 150–151
Koslowski, B., 262
Krakauer, J. W., 63, 154, 281
Krause, M. A., 325, 326
Krauss, R. M., 333
Kravitz, A. V., 180
Kravitz, H., 261
Krubitzer, L. A., 225, 227–229, 249–250, 269–270
Krukenberg, H., 91
Kuehn, E., 21–22
Kuhtz-Buschbeck, J. P., 267
Kuman, K., 240
Kurjak, A., 319
Kuypers, H. G. J. M., 51, 245, 274, 283

Lackner, J. R., 290–291
Lacoste, A. M. B., 224
Làdavas, E., 132, 212
Lahr, M. M., 200
LaMotte, R. H., 80, 214
Langley, A. L., 261
Lashley, K. S., 46–48, 92, 165, 186, 281
Laskin, S. E., 22, 143, 298
Laszlo, J. I., 290
Lawrence, D. G., 51, 245, 267, 274, 275
Lawson, R., 93, 313
Leakey, L. S. B., 193, 195–196, 242
Leakey, M. D., 196
Leavens, D. A., 325–326
Lebel, C., 276
Le Cornu Knight, F., 42
Lederman, S. J., 92–95, 214, 313, 328
Le Gros Clark, W. E., 231, 236
Lee, H., 251
Lee, M., 291–292
Lee, M.-H., 262
Lee, R. B., 342
Legerstee, M., 324
Lehmann, S. J., 69
Leiguarda, C., 186, 188
Leinonen, L., 128–129, 134

Lemelin, P., 4, 232
Lemon, R. N., 51, 226, 245–246
Lende, R. A., 226
Lenka, A., 171
Leo, A., 50
Leont'ev, A. N., 91–94
Lepre, C. J., 198, 242
Leroi-Gourhan, A., 342
Leung, E. H. L., 322
Lewis, J. E., 196, 204
Lewis, J. W., 209
Lewis, O. J., 204, 238
Ley, P., 41
Leyton, A. S. F., 13–14, 54, 186, 281
Lhermitte, F., 73–74
Li, Q., 231
Libertus, K., 265
Liddell, E. G. T., 85
Lie, E., 136
Lieberman, D. E., 253
Liepmann, H., 186–192, 312, 340, 346
Lindhagen, K., 263
Lipsitt, L. P., 261
Liszkowski, U., 324–325
Liu, Y., 31
Livingstone, M. S., 139–140
Llorente, M., 316
Lloyd, D. M., 112
Locke, J., 98–99
Lockman, J. J., 267
Logan, G. D., 152, 164–165
Lombard, M., 200
Long, C., 4
Long, X., 271
Longo, M. R., 12, 20, 28, 30, 35–37, 41–44, 110–111, 118–119, 124, 143–144, 175, 213–214, 249
Longoni, A. M., 309
Lorch, M. P., 160
Lorenzo, C., 241
Lotze, H., 214
Louis, E. D., 280
Lourenco, S. F., 213–214

Lovejoy, C. O., 237–239
Lozano, M., 317–318
Lukos, J. R., 184
Luppino, G., 64, 67, 134
Lyon, I. N., 85

MacIver, M. A., 223–224
MacNeilage, P. F., 313–314
MacSweeney, M., 330
Maeda, R. S., 87
Magnus, C. R. A., 292
Maier, M. A., 244–245
Mainka, T., 175
Maister, L., 112
Majsak, M. J., 184
Makin, T. R., 135, 137–138
Malaivijitnond, S., 194, 249
Malfait, N., 290
Maloney, T. R., 106
Manca, A., 289
Mancini, F., 20, 101, 144
Manger, P., 22
Manni, E., 21
Mannu, M., 194, 233, 244
Manohar, S. G., 183
Manzoni, T., 293
Maravita, A., 32, 116–117, 133, 212
Marcel, A., 113, 301–302
Marchant, L. F., 306, 314
Marchi, D., 240
Marentette, P. F., 329
Margiotoudi, K., 315
Margolin, D. I., 161
Margolis, A. N., 20
Marois, R., 251
Maros, K., 326
Marsden, C. D., 86, 169–170, 186, 188
Marsh, A. P. L., 287–288
Marshall, J., 330
Marshall, J. C., 123, 130–131
Martin, A., 209
Martin, W. L. B., 309
Martinaud, O., 120

Marzke, M. W., 9, 199, 202–204, 237, 239–240
Masataka, N., 324
Maschke, M., 185
Mason, J., 290
Mason, W. A., 264
Masterton, R. B., 51, 226, 244–245
Matelli, M., 64, 67, 254–255
Matiello, M., 162
Matthews, P. B. C., 287
Matthews, P. M., 149
Mattingley, J. B., 132
Mattout, J., 108–109
Mauguiére, F., 94,
Mayberry, R. I., 334
Maynard, I., 171–172
Mayston, M. J., 287
Mazza, P. P. A., 201
Mazzoni, P., 154, 183
McBride, J., 75–76
McCabe, C. S., 117–118
McCandliss, B. D., 331
McCarroll, H. R., 3, 120
McDaniel, K. D., 171
McDonnell, P. M., 262
McGonigle, D. J., 122
McGraw, M. B., 72, 73, 257–259
McGrew, W. C., 194, 197, 205, 314
McIntyre, S., 328
McLean, D. A., 210
McLennan, J. E., 181–182
McManus, I. C., 306, 309–310, 318
McNeill, D., 333–335
McPherron, S. P., 195, 238–239
Meador, K. J., 297
Mechelli, A., 251
Medina, J., 28, 30, 299–300, 302
Medina, S., 141
Medland, S. E., 319
Meguerditchian, A., 315–316, 341
Mehring, C., 3, 120, 273
Meltzoff, A. N., 271, 322
Melzack, R., 109–110
Menichelli, A., 162

Mercader, J., 205
Merchant, S. H. I., 175
Mercier, C., 108
Merrick, C. M., 192, 285
Merton, P. A., 47, 63, 281
Merzenich, M. M., 16–17, 20, 25, 30, 248–250
Messina, C., 106
Meyerbeer, G., 145
Miall, R. C., 36
Michel, G. F., 319
Midgette, A., 170
Miklósi, Á., 326
Miles, L., 340
Miller, G. A., 45
Miller, L. E., 32, 41–42, 215–216, 220
Milner, A. D, 27, 61, 80, 82–83, 94, 139, 152
Milner, B., 152
Mishkin, M., 27, 61, 139
Mitani, J., 325, 339
Mitchell, S. W., 107
Mitchison, G., 273
Mithen, S., 194–195
Mittra, E. S., 237
Moffett, A., 97
Møller-Christensen, V., 106
Molyneux, W., 98–99
Monaco, S., 102
Moore, C., 324
Mooshagian, E., 285–286
Mora, L., 36
Mora, R., 198
Morgagni, G., 280
Moro, V., 120
Morrel-Samuels, P., 334
Morton, H. B., 281
Morton, S. M., 290
Moscovitch, M., 39
Moseley, G. L., 118
Mosquera, M., 316
Moss, A. D., 114
Mott, F. W., 15, 300–301
Mountcastle, V. B., xii, 20, 22–24, 43, 57, 61–62, 64, 285

Moyà-Solà, S., 237
Mruczek, R. E. B., 209–210, 251
Muir, R. B., 51
Mukamel, R., 290
Müller, F., 282
Müller, K., 276
Mulliken, G. H., 90
Mumford, K. H., 337–338
Murata, A., 64–65
Muret, D., 303
Murphy, W. F., 261
Musgrave, J. H., 241
Mushiake, H., 77
Myers, R. E., 303
Mylander, C., 329
Myowa-Yamakoshi, M., 261

Nagarajan, S. S., 303
Nakajima, K., 245–246
Nakamichi, M., 260
Nakamura, T., 222
Napier, J. R., x, xii, 3–4, 7–9, 231–234, 236–238, 242, 346
Nashed, J. Y., 86
Nathan, P. W., 300–301
Needham, A., 265
Negrotti, A., 184
Nelson, E. L., 315
Nelson, R. J., 250
Neubauer, S., 244, 254
Neville, H. J., 330
Newell, K. M., 267
Newport, R., 119, 124
Newton, I., ix
Ni, X., 231
Nicoladis, E., 342
Niekus, M. J. L., 200–201
Nielsen, J. M., 114, 116
Niewoehner, W. A., 241
Nightingale, S., 114
Nirkko, A. C., 284–285
Nishimura, Y., 245
Noachtar, S., 295

Nonaka, T., 205
Norman, D., 164–165
North, N., 135
Nuccio, J., 332
Nudo, R. J., 226, 245

Obayashi, S., 208
Obersteiner, H., 298–299
Ochoa, J., 29
Ocklenburg, S., 319
Ogawa, K., 158
Ohsaki, K., 273–274
Ohtake, H., 103
Oliveri, M., 297–298
Olivier, A., 52
Olivier, E., 276
Omrani, M., 90
Onimus, E., 167
Orban, G. A., 209, 251
Orgass, B., 29
Osiurak, F., 190, 207, 211
Ossmy, O., 267, 290
Ostry, D. J., 290
Osumi, M., 118–119
Otsuru, N., 118
Ottoni, E. B., 194, 233, 244
Over, R., 101
Overduin, S. A., 50
Owen, D., 171–172
Owen, R., 2, 221
Oxnard, C. E., 243–244

Padberg, J., 21, 234, 250
Paillard, J., 32
Pammi, V. S. C., 151–152
Pan, B. A., 322
Papadatou-Pastou, M., 311
Paré, A., 106
Pargeter, J., 200, 208
Parins-Fukuchi, C., 234–235
Parkinson, J., 177–178
Parlitz, D., 156
Parlow, S. E., 291

Pascual-Leone, A., 30–31, 148, 153
Passingham, R. E., 77, 134, 184, 242, 248, 254, 275
Patané, I., 134
Paton, R. L., 3, 120, 222
Patterson, F. G., 340
Patterson, K., 162
Patterson, N., 234–235
Paul, R. L., 23
Paulignan, Y., 84
Pause, M., 97
Pavani, F., 133
Pearce, J. M. S., 168
Pearcey, G. E. P., 290
Peckre, L., 260
Pedrazzini, A., 147
Peeters, R. R., 209–210, 251
Pei, Y.-C., 24
Peignot, P., 325
Pelgrims, B., 211
Pellijeff, A., 34
Penfield, W., 13–16, 21, 41, 58–59, 63, 281, 344
Pereira-Pedro, A. S., 254
Perelle, I. B., 306, 308
Perenin, M.-T., 78–81
Perez, M. A., 286, 291–292
Perris, E. E., 264
Perry, C. J., 136
Pesaran, B., 68
Petitto, L. A., 329
Petkova, V. I., 301–302
Petrides, M., 255
Petrosini, L., 21
Peviani, V., 37
Phillips, K. A., 315
Pia, L., 123–124
Piaget, J., 277, 335, 338
Piazza, M., 251
Picard, N., 148, 254
Pickering, T. R., 239
Piedimonte, A., 113
Pietrini, P., 103

Author Index

Pika, S., 325, 339
Pike, A. A., 272
Pilbeam, D., 243
Pisella, L., 83–84
Pitres, A., 281
Pitzalis, S., 69
Planton, S., 160
Plato, 307
Platz, T., 95
Playford, E. D., 77, 184
Pliny the Elder, 3, 106
Plummer, T. W., 195
Poeck, K., 29, 109
Poizner, H., 330
Ponce de León, M. S., 254, 256
Pons, T. P., 27, 30, 250
Pontier, D., 308
Pontzer, H., 283
Pool, E.-M., 312
Popelka, G. R., 337
Popova, T. S., 156–157
Porac, C., 307
Poranen, A., 24–25
Porro, C. A., 31
Porter, R., 226
Posnansky, M., 307
Postma, P., 301–302
Potts, R., 197
Pourrier, S. D., 299
Povinelli, D. J., 218–219, 346
Poza-Rey, E. M., 255
Prado, J., 69, 78
Prang, T. C., 237–238
Pratt, R. T. C., 114
Prechtl, H. F. R., 261–262
Press, C., 141
Preston, C., 119
Preston, J. B., 53, 246
Preuss, T. M., 254–255
Previc, F. H., 319
Price, C., 331
Proffitt, T., 205
Proske, U., 35

Pruszynski, J. A., 85–86, 90
Pubols, B. H., 249
Pubols, L. M., 249
Puts, N. A. J., 142
Putt, S. S., 207

Quallo, M. M., 51, 208
Quartarone, A., 174
Quine, W. V. O., 321

Rabbitt, P., 87
Rabin, E., 165
Rader, N., 262
Radinsky, L., 22, 247–248
Raffin, E., 108
Ragert, P., 159
Rakic, P., 229
Ramachandran, V. S., 107–109, 283, 299
Randerath, J., 340
Randolph, M., 26
Ranhorn, K., 201
Raos, V., 21
Rapp, B., 29–30, 162–163, 299–300, 302
Rasmussen, D. D., 143
Rasmussen, T., 13–15, 59, 186, 282
Ratcheson, R., 283
Ratcliff, G., 78
Rathelot, J.-A., 54–55, 62, 90
Rauscher, F. H., 335
Raymond, M., 307–308
Reales, J. M., 102
Reed, C. L., 94–97, 104, 137, 139
Reed, C. M., 332
Reed, J. L., 22–23
Regaiolli, B., 315
Reich, L., 331
Reichenbach, A., 89
Reilly, K. T., 108
Reissland, N., 261
Reitan, R. M., 303
Remple, F. E., 225, 227
Remy, P., 297
Révész, G., 101

Rheingold, H. L., 322
Ricci, R., 300, 302
Richmond, B. G., 4, 236
Richter, C. P., 259
Ricklan, D. E., 237, 240
Riddoch, G., 78, 107, 122, 345
Riddoch, M. J., 76, 103
Riedel, J., 326
Rijntjes, M., 47, 163
Rilling, J. K., 234
Rimé, B., 335, 337
Riseborough, M. G., 337
Rivers, W. H. R., 11–12
Rizzolatti, G., 64–66, 128–130, 134, 255
Ro, T., 33
Robertson, I. H., 135
Robertson, L. C., 135–136
Robinson, L., 258, 260
Rochat, P., 262
Roche, H., 205
Rock, I., 99
Röder, B., 40–41
Rodgers, B., 87
Roediger, H. L., 101
Roel Lesur, M., 119
Roffman, I., 206
Rogers, M. J., 206
Rohde, M., 111
Rohlfing, K. J., 323
Roland, P. E., 27–28, 95
Rolian, C., 198, 204, 234, 239
Romano, D., 116–117, 220
Romo, R., 24
Ronchi, R., 115–117
Rondot, P., 78
Rönnqvist, L., 267
Rosenblum, S., 181
Rosenfield, D. B., 161
Rosenzopf, H., 192
Ross, C., 260
Ross, P., 327
Rossetti, Y., 32, 80, 89
Ross Russell, R. W., 78

Rothi, L. J. G., 188
Rothwell, J. C., 34, 86
Roux, F.-E., 15, 163–164
Rowe, M. L., 322
Rowe, T. B., 225
Rubel, E. W., 269
Ruck, L., 307, 316–317
Rumelhart, D. E., 164–165
Russell, J. R., 303
Russo, A. A., 59
Rutz, C., 194

Saadah, E. S. M., 109–110
Saadon-Grosman, N., 21
Saal, H. P., 24, 215
Sabater-Pi, J., 325
Saby, J. N., 271
Sacrey, L. R., 261
Sadato, N., 100, 331
Sadnicka, A., 151, 168, 172, 174, 176
Saetta, G., 108, 117
Sahnouni, M., 196
Sainburg, R. L., 290
Sakai, K. L., 330
Sakata, H., 26, 34, 64, 76
Sakurada, M., 159
Saleh, M., 50
Salmaso, D., 309
Sambo, C. F., 143
Sánchez-Panchuelo, R. M., 16, 20, 24, 250
Sand, A., 223
Sandler, W., 329
Sanes, J. N., 53
Sansavini, A., 334
Santello, M., 49–50
Sasaki, S., 245
Sathian, K., 97, 184, 299–300, 303
Savage-Rumbaugh, S., 325, 340
Saxe, R., 251
Scaliti, E., 108
Schacter, D. L., 101
Schady, W. J. L., 29
Schäfer, E. A., 14, 53–54

Schaffelhofer, S., 66
Schauble, B. S., 115
Schellekens, W., 25, 249
Schendel, K., 135–136
Schenker, N. M., 255, 341
Scherberger, H., 66–67, 69
Scherer, R., 285
Schettino, L. F., 184
Schick, K. D., 196–197, 199, 206
Schieber, M. H., 53
Schilder, P., 37–38, 115–116, 299–300
Schlereth, T., 20
Schluter, N. D., 284
Schmidt, P., 200–201
Schmitt, D., 236
Schneiberg, S., 263
Schneider, I., 222
Schoenemann, P. T., 242, 248, 254
Schoenfeld, M. R., 147
Schott, G. D., 287–288
Schwartz, A. S., 16, 301
Schwarz, B., 145–147
Schweisfurth, M. A., 313
Schweizer, R., 30
Schwenkreis, P., 144
Schwoebel, J., 28
Scocchia, L., 141
Scott, S. H., 87–88
Scripture, E. W., 289–290
Sebeok, T. A., 342
Sechzer, J. A., 303
Seelke, A. M. H., 21, 269
Seidenstein, S., 227
Semaw, S., 196, 206
Semmes, J., 26, 95–97
Senghas, A., 329
Senut, B., 237
Sereno, M. I., 16, 20, 230
Serino, A., 141
Seyal, M., 31, 298
Seyffarth, H., 72, 260
Shackley, M. S., 202–203
Shadmehr, R., 87

Shafer, D. D., 316
Shafritz, A. B., 91
Shallice, T., 73, 161
Shamim, E. A., 170
Shea, J. J., 198, 200
Sheehy, M. P., 169
Sherrick, C. E., 298
Sherrington, C. S., 13–14, 51, 54, 186, 281
Sherwood, C. C., 244, 315
Shimada, S., 111
Shimojo, S., 228
Shipman, P., 197
Shoham, D., 25
Shrewsbury, M. M., 239
Shubin, N. H., 2, 222–224
Shumaker, R. W., 194
Silver, M. A., 12
Simmel, M. L., 109
Simonyan, K., 175, 179
Singer, M. A., 339
Sinha, N., 183
Sirianni, G., 194
Sirigu, A., 15, 63, 103
Siuda-Krzywicka, K., 331
Skinner, M. M., 204, 240
Skoyles, J. R., 109
Sloan, T. B., 143
Smaers, J. B., 244
Smetacek, V., 288
Smit, M., 116
Smith, G. E., 252
Smith, K. U., 303
Smith, T. L., 289–290
Snow, J. C., 100, 103–104
Snyder, L. H., 67–68
So, E. L., 115
So, W. C., 336
Soechting, J. F., 165
Sohn, Y. H., 174, 287
Soligo, C., 244
Solly, S., 168
Solvi, C., 98
Sommer, R., 127

Song, J.-H., 137
Soto-Faraco, S., 40
Southey, R., 105
Sparling, J. W., 261
Spatt, J., 190
Spelke, E. S., 264
Spence, C., 40, 132–133
Spencer, J. P., 263
Spencer, W. A., 22, 143, 298
Sperry, R. W., 272, 282, 296, 303
Spinazzola, L., 113
Spinozzi, G., 316
Sposito, A., 220
Sretavan, D., 23
Srour, M., 287
Stack, D. M., 264
Stamelou, M., 172
St. Clair, J. J. H., 194
Stamm, J. S., 303
Stankevicius, A., 107
Starkovich, B. M., 197
Starkstein, S. E., 114
Stepniewska, I., 66, 246, 250
Stern, J. D., 262
Sterr, A., 30
Sterzi, R., 35
Stokoe, W. C., 329
Stoll, S. E. M., 192, 312
Stout, D., 205, 207
Strait, D. S., 236
Straus, W. L., 236
Strick, P. L., 51, 53–55, 69, 90, 179, 234, 244–246, 254–255
Striedter, G. F., 225–226, 228, 242–243
Striemer, C. L., 85
Ströckens, F., 313
Summers, D. C., 313
Sun, T., 319
Sur, M., 17–19, 23, 249, 270
Suresh, A. K., 60
Susman, R. L., 5, 204, 237–240
Suzuki, K., 101
Szalay, F. S., 232

Taglialatela, J. P., 341
Tajadura-Jiménez, A., 41–42
Takahashi, K., 69
Takei, T., 90
Takeshita, H., 261
Tal, Z., 102
Tamburin, S., 173
Tamè, L., 28, 44, 172, 294–295, 298, 303
Tanaka, S., 292
Tanaka, Y., 73
Tang, R., 217
Tanji, J., 24, 66, 90, 283–284
Tapley, S. M., 310
Tastevin, J., 38
Taub, E., 290
Tavare, S., 230
Taylor, H. G., 291
Taylor, J. A., 290
Taylor-Clarke, M., 41, 143, 220
Tazoe, T., 286
Tegnér, R., 113
Teng, E. L., 309
Teuber, H.-L., 95
Thaler, D., 77
Thelen, E., 258, 263, 323
Thibault, S., 342
Thieme, H., 195
Thomas, L. E., 139
Tian, M., 331
Tinazzi, M., 143, 174, 286
Tipper, S. P., 141
Tobias, P. V., 255
Tocheri, M. W., 236, 239
Todd, J. J., 251
Todorov, E., 87
Tohyama, T., 245
Tomasello, M., 324–326
Topham, J. R., 1
Torebjörk, E., 29
Torigoe, T., 233, 249
Torres, F., 30–31, 148
Tosi, G., 111
Toth, N., 196–199, 206, 316–317

Touwen, B., 265
Traub, M. M., 86–88
Trevarthen, C., 324
Trillenberg, P., 79
Trinkaus, E., 106, 241
Trut, L. N., 326
Tryon, C. A., 201
Tsakiris, M., 110–112, 120
Tseng, P., 139
Tucker, M., 74–75
Tulving, E., 101
Tunik, E., 66, 89
Turella, L., 62–63
Turnbull, O. H., 114
Turvey, M. T., 214
Tuttle, R. H., 5, 234, 239
Twitchell, T. E., 258–260, 265, 267
Tzourio-Mazoyer, N., 311–312

Umiker-Sebeok, J., 342
Umiltà, M. A., 216, 218
Ungerleider, L., 27, 61, 139
Uno, Y., 88
Uomini, N. T., 307, 316–317

Vail, A. L., 326
Vaina, L. M., 211
Valenza, N., 97–98
Valenzeno, L., 338
Vallar, G., 35, 115–117, 297
Vallbo, Å. B., 5
Valsalva, A., 280
Valyear, K. F., 209–210
Van Boven, R. W., 31
van den Heuvel, M. P., 271
Van der Loos, H., 21, 273
van der Meer, A. L. H., 262–263
Vanderpool, D., 230–231
Van Essen, D. C., 251
Van Galen, G. P., 160
Vasari, G., 194
Vauclair, J., 314–316
Vaught, G. M., 214

Veà, J. J., 325
Vecera, S. P., 139
Vega-Bermudez, F., 20, 249, 312
Venditti, F., 199
Verhagen, L., 102
Veronelli, L., 95
Verstynen, T., 192, 284–285, 311
Vetter, R. J., 109
Victor, J., 99
Vighetto, A., 78–80
Villemeur, I., 241
Vingerhoets, G., 341
Virányi, Z., 326
Vocat, R., 113
Volkmann, A. W., 302
von Bayern, A. M. P., 194
von Hofsten, C., 262–264, 267
von Reis, G., 169
Vuilleumier, P., 121–122, 130

Wada, J., 186, 281–282, 297
Wade, D. T., 123
Wadley, L., 201, 207
Wagstyl, K., 24
Wakefield, E. M., 337
Walker, A. C., 233
Wallace, P. S., 276–277
Walshe, F. M. R., 345
Wang, J., 290
Ward, C. V., 237, 239, 242
Ward, J. P., 314
Warren, W. C., 227
Warrington, E. K., 161–162
Weaver, D. S., 106
Weber, E. H., 41–43
Weder, B. J., 185
Weidenreich, F., 252
Weidler, B. J., 140
Weinstein, E. A., 121
Weinstein, S., 109
Weir, A. A. S., 194
Weiskrantz, L., 32, 81–82, 135
Weiss, E. J., 49–50

Welker, E., 273
Welker, W. I., 227
Wenger, E., 153–154
Wernicke, K., 94, 186, 188
Wesselink, D. B., 108
Whishaw, I. Q., 227, 276–277
White, B. L., 263
White, R. C., 29, 31
White, T. D., 236–237
Whitehead, K., 272
Whiten, A., 194
Whitwell, R. L., 81–82
Wiesel, T. N., 20, 272
Wiestler, T., 148–149
Wilkins, J., 200
Willett, F. R., 285
Williams, R., 223
Williams-Hatala, E. M., 198–199, 204
Willingham, D. B., 153
Wimmers, R. H., 265
Wing, A. M., 162–163
Wise, S. P., 24, 254, 291
Witham, C. L., 55
Witherington, D. C., 267
Wittgenstein, L., 45, 346
Wolbers, T., 103
Wolff, P., 337
Wolpert, D. M., 34, 87, 90, 113, 218
Wong, P., 242, 249
Wood Jones, F., x, xii, 9, 231–234
Woodward, A. L., 322–323
Woodworth, R. S., 60
Woolsey, T. A., 21, 273
Wright, R. V. S., 206
Wu, C. W., 254–255
Wu, T., 181
Wullstein, K. L., 202
Wyke, M. A., 287
Wymbs, N. F., 149–150, 152, 342
Wynn, T., 199, 205
Wysocki, C. J., 309, 311

Xitco, M. J., 326

Yamada, K., 260
Yamamoto, S., 39–40, 214–215
Yan, Y., 50
Yeo, G., 13, 186
Yeo, S. S., 276
Young, R. R., 299
Yousry, T. A., 52, 248

Zaidel, E., 296
Zaki, M. E., 106
Zanette, G., 286
Zangaladze, A., 100, 303
Zangwill, O. L., 117
Zaporozhets, A. V., 91–94
Zeharia, N., 21
Zehr, E. P., 292
Zeller, D., 120
Zhang, X., 106
Zhao, D., 315
Zimmerman, M. R., 106
Zingerle, H., 114
Zoia, S., 261
Zollikofer, C. P. E., 254
Zopf, R., 141
Zuberbühler, K., 325, 339

Subject Index

Page numbers in italics refer to figures.

Abductor digiti minimi (ADM) muscle, 286
Acanthostega, 223–*224*
Acheulean industry, 195, 198–*199*, 208–208, 242, 316–317
Adaptation
    fMRI, 135, 294–295
    MEG, 295
    slow vs. rapid, 6–7, 23
    tactile distance, 42–43
    visuomotor, 86–87, 217, 290–291
Aegyptopithecus, *231*, 248
Affordances, 71–72, 74–76, 267
Affordance competition hypothesis, 76
Agnosia
    astereognosis, 94–95
    multimodal, 103–104
    for object use, 189
    tactile, 91, 94–97
    visual object, 78, 82–83
Agraphia, 160–164
    and Exner's area, 163–164
    ideational, 161–*162*
    lexical, 161
    phonological, 161
Akinesia (in Parkinson's disease), 178
Allochiria, 15, 298–302
Allometry (in brain evolution), 228, 242–244, *243*, 254

American Sign Language (ASL). *See* Sign language
Amnesia, 152
Amputation, antiquity of, 106–107
Anarchic hand sign, 72–73, 76, 122, 260
Anisotropy
    of receptive field (RF) geometry, 20, 43–44
    of tactile distance perception (*see* Tactile distance perception, anisotropy of)
Anosognosia
    for hemianesthesia, 113
    for hemiplegia, 113–114, 120, 122, 123, 186, 283
Anterior intraparietal area (AIP), 64–69, *65*, 76, 102, 134, 137, 179, 209, 251. *See also* Grasping network
Anterior supramarginal gyrus (aSMG), 209, *210*, 251
Anticipatory control, 261, 267
Anvil technique, 196–197
Ape. *See* Bonobo, Chimpanzee, Gibbon, Gorilla, and Orangutan
Aphasia, 160, 186–187, 280–281, 329–330, 335, 340
    Broca's, 160, 186–187, 280–281, 329–330, 335, 340
    sign language, 329–330
    Wernicke's, 186–187, 329–330, 335

Apraxia, 63, 72–74, 97–98, 162, 167, 186–192, 207, 285, 312, 340
   ideational, 63, 189–191
   ideomotor, 187–191, 346
   left hemisphere and, 191–192, 285, 312, 340
   limb–kinetic, 188
   magnetic, 72–74
   repellent, 72–74
   tactile, 97–98
   taxonomy of, 187–191
   tool use in, 188–190, 207, 211
Arachnodactyly ("spider hand"), 147
Archetype
   of mammalian cerebral cortex, 225
   of vertebrate limb, 2–3, 221, 225
Ardi. See *Ardipithecus ramidus*
Ardipithecus ramidus, 235–239
Area duplication hypothesis, 229–230
Area F5. *See* Premotor cortex, ventral (vPMC)
Area V6a, 68–69, 134. *See also* Parietal reach region (PRR)
Aristotle illusion, 37–38
Armour, Tommy, 171, 345
Arm-projection neurons, 61, 64
Asomatognosia, 114–115, 120
Associated movements. *See* Mirror movements
Astereognosis. *See* Agnosia, astereognosis
Asymmetric tonic neck reflex, 262–263
Atapuerca, Spain. *See* Sierra de Atapuerca
Ataxia. *See* Optic ataxia
Atopognosia. *See* Tactile localization, deficits in
Attentional wand, 136–138
Aurignacian Industry, 196, 201–202
Australopithecus afarensis, 195, 235, 239, 255
Australopithecus africanus, 204, 235, 237, 239–240, 252–253, 255
Australopithecus anamensis, 236
Australopithecus sediba, 235, 237, 240, 255
Autism, 322
Automatic pilot, 83–85
Automatization, and development of expertise, 152–154
Awar, Ethiopia, 236

Babbling. *See* Hand babbling
Bálint's syndrome. *See* Optic ataxia
Barrel cortex, 21, 229–230, 273–274
Basal ganglia, 21, 63, 64, 76, 149, 151–152, 175, 177, 179–181, *180*, 185–186, 270, 342
Bat, 2, 221, 243, 326
Beat gesture, 334
Beckoning gestures (in chimpanzees), 325, 339
Beni Hasan (ancient Egyptian tomb), 307
Bilateral access theory, 291–292
Bilateral representation
   in motor system, 279–293, *284*
   in somatosensory system, 293–303, *294*
Bilateral symmetry, 279, 305–306
Bipedalism, 234, 236, 238
Bipolar technique, 196
Bird, 3, 194, 222, 225, 326
Blass, Steve, 172. *See also* Dystonia and Yips
Blindness, 32, 40–41, 81–82, 91, 98–99, 100, 135–136, 141, 214, 225, 330–333
Blindsight, 32, 81–82, 135–136
Bobbin lace, 155
Body integrity dysphoria, 117
Body representations, 28–44, 219–220
   and somatosensory function, 28–44
   and tool use, 219–220
Body schema. *See* Postural schema
Bonobo, 206, 234, 236, 248, 316, 325, 339–341
   gesture, 325, 339
   handedness, 316, 341
   pointing, 325
   sign language, 340
   tool use, 206
Brachiation, 233–234, 236–238
Bradykinesia (in Parkinson's disease), 178, 180, 181, 183–184
Braille reading, 30, 100, 148, 330–332
Brain size (in primate and hominin evolution), 242–244, *243*
Brake-accelerator model, 179–*180*
Bridgewater Treatises, 1
British Sign Language (BSL). *See* Sign language
Broca's aphasia. *See* Aphasia, Broca's

Broca's area, 152, 164, 207, 254–256, 330, 341
Bumble bees, 98
Butchery (with stone tools), 160, 197, 199, 202

C. Horatius, 3
Canonical visuomotor neurons, 64, 76
Carpal tunnel syndrome, 143
Carpolestes simpsoni, *231*, 232
Casineria, 3, 222
Castorocauda, 225
Cat, 3, 13, 20, 22–23, 85, 143, 227, 245, 269, 303
Cave of the Hands. *See* Cueva de las Manos
Cave paintings, 201, 305–*308*, *306*, 316, 342
Central sulcus, 12–13, *18*, 22, 52–55, *53*, 58, 247–248, 294, 311
Cerebellum, 21, 64, 90, 149, 175–176, 195, 218, 230, 244
Cerebral palsy, 287
Chantek (orangutan), 325–326, 340
Chimpanzee, 13, 193–195, 197, 204–206, 218–219, 222, 230–*231*, 234–238, 247–248, 252–253, 255, 303, 314–316, 325–326, 339–341
   gestures, 339–341
   handedness, 314–316, 341
   pointing, 325–326
   tool use, 193–195, 197, 204–206, 218–219
Chimpanzee Stone Age, 205
Chopper (stone tool), 197
Chopsticks, 160, 309
Chunking. *See* Motor chunking
Circle-marking task, 310
Circular reaction, 277
Coelacanth, 223–*224*
Columnal organization. *See* Cortical columns
Computed tomography (CT), 131
Concatenation (in motor chunking), 152
Conservation (Piagetian), 335–336, 338
Contralateral control of motor function, 280–282
Contralateral transfer
   of motor learning (*see* Cross–education)
   of tactile learning, 302–303

Control policy, 86–88, 90
Convergence, 232
Corpus callosum, 133, 191, 282–283, 287–288, 292–293, 295–296, 303, 312
Cortical columns, 22–24, *23*, 57
Cortical magnification, 15, 17–21, *19*, 25, 41, 52–53, 227–228, 247–249, 271
Corticolimbic pathway, 27, 97
Corticomotoneuronal (CM) projections, 50–52, 54–55, 90, 208, 226, 244–*246*, 265–267, 274–277
   development of, 265–267, 274–277, *275*
   Old M1 vs. New M1, 54–55
   and precision grips, 51, 226, 244–245, *275*
   and tool use, 51, 208
Corticospinal tract (pyramidal tract), 45, 50–51, 226, 245–246, 275–276, 281, 319
Cross–activation theory, 291–292
Crossed hands deficit, 39–41, 214–215
Cross–education, 289–293, *290*, 302
Crossmodal congruency effect (CCE), 132–134
Cueva de las Manos (Cave of the Hands), 305–307, *306*, 309
Cutting (with stone tools), 197–199, 204, 317

Deictic gesture, 333. *See also* Pointing
Denisovan, 241, 244
Der Schwanendreher (the swan turner), 155
Development, 257–277, 318–319
   of corticomotoneuronal (CM) projections, 274–277, *275*
   of grasping, 265–268, *266*
   of handedness, 318–319
   of mirror movements, 287
   prenatal, 261, 319
   prereaching, 262–263
   of reaching, 263–265
   of somatotopic maps, 268–274, *270*
Diagnostic dyspraxia, 73, 122
Diffusion tensor imaging (DTI), 207, 276
Diplesthesia. *See* Aristotle illusion
Directed scratches (in chimpanzees), 325, 339

Disembodiment, experimental induction of, 117–119, *118*
Distal type neurons. *See* Pericutaneous neurons
Distant peripersonal neurons, 128–129
Disturbed sensation of limb ownership, 114–117
Dmanisi, Georgia, 256
Dog, 13, 227, 281, 326–327
Dolphin, 2, 221, 313, 326
Dopamine, 175, 177, 179–180, 184
Doppler sonography, 340–341
Dorsal visual stream, 27, 61, 64, 81–83, 85, 131, 139–140, 251
Dorsolateral pathway. *See* Grasping network
Dorsomedial pathway. *See* Reaching network
Double consciousness, 153
Double-stopping technique, 155–*156*
Dual mode principle, 153
Duck-billed platypus, 227, 248
du Pré, Jacqueline, ix
Dystonia, 143, 167–177, *169*, 179–180, 287, 345–346
   mechanisms underlying, 172–177
   and motor chunking, 176
   in musicians (*see* Musician's dystonia)
   in sports (*see* Yips)
   Telegraphists' cramp, 167
   of writing (*see* Writer's cramp)

Earl of Bridgewater, 1
Egerton, Francis Henry. *See* Earl of Bridgewater
Eglon (King of Moab), 307
Ehud, 307
Electroencephalography (EEG), 32, 67, 141, 143–144, 159, 173–174, 215–216, 268, 271–272, 323
Electromyography (EMG), 5, 108, 110, 113, 157, 169, 181, 203, 276, 287
Elgar *Cello Concerto*, ix
Emotion (communication by hand), 327–329
Estadio Azteca, ix
Ethological action maps, 50, 58–59, 261
Eusthenopteron, 224

Exner's area, 163–164
Experimental finger dyspraxia, 117
Expertise, 145–165
   automatic vs. effortful processing, 152–154
   musical (*see* Musical expertise)
   principles of motor skill acquisition, 147–149
   timescales of motor skill acquisition, 149–150
   typing (*see* Typing)
   writing (*see* Writing)
Extinction, 123, 132, 212–213, 296–298
   multimodal, 132, 297
   tactile, 296–298
   and tool use, 212–213
Extrastriate body area, 103

Fencer's reflex, 262–263
Fetal arm movements, 261, 319
Fish, 3, 13, 73, 221–224, 326
Fish fins, 222–*224*
   genetic similarities to limbs, 222
   sensory functions, 223
Fitts's law, 108
Flanker task, 137
Fleisher, Leon, 170, 176
Flexor pollicis longus (FPL), 4–5, 204, 237, 240
Foreign hand. *See* Main étrangère
Forward model. *See* Internal models
Fossey, Dian, 193
Foveal vision, 68–69, 78
Fox, 326–327
Fox, Michael J., 177–178
Frontal eye fields (FEF), 130, 134
Frontal operculum. *See* Broca's area
Fronto-orbital sulcus, 255
Functional magnetic resonance imaging (fMRI). *See* Magnetic resonance imaging (MRI)
Functional transcranial Doppler sonography. *See* Doppler sonography
Furze Platt, England, 307
Fusiform face area, 103

Galago. *See* Prosimian
Galdikas, Birutė, 193

Subject Index

Gamma–aminobutyric acid (GABA), 64, 142–144
Gargas cave, France, 342
Gavagai, 321–322, 333
Geometric illusions, 101
Gesture, 332–342
   in blind people, 333
   classification of, 333–334
   communicative effectiveness, 337–339
   integration with speech, 334–335
   and language evolution, 339–342
   and learning, 337–339
   as reflection of thought, 335–337
Gibbon, 3, 4–5, *231*, 325
Giotto, ix, 194
Globularity (of human brain), 253–254
Globus pallidus, 179–*180*, 186. *See also* Basal ganglia
   external segment (GP_e), 179–*180*
   internal segment (GP_i), 179–*180*
Goldberg Variations, ix
Gombe Stream Chimpanzee Reserve, Tanzania, 193, 339
Gona, Ethiopia, 196–197, 205–206
Gorilla, 13, *54*, 193–194, *231*, 235–239, *238*, 248, 315–316, 325, 339–341
   gestures, 339–340
   handedness, *315*–316, 341
   pointing, 325
   tool use, 194
Gould, Glenn, ix
Gran Dolina cave, Atapuerca, Spain, 241
Grasp phase (of reaching), 46, 60–61
Grasping. *See* Power grip and precision grip
Grasping network, 64–67, *65*, 69, 250. *See also* Anterior intraparietal area (AIP) and Premotor cortex (PMC), ventral (vPMC)
Grasp reflex, 72, 257–260, *259*, 267
Great ape. *See* Bonobo, Chimpanzee, Gorilla, and Orangutan
Greybeard, David, 193
Grips
   cradle, 9, 202–*203*
   power (*see* Power grip)
   precision (*see* Precision grip)
   three-jaw chuck, 202–*203*
   two-jaw pad-to-side, 202–*203*

Hadar, Ethiopia, 239
Hafted tools, 200–201, 207, 241, 341–342
Hand-assisted blindsight, 135–136
Hand axe, 195, 198–200, *199*, 206, 208, 242, 307. *See also* Acheulean industry
Hand babbling, 277, 329
Handedness, 305–319
   in animals, 313–316
   antiquity of, 305–*308*, *306*, 316–318
   consistency in human populations, 306–309
   development of, 318–319
   evolution of, 313–318
   in fetal behavior, 319
   genetic models of, 318–319
   hand skill, 309–311, *310*
   mixed, 310–311
   neural bases, 311–312
   preference vs. skill, 308–*310*
   for somatosensory function, 312–313
Hand-manipulation neurons, 61–62, 64, 285
Hand paintings (in caves), 305–*308*, *306*, 309, 316, 342
Haptics, 92–94, 97–98, 100–103, 214–215
   common mechanisms with vision, 100–103
   exploratory procedures, 93–*95*, 97–98
   and tool use, 214–215
Hard-hammer percussion, 196
Hemianesthesia, 28, 113, 186
Hemiparesis, 113–114, 186–188, 191, 282, 288, 293
Hemiplegia, 113–114, 123, 125, 130, 186–188, 191, 281, 283, 287–288, 297, 346
Hemispheric rivalry, 297–298
Hierarchical processing
   and hafted tools, 200–201, 207, 341–342
   in motor system, 45–46, 49, 84, 164, 188, 284, 341–342
   in somatosensory system, 24–28, 61, 272

Hindemith, Paul, 155, 170
Hogan, Ben, 171
Hominins. *See also Australopithecus,* Denisovan, *Homo, Kenyanthropus,* Neanderthal, and *Paranthropus*
  brains of, 242–244
  evolution of, 230–231, 234–244, *235*
  handedness, 316–318
  hands of, 234–242
  tool use, 160, 195–208
Homo antecessor, 235, 241, 317
Homo erectus, 195, 198, *235,* 252, 256
Homo habilis, 195, *235,* 237, 238, 242, 255, 317–318
Homo heidelbergensis, 195, *235,* 241, 316–317
Homology of vertebrate limbs. *See* Archetype of vertebrate limb
Homo naledi, 237, 240–241, 255, 344
Homo neanderthalensis. See Neanderthal
Homo rudolfensis, 235, 255
Homunculus. *See* Somatotopic maps
Horizontal-vertical illusion. *See* Geometric illusions
Horse, 2, 106, 107, 109, 197, 221, 326
Hox genes, 222, 229
Hypokinesia (in Parkinson's disease), 178, 183

Ichthyostega, 3
Iconic gesture, 333
Ideational agraphia. *See* Agraphia, ideational
Ideational apraxia. *See* Apraxia, ideational
Identification anosognosia, 123–124
Ideomotor apraxia. *See* Apraxia, ideomotor
Imagery. *See* Motor imagery and Proprioception, imagery
Indeterminacy of translation, 321
Inferior frontal gyrus, 207
Inhibition, 66, 142–144, 159, 174–175, 179, 286–287, 291, 298, 311–312
  interhemispheric, 286–287, 291, 298, 311–312
  lateral, 142–144
  surround, 287

Insight learning, 204
Instinctive grasp reaction, 259–260
Insula, 27–28, 97, 102, 120, 255, 270
Intention. *See* Motor intention
Intermediate-level representations, 150–152, *151*
Intermittently recursive mapping, 22
Internally generated movements, 76–77, 183–184
Internal models, 87–90, *88,* 113–114, 217–219
Intraparietal sulcus (IPS), 18, 47, 64, 68, 80, 110, 120, 128–129, 134–135, 152, 160, 209–211, 217, 251, 294. *See also* Anterior intraparietal area (AIP); Lateral intraparietal area (LIP); Medial intraparietal area (MIP); Ventral intraparietal area (VIP)
Inverse model. *See* Internal models
Ipsilateral representations
  in motor system, 66–67, 282–286
  in somatosensory system, 293–296

Joint, 3–4
  as category boundary, 42
  evolution of, 204, 232–233, 238–240
  hyper–flexibility (*see* Marfan syndrome)
  neurons sensitive to movement of, 23–26, 35, 50, 57–58, 90, 186, 293–294
  saddle, 4, 232, 233, 238
  universal, 3

Kaitio, Kenya, 242
Kanzi (bonobo)
  sign language, 340
  tool making, 206
Kathu Pan, South Africa, 200
Kenyanthropus platyops, 195
Klaviermusik mit Orchester (piano music with orchestra), 170
Klippel-Feil syndrome, 287
Knob. *See* Pli de passage fronto–pariétal moyen (PPFM)
Knoblauch, Chuck, 172
Knuckle walking, 234–237

Subject Index

Koko (gorilla), 340
Konso, Ethiopia, 198–*199*
Koobi Fora, Kenya, 202, 316–317
Krapina cave, Croatia, 317–318
Krukenberg procedure, 91–94

Language
   aphasia (*see* Aphasia)
   evolution, 339–342
   laterality, 340–341
   sign language (*see* Sign language)
Lateral intraparietal area (LIP), 67–68, 134
Lateral occipital complex (LOC), 100, 102, 104
Lateral occipital tactile–visual region, 102
Lateral reticular nucleus (LRN), 245–246
Later Stone Age, 201–202
Latimeria, 223–*224*
Ledi-Geraru, Ethiopia, 196
Le Moustier, France, 200
Lemur. *See* Prosimian
Le Pech-Merle, France, *308*
Lepidosiren, 221
Levallois technique, 196, 200–201, 207, 241
Limb-kinetic apraxia. *See* Apraxia, limb–kinetic
Limb preferences in animals, 313
Little Foot, 239
Loango National Park, Gabon, 195
Locomotion, 232–234
Lokalalei 2C, Kenya, 205
Lomekwi 3, 195–196
Lomekwian culture, 196
Long-latency stretch reflex, 85–90
Loris. *See* Prosimian
Lucy. See *Australopithecus afarensis*
Lunate sulcus, 252
Lungfish, 221, 223–*224*

M1. *See* Primary motor cortex (M1)
Macrogeometry, 27–28
Magdalenian Industry, 201–202
Magnetic apraxia, 72–74

Magnetic resonance imaging (MRI)
   functional (fMRI), 16, 20–21, 25, 31, 34–35, 44, 47–48, 50, 57, 62, 68–69, 76, 100, 102–104, 108, 110, 122–123, 125, 135, 137, 144, 148–149, 152–153, 158, 173, 174–175, 192, 209, *210*, 217–218, 249, 251–252, 270–273, 284–285, 292, 294–295, 311–313, 331, 341, 342
   structural, 24, 115, 207–208, 248, 251, 311, 315
Magnetic touch illusion, 134
Magnetoencephalography (MEG), 57–58, 147–148, 173, 295
Magnification. *See* Cortical magnification
Magnocellular system, and effect of hand proximity, 139–140
Main étrangère, 116
Malapa, South Africa, 240
Maltravieso cave, Spain, 316
Mammals (evolution of), 224–230
   brain archetype, 225
   sensorimotor cortex, 226–228
Maradona, Diego, ix
Marfan Syndrome, 147
Marrow extraction (with stone tools), 197–199, 204
Marsupials, 225–226, 232
Means-end reasoning, 45
Mechanoreceptors, 5–7
   Meissner corpuscle, 6, 23
   Merkel cell, 6, 23
   Pacinian corpuscle, 7, 215
   rapidly adapting type 1 (RA1), 6, 23
   rapidly adapting type 2 (RA2), 7
   Ruffini corpuscle, 6–7
   slowly adapting type 1 (SA1), 6, 23
   slowly adapting type 2 (SA2), 6
Medial intraparietal area (MIP), 68–69, 134.
   See also Parietal reach region (PRR)
Metaphoric gesture, 333–334
Michelangelo, ix
Microgeometry, 27–28
Micrographia, 181–*182*

Microliths, 201
Microneurography, 28–29
Middle Paleolithic. *See* Mousterian Industry and Levallois technique
Middle Stone Age. *See* Mousterian Industry and Levallois technique
Midline fusion, 293
Migration (of tactile sensation), 301–302
Minimum intervention principle, 89
MIRAGE (augmented reality system), 119
Mirror movements, 268, 288–289, 291, 311
Mirror writing, 48, 279, 288–*289*, 291
Misoplegia, 114, 122
Modality specificity, 22–24
Modes (of lithic technology), 195–196
Modularity. *See* Cortical columns
Module aggregation hypothesis, 228–229
Mole, 2, 225, 227–229, *228*, 248, 273
Molyneux question, 98–99
Monkey, 6, 13, 14, 16–*26*, *17*, *18*, *19*, 34–35, 50–51, 53–69, *54*, *58*, 73, 76–77, 80–81, 90, 97, 120, 128–130, *129*, 134, 136, 148, 172–*173*, 180, 194, 205, 208–209, 211–*212*, 216, 228, 230–237, *231*, 242–252, *243*, *246*, *247*, 254–255, 258–*259*, 264–265, 269–271, *270*, 274–276, *275*, 282–286, *284*, 293–*294*, 300–301, 303, *315*–316
  baboon, *231*, 248, *315*–316
  capuchin, 55, 69, 205, *231*, 233, 244–246, 249–251, 254–255, 315–316
  macaque, *18*, 25–*26*, 35, 51, 54, 62, *65*–68, 77, 90, 134, 186, 208–209, 211, 216, 228–229, *231*, 244–251, *246*, *247*, 254–255, *259*, 264, 269–270, 274–276, 283–284, *294*, 315
  marmoset, *231*, 249–251, 269
  New World, 16–17, 194, 230, *231*, 233–234, 242–244, 246–247, 249–251, 254, 269–270
  Old World, 194, 230, *231*, 233, 235–237, 242–244, 247–251, 254–255, 269–270
  owl, 16–*17*, *231*, 248
  squirrel, 20, 244–*246*, 249–250, 269–*270*
  snub–nosed, *231*, *315*
  titi, 250
Monotremes, 225–228, 248
Morganucodon, 225
Motoneuron, 51, 55, 62, 90, 226, 245, 274
Motor chunking, 150–152, *151*, 176, 341
Motor cortex. *See* Primary motor cortex (M1)
Motor equivalence, 46–*48*, 66–67, 78, 93, 125, 163, 170, 217, 286, 288, 303, 309
Motor evoked potential (MEP), 66, 153, 159, 173–175, 286–287, 290–291, 293
Motor imagery, 63, 108–109, 178, 283
Motor intention, 45, 61, 63, 67–68, 72–73, 77, 85–86, 90, 108, 113–114, 260, 265, 267, 324–325
Motor synergies. *See* Synergies
Mousterian industry, 196, 200–201, 206
Müller-Lyer illusion. *See* Geometric illusions
Multidimensional scaling (MDS), 44
Multivoxel pattern analysis (MVPA), 62, 69, 137, 148, 210–211
Muscimol, 64, 66, 81, 250
Muscles, 4–5
  accessory adductor pollicis, 5
  biceps brachii, 290
  extensor pollicis brevis, 4
  first volar interosseous muscle of Henle, 5
  flexor carpi radialis, 290
  flexor digitorum profundus, 4–5
  flexor pollicis longus (FLP), 4–5, 204, 237, 240
  intrinsic hand muscles, 4
Musical expertise, 155–160, *156*
  changes in somatosensation, 159–160
  dystonia, 170–171
  efficiency of movement, 156–157
  formation of new motor synergies, 157–158
Musician's dystonia, 170–171, 172, 174, 175. *See also* Dystonia

Subject Index

Neanderthal, 106, 160, 195, 198, 200, *235*, 240–241, 244, 254–255, 316–*318*, 341
Neck (evolution of), 223–224
Neglect, 39, *121*, 130–132, *131*, 135, 138, *213*, 297–298
   personal, 130
   selective for peripersonal vs. far space, 130–132, *131*, *213*
   tactile, 39
   and tool use, *213*
Nelson, Admiral Horatio, 105, 107
Neoceratodus, 224
New World monkey. *See* Monkey, New World
Nissl stains, 22
Novel tool test, 189–*190*
Numbsense, 32–33
Nutcracker man. *See Paranthropus boisei*

Olbermann, Keith, 172
Oldowan industry, 196–198
Olduvai Gorge, Tanzania, 195–200, 202–207, *238*, 242, 316, *318*
Old World monkey. *See* Monkey, Old World
Operation (board game), 290
Optic ataxia, 66, 78–81, 83, 85, 185
Optimal feedback control (OFC), 86–90, *88*
Optimal integration of vision and touch, 99–100
Orangutan, 13, 193–194, 206, *231*, 236, *315*–316, 325–326, 339–340
   handedness, *315*–316
   pointing, 325–326, 339–340
   tool use, 194, 206
Oreopithecus bambolii, 237
Orpheus, 145–*146*
Orrorin tugenensis, *231*, 237–238
Owl monkey. *See* Monkey, owl

Paganini, Niccolò, 145–147, *146*, 154
Paleolithic tools. *See* Stone tools
Paley, William, 1
Pan-Banisha (bonobo), 206
Panda's thumb, 3
Panderichthys, 223–224
Papillary ridges, 5
Parahippocampal place area, 103
Paranthropus, 195, *235*, 237, 242
Paranthropus boisei, *235*, 242
Paranthropus robustus, *235*, 237
Parietal lobe. *See* Posterior parietal cortex (PPC) and Somatosensory cortex
Parietal reach region (PRR), 67–68, 78–81. *See also* Area V6a and Medial intraparietal area (MIP)
Parkinson's disease, 177–186, *180*, *182*, *185*, 287
   and basal ganglia, 179–*180*
   clinical characteristics, 177–179
   deficits in hand function, 181–183, *182*
   deficits in internally generated movements, 183–184
   deficits in somatosensory function, 184–186, *185*
   mirror movements, 287
Parmastega, 223
Patient,
   A.H. (impaired tactile localization), 31
   A.P. (removal of right cerebral hemisphere), 282
   Armour, Tommy (yips), 171, 345
   A.T. (optic ataxia), 79–80
   A.Z. (phantoms of congenitally absent limbs), 110
   Barbara M. (synchiria), 300
   Blass, Steve (yips), 172
   Charles (sign language aphasia), 330
   C.P. (M1 damage, slowed motor imagery), 63
   D.B. (action blindsight), 81–*82*
   D.D.C. (split brain patient, ipsilateral tactile projections), 296
   D.F. (visual agnosia, preserved action ability), 82–83
   D.K. (agraphia specific to lowercase letters), 162
   D.L.E. (synchiria), 299–300
   D.S. (impaired tactile localization), 31

Patient, (*continued*)
　E.C. (tactile agnosia), 96–97
　E.P. (misoplegia and supernumerary phantom limb), 114, *121*–*122*
　F.B. (multimodal agnosia), 103
　F.B. (somatoparaphrenia), 116
　Fleisher, Leon (musician's dystonia), 170, 176
　Fox, Michael J. (Parkinson's disease), 177–178
　G.O. (loss of proprioception and touch), 34
　G.O.S. (agraphia specific to cursive), 162
　G.P. (anarchic hand sign), 73
　G.S. (multimodal extinction), 132
　Harriet C. (somatoparaphrenia), 115–116
　H.J.A. (visual agnosia, without tactile agnosia), 103
　H.L. (agraphia), 162–163
　H.M. (anterograde amnesia, retained motor skill learning), 152
　H.P. (neglect for far, but not peripersonal space), 130–*131*
　H.S. (cataracts, tactile spatial remapping), 41
　H.W. (tactile extinction), 296–297
　I.D.T. (ideational agraphia), 161–*162*
　I.G. (impaired online action correction), 89
　I.W. (loss of proprioception and touch), 33–34, 36, 89, 333
　J.A. (numbsense), 32
　James S. (allochiria), 298–299
　J.G.E. (agraphia), 163
　J.H. (tactile agnosia), 95–96
　J.H. (mirror writing), 288
　J.K.P. (impaired visual localization in peripersonal space), 130
　J.M. (identification anosognosia), 123
　J.W. (split brain, visual-tactile interaction near hand), 133
　Knoblauch, Chuck (yips), 172
　L.M. (cataracts, tactile spatial remapping), 41
　M.C. (visual agnosia, without tactile agnosia), 104
　McGrath, John (phantom limb), 107–108
　M.N. (agraphia specific to uppercase print), 162
　Mrs. H. (disturbed sensation of limb ownership), 115
　Mus, a.k.a "Professor Krukenberg" (Krukenberg hand), 91
　Nelson, Horatio (phantom limb), 105
　P.J.G. (action blindsight), 81
　P.P. (projection of neglect into far space by tool use), *213*
　P.R. (phonological agraphia), 161
　Regierungsrat (ideomotor apraxia), 187
　R.G. (lexical agraphia), 161
　R.S.B. (tactile mislocalization), 29–*30*
　Schumann, Robert (musician's dystonia), 170–171
　Sergius, Marcus (amputee), 106–107
　Shanidar 1 (Neanderthal amputee), 106–107
　S.J. (action blindsight, grip constancy), 81–82
　T.M. (neglect for peripersonal, but not far space), 130–*131*
　T.N. (object blindsight), 81
　Ukr (Krukenberg hand, tactile agnosia), 91
　W.J. (split-brain patient with left hand apraxia), 191–192
　W.M. (hand-assisted blindsight), 135–136
Pebble tools, 196, 205–206
Pegboard task, 310
Penfield homunculus. *See* Somatotopic maps
Pentadactyly, 2–3, 120, 222
Percussion (with stone tools), 196–197, 204, 206
Pericutaneous neurons, 128–*129*
Peripersonal space, 127–135
　in healthy humans, 132–134
　lesion studies in monkeys, 130
　network underling, 134–135
　neuropsychological studies in humans, 130–132
　receptive fields of single neurons, 128–129
　and tool use, 211–214

Subject Index

Peripheral vision, 68–69, 78, 89
Phantom limbs, 105–110, 117, *121*–124, 186, 283, 299
  antiquity of, 106–107
  in congenital limb absence, 109–110
  movement of, 107–109, 283
  supernumerary, *121*–124, 186
Phase-encoded mapping, 16, 20, 108, 230
Pill-rolling tremor, 178
Pixel model (of tactile distance perception), 43–*44*
Place specificity, 22–24, *23*
Platypus. *See* Duck-billed platypus
Pli de passage fronto-pariétal moyen (PPFM), 52–53, 248
Pointing, 321–327
  in animals, 325–327
    communicative function, 321–322
    development of, 322–324
Ponzo illusion. *See* Geometric illusions
Population receptive field (pRF), 25
Population vector, 56, 59
Porpoise, 2
Positron emission tomography (PET), 27, 77, 153, 175, 185, 208, 297
Posner cueing task, 137, 323
Posterior parietal cortex (PPC), 21, 25–28, *26*, 32, 35, 40, 47–48, 55, 61–63, 67, 80–81, 84, 89, 97, 100, 110, 115, 117, 120–121, 128, 134–134, 142, 144, 163, 187, 189, 192, 207–209, 216, 234, 250–256, 270, 272, 274, 285–286, 297–298
  anterior intraparietal area (AIP) (*see* Anterior intraparietal area (AIP))
  Area 5, *26*, 34, 55, 62, 90, 250, 294
  Area 7, 80, 128, 285
  as command apparatus, 61–63
  evolution of, 250–256
  inferior parietal lobe, 63, 96, 120, 175, 190, 207–211, 268, 297, 301
  intraparietal sulcus. (*see* Intraparietal sulcus)
  lateral intraparietal area (AIP) (*see* Lateral intraparietal area (LIP))
  medial intraparietal area (MIP) (*see* Medial intraparietal area (MIP))
  parietal reach region (*see* Parietal reach region (PRR))
  superior parietal lobe, 34, 68–69, 80, 90, 254
  supramarginal gyrus, 120, 187, 209–211, 251 (*see also* Anterior supramarginal gyrus (aSMG))
  ventral intraparietal area (VIP) (*see* Ventral intraparietal area (VIP))
Postural origins theory, 314
Postural schema, 34
Power grip, 7–9, *8*, 46, 62, *65*–66, 75, 139, 202–204, *203*, 239, 241, 265–*266*, 346
  development of, 265–*266*
  evolution of, 241, 249
  neural representation of, 62, 65–66
  tool use and, 202–204, 239
Precision grip, 7–9, *8*, 25, 51, 62, 65–67, 75, 79, 80, 128, 139, 202–204, *203*, 217, 233, 237–241, 244, 249, 265–268, *266*, 274–277, 316, 346
  corticomotoneuronal (CM) projections and, 51, 274–*275*
  development of, 67, 265–268, *266*, 274–*275*
  evolution of, 233, 237–241, 249
  handedness of, 316, 341
  neural representation of, 25, 62, 65–67, 128, 217
  tool use and, 202–204, *203*
Premotor cortex (PMC), 21, 48, 55, 57–59, *58*, 63–69, *65*, 72–73, 76–77, 90, 110, 115, 117, 120, 122–123, 128–130, 134–135, 148–152, 160, 163–164, 175, 179, 184, 187–188, 195, 207–210, 216–218, 226, 244, 254–255, 268, 270, 275, *284*–*285*, 292, 312
  Broca's area (*see* Broca's area)
  dorsal (dPMC), 55, 67–69, 90, 110, 134, 299, 292
  medial, 21, 55, 72–73, 76–77, 122, 181, 184, 254

Premotor cortex (PMC) (*continued*)
  ventral (vPMC), 55, 57–*58*, 64–69, *65*, 76, 117, 120, 128–129, 134–135, 207, 209, 216–217, 254–255, 268
  supplementary motor area (SMA) (*see* supplementary motor area (SMA))
Prepared core methods. *See* Levallois technique
Prereaching, 262–263
Primary motor cortex (M1), 13–15, *14*, 21, 32, 35, 44, 49–63, *53*, 66–67, 69, 77, 89–90, 108–109, 122, 125, 148–150, 153, 158–159, 173–175, 179, 184–186, 188, 192, 208, 216, 226, 244–249, 254–255, 265–266, 274–276, 283–287, 291–293, 311–312
  bilateral representation, 283–286
  contralateral control, 280–282
  corticomotoneuronal (CM) projections (*see* Corticomotoneuronal (CM) projections)
  direction selectivity, 56–57
  evolution of, 244–249
  force selectivity, 56
  intrinsic vs. extrinsic representation, 57–58
  muscle vs. movement representation, 56–60
  Old M1 vs. New M1, 53–55, 58, 90, 244–246, 265–266
  population vector, 56
  role in learning, 153–154
  somatotopic organization of, 12–15, *14*, 52–54
Primary somatosensory cortex (S1). *See* Somatosensory cortex, primary
Primary visual cortex (V1), 16, 20, 81, 100–101, 136–139, 230, 229, 252, 272, 331
Primates. *See also* Bonobo, Chimpanzee, Orangutan, Gorilla, and Gibbon
  central sulcus in, *247*–248
  evolution of, *231*–232
  hands of, 232–234
  maternal carrying of infants, 260
  Old and New motor cortex, 244–246
  somatosensory cortex in, 248–250

Priming
  of affordances, 74–76
  of tactile and visual form, 101–102
Primitiveness of hand, x, 232, 236–237, 343
Principal components analysis, 44, 49–50, 59, 111, 119, 157, 237–238, 243, 285, 308
Proconsul, *231*, 235, 237
Proprioception, 24–25, 33–37, *36*, 112, 117–118, 122, 130, 165, 208, 223, 333. *See also* Patient, I.W.
  deficits in, 33–34
  and disturbed limb ownership, 116–117
  imagery, 36
  and motor expertise, 159–160
  neural representation of, 34–35
  perceptual maps of, 35–37, *36*
  and supernumerary phantom limbs, 122
Proprioceptive constructions, 332
Proprioceptive drift (in rubber hand illusion), 110–111
Prosimian, 22, 230–*231*, 233, 242–*243*, *247*–251, 254–255, 260, 314
Protactile, 332
Protolanguage, 340
Proximal type neurons. *See* Distant peripersonal neurons
Pseudoneglect, *213*–214
Pseudo–opposability, 233
Pyramidal tract. *See* Cortico–spinal tract

Quadrupedalism, 233

Raccoon, 143, 227–228
Rafferty, Mary, 281
Reaching network, 64, 67–69, 73, 76, 78–83, 120, 134, 207, 209, 219, 250, 311. *See also* Area V6a; Medial intraparietal area (MIP); Parietal reach region (PRR); Premotor cortex (PMC), dorsal (dPMC)
Receptive fields (RFs), 16, 18–20, *19*, 22, 24–25, 29–30, 43, 62, 128–*129*, 134–136, 142–143, 172–173, 211–212, 223, 269–270, 287, *293*–294, 303

Subject Index

bilateral, 25, 293–*294*, 303
and cortical magnification, *19*–20
and dystonia, 172–173
elongated shape of (*see* Anisotropy, of receptive field (RF) geometry)
hierarchical organization, 25
multi-finger, 25, 249, 293–*294*, 303
peripersonal, 128–*129*, 134–135, 211–*212*
population (*see* Population receptive field (pRF))
and tool use, 211–*212*
visual, 128–*129*, 134, 136, 211–*212*
Regierungsrat, 187–188, 346
Repellent apraxia, 72
Repetitive transcranial magnetic stimulation (rTMS). *See* Transcranial magnetic stimulation (TMS)
Revadim, Israel, 199
Rising Star cave, South Africa, 240
Rotational dynamics, 59–60
Rubber hand illusion, 110–112, 120, 124, 134, 301–*302*

Sassoon, Siegfried, 11
Scandentia. *See* Tree shrew
Schemata, 28–29, 34, 60, 64
Schumann, Robert, 170–171
Scraper (stone tool), 197, 201
Scratches on teeth (as measure of hominin handedness), 317–*318*
Scrivener's palsy. *See* Writer's cramp
Secondary somatosensory cortex (S2). *See* Somatosensory cortex, secondary (S2)
Sedigitae, 3
Segmentation (in motor chunking), 152
Septum, 21–22
Sergius, Marcus, 106–107
Shanidar 1, 106–107
Shrew. *See* Tree shrew
Sibudu Cave, South Africa, 201

Sierra de Atapuerca, Spain, 241
Sign language, 36, 49–50, 325–326, 329–330, 332, 339–342
in animals, 325–326, 339–340
and language evolution, 339–342
tactile, 332
Sima de los Huesos (Pit of Bones), Spain, 241
Sima del Elefante (Pit of the Elephant), Spain, 241
Sistine Chapel, ix
Skin, 5–6
glabrous, 5–6, 16–17
hairy, 5–6, 17
Somatoparaphrenia, 115–116, 122, 186
Somatoperception, 12, 28–44
Somatosensory cortex
bilateral representation, 293–298, *294*
Brodmann's area 3a, 16–17, 24–27, 35, 225, 249
Brodmann's area 3b, 16–27, *17*, *19*, *23*, 62, 143, 172–*173*, 226, 249–250, 269–272, *270*, 293–294
Brodmann's area 1, 16–19, *17*, *18*, *23*–25, 27, 62, 143, 249–250, 269–271, *270*, 293–294
Brodmann's area 2, 16, *18*, *23*–25, 27, 35, 62, 249–250, 272, 293–294
development of, 269–275, *270*
evolution of, 226–228, 247–250
primary (S1), 13–27, *14*, *17*, *18*, *19*, *23*, 31–33, 35, 44, 52–55, 62, 89, 97, 102, 108–109, 125, 143–144, 148–149, 172–174, *173*, 186, 216, 226–230, 247–250, 269–275, *270*, 293–298, *294*, 313
secondary (S2), 21, 27–29, 97, 102, 144, 208, 227, 295–297
serial organization of, 26–27
somatotopic organization of, 13–21, *14*, *17*, 53–*54*, 269–274, *270*
Somatosensory evoked potential (SEP), 141, 143–144, 159, 174, 272

Somatotopic maps, 12–21, *14, 17,* 23–24, 29–31, 43, 52–*54,* 59, 108–109, 124–125, 142–143, 172–175, 227–230, 244, 248–249, 268–274, *270,* 303, 313
  bilateral symmetry of, 313
  comparison of M1 and S1, 53–*54*
  in dystonia, 172–175
  development of, 268–274, *270*
  evolution of, 227–230, 244, 248–249
  magnification of (*see* Cortical magnification)
  reversal at area boundaries, 16–18, 230, 249–250
Somatotopy. *See* Somatotopic maps
Spider hand. *See* Arachnodactyly
Spinal cord, 1, 6, 11, 15, 20, 33–34, 45–46, 50–51, 55, 62, 85, 90, 195, 226, 244–*246,* 265, 274–276, 291, 300–301, 319
Spinal interneurons, 51, 55, 226, 245–*246*
Split-brain patients, 133, 191–192, 282–284, 296, 303, 312
St. Acheul, France, 198
Star-nosed mole, 227–*228,* 248, 273
Stepping reflex, 258
Sterkfontein, South Africa, 239–240
Steve Blass Disease, 172. *See also* Dystonia and Yips
Sticky mittens, 265
Stone tools, 195–208, *199, 203*
  brain regions involved in, 207
  cognitive requirements of, 204–208
  manual requirements of, 202–204, *203*
  Mode 1 stone tools (Oldowan Industry), 196–198
  Mode 2 stone tools (Acheulean Industry), 198–*199*
  Mode 3 stone tools (Mousterian Industry), 200–201
  Modes 4 and 5 stone tools (Aurignacian and Magdalenian Industries), 201–202
  use by animals, 205–206
Stretch reflex. *See* Long-latency stretch reflex

Striatum, 175, 179–*180*. *See also* Basal ganglia
Stump Hospital. *See* Turner's Lane Military Hospital
Stuttering, 334–335
Substantia nigra pars reticulata (SNr), 179–*180*. *See also* Basal ganglia
Subthalamic nucleus (STN), 179–*180*. *See also* Basal ganglia
Sucking (in infants), 261–262, 319
Superficial schema, 29–34
Supernumerary body parts, experimental induction of, 124–125
Supernumerary finger, 3, 120, 273
Supernumerary phantom limb, 120–123, *121,* 186
Supplementary motor area (SMA), 21, 73, 77, 122, 148, 150, 160, 179, 181, 184, 254, 268, 274, 281, 284, 292, 312
Supramarginal gyrus (SMG), 120, 187, 209–211, 251
Swartkrans, South Africa, 240
Synchiria, 16, 299–*302*
Synergies, 49–52, 150–*151,* 157–158, 188, 259–260, 265

Tactile agnosia. *See* Agnosia, tactile
Tactile distance perception, 41–*44*
  adaptation, 42–43
  anisotropy of, 43–*44*
  categorical perception, 42
  equivalence, 303
  and perceived body size, 41–42
  pixel model of, 43–*44*
  and tool use, 42, 220
Tactile Italian Sign Language, 332
Tactile localization, 28–33, *30*
  allochiria, 298–302
  deficits in, 29–31, *30*
  in external space (*see* Tactile spatial remapping)
  synchiria, 299–*302*
  on tools, 215–216
Tactile Quadrant Stimulation test, 300

Subject Index

Tactile spatial remapping, 26, 37–41, *38*, 214–215
  in blindness, 40–41
  neural basis of, 40
  time-course of, 40
  and tool use, 214–215
Tarsier, 230–*231*, 233
Task complexity hypothesis, 314
Taung child, 252–*253*. See also *Australopithecus africanus*
Technition, 190, 207, 211
Telegraphist's Cramp, 167–168, 172. *See* also Dystonia
Tetrapod, 2–3, 120, 222–*224*, 344
Tetris, 155
Thebes (ancient Egyptian tomb), 307
Throwing (technique for making stone tools), 197
TicTac, 330
Tiktaalik, 223–224, 344
Toccata in C Major (Op. 7), 171, 177
Tool use, ix–x, 5, 7, 9, 42, 51, 133, 189–191, *190*, 193–220, *199*, *203*, *210*, *212*, 233, 238–241, 244, 249–252, 305, 307, 314, 316–*318*, 321, 341–343
  in animals, 193–195
  in apraxia, 189–191, *190*
  brain regions involved in, 208–211, *210*
  in capuchin monkeys, 233, 244, 250
  in chimpanzees, 193–194, 205–206
  evolution of, 238–241
  in gorillas, 194
  and handedness, 305, 307, 314, 316–*318*
  and haptic exploration, 214–215
  integration with motor programs, 216–219
  in New Caledonian crows, 194
  in New World monkeys, 195, 233, 244, 250
  in Old World monkeys, 51, 195, 244, 249, 250
  and peripersonal space, 133, 211–214, *212*
  and tactile distance perception, 42, 220
  and tactile localization, 215–216

Touch, 5–7, 11–33, *14*, *26*, *30*, 37–44, *38*, 91–104, *95*, *98*, 110–112, 116–119, 123–124, 130, 132–135, 137, 140–144, *142*, 165, 175, *185*, 208, 212, 214–216, 219–220, 223, 227, 269–274, 293–303, *294*, *302*, 312–313, 328–329, 332, 345
  acuity, 20, 312
  agnosia (*see* Agnosia, tactile)
  apraxia (*see* Apraxia, tactile)
  bilateral representation of, 293–303, *294*
  communication of emotion by, 328–329
  distance perception (*see* Tactile distance perception)
  handedness, 312–313
  haptic (*see* Haptics)
  localization (*see* Tactile localization)
  remapping (*see* Tactile spatial remapping)
  and sign language, 332
  and tool use, 214–216, 219–220
  visual enhancement of (*see* Visual enhancement of touch)
Traction response, 259–260
Transcranial direct current stimulation (tDCS), 192
Transcranial magnetic stimulation (TMS), 31, 33, 40, 50, 52, 63–64, 66, 89, 100, 108, 110, 131, 137, 144, 153, 158–159, 173–175, 211, 276, 281, 286, 290–293, 297–298, 331
Transport phase (of reaching), 46, 60–61, 64, 67, 69, 79, 219
Traveling wave. *See* Phase–encoded mapping
Tree shrew, *231*–232, 236, 242, 250
Tremor, 168, 171, 178
Trent Valley, England, 307
Troglodytian stage, 236
Tube task, *315*–316
Tulerpeton, 3, *224*
Turner's Lane Military Hospital, 107
Typing, 164–165

Ultrasound recording of fetal behavior, 261, 319
Unity of phenomenal experience, 37

Unity of type. *See* Archetype of vertebrate limb
Upper Paleolithic, 201–202
Utilization behavior, 72–*74*, 260

V1. *See* Primary visual cortex V1)
V6a. *See* Area V6a
Valsalva doctrine, 280
Ventral intraparietal area (VIP), 128, 134
Ventral tactile pathway. *See* Corticolimbic pathway
Ventral visual stream, 27, 64, 83, 97, 100, 102–103, 131, 251, 331
Vertical clinging and leaping, 233
Vindija cave, Croatia, *318*
Visual cortex. *See* Primary visual cortex (V1)
Visual dominance over touch, 99
Visual enhancement of touch, 140–*142*
    and modulation of intracortical inhibition, 142–144
Visually induced analgesia, 144
Visual word form area, 331
Voxel-based morphometry (VBM), 153–154, 207

Washoe (chimpanzee), 340
Watchmaker analogy, 1
Watson, Tom, 171
Weber's illusion, 41
Window of temporal binding, 111–112
Wittgenstein, Paul, 170
Wolf, 326
Woods, Jimmy, 257
Woods, Johnny, 257
Woods, Tiger, 171
Writer's cramp, 168–170, *169*, 175. *See* also Dystonia
Writing,
    agraphia, 160–164
    dystonia of (*see* Writer's cramp)
    Exner's area, 163–164
    motor equivalence, 47–*48*
    orthographic buffer, 161
    slips, 163

Yips, 171–172. *See* also Dystonia